The Story of Knife Steel: Innovators Behind Modern Damascus and Super Steels

Dr. Larrin Thomas

ACKNOWLEDGMENTS

Thanks to everyone who helped with this book, whether by providing suggestions or agreeing to be interviewed. The list is long: Devin Thomas, Shawn Houston, Matt Gregory, Phil Wilson, Bernard Levine, Mark Zalesky, Dennis Ellingsen, Michael Walker, Rick Dunkerley, Steve Schwarzer, Ron Lake, Jim Sornberger, Brad Stallsmith, Tinnell Hickory, Don Fogg, Steve Dunn, Daryl Meier, Greg Gottschalk, Tim Zowada, Ed Schempp, Joel Davis, Tom Ferry, Phil Baldwin, Mike Norris, Bob Eggerling, Salem Straub, Mareko Maumasi, Don Hanson, Ron Hardman, Sean McWilliams, Beren McKay, Tim Wright, James Batson, Roman Landes, Tobias Hangler, Scott Devanna, Dick Barber, Maria Sawford, Alojz Kajinic, Garry Maddock, Ed Severson, Sid Suedmeier, Ric Furrer, Shane Taylor, Barry Gallagher, Josh Smith, Ed Tarney, Steve Filicietti, Ken Pinnow, Daniel O'Malley, Bill Harsey, Tim and Anne Reeve, Peter Hjortberger, Sal Glesser, Doug Flagg, Ken Onion, Jimmy Fikes, A.G. Russell, Gil Hibben, Linda Hibben, Wes Hibben, RJ Martin, Beverly Imel, Ray Rybar, Gary Creely, Tai Goo, Robert Terzuola, and Malachi Chou-Green.

CONTENTS

INTRODUCTION

I've never thought of myself as a historian, or even someone particularly interested in history. But when writing profiles on different steels for knifesteelnerds.com, I wanted to add a history of the steel at the start of these articles. I thought, surely, a quick google search would reveal the history of each steel, and boy was I wrong. There would be no results for most steels, and finding some history online was no guarantee of accuracy. For example, the top result for D2 steel history says that the steel's "popularity began during the 'Rosey the Riveter' days of World War II." Except the steel predated WWII by quite a few years and was popular as soon as it came out. The internet is not a good place to learn steel history.

Instead, I began the search through scientific journal articles. The problem is that scientists and engineers are not historians; recording history in a journal article is rarely a top priority. Sometimes the "background" section has some tantalizing clues, such as referring to a steel being "recent" or having been used for "many years." However, one man's "many years" can also be the same length of time as another man's "recent." The search is doubly difficult because the standard names by which we know steels: A2, D2, 440C, etc., were not named until several decades after they were introduced, and instead used different brand names. Journal articles are not a good place to learn steel history.

So, while history has never been a significant interest of mine, learning how people invented things, how new developments were discovered, and why someone might think to try "this" has fascinated me. So, I began the long and arduous task of going through patents, advertisements, trade journals, and biographies, and interviewing many people to piece together what happened in the history of tool steel and stainless steel. Some exciting elements to me were when there was an intersection between steel metallurgists and knifemakers. The first experiments with adding different alloying elements to steel happened when a scientist, Michael Faraday, worked with cutlery manufacturer James Stodart, to try to recreate the properties of the legendary sword steel, Wootz Damascus. This has been repeated several times, such as through the experiments of metallurgist Dr. John Verhoeven and bladesmith Al Pendray, who recreated Wootz Damascus. Or when Crucible Steel metallurgists like Dick Barber worked with knifemaker Chris Reeve on the development of the popular knife steels CPM-S30V and CPM-S35VN. And in 1912, the very first commercialized stainless steel was made to be used in table knives!

While this history is certainly not about me, it is about the long line of metallurgists, engineers, and knifemakers who innovated to create the beautiful world of knife steel, where I now work. And in my position as a metallurgist who develops new steels for knives, their history is particularly relevant to my work. In some ways, I have searched their histories to find their secrets in how to push the envelope of steel. And it has been fruitful. For example, MagnaCut, a knife steel I designed, was developed not just from knowing how the different elements work, but also from understanding how previous steels were developed. I knew that stainless steels like S90V and S30V improved the previous generation by reducing chromium, and part of MagnaCut's development was to reduce the chromium further, along with a few other tricks. So it wasn't just the effects of chromium on steel that helped to guide the development, it was the path that those previous metallurgists had followed; one that I attempted to continue.

I grew up surrounded by pattern-welded Damascus steel. My father, Devin Thomas, ran a Damascus business as far back as I remember. He would sometimes talk about the history of who introduced different steels, patterning techniques, and ideas into the Damascus world, but I never did get a complete picture of how it all happened. I had heard from many of the old Damascus guys that names like Daryl Meier and Don Fogg had contributed much to pattern welding, but what that meant beyond the patterns didn't mean much to me then. I felt this book was an excellent excuse to finally figure it out. All I had to do was interview every significant living Damascus smith and piece together the timeline with magazines and books that published their work. Without Damascus, the American Bladesmith Society may not even exist today. This legendary steel inspires many metallurgists and knifemakers to start their craft.

For these great metallurgists and knifemakers I can only introduce their lives and careers. There are full books on their histories for a few of them, and for many others, there are individual magazine articles. The *Knife World/Knife Magazine* and *Blade Magazine* back issues are available to read online for a nominal fee. I have my favorite sources in bold in the References section if you want to learn more. And I hope you do.

-1-

THE INVENTION OF TOOL STEEL

Knives Before 1900

Before about the year 1900, all knives were made with carbon steel. I don't mean just in the colloquial sense of "non-stainless," but rather, the steel was essentially just carbon and iron. Bowie knives, razors, Viking swords, Wootz steel, etc. You name it, the steel was iron and carbon. These various types of steel were different manufacturing paths to a similar result - simple, unalloyed steel. Until then, the development of steel was finding ways to make steel cheaper, faster, and of better quality.

Wootz, European Steel Production, and Early Steel Alloying Experiments

At the end of the 18th century, European scientists were fascinated by Wootz steel[1]. They attempted to analyze the steel, but the knowledge of metallurgy was in its infancy; even the knowledge that carbon is the difference between iron and steel had only recently been discovered[2]. Wootz was a steel product that originated in India, famous for its excellent properties and "Damascus" pattern. It was made as early as the 3rd century BC by melting iron with carbon in a Crucible. Cutlery manufacturer James Stodart in the 1790s made a small knife out of the Wootz steel and reported[1] that "Wootz is superior for many purposes, to any steel used in this country ... it would carry a finer, stronger, and more durable edge, and point. Hence, it might be particularly valuable for lancets, and other surgical instruments."

The most common type of steel in Europe from around 1600 was "shear steel," produced using "blister steel." Blister steel[3,4] was made by cementation, where they stacked alternating layers of wrought iron and charcoal and heated to high temperature to allow the carbon to diffuse into the iron. Then shear steel was produced using layers of blister steel which were forge welded together and hot worked to homogenize the carbon content. However, this did not result in perfectly homogenous steel. Benjamin Huntsman in 1740 developed a process whereby the

1

blister steel was melted and cast in a crucible so that the carbon would be evenly distributed throughout.

Benjamin Huntsman's "Crucible" method for manufacturing steel was developed independently of the Wootz process, but fundamentally they are very similar. However, future improvements to the crucible process were often based on observations of Europeans from craftsmen making Wootz in India[5]. For example, David Mushet improved Huntsman's process in 1800 by using wrought iron and carbon to make steel, skipping the time-consuming cementation process to make blister steel. David Mushet published a paper in 1805[6] of his analysis of Wootz, concluding that it had been produced with a fusion process using iron and carbon, similar to his own process. Some have speculated that Mushet was aware of the Wootz process beforehand, such as John Percy in 1864[7]: "It is curious that Mushet's process so far as it relates to the use of malleable iron in the production of cast steel should in principle, and I may add even in practice too, be identical with that by which the Hindoos (sic) have from ancient times prepared their Wootz. I cannot discover any essential difference between the two."

James Stodart continued his analysis of Wootz steel along with Michael Faraday, later famous for his discoveries in electricity and magnetism. Faraday reported[8] that their experiments found Wootz to contain silica and aluminum oxide, so he and Stodart attempted to replicate Wootz by making steel with the same additions. After etching their steel in acid, they did not find the same pattern as in the Wootz steel. Eventually, with a modified process, they were able to recreate a pattern similar to Wootz[9].

Stodart and Faraday extended their research to look at other alloy additions; these were the first experiments on steel to add different elements. Therefore, Wootz inspired the study of steel alloy design. They stated[9], "In proposing a series of experiments on the alloys of iron and steel, with various other metals, the object in view was two-fold; first, to ascertain whether any alloy could be artificially formed, better for the purpose of making cutting instruments than steel in its purest state; and, secondly, whether any such alloy would, under similar circumstances, prove less susceptible to corrosion." So, their goal was also to make an improved knife steel, preferably one less prone to rusting.

Since this was the infancy of steel alloying, their experiments were largely unsuccessful. Many elements they tried have had no commercial use in steel, such as palladium, platinum, silver, and rhodium. However, they believed that the alloy with silver was "the most valuable of those we have made."[9] "This alloy is decidedly superior to the very best steel... Various cutting tools have been made from it of the best quality."[9] While silver did not become a standard alloy addition, this is the source of the term "silver steel," often used to refer to English steel used in knives and razors, though the steel did not contain silver[10].

Robert Forester Mushet and The First Tool Steel

Robert Mushet followed in his father David's footsteps and became a metallurgist. Also like his father, he contributed to a new steelmaking process, in this case the Bessemer process. While the Bessemer process was unsuitable for making high carbon tool steels used in knives, it was important in greatly reducing the cost of steel production and was the first method for mass production of steel. It was invented by Henry Bessemer in 1856[11], but Bessemer had been unsuccessful in making useful steel from it. The process involved melting pig iron (the intermediate iron product resulting from smelting iron ore) in a blast furnace, and a flow of oxygen removed the impurities from the steel. Bessemer attempted to stop the flow of oxygen when an appropriate amount of carbon remained, but this meant impurities still remained. Iron sources had different amounts of carbon and impurities; therefore, the point at which to turn off the oxygen varied by source. Mushet instead would remove all carbon and impurities[12], then added carbon through the addition of spiegeleisen, an alloy of iron, manganese, and carbon. Mushet had previously tried the addition of spiegeleisen with crucible-made steel and found it to increase hardness, ductility, and made it more workable by hammering. Adding carbon and manganese through the spiegeleisen after all of the impurities and carbon were removed with the Bessemer process made it a viable product. Mushet said, "the Bessemer process was perfected and that, with fair play, untold wealth would reward Mr. Bessemer and myself"[13].

Mushet patented his improved Bessemer process with Thomas Brown of the Ebbw Vale Ironworks and S.H. Blackwell. However, after the patent had been accepted and in place for a year, Blackwell and Brown failed to pay the annual fee to maintain it, and it expired[12]. Bessemer was free to use the addition of spiegeleisen in his process without payment. However, Bessemer claimed that he had independently found the benefits of a manganese addition just before Mushet's patent had been published[14]. Mushet produced a large ingot of steel, and it was rolled at the Ebbw Vale Ironworks into double-headed steel rail. This was the first commercial use of the Bessemer-Mushet process and the first steel rail ever produced.

Robert Mushet was greatly disappointed that he had failed to gain any money or recognition for his contributions to the Bessemer process. In the years following, he performed many experiments on alloy additions to steel, such as titanium, manganese, chromium, and tungsten. His father, David Mushet, had previously experimented with alloying steel, particularly with titanium[12]. Robert took out 48 patents in the following years on various small alloying additions for improving steel's casting process and quality. His experiments with titanium resulted in a new product he called Titanic Cast Steel. The development of that product led to a new company, Titanic Steel & Iron Co, Ltd, with the backing of three businessmen. A manufacturer named J.P. Smith asked Robert Mushet to help him manufacture machine tools using Smith's specifications. The tools failed, probably because of too much tin in the steel. But this led to Robert Mushet experimenting with different alloys for machine tools. In 1868, he pulverized and melted a lump of manganese-rich pig iron and added finely powdered tungsten ore. This new steel did not require a quench to harden but would harden in the air after forging, requiring no separate heat treatment, being called a "self-hardening" steel (later, the term "air hardening" became more accepted). Titanic company secretary Robert Woodward convinced Mushet not to patent the process but to keep it a secret. So, despite the many patents previously taken out by Mushet, the first alloy tool steel ever commercialized was not patented.

A piece of the original ingot of R. Mushet's Special Steel[15].

R Mushet's Special Steel was marketed for lathe and planer tools. Despite the exciting new product, the Titanic Steel company was failing. The secretary Woodward advised Mushet to find a Sheffield steel company to produce and market the steel instead. Robert Mushet met Samuel Osborn, who was interested in making the Special Steel and paying Mushet a royalty on every ton sold. The steel became popular because it didn't need to be quenched in water, which was risky because of the chance of cracking. The tungsten addition also meant that many hard tungsten carbides formed in the steel for improved wear resistance. This meant that the Special Steel was prized for machining difficult materials like cast iron. In the *Engineer* journal in 1868, it was reported[16], "A tool of the best ordinary Cast Steel required to be ground three times in planing a given area of hard cast iron, whereas a tool of Mr. Mushet's steel not only planed a similar area without regrinding, but remained to all intents and purposes nearly as sharp as when it began."

To obtain
IMPROVED FEEDS and SPEEDS
"R. Mushet's
Special
Steel"
(SELF-HARDENING)

should be heated Gradually but Thoroughly, so as to be quite ductile.

The Tool should then be forged at a Good Heat.

Avoid Forging it too Cold.

When forged the cutting end of Tool should be REHEATED TO A GOOD SCALING HEAT, and, if possible, then BLOWN COLD.

Whilst Hot this Steel must be kept from Water.

DURABLE, RELIABLE and UNIFORM.

MADE ONLY BY
SAMUEL OSBORN & CO.,
Clyde Steel and Iron Works,
SHEFFIELD.

Leaflet advertising 'R. Mushet's Special Steel' circulated during the 1890s

Mushet's steel saw some evolution in composition in the early years. The carbon, silicon, and manganese were increased, and the tungsten decreased. For many years, it was believed that the tungsten gave the steel its air-hardening capability; it wasn't until 1892 when it was recognized that the high manganese content gave the steel that property[18]. Julius Baur of New York was the first recorded metallurgist to alloy steel with chromium in about 1865[18], which he patented[19]. This resulted in "Chrome steel," branded as Adamantine by the Brooklyn Steel Works in 1867. The steel had 1% carbon and 0.5% chromium. This steel would inspire many metallurgists to try chromium additions, including Brustlein of Holtzer Steel, who will be discussed later. A small chromium addition was made to Mushet Special Steel from about 1870[16]. Henry Gladwin, an assistant to Samuel Osborn, found that the performance of the steel was better if given a separate heat treatment (rather than "self-hardened" after forging). The steel was re-heated to have more control over the temperature and to remove strains from forging[16]. He also found that when the steel was given an air blast (instead of cooling in still air) it would harden more fully and therefore improve performance.

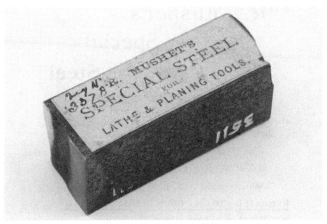

1872 Special Steel[17].

Steel	Year	C (%)	Si (%)	Mn (%)	W (%)	Cr (%)
Adamantine	1867	1				0.5
Mushet	1868-1869	1.55	0.7	1.2	9	
Mushet	1872-1882	2.3	1.2	1.8	6	0.5

The Discovery of High Speed Steel

Some steel manufacturers began to add yet higher additions of chromium, such as 1.25% or even higher. Chromium would also contribute to air hardening in addition to the manganese. Over time the manufacturers further increased the chromium up to 2.5-4% while reducing the manganese to standard carbon steel levels (~0.35%). Two companies that had replaced manganese with chromium were Sanderson Works and Midvale Steel. Midvale partnered with metallurgist Henri Brustlein of Holtzer Steel Works to make metal-piercing projectiles. Brustlein sent samples of two self-hardening steels in 1893 to Midvale Steel Co. So, by 1894 Midvale was producing a self-hardening steel that had replaced the high manganese with chromium.

Steel	Year	C (%)	Si (%)	Mn (%)	W (%)	Cr (%)	V (%)
Holtzer	1893	1.49		0.34	1.51	4.76	
Holtzer	1893	1.66		0.64	2.4	6.59	
Midvale	1894	1.14	0.25	0.18	7.72	1.83	
Taylor-White	1898	1.85	0.15	0.3	8	3.8	
High Speed	1902	0.42			10.75	4.95	
High Speed	1903	0.7	0.01	0.02	14.7	2.9	
High Speed	1906	0.67	0.04	0.07	18.2	5.47	0.3
High Speed	1910	0.7			18	4	1

Frederick Winslow Taylor began working for Midvale steel in 1878. He started as a machinist but was eventually promoted to "gang boss" over the lathe hands, then machine shop foreman, research director, and chief engineer. While at Midvale, Taylor noted that many machinists did not work as quickly as they could[20]: "Hardly a competent workman can be found in a large establishment who does not devote a considerable part of his time to studying just how slowly he can work and still convince his employer that he is going at a good pace." The issue with asking machinists to work more quickly is that if the machine was run at too high a speed, the steel tooling would be overheated and need to be replaced. Therefore, it was easier for the machinist to run at a conservative speed and avoid a quick breakdown. So, Taylor experimented with machining different materials to dial in the optimum speeds and feeds for each job, thus avoiding arguments with the veteran machinists. He created a table listing the best speeds and feeds for each condition which improved efficiency by 30% and no longer required highly experienced machinists to do the work, as relatively new employees could follow the table. Taylor eventually published *The Principles of Scientific Management* in 1911, systematizing his thoughts about the efficiency and management of workers.

Taylor also experimented with changes to the machining process so that efficiency and productivity could be further improved[21]. He found in 1881 that a round-nosed tool could be used at a higher speed than the previous diamond-shaped tools. In 1883, he discovered that a stream of water poured where the tool was removing steel would allow an increase in cutting speed, and they upgraded the shop in 1884 to accommodate this change. In 1894-1895 he performed more experiments with self-hardening steels and found that they were 90% better than typical carbon steels when cutting soft metals, but only 45% better when cutting hard steel or cast iron. In the past, self-hardening steels were only used on hard metals. He also found that a stream of water on the self-hardening steel allowed a 33% increase in cutting speed. At that time, machinists only used water cooling on the standard steel tools. The advice from Samuel Osborn & Co. that "whilst hot this steel must be kept from water" (see the advertisement on the previous page), was made to convince the heat treaters not to quench the self-hardening steel in water. But this had been understood by machinists that it must not be cooled with water during machining either.

In 1898 Fred Taylor was hired at Bethlehem Steel to manage their machine shop. Machine Shop No. 2 was "the largest building of its kind under one roof in the country, and possibly in the world" in 1891[22], but it was also Bethlehem Steel's biggest bottleneck. Taylor made immediate changes to the machine shop, such as no longer having machinists grind their own tools; different employees sharpened them and delivered them to the machinists ready to go. This meant machines would not be left idle while waiting for machinists to sharpen. While Taylor was already convinced that self-hardening steel should be used for all "roughing cuts," he still was not sure which steel was best, so he had an experimental lathe built for comparing different steels and other machining parameters. He determined that the steel labeled "Midvale" in the previous table was best and then invited company superintendents and foremen to demonstrate the superiority of the steel. Taylor said, "In this test, however, the Midvale tools proved to be worse than those of any other make... This result was rather humiliating to us as experimenters who had spent several weeks in the investigation... It was of course the first impression of the writer that these tools had been overheated in the smith shop [prior to quenching]. Upon careful inquiry among the smiths, however, it seemed as though they had taken special pains to dress them at a low heat."[21]

Taylor and metallurgist Maunsel White decided "to discover if possible some heat treatment which should restore Midvale tools injured in their heating to their original condition... For this purpose Mr. White and [I] started a carefully laid out series of experiments, in which tools were to be heated at temperatures increasing, say, by about 50 degrees all the way from a black heat to the melting point."[21] They found that the hotter the steel was heated before quenching, the better the steel performed when machining. This was "directly the opposite of all previous heat treatment of tools,"[21] as conventional wisdom was that heating tools above 1700°F or so would only lead to a large grain size, poor toughness, and overall poor properties. Only the fact that they had decided to cover the entire temperature range allowed them to make this counterintuitive finding. Taylor reported, "Neither Mr. White nor [I had] the slightest idea that overheating beyond the bright cherry red would do anything except injure the tool more and more the higher it was heated"[21].

What Taylor and White had discovered was the occurrence of "secondary hardening." Typically, when quenched steel is heated, it becomes softer the higher the temperature it reaches from tempering. Almost all steel is tempered after quenching to reduce hardness somewhat and increase toughness. When operating tools at high speeds, the steel would heat and soften, therefore no longer having the strength to maintain a sharp cutting edge for machining. However, for steels that have certain alloying elements added to them, especially tungsten and molybdenum, the hardness will drop when tempering up to 600°F or so, and then increase again, as shown in the chart on the following page.

During tempering at 900-1220°F tungsten, molybdenum, etc. will form a very fine array of carbides, increasing the steel's hardness. That gives the steel high hardness even when tooling is operated at high speeds, and therefore reaches high temperature; the tool maintains its strength. However, before austenitizing, the tungsten is already locked up in a carbide, and therefore is not free to form fine carbides during tempering. The high temperature utilized in the heat treatment of high speed steels is to dissolve the tungsten/molybdenum carbides so that those alloys are "in solution" before tempering. Then, new, much finer carbides form during tempering to provide hot hardness. Taylor and White would eventually find that the tool would have a higher performance if the steel was tempered in this elevated temperature range. The fine array of carbides formed at these temperatures (~900-1200°F) could be better controlled than simply allowing the tool to reheat during the machining operation.

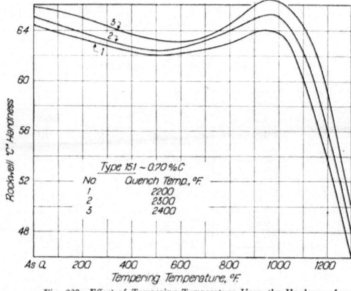

Fig. 228—Effect of Tempering Temperature Upon the Hardness of a 0.70% C Type 151 High Speed Steel After Quenching from 2200, 2300, and 2400 °F. For this steel 2350 °F. is the usual austenitizing temperature. Tempering time—2½ hours. Specimen size: 1¼ x 1 x ¾ inches. Vanadium-Alloys Steel Company.

Image from *Tool Steels*[23].

Taylor and White would patent their new process and the composition of the best steel they had yet tested (labeled "Taylor-White" in the table) in 1900. The patent would later be rescinded because they had not developed the composition but had only heat treated it with a new type of heat treatment. They presented their findings to the world at the Paris Exhibition in 1900, a world's fair that celebrated the previous century's achievements. Other new inventions displayed at the fair included the escalator and the diesel engine. People were amazed by the discovery of high speed steel. J. Hartley Wicksteed, a British Engineer, stated that the discovery was as incredible as the electric lightbulb[20] and, "The whole idea of a tool preserving its edge when in a state of dull red heat was, I think, to all of us a revelation... I consider that it is by far the greatest revolution that has taken place in my life."

High Speed Development of High Speed Steel

Between about 1901 and 1903, there was a shift from the relatively high carbon (1.2-2.3%) and intermediate tungsten (~8%) of the Mushet Special Steels, to lower carbon (~0.4-0.9%) and higher tungsten (10-20%). John A. Mathews reported in 1919[24] that they had analyzed commercial high speed steels in 1901 and then again in 1902 and found this radical shift in composition. It is unknown who first started this change in carbon and tungsten, but it proliferated relatively quickly. The higher tungsten gave higher red hardness from more precipitated tungsten carbides, and the lower carbon gave improved toughness. In 1906[21], Taylor presented compositions of steels Taylor and White had analyzed at different points, including a range of steels analyzed in 1903, which also showed the growing consensus among steel companies that the optimal composition was approximately 0.65% carbon, 17.5% tungsten, and 3% chromium.

1901 Self-Hardening Steels[24].

C (%)	Mn (%)	Cr (%)	W (%)	Mo (%)
2.19	1.32	0.5	5.63	
1.69	0.45	3.73	7.63	
1.14	0.33	2.09	7.98	
1.79	0.5	3.95		4.54
1.55	0.94	3.22	7.8	
1.55	0.21	3.67	9.42	1.1
1.78	1.18		7.22	
1.4	1.65	3.69		4.59
1.75	3.92		6.61	

1902 High Speed Steels[24].

C (%)	Mn (%)	Cr (%)	W (%)	Mo (%)
0.63		4		6
0.42		4.95	10.75	
0.57	0.43	3.3	11.58	
0.75			19.5	
0.87		5.1	13.83	
0.62		6.5	21.06	
0.84	0.07	2.76	11.25	
0.56		2.95	9.74	
0.6	0.3			9.25

1903 High Speed Steels[21].

C (%)	Mn (%)	Cr (%)	W (%)	Mo (%)
0.65	0.12	2.84	17.79	0.48
0.76	0.3	2.85	19.64	
0.67	0.2	2.61	18.99	
0.8	0.11	2.8	23.28	
0.58	0.19	3.52	18.92	2.03
0.7	0.12	2.9	14.71	
0.54	0.12	2.88	15.31	
0.45	0.1	2.8	14.91	0.75
0.6	0.18	2.81	14.62	

Studying Red Hardness

The discovery of the mechanisms by which tungsten creates the property of red hardness took much longer than it did to develop high speed steel. A summary of the different discoveries can be found in Ref. 25. There were a series of half-steps and discoveries by investigators such as Frederick Taylor, Henry Cort Harold Carpenter[26,27], Charles Alfred Edwards, Edgar Bain, and Zay Jeffries. Such as the necessity of high austenitizing temperatures to dissolve the tungsten carbide[28], that "red hardness" was not a separate property from "hardness at high temperature"[29], chromium allows more tungsten carbide to dissolve in the steel during the high austenitizing temperature treatment[29], and the effect of small precipitated tungsten carbides on strength after tempering[30,31]. Eventually, researchers Morris Cohen and P.K. Koh showed more conclusively in 1939 that secondary hardening that occurs during tempering is primarily from the precipitation of carbides and not other potential mechanisms such as the reduction of retained austenite[32].

The Alloy Steel Age

After 1900, the development of new tool steels was rapid. Taylor and White effectively kickstarted modern steel development. In 1930, John A. Mathews stated[33], "The year 1900 may well be accepted as a milepost in the evolution of metallurgy. We may call the period since then the 'Alloy Steel Age.'" He stated that 1900 was the "dividing point" because of "the introduction of high speed steel in that year and the results attained by this epoch-making event."[33] He clarified that many good tools and much progress were made before 1900, but that the period before 1900 was "an era of process development" while post-1900 was "an era of products development."[33]

-2-

CARBON STEEL, ALLOYS, AND SPECIFICATIONS

The most significant effect of carbon content on steel is to increase hardness (see chart below). However, 0.6% carbon results in about 64 Rc, a relatively high hardness, and further additions of carbon result in only a small hardness increase. Why then do many knife steels have greater than 0.6% carbon?

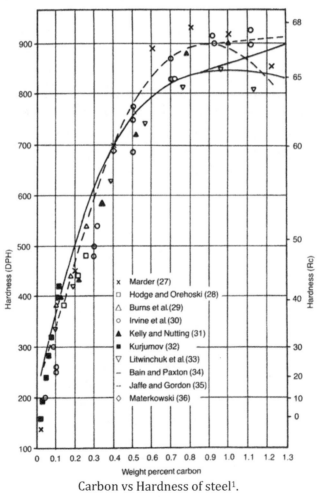

Carbon vs Hardness of steel[1].

In some cases, the higher carbon is for very high hardness, but generally, these higher carbon additions are for increased wear resistance. The higher carbon leads to more iron carbide, also called cementite, which are hard particles that resist wear. So, higher carbon steel can be used for applications that fail due to wear. Or in applications where toughness is not critical, a higher carbon, more wear resistant, steel will last longer in service.

Before standard steel specifications like 1075, 1084, 1095, W1, etc., carbon steels were sold by individual steel companies by "temper," which indicated the carbon content of the steel rather than the heat treatment. For example, Henry Seebohm, in 1881[2] wrote, "When I first began business, the 'temper' of steel, or the percentage of carbon which it contained, was concealed from the consumer." He then laid out some common "tempers" of steel and what they are commonly used for, including 1.5% carbon razor steel, 1.375% carbon saw file steel, 1.25% carbon "tool temper" steel, 1.125% carbon spindle steel, 1% carbon chisel steel, 0.875% carbon "Sett" steel, and 0.75% carbon die steel.

Different companies designated these "tempers" differently. Thallner, in 1902, wrote[3], "In some manufactories (sic) of tool-steel to the corresponding to the content of carbon, for instance, degree of hardness 7 - containing 0.70 per cent. carbon - while in others it is expressed in such allegorical designations as: very hard, super-hard, extra hard, hard-hard, medium hard, tenaciously hard, tenacious, soft." He had the following table for carbon contents and their typical uses:

Degree of hard-ness.	Average per cent. of carbon.	Purposes for which Employed.
Very hard.......	1.5	For turning and planing knives, drills, turning gravers, etc., for very hard materials.
Hard	1.25	For ordinary turning and planing knives, rock drills, mill picks, knife picks, scrapers, etc., and for cutting tools for hard metals.
Medium hard....	1.0	For screw-taps, broaches, cutters, tools for stamping presses and for various tools used by locksmiths and blacksmiths.
Tenaciously hard.	0.85	For screw-taps, cutters, broaches, matrices, swages, pins, bearings, chisels, gouges, etc.
Tough	0.75	For chisels and gouges, shear-blades, drifts, springs, hammers, etc.
Soft	0.65	For various blacksmith tools, as weld-steel for steeling finer tools and larger surfaces, etc.

In 1909, J.M. Darke proposed a specification system for carbon steels[4], which doesn't seem to have caught on, or at least not in the long term. He grouped them into five classes based on carbon content from 0.75 to 1.4%. He also proposed they be grouped into Grades A, B, and C according to how tight the tolerances were for manganese, silicon, phosphorus, and sulfur.

	Carbon (%)	Grade A	Grade B	Grade C
Manganese		0.25-0.42	0.25-0.35	0.25-0.35
Silicon max		0.2	0.15	0.12
Phosphorus max		0.03	0.02	0.01
Sulfur max		0.04	0.02	0.01
Class	Carbon (%)	Uses		
Class 1	0.75-0.85	Drop dies, Dies for hammers	Hammers, Drifts, Punches, Hot chisels	Cold chisels, Cold sets
Class 2	0.85-0.95	Cams		
Class 3	0.95-1.10	Sheet steel for springs	Lathe centers, jigs, templates	Misc. steel for blacksmiths
Class 4	1.10-1.25	Coil springs	Springs	Misc. steel for die room, large reamers, saws, large dies, large taps, letters, figures, milling cutters, drill rod
Class 5	1.25-1.40			Milling cutters, small dies, chases, gravers, dies for heading machines

Development of Atkins Silver Steel

Of course, the relative simplicity of carbon steels didn't stop people from telling stories about their development. For example, E.C. Atkins & Co. was a saw company with a proprietary steel called "silver steel" trademarked in 1875[5]. The Atkins Silver Steel was made by Jessop in England, and in 1911 the Atkins company was sharing the story about how the steel was invented:

> E.C. Atkins, founder of E.C. Atkins & Co., was also the discoverer and inventor of the formula and process for making silver steel. When the new diamond point tooth, as used in the Perfection, Peerless and Rex cross-cut saws, was invented, Mr. Atkins realized at once that this style of tooth, being required to do much more cutting than had been necessary with the old-style tooth, must be made of finer steel than had ever been used before. No manufacturer of saw plate in the United States had, at that time, the facilities for making as high-grade steel as Mr. Atkins determined to use in his saws. And so he went to England. He went to see Jessops, who were and still are regarded as the very foremost manufacturers of crucible steel in the world. Being an expert metallurgist, he prepared his formula and handed it to Mr. Jessop with the request that a few plates be made according to specifications. Mr. Jessop was astonished and exclaimed: "Do you mean to say that you expect to use such steel as this in saws?" "I will either use that steel or something finer," was Mr. Atkins' reply.

Experimental quarters were erected on the grounds at the Jessop plant and it was here that silver steel saws were born. The first batch of plates were made and turned into saws by Mr. Atkins' own hands and carefully tested by him. "Not right yet," was the verdict, and so Mr. Jessop was again directed to change the formula. "Put in more carbon and add more of this and that," said Mr. Atkins. "Why, man," said Mr. Jessop, "that's razor steel. It's too good for saws." "It's not too good for Atkins saws," was the reply. "You make the kind of steel I tell you to. I know what I am doing." The Jessops shook their heads and tapped their foreheads and thought – "poor man, too bad." And so the new batch was made, and then another and another, until at last the proper quality was secured. A steel that had the quality of receiving a hard, tough temper, stiff and firm, but not brittle. That would receive a very keen, sharp cutting edge and hold it for a remarkably long time with but little filing. "That's silver steel," said Mr. Atkins, "you may enter my order for 5,000 plates." Contrary to their own judgment, and with reluctance, the order was filled and the plates were shipped. The wisdom shown is best demonstrated by the immediate and constantly growing

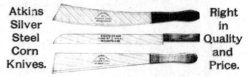

demand for "silver steel saws." From that day to this, everywhere a silver steel saw is used it makes a friend, and each friend makes another, until today Atkins' silver steel saws are recognized the world over as the standard for everything a perfect saw should be.[6]

Steel Names, Specifications, and Standards

One difficulty in researching the historical development of steels is that the common names by which we know them did not come until years or even decades after they were invented. So, searching for "AISI W1" or "W1" steel in old trade magazines, books, and journal articles would not result in anything useful because that designation did not exist until the 1950s when tool steel specifications were standardized. So, for unpatented new tool steels, it is necessary to find the brand name which appeared first in those early sources. Because steel development and copying were so rapid at that time, there is no guarantee that the first brand to appear was the first example of that alloy. Even a company claiming to be the first to make something is not a guarantee, as there are many examples where these were false claims. While it is generally possible to narrow the appearance of steels to close to when they were first released, for some there is a 1-5 year delay until they appear in print. And even then, whether the first one to appear is actually the first to be released is often in question. I have noted when a steel is the first example I have been able to find vs. when I know that it was the first such example. Frequently the earliest available composition for a steel is many years after it was released, so some steels may have been released with a somewhat different composition.

Said the Carpenter:

"I've heard all these wonderful tales about that fellow, Saladin and his Damascus blade—how he could bend it double and cut a silk thread with it.

"That's nothing! I can bend my Atkins saw double and twist it all around, and it will spring back into shape without a kink. Wish my wife had that kind of temper! She gets all kinds of kinks and wrinkles when she gets out of shape.

"Then I've heard of these wise lads that write books telling about the 'lost art' of making such steel. I guess they have never heard of SILVER STEEL, made on the Atkins formula, nor about the Atkins secret process of tempering and hardening! Cut a silk thread! Huh! I can cut bar iron and never hurt a tooth! Can't tell me the old Arabs could make as fine steel as that!

"You can't see the way this saw-blade tapers, but I can FEEL it when I saw. That means that the saw makes leeway for itself, runs easy and never buckles. Mebbe you don't know what that means to a carpenter, but I do. Saves a lot of time and wicked words.

"Then that perfection handle! They have the right name for it, all right. Don't have to hump over and pull straight up and down and break your back. You can stand up and get the best angle for cutting.

"Pay more for it? Of course! Isn't it worth it? When you do anything you want to do it right, and it takes a first-class man with first-class tools to do a first-class job.

"Ask your dealer for Atkins saws, and don't let him substitute a 'just as good.' There are none as good as the Atkins."

Ask for our Carpenters' Universal Time Book. It is free.

SILVER-STEEL

This Trade Mark, indicating the finest steel ever made in modern or ancient times, is found only on the saws—all types and sizes—made by

E. C. ATKINS & CO., Inc.

LARGEST SAW MANUFACTURERS IN THE WORLD.

Factory and Executive Offices, INDIANAPOLIS, INDIANA.

BRANCHES: New York City, Chicago, Minneapolis, Portland, Ore., Seattle, San Francisco, Memphis, Atlanta, and Toronto, Can.

Both Atkins Silver Steel ads from 1904-1905.

Carbon Steel Specifications

W1 is a steel that has been used in many knives though only rarely advertised as being W1. In part because many of the knives made with it would have been produced before the W1 name existed. For example, many carbon steel pocket knives were produced in what could be classified as W1 even though many manufacturers eventually switched to the similar 1095 or a "chromium vanadium" alloy steel. These steels were sold to knife companies simply as "cutlery steel" or "pocket blade steel." For example, William Scagel, considered the earliest American custom knifemaker, made knives with simple high carbon steel such as SKF "Carbon Cutlery" or Jessop's steel[7] which would have been W1. Camillus used a simple cutlery carbon steel with 1-1.1% carbon made by Crucible until at least the 1950s[8,9]. As noted by Seebohm in 1881, razor steel was historically a simple, very high carbon steel, with early sources reporting 1.5% carbon. However, this carbon content was reduced somewhat by the 1920s to 1.25%[10] or 1.3%[11]. Some razor steels began to have a small chromium addition by the 1930s[11]. The Japanese Hitachi White/Shirogami series used by bladesmiths are also W1-type steels with 0.8-0.9% C (White #3), 1.05-1.15% C (White #2), or 1.25-1.35% C (White #1)[12]. The Hitachi Yellow series steels are similar but with higher allowable impurities.

Steel	C (%)	Mn (%)	Si (%)
W1	0.60-1.40	0.1-0.4	0.1-0.4
1095	0.90-1.03	0.3-0.5	
1084	0.80-0.93	0.6-0.9	
1075	0.70-0.80	0.4-0.7	
1050	0.48-0.55	0.6-0.9	

W1 has a very wide carbon range (0.6-1.4%), unusually wide compared to other steel specifications, because of the history of offering carbon steels at different carbon contents as described above. The wide specification isn't because the steel can be at any given carbon content and still "meet" the specification, but because it was common for one brand name of carbon steel to be used along with different "temper" or carbon content targets. W1 most typically refers to 0.95-1.1% carbon steel, while other carbon levels of W1 were unused primarily in favor of the SAE "10" designation, such as 1075 steel. Note: the specifications for the "10" series steels do not have a Si requirement, but all commercially produced "10" series steels have a small Si addition like W1.

Early Attempts to Classify Tool Steels

There were several attempts at steel specifications, such as the one for carbon steel by Darke described previously. The United States Navy first provided specifications for tool steels in 1908-1909[13]. They specified high speed steel as roughly similar to the standard 18% tungsten high speed steel, later named 18-4-1 or T1, though the vanadium content was lower at ~0.3% rather than the increasingly common 1%.

"Grade A" was a grade F2 steel described in the next chapter, "Grade B" was a simple carbon steel with more than 1% carbon, and "Grade C" was a simple carbon steel with less than 0.9% carbon.

Alloy Steel Specifications

When we started to get specifications that are still relevant today, there is the Society of Automotive Engineers (SAE) which created the alloy steel specifications we still use. The original specifications in 1911[14] included only sequential numbers, i.e., "Specification No. 1," "Specification No. 2," etc., along with a simple description like "0.20 Carbon Steel," "0.45 Carbon, Chrome Vanadium Steel," or "Silico-Manganese" steel. In 1912[15], they introduced the two number prefixes, such as "61" for chrome vanadium steels. For example, 6150 refers to chrome vanadium steel (61) with 0.50% carbon (50). The prefix "92" referred to silico-manganese such as 9250 or 9260. "51" and "52" referred to chromium steels, with 52 having more chromium than 51, such as 5295 in 1912, later changed to 52100. The original 1912 report included dashes between the prefix and carbon content, such as 61-50, but those were removed in the 1914 report[16]. Simple carbon steels were given the "10" prefix, such as 1075, 1084, and 1095.

Grade	C (%)	Mn (%)	Si (%)	Cr (%)	V (%)
5160	0.56-0.64	0.75-1.0	0.15-0.30	0.70-0.90	
6150	0.48-0.53	0.70-0.90	0.15-0.30	0.80-1.10	0.15 min
9260	0.56-0.64	0.75-1.0	1.80-2.20		
52100	0.98-1.10	0.25-0.45	0.15-0.30	1.3-1.6	

Tool Steel Specifications

The standard tool steel specifications came much later. Tool steel manufacturers generally opposed standard specifications because they felt this would mean that consumers viewed steel as a commodity rather than seeing the superior quality, or at least unique attributes, that their brand provided[17]. Of course, this did not stop publications from grouping steels into general categories such as "high speed steel," "finishing," "oil hardening," etc., even if the exact compositions were not given[18,19]. Typically the compositions of various brand names were not advertised. This meant that in journal articles they often reported experiments while giving a composition without providing the attached brand name, or reports were made on a specific brand of steel without providing the composition.

The tool steel designations began with the Gorham Tool Company in Detroit, which came up with its internal names for high speed steel grades to categorize them for use despite the different manufacturer brand names[20]. In 1937 a customer asked for a copy of Gorham's classifications which led the president of the company L. Clayton Gorham to publish a booklet on the classification system. By 1945 General Motors, Chrysler, and Ford agreed to adopt the system as well, which was in its 4th edition[20]. This categorized the steels as either a tungsten-alloyed high speed steel "T" or a molybdenum high speed steel "M." Then the grades were assigned a number such as T1, M2, etc. The T-series includes steels almost exclusively alloyed with tungsten though the M-series includes many steels alloyed with both molybdenum and tungsten, such as M2 with ~6% W and ~5% Mo. By 1954 this general classification system was applied to other tool steels by the American Iron and Steel Institute (AISI) along with help from SAE and ASM[21]. Along with the "T" and "M" high speed groups, they added water hardening steels (W), shock resisting steels (S), oil hardening steels (O), air hardening cold work steels (A), high carbon, high chromium cold work die steels (D), hot work steels (H), low alloy chromium steels (L), carbon-tungsten low alloy "finishing" steels (F), and mold steels (P).

Stainless Steel Specifications

The AISI number system for stainless steels was introduced in 1932[22], which included the "300" series of austenitic stainless steels alloyed with chromium and nickel like 302 and 304. And the "400" series for chromium-alloyed ferritic and martensitic alloys (tool steels and knife steels are martensitic). These specifications included the original 420 stainless steel (13% Cr). There was also a specification for "440" stainless steel (17% Cr), but this was a very low carbon steel (0.13-0.20% C). The specifications for the common high carbon knife steels 440A, 440B, and 440C were added in 1943, and the original 440 specification was removed[23].

Grade	C (%)	Cr (%)	Mo (%)
420	0.15 min	12-14	
440A	0.68-0.75	16-18	0.75 max
440B	0.75-0.95	16-18	0.75 max
440C	0.95-1.20	16-18	0.75 max

Robert A. Hadfield and Early Steel Alloying

Robert A. Hadfield added high amounts of manganese to steel in 1882[24], discovering a steel referred to as "manganese steel" or "Hadfield steel." Hadfield steel is austenitic at room temperature for an excellent combination of strength, ductility, and wear resistance. He patented this manganese steel in 1883-1884[25]. He also developed 1.5-5% silicon steels with excellent magnetic and electrical properties[26]. Neither of these steels are useful for knives, but these products increased interest in alloy steel design. Hadfield was a prominent metallurgist who published many papers on alloying steel, including manganese[27], silicon[28], chromium[29], tungsten[30], and nickel[31]. So, his influence was not restricted solely to the development of alloys, but also summarizing his experiments and those of others. The next chapter will cover several early alloy steels that have been used in knives and tools.

-3-

NEW ALLOY ADDITIONS IN STEELS

Early Bearing Steels - 52100

Early bearing steels were simple high carbon steels with 0.80-1.0% carbon. In the early 1900s, chromium additions were being tested for bearings, and by 1903 a steel with 1% chromium was produced by the German company Deutsche Waffen- und Munitionfabrik[1]. The first steel to have the 1% carbon and 1.5% chromium of 52100 was produced by Sachs AG Schweinfurt in 1905. This company was then taken over by Svenska Kullagerfabriken AB (SKF), a Swedish company, in 1929, which sold 52100 under the name SKF3. The chromium additions were made for enhanced "hardenability" so that the ball bearings could be hardened all the way through. In fact, some companies added different amounts of chromium depending on how large the ball bearings were. The chromium additions also led to finer and more numerous carbides than in plain carbon steel[2], providing higher toughness and fatigue resistance. There have been some modified grades of 52100 for different reasons in the years after. However, 52100 is typically the base choice for those applications that do not require corrosive or high temperature environments. Molybdenum-alloyed and vanadium-alloyed versions were tested by 1927[3,4]. Molybdenum allows the heat treatment of larger bearings, while vanadium provides grain size control. Higher manganese versions were also developed for improved hardenability for larger bearings, including SKF1 introduced in the 1930s, containing 1% carbon, 1% chromium, and 1% manganese[1]. Germany has a 52100 grade with 1.5% Cr and 1% manganese introduced for the same reason in 1934, designated 100CrMn6[1]. The vanadium-alloyed version was called "Type BB" or "Vasco BB" by its inventor, Vanadium Alloys Steel Company (VASCO)[5]. For the tool steel naming convention, 52100 is called L1, the vanadium-alloyed version is L3, and the molybdenum-alloyed version is L7.

William Scagel is the first reported knifemaker to use 52100, using bearing races (also made from 52100) during WWII when steel was less readily available[6]. Ed Fowler is due some credit for popularizing 52100 as a knife steel. He says that Wayne Goddard first suggested the use of ball bearings to him[7]. D.E. Henry was the first knifemaker to use Vasco BB/L3 in 1960[8],[9].

Steel	C (%)	Cr (%)	V (%)	Mo (%)
52100	1.0	1.5		
L3	1.0	1.5	0.2	
L7	1.0	1.5		0.4

Low Alloy Tungsten Steels

F2, F3, and "Finishing Steels"

While Mushet Special Steel was getting a lot of attention for its air hardening characteristics, other tungsten steels were also developed. A category of steels called "finishing steels" were developed that were high in carbon (~1.3%) and high in tungsten (6-7%) like Mushet steel but were missing the large additions of manganese or chromium[10-12]. These steels tended to graphitize during prolonged heating cycles, but it was discovered that a small addition of chromium (0.3-0.5%) helped suppress this problem. The tungsten content was lowered to 3-5% for future finishing steels. The steels were often named "Double Special"[13-16] by the various manufacturers and nicknamed "finishing steels," and were later given the AISI designations F2 and F3. These water hardening steels could reach very high hardness (66-70 Rc), which gave machining tools the ability to hold a very keen cutting edge for the final, or finishing, cuts. The formation of tungsten carbides (WC or W_2C) means they have high wear resistance, particularly for water hardening steels. These steels were also known for their extremely fine grain size.

These finishing steels have never seen wide use in knives, as air hardening and oil hardening steels largely replaced them. Though in recent years, the European version 1.2562 has seen some use.

Steel	C (%)	Cr (%)	W (%)	V (%)
F2	1.3	0.3	3.5	
F3	1.25	0.75	3.5	
1.2562	1.4	0.3	3	0.2

O1, O2 and "Oil Hardening Steels"

A new category of steels referred to as "non-deforming," "non-shrinking," or "oil hardening" was created with the introduction of Halcomb Ketos steel in 1905[17-18], later given the standard designation of O1. Halcomb Steel Co. was created when Charles Herbert Halcomb left as the President of Crucible Steel to form his own competing company. Crucible had started only a few years prior in 1900 as a consolidation of the 13 largest steel companies producing steel with the crucible method, accounting for 95% of production at that time. Halcomb had become president of the largest, Sanderson Brothers, after immigrating to the United States to work there at 22 years old in 1881. When Crucible was formed, Halcomb was made the president of the entire company. Several business people, and US Steel, had purchased significant shares in Crucible Steel and wanted Halcomb out and a new President put in place[19]. One rumor was that US Steel wanted to absorb Crucible and Halcomb opposed this[19]. Another rumor was that these new "stockholders objected to the lavish furnishment of his new offices"[20]. Halcomb Steel had the first electric furnace for steelmaking in 1906[17] and built the largest high alloy steel plant in the country[21]. The new plant he made was right next to Crucible's Sanderson Works in Syracuse, NY. However, after all this investment, Halcomb Steel was relatively short-lived, at least as an independent company, as it was purchased by Crucible in 1911[22]. It was reported then that Crucible wanted the Halcomb company to obtain the electric furnace because of its superior capabilities[23].

John Mathews was the metallurgist who developed Ketos/O1. Mathews[24] was born in Washington, PA in 1872 and got a bachelor's and master's degree from Washington and Jefferson College. He went to Columbia University and earned a Ph.D. in Chemistry in 1898, and he was given a fellowship at the Royal School of Mines in London. He studied metallography under Sir William Roberts-Austen, famous for his research on metals. The iron/steel phase austenite was named after Roberts-Austen. Mathews returned to Columbia in 1901 and worked on a series of alloy steels that were made for him at the Sanderson Works of Crucible, and the Iron and Steel Institute gave him the Carnegie Gold Medal for his published work. He then began working for Crucible Steel at the Sanderson Works before leaving to work for Halcomb where he developed O1 (and other grades, as will be discussed in the section on vanadium alloys). Mathews was well known for his metallurgy contributions worldwide and published many technical papers and served on many committees. His obituary writer reported he had a "kindly personality and the charm of his company endeared him to all."[24] Those that worked with him "always had a real friend, who combined those rare qualities of leadership in both the practical and scientific side of their work, of a patient teacher and a reliable guide"[24]. He continued working for Crucible after the company purchased Halcomb Steel, where he was the general manager from 1913-1920, president of the company from 1920-1923, and then left upper management to be VP and Director of research because "Three years of being a president of a steel company was enough for any real scientist"[24].

Though Mathews did not explain how he developed the steel, O1 was a cut-down version of Mushet Special Steel, with reduced Mn so that it hardens easily in oil, even in large cross-sections, but does not harden in air. It also had small amounts of chromium and manganese like many other low alloy tool steels in 1890-1910.

Mathews talked about the development of O1 in 1930:

> It may be a surprise to you to know that prior to 1905 there was no oil-hardening die steel. Some tool steel, particularly very light gages, and edged tools, were from time to time hardened in oil, but the introduction of Ketos steel at the beginning of Halcomb's activities was the first time that any steel had been available which would harden sufficiently for dies when quenched in oil in fairly large masses. This not only was of great advantage from the standpoint of safety in hardening of intricate tools but a secondary advantage arose from the minimum of distortion or change in shape which was characteristic of their oil-hardening Ketos steel. While this product was not of as great primary interest to tool steel users as high speed steel, yet it was enthusiastically received and filled a real need in tool and die shops.[17]

A simpler version of the steel was developed, later designated O2, which did not have Cr and W additions. The earliest version of this grade I have found was Carpenter's Stentor, trademarked in 1908[25].

Steel	Year	C (%)	Mn (%)	Cr (%)	W (%)	V (%)
O1	1905	0.9	1.2	0.5	0.5	
O2	1908	0.9	1.6			
1.2842		0.9	2	0.3		0.1

The earliest use of O1 in knives I know of was by W.D. "Bo" Randall by 1940[26-28]. In terms of production knives, Marble was marketing knives in the 1950s as using "high carbon manganese steel"[29], but it might require too many assumptions to call that O1 or O2 as the description is very generic. More recently, the European version of O2, 1.2842, has seen use in custom knives in that region, particularly in pattern-welded Damascus.

O7, Intra Steels - Blue Super, V-Toku 1, 1.2519

Another series of steels which were essentially cut-down Mushet Special steels, started with Böker's Intra steel introduced around 1907 and was advertised as "between high speed steel and carbon steel"[30]. It had lower tungsten than the finishing steels at 1.3%, which gave it some wear resistance compared to carbon steels but not as extreme as F2.

These steels were given higher hardenability with 0.5% chromium and 0.2% vanadium for grain size control, such as Ludlum Utica steel, which was advertised in 1921[31], and similar steels were later given the designation O7. Steels that fit this category have been used in knives under various names, such as the Aogami/Blue series of steels and V-Toku series sold in Japan. And in Europe as 1.2516, 1.2442, and 1.2519.

1921 advertisement for Utica steel.

Steel	C (%)	Cr (%)	W (%)	V (%)
Intra	1.2		1.3	
O7	1.2	0.5	1.5	0.2
1.2516	1.2	0.2	1	0.1
1.2519	1.1	1.2	1.3	0.2
1.2442	1.15	0.2	2	
Blue #1	1.3	0.4	1.75	
Blue #2	1.1	0.3	1.25	
Blue Super	1.45	0.4	2.25	0.4
V-Toku 1	1.15	0.35	2.25	0.1
V-Toku 2	1.05	0.25	1.25	0.1

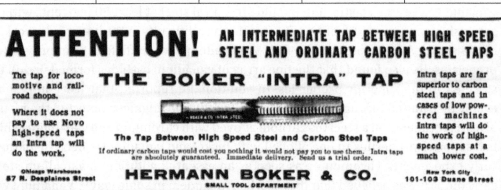

1908 Boker "Intra" advertisement.

Vanadium Additions to Steel

J. Oliver Arnold of Sheffield was the metallurgist who performed the early research on vanadium additions to steel. At 20 years old, he began working in the engineering department of steel company Brown, Bayley & Dixon, where he eventually became chief chemist and test master[32]. He became a consulting chemist for the Farnley Iron Company and then a chemist and manager of the Spanish Steel Works in Sheffield and then a private consultant. In 1889 he became Professor of Metallurgy at Sir Frederick Mappin's Technical School, which combined with Firth College in 1905 to become the University of Sheffield. And about 40 of the biggest Sheffield steel companies formed a "more or less secret tribunal" called the Sheffield Steel Makers Ltd, which retained Arnold as a scientific expert. Despite these many ventures, Arnold was not paid very much for these varied jobs, and when he died, he had only 115 British pounds. "Arnold seemed to have been driven more by the scientific interest of his work, reveling in his close contact with firms and industrialists, rather than by any concern to amass riches for himself. Accordingly, the cheapness and availability of his scientific services was one of the hidden strengths on which the steel industry could rely."[32]

Studying Vanadium Additions

In 1899 a sample of ferrovanadium was given to Arnold by Augustus F. Wiener, and Arnold researched vanadium additions at the Vickers's River Don Works. Wiener said that "research work has continued uninterruptedly by Professor Arnold as regards the properties of vanadium and his name must be connected for all time with the practical application of vanadium in the steel industry"[33].

Arnold reported that[33] "It is evident that as a steel-making element vanadium will place in the hands of metallurgists and engineers a very powerful weapon, because it is now demonstrated beyond doubt that the addition of a few tenths per cent. of vanadium raises the elastic limit of mild structural steel at least 50 per cent., without seriously impairing its ductility, or presenting any difficulty in the hot or cold working of the steel." Arnold found that ~0.25% vanadium was sufficient to improve steel properties. He also found that the fine array of vanadium carbides formed within the steel led to the superior properties. In addition, he reported that "carbon is more evenly distributed throughout the mass."[33] In 1902, an engineer reported the results of his experiments on a steel that Arnold alloyed with chromium and vanadium, which had superior properties to one alloyed with nickel and vanadium[33]. Because of these early experiments, chromium was seen as an essential addition in vanadium steels early on, as it was believed that they worked in tandem in some way[34]. Vanadium was found to be effective at "deoxidizing" molten steel by forming a compound with undesirable oxygen, which is then removed with the slag. Vanadium-alloyed steel was also promoted as a better-performing alternative to nickel-alloyed steel, for applications outside of tool steel.

Chrome-Vanadium Spring Steel - 6150

Arnold developed a steel with 0.5% carbon, 1% chromium, and 0.25% vanadium, later called 6150, manufactured by Vickers for automobiles. A large paper was published in 1904 in the *Proceedings of the Institution of Mechanical Engineers* by J. Kent Smith[35]. They reported experiments on a range of vanadium-alloyed steels, including Arnold's steel for springs with excellent properties. After reading the report from J. Kent Smith, Ford hired him to produce the steel commercially[36]. 6150 became a standard spring steel used by Ford and was included in the early SAE steel specifications.

Steel	C (%)	Mn (%)	Cr (%)	V (%)
6150	0.5	0.8	1	0.2

Chrome-Vanadium Cutlery Steel

Steels have been developed for cutlery and tool and die applications with high carbon plus vanadium. Steel with a composition of 0.82% carbon and 0.25% vanadium were reported as early as 1903, which would be W2 steel[37]. W2 became a standard tool steel and has been used in many knives.

1907 Ford ad[38].

The American Vanadium Company found a deposit of vanadium in Peru in the Andes, which led to much greater availability of vanadium and therefore lower cost[39]. At the high elevation the mines were located, the ore had to be carried on the backs of llamas, then transferred to boats, railway, and steamships, and then processed in Bridgeville, PA, near Pittsburgh. The American Vanadium Company freely published compositions of various steels that could be produced with vanadium in 1912[39]. Among those published included a "type H" grade (unrelated to later H-series tool steels) with two example compositions: 0.8% C, 0.9% Cr, 0.2% V (later called L2) or 0.9% carbon, 0.5% chromium, and 0.2% vanadium (the common chrome-vanadium knife steel). This grade was listed as being suitable for cutlery, among other uses. This grade was also previously published as a die steel in 1909, so it existed at least as early as then[40]. In a 1926 publication[41], "Typical Analyses of Cutlery Steel" were given, which included "straight carbon" grades with 0.85-1.05% carbon with small amounts of manganese and silicon, "chrome" grades which were 0.9-1.1% carbon with the same level of Mn and Si but also 0.75% chromium, and "chrome-vanadium" types with 0.85-1.05% carbon but with 0.5% chromium and 0.2% vanadium.

Grade	C (%)	Mn (%)	Si (%)	Cr (%)	V (%)
Carbon #1	0.85-0.95	0.25-0.30	0.15-0.20	-	-
Carbon #2	0.95-1.05	0.25-0.30	0.15-0.20	-	-
Chrome #1	0.90-1.00	0.25-0.30	0.15-0.20	0.70-0.80	-
Chrome #2	1.00-1.10	0.25-0.30	0.15-0.20	0.70-0.80	-
Chrome-vanadium #1	0.85-0.95	0.25-0.35	0.15-0.20	0.45-0.55	0.15-0.25
Chrome-vanadium #2	0.95-1.05	0.25-0.35	0.15-0.20	0.45-0.55	0.15-0.25

1926 compositions of cutlery steels[41].

Vanadium-Alloyed High Speed Steel - T1

Concurrent with the research into vanadium-alloying to improve spring and structural steels, experiments were also performed on the development of tool steels. In 1900 Professor Arnold analyzed a steel with 1.25% C and 3% vanadium which was found to cut 75% longer than the same steel with 3% tungsten[39]. John A. Mathews of Crucible Steel patented vanadium additions to high speed steel of up to 1% in 1905[42]. He recalled in 1919:

> [I] began experimenting with the use of vanadium in 1903, and it is well to bear in mind that at that time vanadium was almost a chemical curiosity. It was worth about $15.00 a pound, and this was some time prior to the formation of the American Vanadium Co. which manufactured and sold vanadium in large quantities. So far as [I am] aware, the entire stock of ferro-vanadium in the country when these experiments were begun consisted of not over 100 lb. in the hands of two different dealers in New York. We purchased one-half of the entire stock of each dealer. As the result of these experiments carried on at the old Sanderson Works, a patent was granted [to me], issued on January 3, 1905. Other experimenters were doubtless working with the same thing, and, in fact Mr. Gledhill referred to its use in 1904, as did also Mr. Taylor. In fact, the composition of tool steel previously referred to as giving Mr. Taylor his best results, showed 0.3 per cent of vanadium. During the year 1905, the Rex AA steel was put upon the market, and other vanadium steels followed shortly, but it was not until three years later that certain foreign makers copied this original steel exactly and made great claims as to originality in regard to their product.[43]

Mathews developed the Crucible Rex AA grade, later designated 18-4-1, or T1, the most common high speed steel for the next few decades. The patent was for vanadium additions to a high speed steel in general, so it is hard to know what the exact composition was of the original Rex AA[43]. But two published compositions in 1919[43,44] show 0.63-0.77% carbon, 16.9-17.8% tungsten, 3-3.5% chromium, and 0.75-0.85% vanadium. And Mathews did say the Navy tested a steel of about the same composition a year after their tool steel recommendations (~1910)[43].

REX "A" HIGH SPEED STEEL IS THE BEST
Crucible Steel Company of America

1905 Rex A ad.

Crucible Steel Company of America
PITTSBURGH, PENNSYLVANIA
MANUFACTURERS OF

REX AA   REX A
HIGH SPEED STEEL
THE HIGHEST DEVELOPMENT IN STEEL MAKING

HEAVIER CUTS   HIGHER SPEED   LONGER SERVICE

1910 Rex AA ad[45].

With 6150 steel being common in early springs, it has been used in quite a few custom knives, such as by Rudy Ruana, though by many custom makers it was often unlabeled, or mislabeled as 5160. The higher carbon "chrome-vanadium" cutlery steel is more widely known for its use in production pocket knives, especially in W.R. Case Knives[46]. T1 high speed steel was used in early Gerber and David Z. Murphy knives. W2 is known for being used by many custom makers, such as Bill Moran. L2 has not been typically used, but 80CrV2 with somewhat less chromium is a descendant of the two "Type H" grades given above. 80CrV2 has seen regular use in recent years; it is either a lower chromium version of L2 or a lower carbon version of the chrome-vanadium knife steel.

Steel	C (%)	Mn (%)	Si (%)	Cr (%)	V (%)
80CrV2	0.8	0.4	0.25	0.5	0.2

-4-

CARBON STEEL USA FACTORY KNIVES

San Francisco Knives and Chrome Steel

San Francisco famously grew thanks to the 1849 gold rush, and many knives were needed[1]. At this time, there were relatively large knife manufacturers in Massachusetts and Connecticut, such as Lamson & Goodnow. As a result, many people in San Francisco would obtain knives made by those companies. But soon, cutlers set up in San Francisco to deliver locally-produced products for the growing city. The earliest of these cutlers was Hugh McConnell, an Irish immigrant who began selling knives in San Francisco in 1852. The most famous cutlery companies in San Francisco were Michael Price and Will & Finck. Will & Finck started in 1863, and cofounder Frederick Will had previously worked for McConnell and for McConnell's chief competitor Frederick Kesmodel. McConnell died in 1863, and Frederick Will set up shop in McConnell's cutlery location. The partner was Julius Finck, a German-born locksmith and bellhanger. The primary competitor for Will & Finck was Michael Price, who set up shop in 1859. Price and his father learned to make cutlery in Limerick, Ireland. These companies made some of everything from razors to carving sets to Bowie knives.

It is perhaps misleading to describe either of these companies as "factories," as both had a relatively small number of employees and did not use automated equipment. The knives were even advertised as "home made." By 1882 there were about 50 employees split between the two cutlery manufacturers[1]. In Sheffield, there were many similar small cutlery companies. Perhaps the closest current analog would be Randall Made Knives. At the 1874 Mechanics' Fair, both companies showed carving sets made with "chrome steel, " presumably referring to the Julius Baur-patented steel made by Brooklyn Steel Works branded as Adamantine with 1% carbon and 0.5% chromium[2]. However, some companies were branding their steel as "chrome steel" when it contained no chromium, as the steel was generating some buzz, and unscrupulous companies wanted to cash in[2,3]. The only "chrome steel" produced in the United States then came from Brooklyn Steel Works[3]. Assuming they were using genuine "chrome steel," this is the first record known of a knife company using alloy steel. The San Francisco newspaper *Alta California* described Price's knives in the steel on September 6, 1874:

> Directly opposite the musicians' stand is a large plate glass show case containing a quantity of California manufactured cutlery which attracts the attention of everyone who passes by. This show case contains one dozen carvers made by M. Price of chrome steel. This steel is generally acknowledged to be superior to every other kind manufactured; it has passed successfully through the severest tests. It is an alloy of the two metals iron and chromium, and besides being far in advance of anything hitherto known for strength, ductility, plasticity, and uniformity, it possesses some features which are peculiarly its own. These carvers were manufactured by Mr. Price on an order received from the East. They are highly polished, with ivory handles, and are considered 'the thing' for the dining table. The show case also contains a number of razors, scissors, pocketknives, etc., manufactured expressly for this exhibition. The exhibitor has already received a number of prizes for his cutlery, and no doubt had there been premiums offered, he would again be the successful competitor.[1]

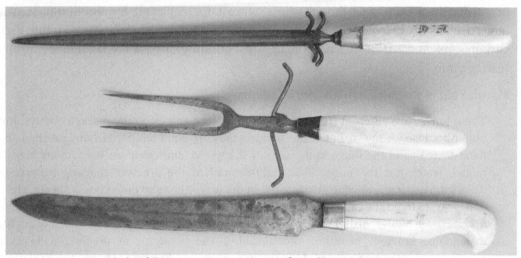

Michael Price carving set. Image from Heritage Auctions.

Hunting Knives - Marble Safety Axe Co.

Before Webster Marble, there was no such thing as a "hunting knife."[4-8]. There were knives used in skinning game, but these were typically kitchen, butcher, or Bowie knives, not specifically designed for general use in hunting and skinning. Around the late 1890s, hunting, fishing, and other outdoor activities began to be viewed as recreation rather than necessary for survival. This led to the introduction of products intended for those purposes. Marble was an avid outdoorsman and worked for two decades in the lumber industry. He said[8]: "I took to the woods like a duck to water. The greatest enjoyment of my life has come in the woods. As a young fellow I got my greatest thrill starting out with a map and a compass to explore new territory, to carry very little equipment, and to stop wherever night overtook me. It was gratifying to feel I was remote from everything civilized, and exploring new regions."

He invented his first gunsight in the early 1890s[9]. He next invented the Marble "Safety Pocket Axe" in 1898[10], after which his company was originally named. This was called a "pocket axe" because it was small, about the size of a hatchet. And it was a "safety axe" because the handle contained a guard that rotated over the blade when not in use. Because of this invention, he built a workshop near his home where the safety axes were constructed, and other inventions such as the waterproof matchbox[11]. In 1900 he would build a 24,000 sq. ft. factory with investments from a new business partner, Frank H. Van Cleve, and a financier named James T. Jones. His first line of hunting knives, introduced around 1900, was Marble's "Ideal" hunting knife. It had a hollow ground "fuller" in the blade, giving the knife a thin edge but a thick spine for strength and reducing weight. The handle was made of either leather washers or stag. Marble brought the leather washer handle into "widespread use and acceptance," according to knife historian Bernard Levine[12]. The knife also had a small guard, and originally horn butt caps though those were later replaced by aluminum. It was advertised as made from "razor steel" or "cutlery steel" which refers to a simple high carbon steel. And a 1906 ad said, "the greatest possible care is taken in tempering and testing them."

Marble said of the design in his 1905 catalog:

> These knives are different from any other one-piece knives on the market - better. Not better because we say so, but better because they are hand tempered, hand finished and hand tested from the finest knife steel in the world, and made on the famous Marble model, which has the unqualified endorsement of the greatest sporting experts of America. They have no wide guard to get in the way - they slip down into a sheath, clean and clear, so that they cannot fall out or catch on a bush and be pulled out, and they hold a keen edge under ordinary use without nicking or bending. They are thoroughly dependable knives - beautiful and durable... The blade, as at present made, is a modification of the two shapes of blades formerly made, known as sticking and skinning points, and is claimed by many expert hunters and woodsmen to combine more of the essential qualities for all-around use than are usually found in one style of knife... The bone chopper at back of point is a valuable feature for rough work.

Different variations of the Ideal were introduced in terms of blade length, including 5, 6, 7, or 8 inches. Marble introduced several other designs of hunting and outdoor knives, including the "Expert" in 1906 and the popular "Woodcraft" knives in 1916[13,14].

He changed the name of his company in 1911 to Marble Arms & Mfg. Co. Apart from introducing the idea of a hunting knife, Marble's knife designs were highly influential and widely copied. Many early 20th century fixed blade knife designs look like descendants of Marble knives, such as the "USMC Fighting Utility," also called the "Mark 2" or "Ka-Bar" knife[4]. Knifemaker Bob Loveless named Webster Marble as one of his three knifemaking heroes along with Rudy Ruana and Harry Morseth because Marble's experience in the woods led him to design knives that really worked[15].

Loveless said[15], "Marble's knives were twice as expensive as anybody else's and four times as good."

Marble Ideal Hunting knife ad from 1901.

MARBLE'S IDEAL HUNTING KNIFE.

5 inch.

These knives are made from the finest cutlery steel, the greatest possible care is taken in tempering and testing them, and they are fully guaranteed. They have bone chopper at back of point, are heavy and keen. The entire knife is hand made and highly polished, and while being the handsomest knives ever produced are also the most reliable. Tne No. 1 handle is made of leather, colored hard fibre and brass washers that are a driving fit on the heavy tang. The handle parts are held in place by a nut countersunk into the end of the stag tip. We are now furnishing, without additional cost, an ALUMINUM SHEATH PROTECTOR as well as the regular tube sheath with every

Blade 5 inches.............................price $2.25. Given for five new subscribers.
Blade 7 inches.............. .... price 2.75. Given for six new subscribers.

1906 ad featuring the cross-section of the fuller grind.

Camillus

Adolph Kastor immigrated to the USA from Germany in 1870 at 14 years old, working for his uncle Aaron Kastor in his hardware supply business[16]. He began an importing business in 1876, selling general hardware made in Germany. In 1883 Kastor met with a German manufacturer of pocketknives and began importing German knives. The McKinley Tariff of 1890 and the Dingley Tariff of 1897 significantly increased the costs of importing knives, so he sought domestic manufacturing. A knife company started by Charles Sherwood in 1894 in Camillus, New York, had twenty employees and fifteen different pocket knife patterns. Kastor felt the company could be ramped up with enough financing. In 1902 Kastor purchased the company and renamed it Camillus Cutlery Co. A second building was constructed, being announced as completed at the end of 1902[17], which was modern, being "supplied with an electric motor and the necessary installation for supplying their own electric light." This announcement also mentioned the steel type used, which was S&C Wardlow Cutlery steel, a simple high carbon steel made in Sheffield. The steel was blanked with hand dies and punch presses and then the blanks were forged[18]. The blades were then[18] "heat-treated in oil-fired furnaces, and here again the workman knew only by eye when the proper temperature had been reached for quenching. One blade at a time. This was standard procedure. Thanks to his experience and the high quality of the steel, he produced a fairly good uniform blade."

Camillus Cutlery Company

CELEBRATED

Salesroom:
109 DUANE ST.
New York City

SWORD ✕ BRAND

Factories:
SYRACUSE, N. Y.
CAMILLUS, N. Y.

AMERICAN POCKET KNIVES

1920 Camillus ad.

Camillus switched to USA-made steel produced by Crucible Steel around 1919. William Wallace of Camillus said[18], "The first cutlery steel produced by Crucible Steel Company of America was developed for Camillus around 1919. Steel is now purchased in coils at a saving of much handling. Progressive dies, working with close tolerances, insure uniformity of parts. Scientific heat treatment and close inspection give long life and proper hardness to the blades." Crucible had sold cutlery steel before that date, so I'm not sure what this refers to, whether they began making steel in more significant quantities or with new dimensions. Wallace may have conflated this story with the start of their use of stainless steel, which also started in 1919, as will be further discussed in Chapter 10. The carbon steels Camillus used were a W1-type simple carbon steel at 1-1.1% carbon for the blades and 0.8-0.95% carbon for the springs[16]. The hardness of a 1943 military-issue pocket knife was specified at 56-61 Rc[19], and in 1980 Camillus knives were 57-58 Rc[20]. In their 1946 catalog[21], Camillus advertised that they used "the newest equipment and by the use of precise electrical controls, instead of the fallible human eye," and heat treatment had changed to lead baths[21] and "electrically controlled furnaces"[22].

The 1946 catalog also expounded on the relationship with Crucible Steel:

> The basic quality of a knife rests in the steel used in the blades and springs. For this reason, Camillus has devoted unceasing research to the development of the finest steel that can be made for pocket knives ... the finest that skilled craftsmen, modern controls and years of experience can produce. The manufacturer of this steel, and partner in the development of it, is the Crucible Steel Company of America, famous as the largest producer of cutting steels in the world. The keen edge of the Camillus blade starts with special ore in Crucible-owned mines in Minnesota. Under utmost care this ore is converted into high-quality pig iron. The pig iron, combined with selected steel scrap and purifying elements, is then refined to a uniformly top-quality steel. This steel is made to specifications developed by Crucible and Camillus after years of experimenting ... specifications so exact that if a housewife were to bake a cake with the same relative care, her measurements would be made in ten-thousandths of an ounce. During the manufacturing process the steel is continually analyzed, chemically and metallurgically. The final analysis is sent to Camillus

to be checked. When the steel arrives at Camillus, it is again analyzed by an independent metallurgical laboratory.[21]

Camillus stayed with simple carbon steels for many decades. A 1952 Camillus Digest said that the steel properties were tested with Rockwell hardness and impact toughness, but stayed with simple carbon steels rather than alloy steel[22]. "Tests such as these keep check on the painstaking which, at Camillus Cutlery Company, begins with the selection of electric-furnace-refined straight carbon steel. This is high carbon steel without 'fancy' alloys or harmful impurities. For many years Camillus searched for alloy steels which might have better characteristics. There is nothing available - nothing better than straight carbon steel for sturdy pocketknife blades that must hold an edge."[22] By 1972, Camillus was using the similar 1095 steel rather than W1[23]. Camillus history with stainless steel is in Chapter 10.

W. Stuart Carnes and Canton Cutlery

In 1879, Henry and Reuben Landis of Canton, Ohio, patented a knife handle made with clear celluloid plastic with a picture or inscription visible through the plastic[24]. The two first knife companies to take advantage of this innovation were Dr. W. Stuart Carnes of Canton Cutlery Company and Major August Vignos of Novelty Cutlery Company, both in Canton, Ohio[25]. Carnes got a copyright for his catalog of knives in 1901[26]. Carnes got his degree in dentistry and became a salesman for dental and surgical equipment before starting his cutlery business.

From Carnes' earliest advertisements, he claimed a special tempering process[27]. His advertising regarding the superiority of his steel ramped up around 1911[28] with the introduction of[29]: "Car-Van Steel Blades, The Super-Master Alloy Blade Steel, claimed by Steel Experts and Metallurgists to be without equal for blade purposes. This Company is the Pioneer in developing an Alloy Blade Steel and the only manufacturer using it in pocket knives." This was the first chrome-vanadium steel used in knives, which Carnes claimed to have developed. Even the American Vanadium company commented on the excellent properties of the steel for cutlery purposes[28]. Canton Cutlery advertised: "This involves not only the development of a formula for Car-Van steel, but the development of special furnaces and methods of heat treatment. The development of Car-Van steel was made possible by the discovery of vanadium some years ago at an elevation of over 15,000 feet in the Andes Mountains of Peru."[30] Carnes also claimed[30] "to be the only one in the United States using an alloy steel made especially for pocket knife blades and other cutlery specialties." Carnes advertised the steel as being the "'Lost Art' of the 'Old Damascus Sword Blade' re-discovered."[31] Carnes would demonstrate his steel and heat treatment by bending a sword back on itself, shaving hair, and hammering through nails and steel.

Canton Cutlery Co. 1915 ad.

The Canton Cutlery Co merged with Brown-McPherson & Co. to become Car Van Steel Products in 1927[32], and ceased production of knives around 1930[25], though Carnes continued to perform his demonstrations for years afterward[33]. He would tell the tale of ancient Damascus blades and how they were heat treated by quenching into slaves, but that his blades could match their properties with vanadium alloying and modern heat treatment. He also had a rectangular blade he tempered to a range of hardness where one end was so hard it could cut glass, three inches away a portion that could shave hair, a further three inches back could cut nails in two, then a portion that would bend like a spring, and the end was so soft it could easily be bent.

OUR goods are made and sold by us only. We save you the middleman's profit which should double your own and give you the benefit of *exclusive patterns* not found in catalogs.
Our goods are guaranteed to give satisfaction to the user.
Quality is our advertisement. A trial will convince you.

W. R. CASE & SONS,
Bradford, Pa.
W.R. Case ad 1908.

W.R. Case

In 1882, the J.B.F. Champlin and Son company was formed by John Brown Francis Champlin and his son Tint as a knife distributor[34,35]. In 1886, the Champlins and four

Case brothers formed the Cattaraugus Cutlery Company making knives. These brothers were William R. (W.R.), Jean, John, and Andrew Case. The Case family left Cattaraugus and formed the Case Brothers Cutlery Company in 1900. John Russell Case, son of W.R., started the W.R. Case & Sons Co. in 1903 and moved to Bradford, Pennsylvania, in 1905. Case Brothers cutlery in Little Valley, NY, was destroyed in a fire, and in 1911 W.R. Case & Sons acquired the equipment and trademarks (including the famous "XX" trademark) remaining from Case Brothers. W.R. Case would become one of the largest knife manufacturers, particularly of pocket knives, and has become the most collected knife brand in the USA. W.R. Case was one of the first users of "chrome vanadium" cutlery steel, with contracts for purchase from Universal Steel in December 1913 and Crucible Steel Company in 1915[34]. A complete history of the Rockwell hardness of Case knives is difficult to come by, though in 1984 it was reported that their blades were 52-55 Rc[36].

Switchblade Knives - Schrade

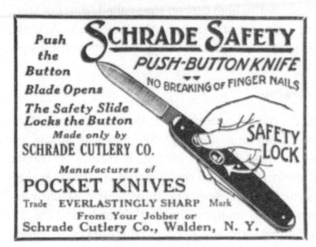

Schrade 1916 ad.

In 1892 George Schrade patented a switchblade with a button in the bolster[37]. His switchblade designs became the dominant form of automatic knives and popularized the style. At this time, switchblade knives were not yet misunderstood and connected with crime, but instead were marketed as saving your fingernails from opening knives with the nail nick. He formed the Press Button Knife Co. but could not raise enough money and sold the rights to Walden Knife Co. George Schrade, along with brothers Louis and William, formed Schrade Cutlery Co. in 1904. George would patent several knife designs and variations on automatic opening mechanisms, including the Safety Pushbutton knife in 1906[38]. George sold his share in the company to his brother Louis in 1910, then later started his own company, Geo. Schrade Knife Co. in 1928 which, after George Schrade died in 1940, was run by son George M. Schrade and grandson Theodore. In 1956 Böker purchased Geo. Schrade Knife Co., but closed it when Congress banned switchblades. In 1941, Albert M. Baer purchased the Ulster Knife Company and merged it with Imperial Knife Company, which had been started in 1914 by brothers Michael and Felix Mirando after leaving the Empire Knife Co. Albert made his brother Henry Baer (Uncle Henry) the president. In 1946, the Baer brothers purchased Schrade Cutlery Co. from Louis Schrade and renamed it Schrade-Walden Cutlery Co. The company was renamed Imperial Schrade Co. in 1985. The series of owners, consolidated companies, and different names are difficult for me to keep track of. Still, Schrade became a relatively large knife company with a wide range of products.

The Schrade Cutlery Co. catalog in 1926 says they had "special cutlery steel used in every blade manufactured by us... Likewise the springs of our pocket knives are manufactured from a special alloy steel, which by long trial has proven to be the toughest and most suitable spring steel known. We especially mention blade steel and spring steel as they are the two vital raw materials used in the manufacture of pocket knives." They also claimed superior heat treatment in the catalog: "Next in importance to selecting the right steels for blades and springs is the designing and heat treatment of the same. We were one of the first manufacturers who years ago adopted the use of scientific heat measuring instruments, so that the manufacture of good blades and springs became an exact science with us." A 1960 catalog for Schrade-Walden shows an illustration of what looks like a man using a lead bath for heat treating. A 1974 article said that Schrade used 1095 at 57-58 Rc[39]. A 1978 article[40] said that the knives were made with a high carbon steel called "Butcher Steel" and had been "for almost 100 years." My test of a carbon steel Schrade knife with a ~1946-1973 stamp revealed it was simple carbon steel, without alloy additions like Cr or V.

Phasing Out Sheffield Steel and Forging

At the beginning of the 20th century, the most common cutlery steels came from Sheffield. In 1899, data was collected from four USA knife companies about the cost of production, and all gave pricing for steel from either S&C Wardlow or Thomas Firths & Sons[41]. However, competition from USA steel companies increased to include cutlery steel from companies like Crucible[42], Jessop[42], and Universal[34]. Wardlow and Thomas Firth & Sons began advertising to maintain market share in 1917[43]. The *American Cutler* reported[43], "The action taken by these two old English houses should furnish some inspiration to American steel houses which have in the past made a bid for the patronage of the cutlery mills and no doubt we shall see some active steps initiated in the competition for cutlery trade business as soon as peace becomes an actuality." Marmaduke Wardlow, head of S&C Wardlow, traveled to the USA in 1920, visiting cutlery plants[44]. However, it seems that these were steps taken to save their export business as the American steel suppliers were taking more of the market. In a USA 1921 tariffs hearing, Arthur Balfour of a Sheffield tool steel company stated that Wardlow "would be practically put out of business" if proposed tariffs were put in place[45]. The increased quality and availability of American-made cutlery steel largely replaced imported Sheffield steel.

Another thing that was slowly phased out, especially in pocket knives, was the forging of blades. Instead, the blades were "blanked," or stamped, out of steel sheet. Cutlery steel sheet thin enough for blanking was introduced in the late 19th century[46]. Blanking of steel sheet grew relatively rapidly to 60% of the pocket knife market in 1931, 80% at the beginning of 1935, and 90% by the end of that year[47].

Cutlery Steel by Sharon Steel

Sharon Steel was advertising their range of cutlery steels by 1946[48], including a "chrome vanadium" steel mentioned in 1949[49]. Sharon gave this steel the designation 0170-6 which had a composition of 0.95% carbon, 0.45% chromium, and 0.2% vanadium[50]. Sharon Steel was also advertising that Camillus used their steels in 1973[51], which roughly coincides with when Camillus was advertising the use of 1095. After Sharon Steel went out of business in 1988, Camillus had 0170-6 steel made by another company but slightly modified it and internally called it 0170-6C, where the "C" referred to Camillus[52]. Reportedly, Case and Schrade had also been using Sharon Steel chrome vanadium knife steel and went in with Camillus when they had the new batch made[53].

Sharon Steel 1949 ad.

-5-

PROJECTILE, SAW, AND SPRING STEELS

Nickel-Steel and Nickel-Chromium Steel

Nickel-alloyed steels were developed concurrently in the 1880s by J. Hall of Jessop Steel in Sheffield and James Riley of the Steel Company of Scotland[1]. Metallurgists found that adding nickel would lead to superior ballistics performance and nickel steels became common for armor plate. Nickel-chromium steels were developed to partially replace nickel with chromium, reducing the amount of expensive nickel. Nickel-chromium steels were used for armor plate, projectiles, and automobile parts. Several nickel and nickel-chromium steels have been used in knives.

Projectile Steels Become Automotive Steels

Carpenter Steel was formed in 1889 by James Henry Carpenter. Carpenter served in the Navy and entered the Naval Academy but left due to poor health. He became interested in steel metallurgy and leased a former Railroad Rail mill in Reading, PA, which became Carpenter Steel. Being a former Naval officer, Carpenter's focus in steel production was making projectiles for the Navy. Carpenter's projectiles were used in the Spanish-American War of 1898, though he died before the war started[2]. There had been a significant demand from the Navy for projectiles during the war, but not many were ultimately used, so Carpenter had a large stock of useless projectiles[3]. In 1902 the automobile was still new, allowing them to find other uses for the steel. Dr. Sargent of Carpenter Steel said in 1907[3]: "The thought was, if the alloy steels make good projectiles, going against the armor plate, and through, being recovered on the other side whole, if this steel is capable of standing this enormous battering, why would it not make elegant material for automobiles which, as you know, are simply locomotives traveling on rough roads instead of on rails, consequently battering there is most severe. So, like the children of Israel, who forged swords into plough-shares, we proceeded to forge our projectiles into bars from which to cut gear wheels, crank-shafts, etc." He also reported[3], "One of the chief steels, or the best steel that we have found for projectiles is this Crucible Chrome Nickel Steel. That is the steel which we have called in our catalogue No. 5-317." This steel 5-317 had 0.5% C, 1.75% Ni, and

1% Cr. In the first SAE steel specifications of 1910 and 1911, there were four "Chrome Nickel" steels[4,5], which match the specifications given by Carpenter for their chromium-nickel steels. "Specification No. 12" was a ".45 Carbon, Chrome Nickel Steel" with 1.75% Ni and 1.0% Cr. These steels were then designated 32- and 33-series steels in the 1913 specifications, including 3250 with 0.5% C, 1.75% Ni, and 1.1% Cr[6]. It was also reported in 1909[7] that "the Wright brothers have used nothing but Carpenter chrome-nickel steel in the vital parts of their flying machines."

Automotive Steel Becomes Tool Steel

In 1906 Carpenter released Reading Double Special (RDS)[8], which would later be given the designation L6, a relatively common knife steel. This steel was a higher carbon tool steel version of 5-317 with 0.75% carbon, 1.75% Ni, and 1% Cr. So L6 had an interesting path to existence, starting as a projectile steel, then an automobile steel for gears and crankshafts, and finally given higher carbon for tool steel use as RDS. RDS was advertised for taps, punches, shear blades, swaging dies, and spindles[9]. However, despite its reputation in the knife industry as a saw steel, I cannot find references to L6 being used in saw blades.

Molybdenum Additions to L6

There was growing evidence that molybdenum would further increase the hardenability of steel and therefore allow larger sections to be oil quenched with full hardness throughout, or to be an "air cooling" grade instead[10]. One such product was A. Finkl's Durodi, marketed as an air-hardening steel in 1922[11], similar to RDS with but with an addition of 0.75% Mo[12]. It was marketed in conjunction with Climax Molybdenum which had recently started a large molybdenum mine in Colorado[13]. Later molybdenum-alloyed versions appeared, though there is no precise timeline. This included VASCO Nikro-M in 1937 with only 0.55% carbon[14], though it was changed to 0.7% C by 1945[15]. Jessop Extra Tough No. 4 existed by 1939 with 1.3% Ni and 0.25% Mo[16]. And finally, Bethlehem Bethalloy and Crucible Champaloy both existed by 1945[17]. Champaloy had a composition of 0.6% C, 2% Ni, 1% Cr, and 0.25% Mo reported in 1945[17] but was modified by 1950 to the final composition shown in the table[18], and Bethalloy is reported to be the same composition in that 1950 publication. The L6 specification says that Mo is optional because of these alternate L6 steels.

Steel	Year	C (%)	Ni (%)	Cr (%)	Mo (%)
No 5-317	1889-1907	0.5	1.75	1	
RDS (L6)	1906	0.75	1.75	1	
Durodi	1922	0.55	1.5	0.75	0.75
Nikro-M	1937-1945	0.7	1.4	0.85	0.25
Extra Tough No. 4	1939	0.7	1.3	0.5	0.25
Bethalloy/Champaloy (L6)	1945-1950	0.75	1.5	0.75	0.3

Saw Steel

Used saw blades have been a common source of knife steel, and many of the steels have nickel additions. James Gill in 1944 wrote that nickel had been added to saw steels around 1890-1895[19], but I haven't found any contemporary sources. In 1916 Simonds Saw and Steel Co. stated in their literature that they had been making nickel saw steels "for a dozen or more years"[20]. They built their first steel plant in 1900, so presumably they wouldn't have produced saw steel earlier than that[21]. Jessop in Sheffield was making simple carbon steel for saws before 1900[22]. The American Jessop Steel Co., in Washington, PA, made nickel-alloyed saw steels by 1936, so it could be that one of the Jessop companies introduced nickel saw steels. The Jessop nickel-alloyed saw steels in 1936 were 96-KC with 1.5% Ni, 0.3% Cr, and 139-B with 2.5% Ni, 0.3% Cr[23].

Simonds "Nickel Steel" ad from 1920[24].

Disston was a crucible steel company and manufacturer of saws that was the first company to produce crucible steel in the United States in 1855 and the first American company to make steel for saws[25]. By the 1870s, they were the largest producer of saws in the world[26]. However, the reputation of American steel at the time was so poor that Henry Disston concealed the fact that they were using their own steel rather than Sheffield steel. They also claim to have produced the first electric steelmaking furnace in 1906[27], which is interesting since Halcomb made the same claim.

Henry B. Allen of Disston & Sons steel of Philadelphia gave a history of saw steel in 1930[28,29]. Unfortunately, he did not provide too many dates in his account, but he did state that carbon steels had been the focus "Up to 15 years ago, or thereabouts". He said that the saw steels started as simple carbon steels with about 0.85% carbon, but these had issues with being too brittle. Next, they tried nickel-alloyed steel, making a stronger saw blade without sacrificing toughness. The nickel ranged from 1.25-3.25% and carbon from 0.6-1.0%. He said, "The plain nickel steel constituted a real advance in saw steels, and it still is used to some extent."

Nickel is known to improve the toughness of steel, particularly at low temperature. This fact was only slowly realized[30], though improved toughness of nickel steels was noted as early as 1897 by William Beardmore[31]. In 1932 Robert Sergeson published experiments showing that 2-5% nickel steels showed superior toughness at low temperatures[32]. In the 1940s, 9% nickel steels were developed for cryogenic pressure vessels with excellent low-temperature toughness[33,34].

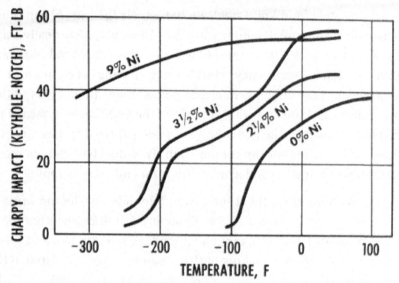

Effect of nickel on low-temperature toughness of steel[35].

After nickel saw steels, next nickel-chromium saw steels were developed[28] which Allen in 1930 said "are used today for many saws where the ductility requirements are not severe. Chromium, too, in small amounts, enhances the hardening power of nickel steels and so aids the saw manufacturer in hardening thicker saws." He also said that vanadium was added to some of the steels. Finally, he said that more recently nickel-molybdenum steels were developed, which also showed excellent ductility, including steel (g) in the table of compositions he provided (see the next page), which had been "used very largely and with good results." Allen patented steel (g) in 1925[36]. Steel (a) is the same as Uddeholm 15N20, which has been used in many knives, especially as one of the steels used in pattern-welded Damascus. Both Sandvik and Uddeholm made band saw steel early on[37,38], though compositions for their steels aren't available until much later. Steel (i) is 80CrV2 (see Chapter 3).

Steel	C (%)	Ni (%)	Cr (%)	V (%)	Mo (%)
(a)	0.7-0.9	2.0	-	-	-
(b)	0.7-0.9	1.25	-	-	-
(c)	0.7-0.9	1.25	0.25	-	-
(d)	0.7-0.9	1.25	0.35	0.20	-
(e)	0.7-0.9	0.70	-	-	-
(f)	0.7-0.9	0.70	0.25	-	-
(g)	0.7-0.9	0.70	-	-	0.15
(h)	0.7-0.9	-	0.40	-	-
(i)	0.7-0.9	-	0.40	0.15	-
(j)	0.55-0.65	1.25	0.35	0.20	-
(k)	0.45-0.55	-	0.80	-	0.20

Saw Steel compositions in 1930.

The standard saw steel (and knife steel) 8670 is steels *(g)* and *(h)* combined to have both the chromium and molybdenum additions. However, historically this steel has not been referred to as 8670, as the 8600 series was developed independently in WWII as a "national emergency steel" since having chromium, nickel, and molybdenum meant that utilizing steel scrap was easier than steels with only two alloy additions like the 4100 and 4600 steels[39,40]. The 8600 series typically topped out at 0.6% carbon[41]. However, the name 8670 has been applied to saw steel recently as it is the closest designation for the composition. This steel has seen increasing use in knives in recent years due to good availability from knife steel suppliers.

Saw steel compositions from the 1960s from Simonds and Jessop are shown in a handbook from the International Nickel Company[42] and *Engineering Alloys*[43]. Both companies have steels like 8670, though Simonds is missing Mo, and Jessop No. 91 lists it as an optional addition, making both the same as steel *(h)* listed in the Disston composition table. They also had a ~2.6% nickel steel, named 139-B by Jessop. Unfortunately, these extra steels beyond 139B and 96KC were not listed in the 1945 *Engineering Alloys* book[44], so I am unsure when Jessop released them. However, Jessop did advertise in 1953[45] that they were expanding their saw steel offerings.

Manufacturer	Steel	C (%)	Ni (%)	Cr (%)	Mo (%)
Simonds		0.75	2.6	0.15	
Simonds		0.8	0.65	0.5	
Jessop	No. 139B	0.8	2.6	0.25	
Jessop	No. 91	0.7	0.7	0.5	0.15*
Jessop	No. 67	0.8	1.35	0.35	
Jessop	No. 96K	1.0	1.4	0.65	
*Optional addition					

1930 Disston ad promoting their "new steel" band saws[46].

Jessop ad 1925.

The document from the International Nickel Company also has some interesting data from Simonds comparing the hardness and toughness of the 8670-type grade vs. the 2.6% Ni-type grade on the next page. The hardness difference comes from carbon rather than nickel. It is hard to say whether the higher toughness of the 2.6% Ni steel comes from the higher nickel, however, as steel toughness is also highly sensitive to carbon.

Bob Loveless used Jessop 139B during the "Delaware Maid" years of 1954 to 1960[47,48]. By the 1970s, it became increasingly common to identify saw steel as L6, such as in a 1974 article by Loveless in *Knife Digest* and David Boye's *Step-by-Step Knifemaking: You Can Do It!* from 1977. L6 is somewhat similar, though the saw steels typically do not fit the L6 specification because L6 requires higher Cr. Whether these distinctions matter is perhaps a matter of perspective, but the steel will behave and heat treat somewhat differently based on these composition differences. Gerber Mark II knives are reported to have used L6 steel from 1966 to 1979 when they were switched to 440C[49], which is the earliest production knife to use L6.

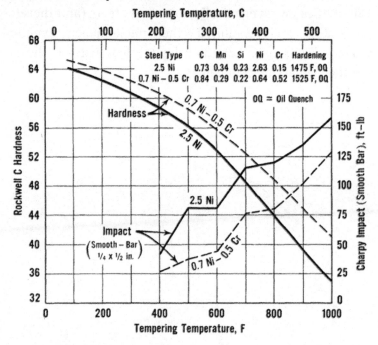

Simonds nickel saw steels compared for hardness and toughness[42].

Silico-Manganese Steels - 9260 and S5

Jacob Holzer (a French steel company) patented a silicon steel in 1892 with 1.8-2.2% silicon along with 0.35-0.45% carbon and 0.45-0.55% manganese[50,51]. This steel was designed for springs. The high silicon addition was intended to increase strength without a decrease in ductility (how far the steel can be stretched without fracture). They also noted that the steel had to be hardened from higher temperatures because of the silicon.

As was learned later[52-55], silicon additions help to suppress "tempered martensite embrittlement," where the toughness is reduced in steels when tempered in the range of about 500-750°F (250-400°C). This phenomenon of tempered martensite embrittlement occurs when large cementite particles form at those tempering temperatures. The addition of substantial amounts of silicon delays the formation of these particles to higher tempering temperatures. In the 1940s, Crucible Steel developed Hy-Tuf steel which relied, in part, on 1.5% Si for its high toughness[52]. Comparisons of Hy-Tuf with low silicon grade 4130 showed a significant change in the tempered martensite embrittlement temperature range[53]. Crucible metallurgists investigated silicon's effects, finding silicon's strong influence in suppressing cementite formation during tempering, which also explained the higher strength of tempered silicon steels[54]. There was debate about the exact mechanism by which silicon delayed TME, however, as retained austenite was also stabilized by silicon to higher temperatures, and the decomposition of that retained austenite also coincided with TME. But studies on cryogenically quenched steels (and therefore little or no retained austenite), still exhibited the phenomena. Eventually, through experimentation, a consensus was developed that instead, the suppression of cementite was the primary mechanism[55].

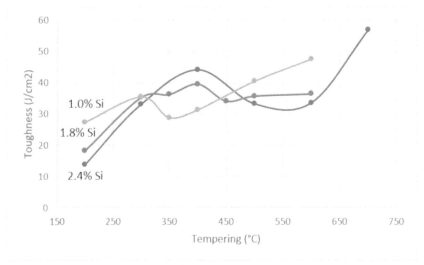

0.6% C steels with different silicon contents showing the difference in tempered martensite embrittlement[56].

This medium carbon 2% silicon steels began to be called "Silico-manganese" steels despite the low manganese content, which is no higher than other steels. These steels were tested in vehicles not long after the Holtzer patent, such as a report from 1906 where springs were tested of silico-manganese steel made by Lemoine of France[57]. This steel was specified in the 1910 SAE specifications as "Silicon Spring Steel" with 0.45-0.55% C, 1.9-2.1% Si, and 0.6-0.8% Mn[58]. The 1912 specification gave the steel designation "92-50" and called it a "silico-manganese steel" and changed the silicon range to 1.5-2.0%[59]. The 1914 specification removed the hyphen to make it 9250 and changed the Si to 1.75-2.0%[60]. A specification for the somewhat higher carbon 9260 was added in 1915[61].

Silico-Manganese Tool Steels

Silico-manganese steels were also developed for tool steel applications. Steels of the S5 type were patented in 1927 by Bethlehem[62], who branded their steel as Omega[63] with 0.6% C, 2.1% Si, 0.45% Mo, and 0.2% V. Often, these steels have a small chromium addition of around 0.25%. They stated[62] that combining silicon and molybdenum "greatly increase the mechanical properties, such as hardness and strength. What is particularly significant is the fact that these steels possess these

This One Tool Steel
handles many railway-shop jobs

RAILWAY shop-men who use Bethlehem Omega handle many different jobs with this single grade of high-performance alloy tool steel.

Omega is primarily an impact-resisting steel, combining extremely high tensile strength with great toughness. It is unmatched for tools subject to the battering of pneumatic hammers. Pneumatic and hand chisels, rivet sets, blacksmith tools, rivet busters, beading tools and punches made of Omega Steel bear up under punishment that no tool steel, lacking Omega's toughness and stamina, could survive.

Omega's high properties make it also an excellent steel for shear blades. Many shops have standardized on it for this purpose.

Using Omega for all these tasks not only provides better tools that last longer and

work faster, but also reduces the number of grades of steel that must be stocked, and simplifies heat-treating. One standard procedure covers a wide range of tools.

Heat-treated, Bethlehem Omega Tool Steel develops a tensile strength of 340,000 lbs. per sq. in. in combination with an Izod value of 7 ft.-lbs. If greater toughness is needed a slightly higher drawing temperature runs the Izod value up to 15 ft.-lbs. with only a trifling reduction of tensile strength, to 320,000 lbs. per sq. in.

Omega Steel is readily forged at about 1750 deg. F. In heat-treatment it responds to a wider temperature range than carbon steels. No expensive equipment is needed.

On the following page are listed other Bethlehem Tool Steels, each an outstanding performer in its field.

Bethlehem Omega (S5) ad from 1935[63].

qualities to a remarkable degree at temperatures which are considerably elevated... Furthermore, these steels effectively resist great stresses, even when suddenly applied and when rapidly varying in intensity, and although these steels are hard and tough, they can be effectively machined." Another effect of the molybdenum not mentioned in the patent is the increase in hardenability vs. 9260 making it more of an "oil hardening" steel. Many companies also made versions of this steel, such as Crucible's LaBelle Silicon #2, which existed by 1932[64]. These steels were made primarily for high-toughness applications like punches and pneumatic chisels. Carpenter's Solar (later designated S2) is a silicon-molybdenum steel developed in 1929[65] which is similar to S5 but has less silicon with 0.55% C, 1% Si, and 0.5% Mo.

Steel	C (%)	Si (%)	Cr (%)	Mo (%)	V (%)
S5	0.6	2.1	0.25*	0.45	0.2
S2	0.55	1.0		0.5	

Silico-Manganese Steels in Knives

Rudy Ruana preferred 9260 leaf springs in his early knives. Bob Loveless used modified S5 steel from around 1966-1971[66]. Probably many knives made from springs were misidentified as 5160 but were instead 9260. For example, Wayne Goddard found a leaf spring knife to have lower than usual hardenability, which is when he learned about 6150 and supposed that is the steel he had, rather than 5160[67]. However, 9260 has significantly lower hardenability than either 6150 or 5160, so it is more likely that he had 9260 or perhaps a simple carbon steel.

5160 Steel

Chromium steels 5120, 5140, and 5165 were added to the 1913 SAE specification[68]. However, these steels were not recommended for springs in the early specifications, with 6150, 9250, and 9260 being recommended for those applications. 5165 was removed in 1922[69]. 5160 as a spring steel was introduced by Carilloy Steels (of US Steel) around 1949[70]. Many knives made of leaf springs were said to be 5160 are likely to be 6150 or 9260 because all three were common.

Steel	C (%)	Mn (%)	Cr (%)
5160	0.6	0.85	0.8

C.Y. Wheeler and Chromium Hot Die Steel

Charles Y. Wheeler was born in 1846 in Berea, Ohio, where his father was a professor at Baldwin University[71,72]. Wheeler attended Baldwin University, after which he became an associate editor of the Burlington Hawkeye newspaper in Burlington, Iowa, in 1963. In 1873, Wheeler left the Hawkeye to work in Chicago for the Hays Steel Company. In 1879-1880 he moved to Pittsburgh to manage the steel company Hussey, Howe, and Co. Wheeler purchased his own company, Sterling Steel Company of Pittsburgh, PA, in 1885[73], and the company was advertising by 1886[74].

In 1889, Wheeler was developing a tool steel with a very high carbon content of 3%, or more, but had been unsuccessful[75]. With the help of Dr. Frank Slocum and James Todd, they successfully made "Sterling Double Special Steel" by adding a small amount of chromium[75]. Wheeler manufactured armor-piercing projectiles using the Double Special Steel, but the tests went very poorly as the high carbon steel was too brittle for projectiles[75,76]. Wheeler and Slocum instead developed a steel that was much higher in chromium (~3.5%) and much lower in carbon (<0.7%). This steel was successful for projectiles and got them a contract with the Navy in 1892[77,78]. The steel was marketed under "C.Y.W.'s Choice" or "C.Y. Wheeler's Choice" steel. Wheeler and Slocum patented the steel in 1895[79]. Todd and Slocum sued Wheeler to be paid a royalty under their contract for C.Y.W.'s Choice, but the court ruled that C.Y.W.'s Choice and Sterling Double Special Steel were very different steels and therefore were not covered under the original royalty contract[75]. C.Y.W.'s Choice was also marketed as a tool steel by 1894[80].

In 1897 Firth & Sons, a steel company of Sheffield, England, and Sterling Steel Company consolidated to make Firth-Sterling Steel Company[81]. Firth had successfully exported steel to the US, but when tariffs were imposed, they lost significant market share, which led to the purchase of steelmaking facilities in the United States[82]. Wheeler became the president of Firth-Sterling while leading the projectile department, and Lewis J. Firth oversaw the steel department in Pittsburgh. C.Y.W. Choice steel became popular for "hot work" applications such as Bull Dies, Gripper Dies, and Header Dies[83].

By 1926 the composition had changed somewhat to around 0.9% carbon along with 3.25-4.1% chromium depending on the manufacturer. It was said to be very common in gripper dies in the bolt and rivet industry[84]. Despite the relatively high carbon content, this grade was often used in large enough cross sections where air cooling and tempering would result in ~43 Rc[85]. But it was also used in smaller cross sections, or oil quenched, to achieve higher hardness levels. So, it was perceived as a versatile steel. By 1934, a version of this steel was made with a small molybdenum addition (~0.5%) which allowed it to fully harden in greater cross sections[85]. While this grade was phased out over time in favor of more optimal products and was never used in knives as far as I know, it acted as a predecessor for other steels, most notably A2, which will be discussed in Chapter 12.

C. Y. WHEELER, Chairman.   WM. P. DE ARMIT, Sec. and Treas.

STERLING STEEL CO., LIMITED.
FINE CRUCIBLE TOOL STEELS ONLY.
BRANCH OFFICE, NO. 41 FIFTH AVE., PITTSBURGH.
WORKS AT DEMMLER.
1886 ad for Sterling Steel.

"All right," *he said,* "give me <u>one</u> good reason why I should try this new spring steel."

1949 ad for 5160 spring steel[70].

-6-

EARLY 20TH CENTURY KNIFEMAKERS

There was a shift in handmade knives around 1900 as factory knives took over. Here is how Bernard Levine, a well-known knife historian, described the change:

> Before 1900, single-author knifemaking was an ancient and long-recognized craft or profession. Its practitioners were called "cutlers." The most skilled were also called "surgical instrument makers." ... As mechanization took over in knife factories (after 1903, thanks to the Hemming grinder) and in the mills that made tool and blade steel (after 1907, thanks to the Heroult electric furnace), knife prices were forced down worldwide. Soon the traditional urban cutlers largely vanished from American cities, no longer able to compete with the mass-producers. The few cutlers who remained in business were reduced to retailing and re-grinding... At the same time, the sharply lowered cost of high-grade tool steel suitable for blades (both new and scrap), and the ready availability of metal-working tools - especially the low-cost surplus tools abundant after World War I - opened the possibility of simple hobby knifemaking to anyone who was so inclined. The newly popular home mechanic type magazines encouraged this pursuit... Premium quality factories such as Marble's, Ulster, New York Knife, and Remington had the high end of the knife market well covered. A $3 Remington or $5 Marble's was equal in quality to a $20 or $30 hand-made hunting knife. Therefore, very few hand knifemakers in this period, certainly fewer than a dozen, even bothered trying to improve upon the most expensive factory knives.[1]

Scagel Knives

William Wales "Bill" Scagel was the first of the major American custom knifemakers of the 20th century. He was Canadian-born but claimed to have been born in Michigan[2]. He made knives from about 1910 to 1962. Scagel was known to be somewhat reclusive and to have few friends. He was extremely independent; because of a "beef" with the power company, his entire house and shop ran on an old gasoline engine. He made all the tools he used for knifemaking, including sanding disks, grinding and polishing wheels, drills, and hammers and tongs. All the equipment was run by a system of pulleys and belts which were driven by the gasoline engine. Scagel

didn't like visiting doctors and even once set his own broken wrist[3]. He was very conservative personally and politically and believed strongly in self-reliance; his saying was, "If you do not work, you do not eat"[2]. He hated mass-produced products, preferring handmade instead. Perhaps such a range of traits and views led to him becoming a maker of handmade knives, a relatively solitary profession requiring a belief that a superior product can be made compared to the "factory knives." Little is known about how he ended up with knifemaking specifically, or even how he learned blacksmithing. His father had a boat-building business, and before being a full-time knifemaker he worked for a lumber camp, the Grand Trunk Railroad as a locomotive troubleshooter, and a construction superintendent at one of the buildings at the end of the route[2].

He sold many knives to sporting goods companies, including Von Lengerke & Detmold, Von Lengerke & Antoine, and Abercrombie & Fitch (at that time a sporting goods store). His knives were first advertised in a 1920 Von Lengerke & Antoine catalog[2]. In 1937 Scagel said in a letter to Bo Randall[2]: "I make an effort to give a man something serviceable and practical ... in all the hundreds of knives I have made, there have never been no two alike. I have several private customers all over the world. Some have goods made that are ivory and silver or gold and platinum trimmed." He wrote to Randall about steel in 1939[2]: "Now about steels for knives. I am using Swedish and Jessops and about out and as the war is on, cannot get more. I will let you know in the near future about other steels I am going to try out, several out of American makers." Jessops refers to Jessop & Sons of Sheffield or Jessop Steel of Washington, PA, which traced its roots back to Jessop & Sons but then became independent. The VL&A ad from the 1920s said that Scagel used Jessop "Silver Steel", a product I cannot find any references to, apart from perhaps the Atkins "Silver Steel" for saws and knives produced by Jessop in Sheffield. Jessop Steel of Pennsylvania did advertise their cutlery steel, however. Scagel's biographer James Lucie said that Scagel purchased his steel from Jessop in England, which was his main steel supply until WWII[2]. Regardless, the Jessop steel that Scagel was using would have been a simple high carbon steel.

The "Swedish" he was referring to is also unclear, as this could refer to several things. Lucie combined the two names calling it "Swedish Silver Steel." Many steels were advertised that were produced using Swedish iron, as the ores in Sweden are known to be very pure. There are also some steels with Swedish in the name, i.e., Swedish Steel No. 4 made by Craine-Schrage in Michigan. In a Knives '83 article[3], Glen Lambert said that Scagel used Ryerson Steel "Swedish Iron," even though that would be iron, not a steel product. There were steel companies with Swedish in the name, such as Swedish American Steel Corp, or Swedish Iron & Steel Corp. I think that Scagel was referring to the Swedish American Steel Company as will be further explained when discussing Randall Made Knives. If Scagel had been purchasing the same steel that Randall eventually used, then Scagel would be the first knifemaker to have used O1 tool steel. Scagel also used simple high carbon steels from other companies, including SKF (he purchased "Carbon Cutlery steel"), Ryerson Steel, and Timken Steel[2]. During World War II (1942-1945), Scagel had difficulty obtaining steel. His friend Max Jester sent him a supply of 52100 bearing races (not the ball bearings) he used during that time[2].

Harry McEvoy reported that Scagel's heat treating had a legendary status: "His tempering art, however, was his greatest secret and it passed on with him. Like with the legendary James Black, who is said to have created the original Bowie knife, the Scagel tempering art is almost a legend too. That is because his blades were always razor sharp and would hold an edge longer, after very tough use, than any others available on the market of his day. Yet they were easily sharpened with a good stone."[4] Harvey Matthews had viewed Scagel performing his tempering process[2]: "He always did his tempering at the same time of the day, usually late afternoons, for this was when daylight was the level that he preferred. He would take the hardened forged martensitic blade and with fine crocus cloth, clean off the scale on just one side of the blade, leaving it as shiny as possible. He would then grasp the blade and with even rhythmic motion, move the blade over the neutral flames of the coal forge. When what he considered the proper tempering color appeared for that particular steel, he allowed the blade to cool, and he repeated this process." Matthews reported that the temper color would range from "straw" (~400°F) to blue (~600°F), depending on which steel he was using. Tempering is a low-temperature process performed after the steel is heated to a high temperature and quenched in oil. James Lucie had tested several of Scagel's blades for hardness and found them to range from 60 to 62 Rc[2].

Bo Randall said the following about Scagel in a 1980 article[5]: "I always felt that he was then - and would be today - the tops of them all in pure ability to make ANYTHING from steel. I think he turned out a greater variety of knives - all top craftsmanship - than any knifemaker in this century. They were well designed to do the jobs for which they were intended. Scagel never compromised on quality or workmanship. If someone brought him a piece of steel out of which to make a knife, he might go ahead and use it, but he would never stamp his name and hallmark into the blade unless he knew it was of the very best quality. Most other modern knifemakers, including myself have drawn great inspiration from cutlery designs that Scagel originated. Although many knife crafters today do not realize it, almost every hunting or fishing knife on the market can be traced back to an old design worked out years ago by Bill Scagel."

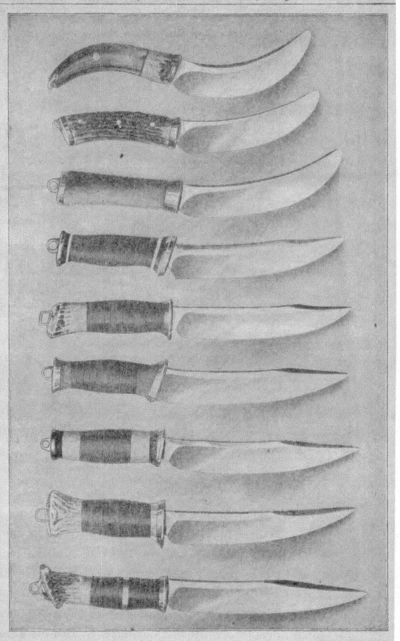

1920s ad for Scagel Knives.

57

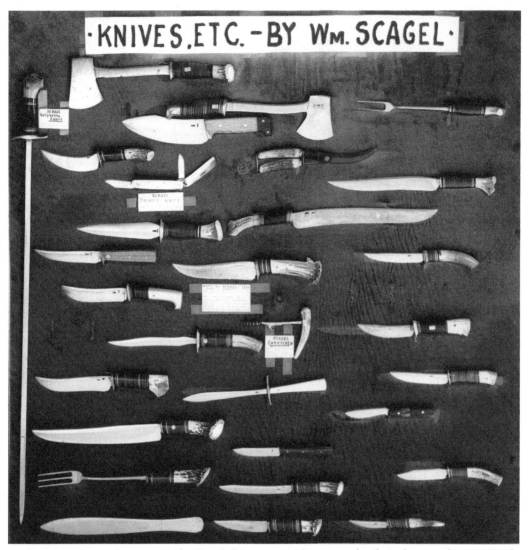

Collection of Scagel Knives in the Randall Museum. The center knife is the one that inspired Bo Randall to make knives. Photo by Leep Zelones.

Frank J. Richtig

Frank J. Richtig sold knives from his Clarkson, Nebraska home until he died in 1977[6,7]. Richtig began as a blacksmith apprentice at 19 years old in 1906. In 1908 he and a friend set up their own blacksmithing shop until 1923, when he sold his share. In about 1916, he began experimenting with a heat treating process for knives. In July 1925, he leased a shop and began producing knives. Despite his previous experience as a blacksmith, his knives were made by stock removal. His typical knives had a cast aluminum handle though some rare knives had leather washer handles. He primarily produced kitchen and butcher knives though also produced some fighting knives for soldiers.

In the early 1930s, Richtig and his wife began traveling to the state fair to demonstrate the superiority of his knives. They would take old pieces of steel like railroad spikes, axles, or horseshoes, and he would hammer the knife through the steel and then slice through newspaper afterward. In 1936 his knives were featured in Ripley's Believe It or Not!, which was published in newspapers across the country. Richtig advertised that his knives could pass these tests due to superior heat treating[8]: "We have discovered and developed a special new process in tempering knives. This method of tempering assures you of a better and more lasting quality of cutlery."

Richtig did, however, say that hammering his knives through iron or steel would void the warranty, and that the knives were ground somewhat thicker for the demonstrations:

> If knife proves to be defective in any way within six months after purchase, return same to us and we will gladly replace it. However, if the knife blade back shows it has been hammered upon or shows abuse in anyway, the guarantee is void and the knife will not be replaced... These knives are made of high-grade steel and are suitable for all ordinary uses in the home or place of business. They will do all we claim for them. In reference to the cutting of cold chisels, axles and many other steel articles, I wish to say that in my steel and iron-cutting demonstrations, I use the same steel and temper you will find in my regular stock knives. The size and thickness of the blade is the same, the only difference you will find is in the grinding of the cutting edge. I do not sell my knives for the purpose of cutting steel and iron. The purpose of my steel-cutting demonstrations is to show the toughness and temper in my knives. Many thousands of people have seen my steel-cutting demonstration and were greatly amused by them. Striking a knife with a heavy hammer in cutting steel requires special skill and experience, both of which I have learned and mastered during the thirty years I have been making and improving these knives. My knives are made for general all-around use where good, substantial knives are required.[6]

Richtig's impressive feats have led many to try to find his heat treatment process. Measurements of Richtig's knives revealed that the steel composition was nothing special, a simple high-carbon steel like 1095 or W1[9]. Richtig used steel from the Ryerson Steel Company in Chicago, branded as Ryolite[10]. The name Ryolite was chosen after a contest in 1915[11]; they liked that it references the name of the steel company while also sounding the same as the rock Rhyolite; obsidian is rhyolitic

volcanic glass. Ryerson was a distributor for Allegheny Steel and Inland Steel[11], so Allegheny Steel would have probably made the simple high-carbon steel Ryolite. Distributors like Ryerson serve as regional distribution and often perform value-added processing like cutting to length or surface grinding, or simply selling in smaller quantities than the large steel companies.

When Glen Lambert asked him about his heat treating process, Richtig answered, "A man is entitled to some secrets, and that's mine."[7] He told Lambert[7], "he had found no one worthy of being told or taught his knifemaking, and until he did, nobody was going to get his 'secret.' During his declining years no one came forward, and when he passed away ... so did the tempering process." Of course, this didn't stop people from trying to recreate his heat treatment. His legend only grew in recent decades as two research papers were published in 2000[9] and 2015[13]. The 2000 paper proposed that Richtig had been using an "austempering" process for a bainite microstructure, a type of heat treatment that wasn't discovered until after Richtig was already famous for his knives' superior heat treatment. In the research reported in 2015[13], they heat treated 1095 steel using an austempering process and could not match the superior properties reported in the 2000[9] research paper!

However, there are problems with this conclusion. One is that the hardness of Richtig's knives varied widely; the knife the researchers attempted to mimic was only 39 Rc, while another knife they tested averaged 50 Rc. A knife I purchased measures 57 Rc. Harlan "Sid" Suedmeier has also tested several Richtig knives and found them to vary in hardness. While Richtig owned a Paragon electric furnace for heat treating, Suedmeier is not sure when he obtained it, which could explain some of the variability. Paragon furnaces did not start until 1948, though perhaps he had another furnace before then.

FRANK
RICHTIG
THE VILLAGE SMITH
OF CLARKSON, Neb.

CUTS COLD STEEL –
TOOL STEEL – AUTO PARTS –
R.R. SPIKES · BUGGY AXLES, ETC
WITH A BUTCHER KNIFE
AND THEN CUTS PAPER
WITH THE SAME KNIFE !

The very low-hardness knife from the research paper is thicker than many of his other blades, and the microstructure shows evidence that it may not have been quenched rapidly enough; in other words, the knife wasn't fully hardened. This "intermediate" microstructure is challenging to mimic with a typical oil quench and temper, or even austempering treatment. The soft phases that form during slow cooling are very ductile and give the overall steel greater ductility. But it could also be that Richtig did not sufficiently heat the knife before quenching because it was thicker than his typical butcher and kitchen knives. Another issue in the 2015 paper where they tried to mimic the 39 Rc knife was that they used a test specimen size that was much larger than what was used when generating the strength-ductility values in the original 2000 paper. The use of a larger test specimen leads to the same strength value but a significantly lower ductility value. One of the researchers from the 2015 paper, Dr. Jared Teague, has since commented that he believes the specimen size is likely the cause of the lower ductility values in that study[14].

So how did Richtig accomplish his steel-cutting feats? He gave us the clues in the information he had previously provided: he left the edge thicker, and he had 30 years of experience cutting iron and steel with them. Sid Suedmeier, an expert on Richtig knives, shared similar thoughts. I asked him what Richtig's secret was, and he gave me a one-word answer: "Practice"[15]. The steel and heat treating that Richtig used were not necessarily extraordinary, but he made good knives and was not afraid to test them. His showmanship in demonstrating his knives gave the buying public an irresistible hook, and he became famous for it.

Knife Crafters - Giles Wetherill

Giles Wetherill (1903-1974) was on a hunting trip in Nevada in 1931. While field-dressing an antelope, his factory-made knife broke where the blade met the handle[16]. Wetherill rode on horseback 18 miles to the nearest ranch and was permitted to use the forge. He took a horseshoe, forged it into a knife, and attached a cross guard and wood handle. Wetherill began making knives as a hobby shortly after and eventually named his company Knife-Crafters.

Wetherill famously sold a knife for $1200 in 1934 (approximately $26,000 in 2022 dollars). The knife was made for a Maharaja of Bhavnagar in India, who wanted a custom knife with rubies, emeralds, and diamonds in a handle made of pure metals. The knife was reportedly sold for that high price even though the Maharaja sent the gems used in the knife. By 1935 Wetherill was selling most of his knives to Abercrombie & Fitch[17]. Wetherill reportedly used "Swedish chromium steel" for a 7-inch "Scimitar" style blade for dressing out a rhinoceros hide[16]. In 1936 he made a reproduction Xerxes dagger for a museum using old Damascus steel left over from a gun company that produced shotgun barrels[17].

In 1937, an article in *Popular Science* magazine about Wetherill's knife business called him the only custom knifemaker in America. The article described Wetherill like so:

> In working out the designs, he tackles the problems involved as an engineer would approach a construction job. He calculates strengths and strains and forces. He carefully balances the knives for work they are to do. He takes into consideration the points where strength is needed most, where the pressure will be exerted in different operations, and how slight variations in the shape and line and grinding of the blade will affect its cutting quality. More than half a dozen different types of steel are used in the knives. To meet special requirements, Wetherill uses steel from England, Sweden, Finland, and Germany, as well as from the mills of the United States. The metal purchased for the knives comes in the form of strips. For use in the Wetherill shop, one American mill produces a special alloy.[18]

Bo Randall sent this article to Scagel, and Scagel was not impressed[2]: "I note the article you sent on the only custom knife maker, but I do not make freaks & daggers. I make an effort to give a man something serviceable and practical. This fellow thinks he is practical and the only one I note in the piece he has, has been working at it for 6 years and doing different knives, 116 in all."

As a reserve officer, by 1940 Giles was away from home much of the time on army business leading up to America's involvement in World War II. In 1941 he became an

Ordnance & Ballistics Officer at the York Safe & Lock Company doing ballistics work, and his knife business effectively ended[19]. For more on Wetherill, and Howard James who stole the name Knifecrafters to make his own knives, read "Custom Maker" Parts I through VII in the April-October 1992 issues of *Knife World*.

Ruana Knives

Rudolph H. "Rudy" Ruana was born in 1903 and made knives near Missoula, Montana, until his retirement in 1983 at 81[20]. Ruana was born in Minnesota, and when he was eight, his family moved to North Dakota. As a boy, he made his first knife out of a nail using his dad's forge to heat it; hammering it to shape and then using a file to finish it. Ruana joined the Army in 1920 and joined the horse-shoeing school. He began making knives in his spare time using 30-06 shells and hammering out small blades. While he was an Army blacksmith, two Native American soldiers came to him asking for a better knife. They were making extra money skinning dead horses, and during the winter their Army-issued knives struggled with the frozen carcasses. In a ditch next to the road, Ruana found a section of leaf spring from a Ford Model-T (6150 steel) and used it to make the knives. In 1923 he left the Army, worked a series of odd jobs, and then moved to Montana. He worked as a welder for the Forest Service at Fort Missoula, and servicemen were asking Ruana for knives. He made knives more regularly as a hobby around 1938 and published his first price list in 1940. He left the welding job in 1944 to make knives full-time. In 1948 due to a war surplus of knives, Ruana returned to working for a welding company, making knives on the weekends and evenings. By 1952 his orders had piled up to the point that he again quit his job, went back to full-time knifemaking, and continued to his retirement. Ruana was known for using aluminum cast handles with an elk antler insert. Ruana viewed his work from a practical point of view[20]: "This ain't no modern knife factory ... This knife-making is dull work and lots of it ... Some of those other knifemakers might feel insulted. They feel it's a craft, but you do 55 years of it and you can shut your eyes."

Ruana's early knives were made with 6150 leaf springs. A 1940 price list described the knives: "The blades are made from the finest Vanadium Spring Steel, hammer forged with a modern power hammer, free hand hollow ground and individually oil tempered and heat treated." Ford was known for its use of vanadium steel, such as 6150. Ruana's heat-treating process included a trade-secret tempering treatment where he would temper the steel according to color to achieve the properties required for the given knife. There are also rare stamps referencing heat treating on some Ruana knives, such as an "S" for "Salt Water Hardened" which presumably refers to a brine quench, and "T" for "Razor Temper" which may refer to a lower tempering temperature for higher hardness. More typical was an "M" stamp which referred to a "Medium Temper" or medium hardness[21]. Around 1941 Ruana began to us 9260 leaf springs, reflected in a 1944 price list that called it "Silica Manganese Spring Steel." Silica refers to silicon dioxide rather than "silico," short for silicon, a mistake in the shorthand name of the steel. This steel reportedly came from Studebaker vehicles. However, Ruana continued to use some 6150 as 1950s price lists said that the knives were "made of vanadium and silico manganese alloy steels." Around 1962 Ruana began using 5160 leaf springs for large knives like Bowies and 1095 for smaller knives like hunting knives.

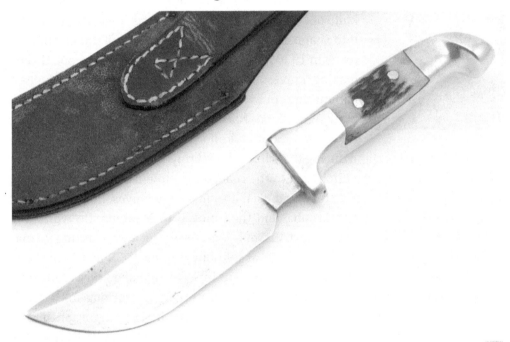

Photo of Ruana knife from Arizona Custom Knives.

Morseth Knives

Harry Morseth was born in 1889 near Selbu, Norway, and emigrated to the United States in 1906[22]. Morseth attempted to make and sell high-priced hunting rifles but found they didn't sell. So instead, he started making knives in the early 1920s. Morseth worked for the Walton Lumber Company, and the mill where he worked had a fully equipped machine shop where he would make the knives in his off hours. At this time, he made his knives using worn-out planer blades. It has been reported that these early planer blades Morseth used were made with A2 steel, but that steel did not exist until the late 1930s; more likely, the planer blades were made with simple high carbon steel. Morseth's early knives had a similar design to those made by Marble Arms Corporation (Marble's), and a 1925 knife he made looks very much like Marble's Ideal hunting knife. Other early knives were also made with recycled steel, such as worn-out files, though planer blades were most common throughout the 1920s and 1930s.

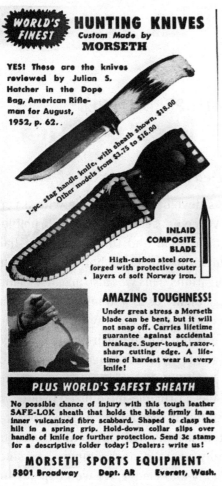
By the 1930s, Morseth was selling knives after positive reviews from local hunters. In 1950 Harry Morseth wrote in his catalog[23], "Frankly, I never intended to make knives on a commercial basis. For over twenty-five years I have made fine hunting knives for my own pleasure and use - then for friends and friends - of friends who had seen and heard about my knives from others. Finally, so many people wanted to own a Morseth knife that I have been forced to manufacture them on a more of a business-like basis." In 1938 Morseth made his first trip back to Norway, and there he was able to visit blade manufacturers, including B. Knudsen and Lars and Ragnar Brusletto. There he learned about laminated steel and brought back test blades from Brusletto and Knudsen that he used to make knives. Brusletto would forge a high-carbon steel between two low-carbon layers of steel and then forge them to shape, and Morseth began importing these forged blades from them soon after World War II. In Morseth's 1950 catalog, he wrote: "For centuries the world's finest, costliest and toughest swords and knives have had carefully forged composite steel-and-iron blades. The famed swords of ancient Damascus, Toledo in old Spain, and feudal Japan were of this type. They used high carbon steel for maximum hardness and keenest edge, then forged it together with soft, ductile iron to give great toughness and resistance to breaking." Morseth would provide Brusletto with the design of the blades so that they could be cut with a 60-ton press using dies to his specifications, and the die-cut blades were delivered to Morseth by 1949. In later years, Brusletto would purchase pre-laminated steel manufactured in Sweden. Morseth specified a hardness of 62-63 Rc, while Brusletto knives were a lower 58-59 Rc. Morseth introduced his first catalog in 1950 and grew his knife business from there. In advertisements, Morseth used a picture of one of his knives being bent in a vice without breaking, showing the unique properties of laminated steel.

-7-

NEW HOT WORK AND HIGH SPEED STEELS

Tungsten Hot Work Steel and "Semi-High Speed Steel"

Around 1910[1,2], another hot work die steel was developed at Crucible with a very different approach than C.Y.W. Choice (C.Y. Wheeler, see Chapter 5). Crucible's "Peerless A" is essentially Rex AA/T1 with the carbon and tungsten cut in half, because hot work die steels do not need as much hardness or hot hardness, and toughness is typically more critical in these applications. This was also an early "semi-high speed steel, " a very loose term that is occasionally used and misused. Over the decades, the definition of semi-high speed steel became a steel with a significant amount of molybdenum/tungsten for hot hardness but not so much that it is used in machining like a high speed steel. But early on, the term was frequently misused for any steel containing tungsten, even the low tungsten "intra" types with no hot hardness. John Mathews wrote about Peerless A as a Crucible-introduced product in 1930[1] though it isn't clear if he was taking credit for its development. This steel was later given the designation H21. There were later variants with different carbon and tungsten contents, of course, but H21 has generally been the most common, and since none of these steels have been used in knives to my knowledge, we won't spend any more time on them.

Steel	Year	C (%)	Cr (%)	W (%)	V (%)
H21	1910	0.3	3.25	9	0.25

Cobalt High Speed Steels, Uranium high speed steels, and Increasing Vanadium

The T1-type high speed steel became the dominant grade used for the next several decades, still covering 80-85% of the high speed steel market even up to 1937[3]. But some variations gained some usage, primarily for special cases. One development that has continued in the design of high speed steels to this day is the addition of cobalt to some steels. This was first done by Reinhold Becker of Germany in 1912. Becker owned his own steel company (also called Reinhold Becker), and he patented a steel

Red Cut Cobalt ad from 1915.

identical to T1 with 0.7% carbon, 18% tungsten, 5% chromium, and 1% vanadium but with a 4% cobalt addition[4]. Many other steel companies copied this composition at the time, such as VASCO's Red Cut Cobalt steel, advertised by 1915[5]. This steel was later given the designation T4.

The benefits of cobalt, or lack thereof, were debated for some time after, as cobalt itself wouldn't give a steel hot hardness but instead would affect the behavior of other elements. With further research, it has been found that cobalt affects the precipitation and growth of the tungsten/molybdenum carbides found in high speed steels and therefore improves hot hardness. It does this without affecting the composition of the carbides themselves. On the negative side, cobalt also reduces toughness. A reasonably complete bibliography of investigations on the effects of cobalt up to 1963 is in Ref. 6.

Cobalt was used in many other types of steels, including molybdenum-alloyed high speed steels in 1915[7] and in high carbon, high chromium alloys in 1916[8]. Molybdenum high speed steels did not catch on until sometime later, and the high carbon, high chromium steel never did catch on for high speed machining.

Steel	Year	C (%)	Cr (%)	W (%)	V (%)	Co (%)
T4	1912	0.7	4	18	1	5

Increasing Vanadium

Vanadium had been added to high speed steel since Rex AA but was increased up to a full 1% or even higher by many manufacturers. The effects of vanadium were not then well understood, but it was found that the tools would last longer with higher vanadium. Around 1913-1914 they found that tungsten could be decreased (relative to T1) to 13-14% as long as the vanadium was increased to 2%[9], sometimes called the 14-4-2 type of high speed steel, and later given the designation T7. These steels gained popularity during WWI due to the high cost of tungsten[3,10]. The earliest version of this steel I have found is Braeburn Gyro high speed steel marketed in 1914[11,12]. Next, there were steels with 18% tungsten and 2% vanadium, the earliest of which I have found was Sanderson Bros' Saben Kerau "super high speed steel" which existed by 1919[13,14]. This steel was later designated T2.

Uranium?

For a short period, there were uranium-alloyed high speed steels. However, the benefits of uranium were debated and rejected by many[15]; the element soon fell out of favor. One example was Latrobe's Electrite Uranium steel, introduced around 1916[9], which was the 14-4-2/T7 type steel with 0.5% uranium added.

ELECTRITE URANIUM STEEL

This planer hand is happy—and with good reason. He has a lot of stock to remove from the plate he is finishing and he has a tool that will carry a good cut—an Electrite Uranium High Speed Steel Tool. He is planing 40—50 point carbon steel with a 1/16 inch speed and a 1/2 inch deep chip. The reason for the staying quality in Electrite Uranium High Speed Steel is due largely to the element Uranium that is introduced by the most modern steel working practice. Uranium gives this steel a toughness that is unattainable in other ways.

Besides Electrite Uranium we make "Mangano" oil hardening non-shrinkable die steel; "Select" die steel for hot work and hot trimming dies; "Renown" die steel, which is a special vanadium steel; also chrome and tungsten magnet steels.

Try us on your next steel order.

"The Smile that Won't Come Off"

LATROBE ELECTRIC STEEL CO., LATROBE, PA.

Latrobe Electrite Uranium ad 1917[16].

Carpenter Gold Star ad 1928[17].

Cobalt and 2% Vanadium

Around the year 1925, there was an acceleration of modified high speed steels with different vanadium and cobalt contents[3]. High cobalt steels and Cobalt+2% vanadium steels were released at various points, including Carpenter Gold Star around 1928, which was the 14-4-2 type plus 5% cobalt (later called T8)[17]. Firth-Sterling Circle C, also in 1928, has 18% tungsten, 2% vanadium, and 8% cobalt (later called T5)[18]. Then T6 by 1930[19] with 18-21% tungsten, 2% vanadium, and 12% cobalt. The earliest version of this steel I found was Crucible's Rex 440 (not related to 440-series stainless steels). This was the maximum amount of alloy possible with high speed steels, especially given the seeming consensus at the time that carbon content should be kept to ~0.7%. Further development of cobalt-alloyed steels had also been occurring in Germany, and by 1928 they had steels with 18% tungsten, 1.5% vanadium, and 5-15% cobalt[20], landing in a similar place to the best USA-produced cobalt high speed steels.

Steel	Year	C (%)	Cr (%)	W (%)	Mo (%)	V (%)	Co (%)
T7	1914	0.7	4	14		2	
T2	1919	0.8	4	18.5	0.75	2	
T8	1928	0.75	4	14	0.75	2	5
T5	1928	0.75	4	18.5	0.75	2	8
T6	1930	0.8	4	20	0.75	2	12

Molybdenum High Speed Steels

Molybdenum was known to be a replacement for tungsten in steel from the early days, such as a self-hardening steel Mo.S.H. by Sanderson Steel in 1898[21]. By 1901 there were some molybdenum high-speed steels (see Chapter 1). Molybdenum is in the same elemental "family" of the periodic table and acts very similarly to tungsten regarding secondary hardening and red hardness. Because of the difference in atomic weight, only half as much Mo must be added for the same effect. However, molybdenum was very expensive, and there was also a belief that molybdenum would oxidize at the surface leading to "demolybdenization" or loss of molybdenum at the surface[22], an issue not seen with tungsten. However, in the USA several factors led to a renewed interest in molybdenum high speed steels. During World War I there had been difficulties in obtaining sufficient tungsten supplies, which greatly increased the price in that period[23]. The United States Government became concerned about consistent access to tungsten because a large percentage of the world's tungsten is present in China[24]. And it was recognized that significant sources of molybdenum existed in the USA with the opening of the Climax mine in 1924.

Capt. Ritchie

Therefore, an investigation into molybdenum high speed steels was initiated at the US Army's Watertown Arsenal in 1927, the results of which were published in 1930[24]. Captain S.B. Ritchie ran this metallurgy research department. They first studied steels where the tungsten was partially replaced with molybdenum and found no reduction in performance, so they then explored steels with molybdenum only. Some important discoveries by this group were that the soft surface layer in molybdenum steels was not from "demolybdenization" but rather decarburization (loss of carbon). Therefore, methods used for other steels to minimize decarburization would work with molybdenum high speed steels. They used a borax coating on the ingots, preventing decarburization during forging and rolling. They also found that molybdenum high speed steels melted from somewhat lower temperatures than tungsten high speed steels, so the typical forging and hardening temperatures established for tungsten steels had to be reduced for the properties of the steels to be similar. Their best-tested steel was 0.7% carbon, 3.5% chromium, 9.5% molybdenum, and 1.3% vanadium. They found it to be 90% as efficient as T1 for heavy high speed tools and equal for smaller tools.

The increased interest in molybdenum high speed steels led to many investigations into new compositions, including by Joseph Emmons of the Cleveland Twist Drill Company[25]. Emmons made a series of heats with molybdenum-only compositions and then used a range of heat treating temperatures to optimize each steel and then

Emmons

compared it against T1, which was assigned a performance rating of 100%. A steel similar to the best of Watertown Arsenal with 0.76% carbon, 3.8% chromium, 9% molybdenum, and 1.2% vanadium had only 62% of the longevity of T1 when tested in drill bits. The best result was with a steel with relatively high carbon (0.8%) and molybdenum (10.7%) which had 91% of the performance of T1. Often comparisons were made in terms of a "tungsten equivalent" where the molybdenum was multiplied by two. So, a 10.7% molybdenum steel would have a tungsten equivalent of 21.4%. So, it was determined that molybdenum high speed steels might require somewhat higher carbon and a higher "tungsten equivalent" than tungsten high speed steels where 18% was optimal. Like the Watertown Arsenal, it was also confirmed that lower hardening temperatures than T1 were required for best performance because the molybdenum carbides dissolve at lower temperatures. Emmons noted that there was significant grain growth in the molybdenum high speed steels at the hardening temperatures found to achieve the best properties, and therefore tried additions of tungsten to help with this issue. A large grain size leads to poor toughness, chipping, and fracture in the drills. He found that steels with 1.25-2% tungsten along with 7.5-11% molybdenum, 1% vanadium, and 0.8% carbon would achieve similar or better performance levels to T1. The small tungsten addition yielded a finer grain size than the molybdenum-only steels from the carbides dissolving at higher temperatures. This investigation led to a patent for the steel, filed in 1932[26]. This steel was trademarked as Mo-Max and licensed to many major steel companies (Cleveland Twist Drill Co. made drills, not steel). This steel was later given the designation M1.

Another approach to solving the grain growth issue with molybdenum high speed steels was to increase the vanadium content up to 2%. Vanadium forms a very stable carbide that only dissolves at very high temperatures. Vanadium carbides are also very high in hardness and therefore improve wear resistance. Arthur Howard Kingsbury of Crucible[27] and James Gill of VASCO[28] independently developed a steel with 0.8% carbon, 8% molybdenum, and 2% vanadium. Kingsbury submitted his patent application earlier in 1934. These were branded as Crucible's Rex VM and VASCO's Van-Lom, later named M10. This composition was found to have significantly higher performance than T1 due to its improved wear resistance.

Despite the good performance achieved with these molybdenum-based high speed steels, there were still desires to create a composition that is less prone to decarburization. Alvin Herzig of Climax Molybdenum patented steels with copper or boron additions[29] which were shown to reduce decarburization. Copper paint was sometimes used instead of borax to coat the steel ingots to prevent decarburization

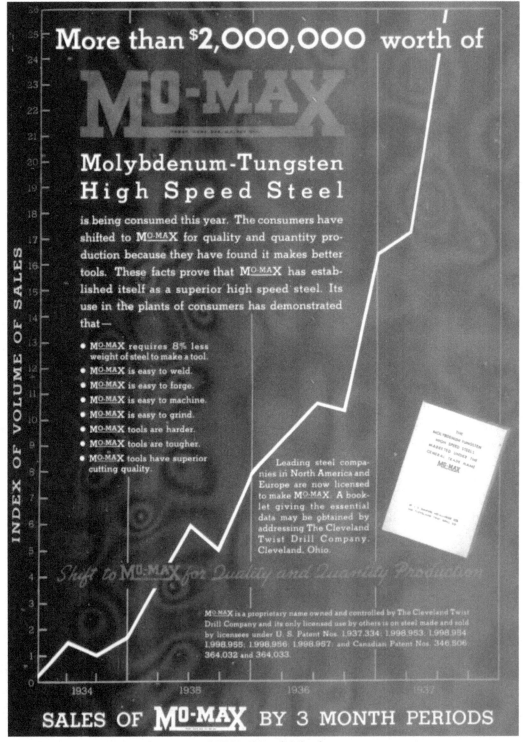

1937 Mo-Max (M1) ad.

1937 Van-Lom (M10) ad.

during forging, and it was thought that alloying the steel with copper may result in a similar effect. However, copper was not entirely effective at preventing decarburization, and there were concerns about putting copper into the high speed steel scrap, making it less desirable for reuse and recycling[23]. As a partial replacement for carbon, Boron was effective at reducing decarburization but led to failures in forging[29]. Borax was suitable for forging and rolling operations, but it was time-consuming to apply and remove, and it would attack furnace refractory, thus reducing the life of the furnace lining[30]. There was an increase in the use of salt baths and atmosphere-controlled furnaces which would limit decarburization in heat treatment. However, there was still a desire to reduce the susceptibility to decarburization during forging. It was instead proposed that molybdenum be reduced to levels where decarburization is not a problem, along with an appropriate amount of tungsten to achieve a good "tungsten equivalent." Climax Molybdenum released data on a steel with 6% molybdenum, 6% tungsten, along with 2.5% copper, and did not apply for a patent so that the steel could be made by any manufacturer[21].

Ralph P. De Vries, of Allegheny Ludlum steel, applied for a patent on a Mo-W steel in 1939 with 0.8% carbon, 4.5% molybdenum, 5.75% tungsten, and 1.5% vanadium[31], though they had been writing about this steel since at least 1938[30]. This steel was branded as "DBL" and later named M2. M2 steel became the dominant high speed steel, taking over T1's position as the most popular choice. It had superior wear resistance and toughness to T1 and used much less tungsten. There were restrictions on tungsten during World War II, which boosted the production of M2, and M2 became even more popular during the Korean War (1950-1953) when supplies of tungsten from China were limited[32]. During WWII there were restrictions on its composition, but in 1945 it became more commonly 0.85% carbon, 6% tungsten, 5% molybdenum, and 2% vanadium[32], higher levels of each element when compared with the original.

Cobalt in Molybdenum High Speed Steels

Of course, it didn't take long for there to be cobalt-modified molybdenum high speed steels. These included "Super MoTung" from Universal-Cyclops, a Mo-Max/M1 with a

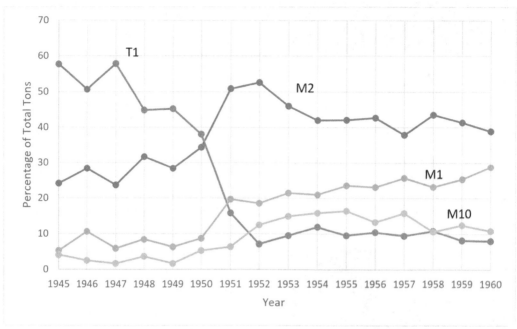

Percentage of sales for popular high speed steels from 1945-1960[32].

Crucible 1941 ad about Mo-alloyed high speed steels and restrictions on tungsten during World War II. It advertises Rex VM (M10), Rex MM (M2), and Rex TMO (M1).

"YES, BUT SUPPOSE THAT SHIP DOESN'T COME IN?"

"DBL High Speed Steel helps to conserve the nation's supply of tungsten. and to protect you against possible shortage, because it contains less than ⅓ as much tungsten as 18-4-1. And...DBL's performance actually equals or beats 18-4-1, and it costs less!"

Another advantage, and just as important—when you change over from 18-4-1 to DBL, you continue to use the same heat-treating equipment — nothing new to learn, nothing new to buy. ● Let us send you full technical data in the form of the "DBL Blue Sheet." Use coupon below.

ALLEGHENY LUDLUM
STEEL CORPORATION   PITTSBURGH, PA.
Tool Steel Division [AL] Watervliet, N.Y.

1941 DBL (M2) ad.

5% cobalt addition which was released by 1937[33], later given the designation M30. Or Mo-Max plus 8% cobalt such as "Super MoTung Special" or Firth-Sterling "Super Mo-Chip," both of which existed by 1940[34], later designated M40. However, M40 has largely been replaced by a version with 2% vanadium called M34[35], and the M40 designation is no longer used. M2 plus 5% cobalt was sold by Bethlehem Steel as "Moco" in 1942 (M35), and M2 plus 8% cobalt was branded as "Super DBL" which existed by 1943[36] (M36).

Steel	Year	C (%)	Cr (%)	Mo (%)	W (%)	V (%)	Co (%)
Watertown Arsenal	1930	0.7	3.5	9.5		1.3	
M1	1932	0.8	4	8.5	1.5	1.2	
M10	1934	0.85	4	8		2	
M30	1937	0.8	4	8.5	1.5	1.2	5
M2	1938	0.8	4	5	6	2	
M34	1940	0.85	4	8	2	2	8
M35	1942	0.85	4	5	6	2	5
M36	1943	0.85	4	5	6	2	8

High Vanadium "Super High Speed Steels"

The vanadium content of high speed steels had gradually increased, first to 1% and then to the proliferation of many 2% vanadium high speed steels. However, increasing vanadium beyond this point would lower the performance of high speed steels rather than improve it. In Germany, Wilhelm Oertel of Becker Steel found that to increase vanadium beyond 2%, or cobalt beyond 10%, would require an increase in carbon as well, which he patented in Germany in 1926[37]. Vanadium takes carbon to form vanadium carbide, necessitating an increase in carbon. The Canadian patent of 1928 is in English:

It is known that standard high speed steel can be considerably improved as regards its working capacity by the addition of vanadium or cobalt or of both elements at the same time. It is also known that tungsten can be replaced partly by other alloy constituents such as for instance by vanadium; further that when a certain vanadium proportion in the high speed steel is exceeded, the steel is no longer hardened, and that owing to the formation of a vanadium carbide with a high percentage of carbon, carbon is withdrawn from the main mass of the steel. The working capacity of a high speed steel is however depended to a very great extent on the presence of a tough hard main mass which must contain numerous particles of very hard carbide. The percentage of carbon in the usual commercial high speed steel with 18-20% W and about 4% Cr amounts, owing to considerations of forging and hardness mostly to no more than 0.75%. An increase of the vanadium percentage in such a steel above 2%, and of the cobalt percentage above 10%, is therefore scarcely possible. It has been found now, that even a steel with a high percentage of cobalt or vanadium, can be again made hard and suitable for working, by increasing the carbon percentage beyond the usual measure, namely in a given relation to the percentage of cobalt or vanadium. For each 1% vanadium starting from carbon percentage of 0.8%, the carbon percentage must be increased by 0.16%, and for each 1% cobalt, by 0.04%, ... Thus for instance a high speed steel with 5% V must contain about 1.4% carbon, a steel with 15% Co about 1.2% C. In the presence of both alloy constituents, the percentage of carbon must be made equal, or approximately equal, to the sum of the carbon percentages required for each of these separately. It has been found that for instance the working capacity of a steel with about 4% vanadium and with a corresponding carbon percentage, can be more than tripled as compared with the same steel with lower carbon percentage without vanadium. Compared with a normally alloyed high speed steel with about 1-2% vanadium, the increase of capacity is about double.[38]

A 1931 book by Oertel and Arthur Grützner listed their best steel in 1928 as 1.4% carbon, 13% tungsten, 5-10% cobalt, and 4% vanadium[20], which was sold as SA900 by Deutschen Edelstahlwerke. And by 1937, there was a steel called Gigant 66[39] from Röchling with a composition of 1.45% carbon, 12% tungsten, 4.75% vanadium, and 4.75% cobalt[40]. The USA was a step behind in high vanadium steels. Allegheny Ludlum introduced a steel in 1926 called "High-Van" with 1% carbon, 18% tungsten, and 3.25% vanadium based on a patent first filed in 1924[41-42]. The inventor William Keen had also determined that increased carbon was necessary for higher vanadium. However, this steel was abandoned due to difficulty in production and erratic performance in the final tools[41]. In 1932 a paper was published by Kinzel and Burgess[41] looking at the effects of vanadium on high speed steel. They found that when maintaining a constant carbon content of 0.7% and increasing the vanadium, the steel would no longer harden above a vanadium content of about 2.3%. They looked at the microstructure of the heat treated steel and found large amounts of "delta ferrite," which means that the soft "iron" phase of steel was still present at the high hardening temperature rather than transforming to the necessary austenite phase before quenching. They solved this issue by increasing the carbon content of the steel along with the vanadium and found that the best properties were achieved with a steel with 1.5% carbon, 17% tungsten, and 5% vanadium, very close to the

VASCO Supreme ad 1951.

carbon and vanadium of the German high vanadium steel.

James Gill of VASCO introduced Neatro in 1937, which was in the same family as Van-Lom (M10) but with 1.25% carbon and 4% vanadium, and the same 8% Mo as M10. Because of the success of M2, this was modified to have 5.5% tungsten and 4.5% Mo in 1939[32]. Neatro was later given the designation M4. In 1946 VASCO introduced VASCO Supreme[43], the same as Gigant 66 (T15). VASCO claimed that they had developed this steel[44], but it was not introduced until Gill published a report on German high speed steels, including Gigant 66, in 1945[40]. However, VASCO developed a modified version called "VASCO Supreme A" which was introduced by 1951[45] and later given the designation M15. This steel was similar to T15 with 1.5% carbon, 5% vanadium, and 5% cobalt, but with 6.5% tungsten and 3% molybdenum (instead of 12.5% tungsten), which was presumably developed with reduced tungsten to avoid the Korean War restrictions.

Steel	Year	C (%)	Cr (%)	Mo (%)	W (%)	V (%)	Co (%)
T15	1928	1.5	4		13	5	5
M4	1939	1.25	4	4.5	5.5	4	
M15	1951	1.55	4.5	3	6.5	5	5

-8-

KUEHNRICH, D2, AND STAIN-FREE RAZORS

Paul Richard Kuehnrich was born in 1871 in Saxony, Germany[1]. He worked for his uncle's steel business in Germany until he was 17 years old, and he moved to Sheffield, England, to work at Marsh Brothers, an old steel and cutlery company owned by Harry Parker Marsh and John Parker Marsh. He was made a clerk and manager of the Marsh Brothers' foreign department. He traveled across Europe to Germany, Austria, Switzerland, and Holland advertising for the Marsh Brothers, helped in part by his fluency with German. Kuehnrich seems also to have been an excellent salesman. According to Kuehnrich, the Marsh Brothers income increased from 6,000 British pounds when he joined to 130,000 in 1897. The Marsh Brothers contended the figures were instead 17,000 to 120,000, but either would be an incredible increase. Kuehnrich became a naturalized citizen of Britain in 1894 after he had been threatened with conscription for German military service. After 1900 Kuehnrich received one-third of the company's profits beyond a particular baseline, and he became the company's Pond Works plant manager. There was a legal battle where Kuehnrich accused the Marsh Brothers of underpaying him, but the court ruled this only amounted to 150 pounds, and Kuehnrich resigned in 1904.

D3 – High-Carbon, High-Chromium Steel

While working for the Marsh Brothers, Kuehnrich began experimenting with different steel alloys:

> In 1895 I was travelling in France on business. At that time the Sheffield firm of Marsh Brothers & Co., whose manager I was, supplied large quantities of steel for metal saw files. While visiting a user of this steel, Ebstein Frères at Jarville, near Nancy, M. Ebstein informed me that he had found another steel which gave much better results. This steel emanated from the firm of Jacob Holtzer. At that time hacksaws with set teeth were not generally known and metal saws were made with a tapered back. The teeth of these saws

were sharpened in the hardened state with small extra hard, three-square files, which naturally had to perform a strenuous task; consequently, all file manufacturers were on the lookout for a better steel. M. Ebstein gave me a piece of the alleged "wonder steel" which I had analyzed in Sheffield. To the astonishment of everyone concerned, it was found that the steel contained 2-1/2 per cent of chromium, which was considered enormously high, because at that time small chromium contents of 1/2 to 1 per cent were the rule. It was feared that more chromium would render the steel brittle, although it was already known that the firm of Jacob Holtzer was manufacturing a steel with a higher percentage of chromium for shells of high penetrating power. This steel was, however, not used for files.

I then caused a trial cast to be made at the Marsh works of a composition similar to that of Holtzer's material, which contained 1.3 per cent carbon and 3.5 per cent chrome. This gave excellent results and was not only quite suitable for files, but also for a number of other purposes, such as hacksaws, twist drills and razors. Articles made from this steel were then put on the market and became favorably known as the Roxo brand. Pocket knives for advertising purposes were also made from this steel, which I showed on my journeys as a curiosity. The blades were exceedingly sharp and had the peculiarity that, although the steel was very hard and would cut glass, the blades could easily be bent to an angle without breaking - a curious combination of great hardness and toughness.

During the period between 1895 and 1903 large quantities of this 3-1/2 per cent chrome steel were manufactured. Trials were made to increase the chrome to 4-1/2 per cent without any difficulties being encountered. Having now conclusively proved that the higher percentage of chromium was possible and useful - in spite of the prevailing opinion that it was dangerous - it occurred to me that it would be interesting to find out up to what limits the chromium content might be brought and yet obtain a material capable of being forged and rolled. In the autumn of 1903 I therefore had a series of ingots melted in our works, containing 6, 9, 15, 18, 21 and 24 per cent of chromium. At that time carbon-free ferrochromium was practically non-existent; Sheffield works generally used ferrochrome with 6 to 8 per cent carbon. Thus, most of the ingots, especially the last four, came out too high in carbon. The ingots with 21 and 24 per cent chromium could not be forged at all. After a few preliminary difficulties it was possible to forge the remaining ingots. Tests with 15 per cent chrome steel led to the surprising discovery that, when made into turning tools, this steel had similar qualities and gave rather better results than the original Mushet self-hardening steel, which was at that time very popular. The above-mentioned trials were, at all events, the first stages of the development and commercializing of high-chromium special steels.[2]

Kuehnrich formed Sheffield Steelmakers Ltd in 1904 with a capital of 30,000 pounds. He began to compete with his former employer, the Marsh Brothers, and even took their customer address books and undercut them on price. He began producing his 15% chromium high carbon steel under the brand name "Unor."[2] Sheffield Steelmakers, however, didn't last very long, and within two years Kuehnrich had formed another company, Darwin & Milner Ltd. Kuehnrich also purchased the former Sybry, Searls, & Co Cannon Steel Works which produced crucible steel which allowed him to produce tool steel and to experiment with new alloys. In 1910 he installed a new electric steelmaking furnace, as that superior technology was slowly growing.

Kuehnrich spent a substantial amount of time and money developing new steels, and he was said to have "carried out more tool steel alloying experiments than any living man."[3]

Kuehnrich went back to high chromium steels, and at that point the steel gained some popularity in France though not in Britain, at least initially[4]. An article in *American Machinist* reported that the steel worked well in "punch and die work." The excellent wear resistance from many chromium carbides led to an impressive number of strokes before the punches needed regrinding. The steel was measured to be about 2.2% carbon and 15% chromium, matching the grade initially tested at Marsh Brothers. This product resulted in a 1911 patent[5] for a steel with 2-3.5% carbon and 13-17% chromium, and the steel was trademarked as "Neor" in 1916[6]. The patent claimed that the steel could be used in high speed machining operations, but this did not end up being the primary use of the steel. Midvale Steel soon copied Kuehnrich's high carbon-high chromium steels in the US, and they produced a steel with 15% chromium and 2.5% carbon as well[7]. Then they patented a steel with 1-2% carbon and 15-20% chromium in 1915 while stating that 12-16% chromium steels with greater than 2% carbon already existed at that time[8]. Kuehnrich did not patent his steel in the USA, leaving the opportunity open for the Midvale patent. The chromium was reduced to about 13% for most high carbon-high chromium steels by 1926[9], which also happened with Neor[10]. This steel was later given the designation D3.

Kuehnrich purchased a large mansion in Sheffield in 1912 and was known for driving to his business in a fancy carriage[1]. He was known to be eccentric, often awake at 4 in the morning, "and in Winter regularly broke the ice of the open-air bathing pool in his grounds. He sun-bathed in a roof garden over his house"[11]. He had a large art collection valued at 24,000 British pounds, and it was said that "[t]he walls of Holly Court, his residence [in Sheffield], are lined with masterpieces which are ingeniously floodlighted"[11]. He was also a musician who would improvise for hours on the piano or organ. However, his eccentricity and German ancestry became a problem during World War I, when distrust of Germans led to many accusations and even imprisonment for some Germans. Stories were circulating that his mansion was full of gun emplacements, but none were found. Questions were asked in Parliament after there were rumors that he was a personal friend of the Kaiser and that he had a secret army at the mansion. While he was never arrested during the war, he was fined twice for seemingly pointless charges, once for "shining a powerful light in the evening," and once for hoarding 69 pounds of bacon (there were food prohibitions during the war)[1].

D5 and D2 tool steels

Kuehnrich next patented high speed steels with cobalt additions that had no tungsten, as it was an expensive element and less available during the war. In 1916 Kuehnrich patented another high carbon-high chromium steel with 1.5% carbon and 12%

chromium, along with 3.5% cobalt[12]. The patent admitted that Neor and other high chromium steels were ineffective for high speed machining tools but that adding cobalt would make them so. Kuehnrich did not state the reason for reducing carbon and chromium vs the original Neor. He marketed the steel as "PRK 33"[13] (Paul Richard Kuehnrich No. 33), or as "Cobaltcrom," or both, later designated D5.

THE NEW STEEL

Cobaltcrom Patent Air Hardening High Speed Steel.

(TUNGSTENLESS HIGH SPEED STEEL)

British Patent No. 106187.

Patents applied for in all Foreign Countries.

THIS STEEL is specially suitable for Milling Cutters, Twist Drills, Reamers, Taps, Automatic Forming Tools, Screw-cutting and Finishing Tools in general, and for these purposes is FULLY GUARAN-TEED to give equal endurance to Best Tungsten High Speed Steel, producing, however, a better finish.

FOR machining Gun Metal and Yellow Metals in general, for High Endurance Drawing and Blanking Dies, Lathe Centres, Shear Blades, Valves for Aeroplane and Automobile Internal Combustion Motors, and all kinds of Gauges and Instruments in general which have to resist abrasion, the Patent Cobaltcrom Steel is clearly superior to Best Tungsten High Speed Steel.

Air Hardening is effected at 1000 Centigrade, a considerably lower and less risky heat than required for Tungsten High Speed Steel.

PROMPT DELIVERIES.

MANUFACTURED UNDER LICENCE BY

DARWIN & MILNER, Ltd.,	SYBRY, SEARLS & Co., Ltd.,
SHEFFIELD.	SHEFFIELD.
Tel. Add.: "URGENT, SHEFFIELD." Tel. No. 4517.	Tel. Add. "CANNON, SHEFFIELD." Tel. No. 165.
BRAND:	*BRAND:*
PRK 33	CANNON

1917 Darwin and Milner Cobaltcrom ad[13].

The Cobaltcrom steel had interesting promotion attached to it, marketed for everything from dies to milling cutters, and was also said to be usable for cast-to-shape parts to save cost vs. machining[14]. Metallurgists recognized that the structure of Cobaltcrom was relatively fine in the cast condition[15], an improvement over the earlier Neor. In 1927 Gregory Comstock of Firth-Sterling took advantage of this improvement when he patented a similar steel called Cromovan[16], also with 1.5% carbon and 12% chromium. Instead of cobalt, it had more typical die steel alloy additions of 0.8% molybdenum and 1% vanadium. This steel was later given the designation D2. Unlike Kuehnrich's patents, this new Firth-Sterling grade was not claimed to work as a high speed steel, but was designed for die steel use. Air hardening of large dies was improved through the molybdenum addition, or at least the patent said that "molybdenum serves as an energetic hardener."[16] The D2 patent also claimed that the hardness is uniform throughout relatively large dies despite faster cooling at the corners, and with "little tendency for the steel to warp during hardening."[16] It was also noted early on that the toughness and machinability were much better with the lower carbon D2/Cromovan vs. the earlier D3/Neor type. Despite the relatively sizeable vanadium addition, the patent makes no statement about why it was added; presumably, it was for microstructure refinement and/or wear resistance. This steel quickly became very popular as a die steel, and many other manufacturers made their own versions[7]. So, despite Kuehnrich's many contributions, the most remembered high carbon, high chromium steel was not directly connected to him, though it was very similar to his earlier Cobaltcrom.

Kuehnrich also patented a molybdenum-based high speed steel with 0.5-0.8% carbon, 4-10% molybdenum, 4% chromium, and 6% cobalt. Molybdenum was known as a potential replacement for tungsten in high speed steels, but none had yet matched the properties of tungsten high speed steels. The patent claimed that the cobalt addition allowed this steel to match the performance of the more popular T1-type high speed steels. This steel was marketed as "Como."

Steel	Year	C (%)	Cr (%)	Co (%)	Mo (%)	V (%)	Ni (%)
Unor	1904	2.2	15				0.75
D3	Before 1926	2.3	13				0.6
D5	1916	1.5	12	3.5			
D2	1927	1.5	12		0.8	1.0	
Crysteel	1929	1.1	18	3.3	0.75	0.25	

D5 and D2 Steel in Knives

D2 has been used in many knives over the years. Loveless said that he first used it in 1957 but didn't like it because it was challenging to finish[17]. D.E. Henry[18], Chubby Hueske[19], and Jimmy Lile[20] were among the first knifemakers to use D2. In recent years, Bob Dozier is perhaps the most well-known custom knifemaker to use D2 regularly in his knives, and he used it at least by 1997[21]. When I asked Mr. Dozier who he knew was making D2 knives early on, he also credited Ted Dowell and Jimmy Lile. D2 has been popular as a "budget steel" in recent years in high-end China-produced production knives. Production knives with D2 didn't show up until the 1990s, such as the Fighting Falcon by Columbia Handcrafted Knives in 1997[22], Knives of Alaska also in 1997[23], and KA-BAR knives in 1999[24,25].

D5 steel (Cobaltcrom PRK-33) also saw some limited use in knives. John Kleinman reported that Joe Funderburg used D5 in 1977[26]: "Wherever applicable he prefers to use D-5 because, in his opinion, he can obtain a better finish and because it has even better edge-holding properties than D-2." Jody Samson was reportedly using D5 by 1978[27]. Wayne Goddard said that he liked D5 steel[28], and he mentioned testing it in 1990[29]; reportedly, it performed similarly to D2 in rope cutting tests he performed[30]. D6 steel, similar to D3/Neor but with a small tungsten addition, has been popular with custom knifemakers in Brazil[28].

Darwin and Milner Razors and Crysteel Cutlery

In 1924 Kuehnrich had the idea to make safety razors and blades. He purchased the Fitzwilliam Sheffield Simplex Motor Works and converted it for razor production. Gillette was a dominant player in the safety razor market, but even they had been making razors with simple carbon steel. He would use the previously described PRK-33[31], also called Cobaltcrom or D5 tool steel. Kuehnrich said in 1924:

> Gillettes made a profit of £10,000,000 on blades alone last year. Their 10-dollar shares are 350 dollars each. I am going to command the market of the world in safety razor blades because I have an infinitely better thing than has ever been known. In 1917 I invented a steel known as Cobalt steel – stainless and rustless, but differing from ordinary stainless steel in possessing super-cutting qualities never before known in razors. I can guarantee that the new blades will last at least three times as long as the ordinary safety blades... Long razors made of cobalt steel have been used for four months without stropping.[32]

Kuehnrich also promoted his steel as being better than the existing stainless steel in the marketplace:

> It is now recognized that stainless steel is very largely a failure for cutlery purposes. Nine people out of ten have the idea that stainless knives are very handy things for keeping clean, but they do not expect them to cut. In the first instance, however, knives should be knives, but stainless knives – with certain exceptions of course – are practically non-cutting or extraordinarily poor for cutting purposes. At Fitzwilliam Works we intend to make cutlery by mass production of a steel that is rustless and stainless, but which has super cutting qualities. It can be used for every kind of knife from a table knife to a cook's knife.[33]

Kuehnrich frequently compared his steels to Brearley's stainless steel (discussed in the next chapter):

> I have followed with a certain feeling of compassion the development of the Firth-Brearley process as far as those industries are concerned which are using the material for the manufacture of cutlery. It was recognized from the beginning that although knives made from this material had their advantages from the point of view of cleanliness, the cutting properties varied very much and generally were uncommonly bad. On the contrary, the introduction of this stainless steel has become a direct calamity for Sheffield, because the world to-day is now inundated with hundreds of millions of these faulty stainless knives, the majority of which possess very poor cutting edges, to such an extent that to-day in nearly every country the world 'stainless' is synonymous with 'bad cutting.' Many cutlery manufacturers have lost their reputation through this unfinished and unreliable steel, and I almost consider it a mischief that it has ever been put on the market...[2]

However, not everything went smoothly, as the razors did not sell well enough to justify all of the investment, especially in the US market. Also, in 1930 there were increasing reports that the blades were rusting. Kuehnrich admitted that[1] "the polished edges of the Darwin blade were rustless, but this did not apply to the other part of the blade (which wasn't polished)." Advertising was modified to state that only the edges were rustless.

1928 Firth-Sterling Cromovan ad.

Kuehnrich was also expanding into cutlery which included a new steel developed for that purpose:

> It is a cause of extreme satisfaction to me that amongst the compositions which I have developed during many years of study and research I have now been able to create a class of steel for stainless cutlery which, in my opinion, will bring forth a revolution of almost fundamental importance on the whole of the world's markets. The P.R.K. Patent cutlery steel, which is being manufactured by Darwins, Limited, in two varieties with different chrome and cobalt contents, makes it possible to produce knives of all kinds with rustless and stainless properties which, as regards keenness of edge and duration of cut, surpass by far any knives made from the very best carbon steels. With one kind of steel, of which considerable quantities are being supplied to one of the largest Solingen cutlery concerns, tests were carried out recently on a machine for testing the cutting edges of knives, with the result that in comparison with a 1 per cent. carbon Swedish steel the P.R.K. Patent steel registered with double feed a quadruple duration of cut. It is intended to make the P.R.K. Patent steel accessible to all the markets of the world from the beginning of the late autumn. All concerns manufacturing cutlery will be able to obtain deliveries without any restrictions as to export. Messrs. Darwins further intend to consider establishing, in the near future, works in America and Germany for the production of this class of steel on a large scale. In this manner a new era will be brought to life for the benefit of the community in the manufacture of cutlery of the highest cutting properties, which will excel in quality the very best cutlery made form carbon steel manufactured before the introduction of stainless steel.[2]

The second variety of cobalt-chromium steel he referred to was used in his "Crysteel" line of cutlery, which also referred to the name of the steel. This stainless steel had a composition of 1.1% carbon and 18% chromium, more similar to 440C-type stainless steel (see Chapter 10) but with a sizeable cobalt addition of 3.3%[10]. The lower carbon, in combination with higher chromium, gave the steel better corrosion resistance than D5. He took out full-page ads in several newspapers in 1932. However, Kuehnrich's debts had been catching up with him. He had restructured businesses several times and declared bankruptcy more than once, having done so again in the spring of 1932. Finally, on April 27, 1932, Kuehnrich locked himself in a room and shot himself.

FREE DIRECT FROM ENGLAND

Making your **regular** *safety razor* **extraordinary**

DARWIN BLADES
with their
Super-Cutting, Rustless
Edges, Revolutionize
Safety Razor Shaving

NOTHING has contributed so much to man's personal convenience as the safety razor. But it has always had one drawback—the nuisance of changing the blade almost daily, and the wastefulness of "dud" blades.

This unreliability is due to the universal use of ordinary carbon steel, with its very definite limitations. The adoption of Sheffield's new Patent Cobalt High-Speed Steel, has completely transformed safety razor shaving. This new steel admits of *mass-hardening with absolute uniformity*—a feature hitherto impossible in safety razor blade production.

DARWIN Blades, with their supercutting *rustless* edges, are as different from the old kind as pneumatic tires are from solids. They give you a degree of shaving comfort never before even thought of. *Every* blade in *every* package is perfect, and lasts *many times longer* than the average blade of the old type. They are a *revelation*.

FREE SAMPLE BLADE

We will mail to anyone in the United States a FREE SAMPLE DARWIN BLADE direct from the Sheffield Works in England. Send us an envelope addressed to yourself and bearing in the top right-hand corner the name of the holder you require the blade for. Enclose this envelope with a *loose* 6-cent stamp to cover mailing expenses, and post to Darwin Ltd., Sheffield, England, and the sample blade will be sent you promptly. Postage 2 cents. Write no letter.

The price of Darwin Blades has been fixed at $1.50 for 10 blades. They are NOW OBTAINABLE exclusively from our American Mailing Department, Darwin & Milner, Inc., 1260-4 West 4th Street, CLEVELAND, Ohio. Send cash $1.50 per package of 10 blades post free, stating type of blade required.

After February next, DARWIN Blades will be purchaseable throughout America from all dealers.

Please note that the free sample blade can be had *only direct from England*, not from Cleveland; also that packages of 10 blades can be bought *only* from Cleveland, *not* from England.

Pass this information on to all your friends. Everybody can verify forthwith the complete change in shaving conditions created by the DARWIN wonder-blade, either by purchasing the blades from Cleveland, or by sending to Sheffield for a free blade. No one need wait to avail themselves of this real boon.

DARWIN
PATENT COBALT HIGH-SPEED STEEL
Safety Razor Blades
DARWINS LTD. FITZWILLIAM WORKS, SHEFFIELD, ENGLAND

1926 advertisement for Darwin razors.

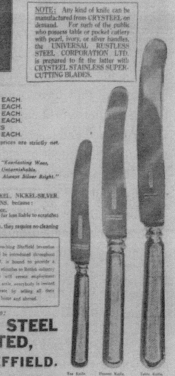
1932 Crysteel ad.

-9-

HARRY BREARLEY AND STAINLESS STEEL

Harry Brearley[1] was born in Sheffield in 1871 in a relatively poor household. His father was a steel-melter, and Harry was the eighth of nine children. Harry described his father as "a good specimen of the country breed spoiled in appearance by the hard brutalising work associated, at that time, with iron and steel-making; and depreciated in manners by the drinking habits commonly practised by, and usually excused in, men who toiled and seated at the hot furnaces. Steel-melters and puddlers were not idealists, they were not encouraged to be. But there was in my father a streak of poetic sentiment."[1] Brearley said that he[1] "was born with some of the instincts of the scholar, but they have come to nothing in particular; they derived, possibly, from a dreamy, other-world, father, and were finally snookered by a regard for matter-of-fact, inherited and learned from a plain-Jane mother." He spoke well of his mother as someone who taught by example though not often with words. "I might have been a home-bird, and remained delicate all my life, if the home had included books, pictures, and toys. But, bar a novelette or a soiled library book, there was no reading matter of any kind."[1] Brearley was not a particular fan of conventional education: "It was curiosity and imagination that made me into a looker and toucher... Apologists for compulsory schooling, who cannot defend the barbarity of compelling children to sit quiet on hard seats and listen to lessons they dislike and misunderstand, find compensation in the good children do each other, and the help they get from games and the team spirit."[1]

Brearley's father and brothers all worked in steel melting furnaces, and he got a job at 11 or 12 years old as a "cellar lad" in a crucible steel melting furnace operation. Harry was slim and frail-looking compared to other cellar lads, but his brother Arthur was a cellar-lad in the furnace next to his and would look after him. The cellar-lads delivered ale, iron, and steel scrap to the workers. Brearley said that the workers required so much ale that it "might have been regarded as one of the raw materials of steel-making."[1] "[O]ne learns naturally from workmen what no professor can teach, and, however extensive a scientific knowledge may ultimately be acquired, the Art of steel-maker can nowhere be learned so well as in the furnace, and nohow so thoroughly as by actually working there. This is rather a song to make about the work of a cellar-lad, but the job was the first in which I was really interested, in spite of the fact that I was very sleepy when called to work at half-past five in the mornings, and very tired when I got home about half-past six in the evenings."[1]

Brearley maintained an appreciation for the practical aspects of steel production. He said that he was "in appreciation of men, typified by my father, who have exerted a greater favourable influence on the practice of steel-making than they are given credit for."[1] "The value of scientific control, with its measurings and weighings, is undeniable, but it is sadly blemished by a strutting vanity and snobbish intellectual pride which smothers with confusions aspects of a subject it does not understand."[1]

However, he also recognized that it could be difficult to get good information from the workers doing the actual job of steelmaking:

> I know how useless the effort may seem, because I tried - too casually - to extract information from my father. I thought my readings in chemistry and physics would take, by another road, far beyond where he stood; but now I know, by inference from what I saw him do, that he knew things I should like to know and never shall know. I nearly made the same mistake with my brother. He is a better workman, a better observer and more resourceful experimentalist than I am; but he has stuck to the furnaces, mills and forges and done things whilst I have read about them, and looked at things other people were doing, and taken some trouble to describe them in the language learned from book.[1]

To illustrate the fact that ordinary workers often have methods that are superior to the scientist, he described the history of toughness testing:

> For a century or two the blacksmith has been in the habit of ending a bar of iron or steel after first nicking it was a slate. When the bar is broken, as intended, through the nick, the effort required to break it, appreciated by the smith's muscles, is registered in his mind - his extremes of judgement are "rotten" and "damned good stuff." There is not the slightest doubt in the smith's mind that the material he intends to make something from is all the worse for being "rotten" and all the better for being "D.G.S." Many years ago, thirty maybe, a young man in Rugby, named Izod, was called upon to explain why of two pieces of steel one failed hopelessly and the other did its job very well. Both chemical analysis and every conventional form of mechanical testing showed that the two pieces were identical; but this young man knew they were not. Bethinking himself of the blacksmith, he notched both pieces, and found the one was "rotten" and the other "D.G.S." In order to state the difference, which was the only difference he could find, quantitatively, he made a pendulum hammer very much like the existing Izod testing machine, and then went to South Africa, blissfully unconscious that his Rugby adventure was the lead to a new kind of official testing.[1]

Brearley next became a bottle washer in a chemical laboratory where he worked for the chemist James Taylor. Brearley described this experience as the greatest "of all the favours bestowed on me by Fate."[1] And "the most important incident in my 'school' life."[1] Taylor gave him assignments from a math book that he required Brearley to purchase himself. When he finished with that book, Taylor gifted him a book about Algebra. "If he had given me one of his old school books I might have understood it ... but a new book, costing seven and sixpence = as much as I could save in a year - given to me by a man who was neither my father nor my uncle, was a shaking up of feeling and perception. If JT's idea was to infect me with that blossoming interest in things which grow out of love for the men who do them, then this action was a stroke of school-masterly genius."[1] Taylor encouraged him to enroll in night school, but Brearley always considered James Taylor to be "my real teacher, the man who inspired me to do my best."[1] So important to Brearley was Taylor that he named his son after him: Leo Taylor Brearley.

Caption from Brearley: "My brother (sees) more with the naked eye than most people see in them using laboratory equipment. On the other hand, I spent about 20 years in a laboratory, mainly occupied in measuring things. The picture is intended to be symbolic: My brother on the right-hand side is looking at a fractured surface of a piece of an ingot, and I on the left-hand side am pretending to be doing something with a pair of calipers."

When Brearley was 20, there was an opening for a laboratory assistant (Brearley had been a bottle-washer from 12 to 20). At the time, it was common to require a payment to start a profession, and this one required 50 British pounds, and Brearley had to arrange to pay in installments to a friend, as his family did not have nearly that much. By then, Brearley had "fallen in love with analytical chemistry and a girl"[1] and married Helen Crank in 1895 when he was 24. Brearley described their relationship: "We were a practical couple, with no expensive habits, very much in love with each other, and delighted to be in the country away from town life. One could do anything in the house, and the other could make or repair anything in reason the house demanded."[1]

"By the time high speed steel was being introduced I had come to feel that analytical chemistry was a calling as imperious as that given to any church officer. I was in love with my work, and could think of few better things than the privilege of living to continue it. My work was indeed my idol, and I have staggered home intoxicated with tiredness in as elated a frame of mind as ever was born of bubbling wine and choice company. I analysed alloys and fancy high speed steels by the score for merchants and manufacturers and for public analysts who might know of no reliable way of doing the job."[1] Brearley wrote a book about analyzing steel chemistry, but the book's publication was essentially the end of his career as a chemical analyst. "I did not realise I had finished looking at everything through a laboratory window, without a care for what was happening in the works - but I had."[1]

Brearley got a job for a steel factory in Riga, then part of Russia, though Thomas Firth & Sons, a Sheffield steel company, owned the factory. The company was making armor-piercing projectiles, and no one at the plant had experience with heat treating. This left Brearley to learn about heat treating of steel, a subject with which he had no previous experience. "I was beginning to feel I could do any job if I fancied it, and this one appeared to be to my liking."[1] Harry and his brother Arthur, who also worked at the Riga plant, hardened the projectiles from different temperatures to find the optimal range. Arthur's "supreme judgment of fractures, and his excellent rule-of-thumb knowledge, were invaluable."[1] They had no way of measuring temperature, so Harry found metal alloys that melted in a range that would show them from what temperature to quench the steel. One alloy would start to melt at the very lower limit of where the steel would harden, another would melt at just the right temperature, and another would melt at the point where the steel was overheated. They purchased a microscope and a thermocouple without knowing how to use them, and measured the melting temperature of the alloys, though they struggled to find uses for the microscope, preferring to look at fracture surfaces instead. "Riga provided, casually, more experience and more kinds of experience than might have been found on a purposed search. Four years had aged the eyes, especially the mind's eye."[1]

Times had been dangerous in Riga; while Brearley was there, he saw the Russo-Japanese War and the 1905 Russian Revolution. So, he was happy when Thomas Firth & Sons sent him to Sheffield to start a research laboratory. The steel company John Brown & Co became a partner in the laboratory as well, and it became the Brown-Firth Research Laboratory. Brearley negotiated an agreement where his discoveries in the laboratory would be co-owned by himself and the steel company, which allowed him to build his consulting business.

> To have a job in a laboratory, meaning a place where one can work without the urge and restraints of routine production, is to be amongst the favoured few. No works laboratory conforms to this description; some are just production workshops, driven as automatically as a Ford operator. But add the word research, and the suggestion is that clever brains are directing unhurried jobs, whose aims, if attained, will lead to fame and fortune: this also is a misleading picture of an industrial research laboratory. The most the awakening dreamer can hope for in a laboratory, however its attractive name may be qualified, is ready-made equipment that he may be allowed to use, without too many hurrying questions being asked; and if he finds also bread and cheese and camp bed, he is in paradise ... the dreamer will dream on in his own way.

From 1907 to 1914 there was no man living better pleased with his job. It was my business to observe, to experiment, to advise, to criticise constructively, to teach, to discover (if the Fates were kind), and above all not to mislead. My connection with Brown's and Firth's gave me every facility that could be wished for in relation to the manufacture of steel.[1]

In this environment, Brearley was tasked with studying erosion and fouling in rifle barrels in 1912. He proposed that they experiment with low carbon, high chromium steels and produced a series of steels with chromium content varying from 6 to 15 percent using the crucible process. Unfortunately, none of the steels had the properties they were looking for, either having too much carbon or too little chromium. So, they used the electric furnace instead, and the second heat they produced had 0.24% carbon and 12.8% chromium. They found this steel to have excellent mechanical properties for rifle barrels, and it was also easily forged, hot rolled, and machined. Brearley set out to characterize this new steel, conducted experiments with different heat treatments, and then observed the steel with microscopy. He found that the typical acids they were using to reveal the microstructure had little or no effect on the new steel, and the degree of etching would vary based on the heat treatment. He also realized that since etching steel is a form of corrosion, this was an indication that the steel had excellent corrosion resistance.

The steel was not getting any interest from the rifle manufacturers, but one of Brearley's reports about the steel indicated that cutlery would be a good use. At the end of 1913, steel was sent to two cutlery companies, James Dixon and George Ibberson. Brearley was not involved, and they didn't like the steel because of the difficulties in forging, grinding, and heat treating which would have been very different than carbon steel (see Chapter 10). Brearley found Ernest Stuart of R.F. Mosley cutlery company and started working with him in 1914. Stuart initially refused help from Brearley but eventually invited him to the factory to help with the selection of temperatures in forging and processing the steel. He bought 125 lbs. of the steel, then called Firth's Aeroplane Steel, and the knives were found to resist tarnishing even when in contact with food, fruit, or condiments. Brearley credited Stuart with believing in the steel: "It is due to him to say that, from the very first trials, he had confidence in the possibilities of the steel. He made unremitting efforts to adapt the process of knife-making to the unusual qualities of the steel and, so far as the initial use of stainless steel for cutlery is concerned, the credit is due to him and his firm."[1] Brearley also credited Ernest Stuart with the name "stainless" for the steel.

After the experience with Stuart, and the difficulties he saw in working with the steel using the historical process for making cutlery, Brearley felt that the steel should be sold as heat treated blanks to the cutlers so that only final grinding and polishing were necessary for production. Firth's management disagreed and wanted to sell the steel and have Brearley offer help as necessary. He was upset about his rejected proposal and refused to offer assistance with the steel. He was reprimanded, and Brearley's relationship with Firth-Brown soured. "I left ... feeling that I had been wronged, and if I was not angry I was sad."[1] The directors at Firth's were unimpressed with Brearley's steel, so the steel was not patented in Britain. The Firth's directors were more impressed by a stainless steel independently developed by Krupp in Germany, and they saw the licensing of the Krupp product as the profitable move. But Brearley met John Maddocks, who encouraged him to patent the steel in North America, and Brearley got patents in both the USA and Canada.

Elwood Haynes, better known for developing the cobalt alloy Stellite, also got a patent for high chromium steel in the USA. Rather than argue over the rights, he and Brearley agreed to share the royalties, and the Firth-Brearley Stainless Steel Syndicate was formed. The name Firth remained because Brearley worked with the Firth-Sterling Steel Company in Pittsburgh, which was related to Firth in Britain but had different management. This led to the American Stainless Steel Company, formed between the Syndicate (40%), Haynes (30%), and the rest split between steel companies Firth-Sterling Steel, Bethlehem Steel, Carpenter Steel, Crucible Steel, and Midvale Steel.

Was Brearley the First to Discover Stainless Steel?

There have been several papers and books on the discovery of stainless steel and how complicated the history is. Brearley's stainless steel was the first to be commercialized, but several other developments occurred in a similar timeframe or even before Brearley. One critical paper presented in 1924 by Percy Armstrong[2] laid out an impressively extensive history of patents and journal articles up to that time on high chromium alloys, both from the United States and Europe. Another paper from Carl Zapffe in 1948[3] includes a somewhat later description of the many people looking at stainless steels early on. In 1904[4] French metallurgist Léon Guillet presented on ferritic and martensitic high chromium steels, and in 1906[5] on austenitic chromium-nickel steels. Several of these steels were very similar to compositions that would be commercialized later as stainless steels, similar to 410, 420, 440C, and austenitic stainless steels. He tested their mechanical properties but did not recognize their high corrosion resistance. The property of stain resistance was discovered in Germany in 1908 by Phillip Monnartz, leading to a patent in 1910[6] and a paper in 1911[7], where he showed that the corrosion of steels dropped greatly as the chromium content approached 12%. Monnartz described this phenomenon as occurring from "passivation," which is indeed the mechanism by which the chromium imparts corrosion resistance. Interestingly, the patent from Monnartz also claimed molybdenum as an element that would enhance the corrosion resistance of stainless steels, which would become an important element later, but was not immediately recognized in early commercial stainless steels.

German Stainless Steels

Maurer   Strauss

Eduard Maurer and Benno Strauss patented austenitic stainless steels for Krupp in Germany in 1912[8] after a series of investigations that began in 1909. Maurer worked on his doctorate in France under Floris Osmond in 1908, and Zapffe speculated that he may have learned about Guillet's work on austenitic chromium steels at that time[3]. Maurer likely also would have been aware of the work of the German Monnartz. Maurer and Strauss made a series of chromium alloys for study. Strauss told the story in 1924:

> Metallurgists for a long time past have been accustomed to alloy iron with nickel with a view to the improvement of the iron, whereas the discovery of the valuable properties of chromium is of comparatively recent date... In seeking for a suitable material from which to make protective tubes for thermo-electric couples (thermocouples), [I] started on the supposition that it ought to be possible to make use of the valuable properties of chromium, that is, its resistance to the action of oxygen, for the purpose of increasing the resistance of iron against the attack of hot gases, in view of the fact that nickel had, shortly before, been rendered extremely heat-proof by the addition of about 10 to 15 per cent of chromium. In 1910, [I] had the following five experimental alloys prepared:

Steel	C (%)	Cr (%)	Ni (%)
C1	0.38	9.81	0.1
C2	0.49	19	0.1
C3	0.6	28.85	0.1
C4	0.22	10.05	1.75
C5	0.31	20.1	5

> These steels had been forged into rods and were to be made into protective tubes for pyrometers. Of the above, the two alloys C4 and C5 containing a higher percentage of nickel, after having been subjected to a simple annealing operation at 800°C, turned out to be unworkable. Polished specimens taken from the above bars, after having been exposed for some time to the atmosphere of the laboratory, revealed the curious fact that particularly those steels which contained 20 per cent of chromium had remained perfectly bright, while other bars, even those containing up to 25 per cent of nickel, had become rusty.[9]

They found that the high nickel steel (C5) was austenitic and had excellent ductility, while the lower nickel steel (C4) was martensitic after heat treatment. These grades were commercialized as V2A and V1M. The V stands for Versuchstahl ("test steel"), the A stands for austenitic, and M for martensitic. So V2A was test steel no. 2 austenitic, and V1M test steel no. 1 martensitic. The compositions for these steels are shown below:

Steel	C (%)	Cr (%)	Ni (%)
V1M	0.15	14	1.8
V2A	0.25	20	7

There became an increasing rivalry between Maurer and Strauss as to who developed stainless steel. The patent and early articles about the steel credited both, but in 1918 only Strauss received a bonus from Krupp for the invention. In 1927 an article was published about the invention without mentioning Maurer. In 1935 Strauss was fired from Krupp because of his Jewish descent, and Maurer reported him to the Gestapo, which led to Strauss's firing as a professor at the University of Münster. Maurer said in 1939 that Strauss had nothing to do with the invention of stainless steel[3], but given their history, I don't think it should be taken seriously.

Steel	C (%)	Cr (%)	Ni (%)
V3M	0.4	14	0.6
V5M	0.15	13	0.6

Austenitic stainless steels are not the focus of this book, being used in applications where high hardness is not necessary and high ductility is required. The composition of V2A is not far from where common austenitic stainless steels ended up, around 18% chromium and 8% nickel. The V1M is close to the modern designations of 414 or 431 stainless, low carbon martensitic stainless steels with 2% nickel. Krupp also developed a somewhat lower nickel version called V5M. Maurer and Strauss attempted to use this steel for cutting tools, as Strauss reported:

> [I]n examining the first surgical knives made from a steel containing 14 per cent chromium, 0.8 per cent nickel and 0.15 per cent carbon, the author found that in the hardened state they were very corrosion-resisting, but not hard enough. Endeavors to impart greater hardness to these knives by means of cementation in powdered coal (carburization) were successful, but at the same time it was found that the knives then rusted very quickly in water containing air. Investigation revealed the presence of free carbide, this having been the cause of the lack of homogeneity and the ease with which rust formed.[9]

In other words, there was insufficient carbon for high hardness of blades. When he carburized the surface, there was an excess of carbon, leading to a loss of corrosion resistance from chromium carbide formation. When chromium forms a carbide, it is no longer present to contribute to corrosion resistance. Krupp also released a martensitic stainless steel closer to Brearley's called V3M by 1922[10], though with somewhat higher carbon. V3M became the standard German cutlery stainless steel 1.4034. Initial versions had 12% chromium[10], which was increased to 14% by the

1930s[11]. The Maurer-Strauss patent was for steels with 0.5-20% nickel which explains why the martensitic steels all have a relatively unnecessary addition of 0.6% nickel.

Percy Armstrong and Neva-Stain Cutlery Steel

Percy Armstrong was born in 1883 in London[12]. Between 1909 and 1914, he was employed by Böhler Brothers of Sheffield, a significant tool steel manufacturer headquartered in Austria. Böhler Brothers had opened a warehouse and small forge in Sheffield in the 1890s to import steel from Austria. Working in Sheffield at the boom of high speed steel and alloy steels meant he learned a great deal about steel, and he even met Harry Brearley during that time. At the start of World War I, he refused to fight the Germans and left for the United States in 1915, leaving his wife and two children for a wealthy divorcée, never seeing his wife or children again.

Armstrong was hired as a vice president of Ludlum Steel Co in Watervliet, New York. He developed a steel called Silcrome with 0.4% carbon, 8% chromium, and 3% silicon. The large addition of silicon was at first accidental. Armstrong was developing a steel for high-temperature automobile valves but had issues with "scaling" at high temperatures. The silica (silicon dioxide) electrode reduced to silicon, and added a large amount to the melt. The electrode is used to melt steel through applied current in electric steelmaking. Tests of those silicon steels revealed excellent resistance to scaling. This steel had relatively low chromium compared to generally accepted definitions of "stainless" steel. However, the chromium was sufficient in combination with the high silicon for high-temperature applications where scale formation is the biggest issue. This steel was patented in the USA in 1919[13-14].

Silcrome was also marketed as a cutlery steel named Neva-Stain with the same composition[15]. A 1920 ad for Neva-Stain said[16]: "A new 'Ludlum' product of high silicon chromium alloy, completely resistant to the affects (sic) of fruit acids, staining, tarnishing, or surface deterioration ... 'Neva-Stain' can be hardened in oil and made **file hard.** It can then be drawn to requirements." Armstrong himself promoted the steel as an answer to inexpensive German stainless steel imports in 1921[17]:

> Rustless and stain-resisting steels have been the subject of keen investigation by American, British and German organization and there has been a race to produce a stain-resisting and rustless steel at sufficiently low price to enable it to be generally used for cutlery ... the Ludlum Steel Company ... has developed the inexpensive low chromium, high silicon steels... It has heretofore been considered by metallurgists that silicon is an alloy which should not be used in quantities but this has been proved a fallacy and at the most only applies to iron-silicon combinations and not to chromium alloys; in fact, peculiarly enough, silicon enables chromium and iron to form a chemical combination ... If it were not for the very high silicon content, this material would be easily rusted and readily stained... German competition in cutlery is being keenly felt at this time and, without a

doubt, the German rustless steel would be very largely employed in the United States; but due to the enterprise of the American manufacturer, this competition has been forestalled, as the cost of the high chromium nickel steel of the Germans could not be sold, even though their rate of exchange is so low, at a price which would compete with the low chromium, high silicon steel. The United States has not employed stain-resisting steel for its cutlery to any large extent. This type of material is very largely used abroad and one of the reasons for this is the lower cost in foreign countries. The new low chromium silicon steel can now be sold at very low cost ... thereby enabling the cutlery manufacturers to offer to the buying public a very keen-edge, stain and rust-resisting cutlery at a greater price than the plated knife. It will be but a short time before the housewife can take her choice in any hardware store of a plated knife or a stain-resisting, rust-resisting sharp steel knife at about the same cost. The cutlery manufacturers realize that this new development assures them a very big market in this county, free from German competition... This new steel is sold under the trade name of Neva-Stain, which is a very apt term and means what it says - that it ... will never stain.[17]

In a 1923 Ludlum document for its steels, Neva-Stain was described:

Stain resisting steels that are to be used for cutlery purposes and for knives generally have been found wanting because of the inability to impart to these steels a keen and lasting edge. It is very well known indeed that to impart a keen and lasting edge high carbon is necessary, and that is equally true when high alloys are employed. Heretofore it has not been possible to produce a high carbon alloy steel which was rustless and stain-resisting. Neva-Stain steel, due to the high silicon employed, will carry sufficiently high carbon to give it the necessary keen and lasting cutting edge... Neva-Stain ... is very similar in these characteristics to high grade, high carbon cutlery steel.[18]

Armstrong and Ludlum were soon sued by the American Stainless Steel Co., the syndicate formed between Brearley, Haynes, and steel companies producing stainless steel. Ludlum was not paying royalties while producing their stainless steels, arguing that the high silicon addition made their steels novel and outside of the Brearley-Haynes patents. An initial lawsuit was dismissed, with the judge agreeing that the high silicon put the steel outside of the previous patents, but after an appeal, the decision was overturned. This verdict was clear about disallowing the use of steel in polished items such as cutlery, but Ludlum and Armstrong thought their valve steel was not affected and continued to sell Silcrome for valves. Of course, this meant that "Neva-Stain" for cutlery was relatively short-lived, though some knives were made. Another lawsuit in 1928 was successfully defended, giving Armstrong credit for inventing Silcrome. By the 1930s, the steel was almost universally used in exhaust valves in the automotive industry.

-10-

DEVELOPING STAINLESS KNIFE STEELS

420 Stainless Pocket Knives and Hunting Knives

Stainless steel of the low carbon 420 type with 0.3% carbon and 13% chromium became relatively popular in "table cutlery," such as carving knives, forks, etc. Unlike silver-plated cutlery, stainless knives could be re-sharpened without removing a necessary coating. Interestingly, while stainless steel was first produced in Britain, that was only in small amounts, and the first industrial production was in the USA at Firth-Sterling in 1915[1]. Initial sales were good; however, there were significant headwinds against stainless steel. In 1917 there were restrictions placed on using stainless steel for cutlery in Britain and the United States due to World War I[2-3]. Even after access to stainless steel was restored in 1919[4], companies were relatively slow to pick it back up[5]. Stainless steel was more challenging to forge, as *American Cutler* reported in 1919[4]: "Small pieces on which not much work required to be done may be forged by hand, but a power hammer is practically indispensable for such articles as table blades or larger pieces."

Heat treating also had much different requirements than carbon steels. The thermal conductivity was much lower and would therefore heat up more slowly, thus requiring longer soaking times[6]. Carbon steels required heating to a red heat of around 1400-1500°F (760-815°C), while stainless steel needed to be much hotter, around 1750-1900°F (950-1040°C). If heat treated at the low temperature used in carbon steels, the steel would not be "stainless" and would be too low in hardness to hold a good cutting edge. "Tempering colors," which develop after the low-temperature tempering process of steel, were also unreliable with stainless steel. The colors develop from an oxide that forms on the steel; stainless steel resists oxidation, so the colors develop at much higher temperatures.

Grinding and polishing were also an issue with stainless steels, since the steel would dissipate heat more slowly, leading to overheating in grinding more often relative to carbon steel blades[6]. Finally, any oxides, pits, or scale needed to be removed because they serve as sites for corrosion to initiate. In addition, some cutlers were "burnishing" rather than polishing the blades, leading to easier staining[6].

Apart from the difficulties in manufacturing stainless cutlery, there were also many rumors and myths about the behavior of stainless steel. One of these rumors was that stainless blades did not hold an edge. Reduced edge retention was blamed on several factors, including poor heat treatment[7], the edge overheated in grinding[7], or that carbon steel knives were more frequently sharpened because they require polishing[8]. Other rumors could be even more outlandish, such as the claims that stainless steel knives should never be placed in hot fat or hot water for more than 1-2 seconds, that the knives should only be stropped and not ground, that you should avoid prolonged exposure to sunlight, and that the knives need to be occasionally rested in a dark place pointing due north because "the weak current of electricity that is always going over the earth's surface between north and south will restore the blades in time."[9]

While stainless steel was known for being used in table cutlery, it was slow to take off in hunting and pocket knives[10]. By the 1930s, 90% of table cutlery was made with stainless steel, but only 10% of pocket knives were[11]. Edwin Jackman of Firth-Sterling said in 1922:

> While the value of Stainless Steel in these household knives is recognized, a tour of New York stores selling sporting goods reveals the fact that Stainless Steel hunting knives are practically unobtainable. The clerks at the counters where hunting knives are sold give amusing reasons as to why they do not sell stainless knives. One informed a potential buyer that the steel was too brittle and would break when submitted to rough usage, such as hunting knives are subjected to. Another gave the opposite opinion that Stainless Steel was too soft. Another thought Stainless Steel was merely a plated blade, and that it could not be sharpened, etc., etc. We are told of one instance where a resident of a gulf city who happened to be visiting in New York, tried to secure a half dozen stainless hunting knives for himself and friends. He had used stainless household knives and had a stainless pocket knife. After several inquiries, he was referred to a steel manufacturer as the only probable source of supply of stainless hunting knives. Unfortunately, the steel company only had a few knives made for exhibit, not for sale. The gentleman insisted on purchasing a part of the exhibit and finally succeeded, saying that he was perfectly willing to pay any price which would cover the cost of making these knives, which were made to order for the exhibit... Why should it be so difficult to secure this type of cutlery outside the household line?[10]

Early 420 Stainless Pocket Knives

Victorinox

Victorinox got its name from an amalgamation of "Victoria" and "inox" in 1921[12]. Inox is short for acier inoxydable, the French term for stainless steel. The company was started by Karl Elsener in Switzerland in 1884. He began producing "Swiss Army Knives" in 1891, a multitool knife for the Swiss Army. The first knives made for the

Swiss Army were made in Solingen, Germany, but Elsener wanted them to be made in Switzerland, so he sought the contract even though he initially produced them at a loss. He patented a new multi-tool knife in 1897, which was internationally marketed and made the company successful. In 1909 Elsener's mother Victoria died, and he began using Victoria as a brand name. This was also the year when he started using the Swiss coat of arms as a logo. When Karl Elsener died in 1918, his sons took over the business. Because of the new and exciting stainless steel, they began using stainless in 1921, and the company was again rebranded as Victorinox.

Empire

Another company that made stainless knives early on was Empire Knife Company in Connecticut. Empire started as Thompson & Gascoigne in 1852, renamed Empire Knife Company in 1856[13]. The company was relocated into a larger factory in 1880, and Charles L. Alvord grew the company starting in 1890 until he left in 1923[13]. The 1890s were the most significant period for Empire when they were one of the largest cutlery companies. Empire began offering stainless grapefruit/orange knives in 1921[14] using a design patented in 1915 by Danforth Fletcher Alvord[15], and in 1922 they were advertising a stainless pocketknife[16].

THE EMPIRE BRAND STAINLESS STEEL PEN KNIFE

This knife is made thruout of Stainless Steel, blades, springs, handle and rivets. The handle is handsomely decorated with engine turning with plain space for monogram. We have similar patterns with pearl handle, Stainless blades and springs. All Stainless parts acid tested before leaving the factory. Not effected by salt water, perspiration, fruit or citrus juices.

Write for prices and further information.

The Empire Knife Co., Winsted, Conn.

Makers of Superior Pocket Cutlery Since 1856

1922 Empire ad.

Schrade

Schrade was also advertising stainless steel knives by the end of 1922[17]. In the 1926 Schrade catalog, they said: "When we began the manufacture of blades and springs out of Stainless Steel, many new and difficult problems appeared, especially in heat treating and polishing. By persevering and selecting only the best Stainless Steel to be had, we have finally succeeded in producing a satisfactory Stainless Steel Knife. We feel confident that an examination and trial of our product will demonstrate the excellence of Schrade Stainless Steel Pocket Knives." Later in the catalog, under the heading of "Advantage of Stainless Steel Pocket Knives" they said:

Schrade Stainless Steel knives are especially suitable for all outdoor work or sport. Farmers, linemen, cattlemen, etc., will welcome a pocket knife which will not rust, even though it is used in the rain, or left in a damp place. For use in outdoor sports such as fishing, hunting, boating, especially on salt water, it has no equal. Stainless Steel knives can be used for cleaning fish, skinning, cutting or

1922 Schrade ad for stainless knives.

peeling vegetables, fruits, etc., without the least danger of stain or rust. If one of our Stainless Steel knives is used as above and not cleaned even if left for a long time, it will not be damaged at all. Simply wash it off and it will be as clean and bright as when new.

Camillus – Stainless Cutlery Company and William D. Wallace

William Wallace

Alfred Kastor hired William D. Wallace in 1917 to work for Camillus. Wallace was a mechanical engineer and was hired because of his prior scientific training, approach, and personality[18]. Wallace was made the factory manager in 1922 after the death of the preceding Mr. Mayer. Wallace systematized each stage of the manufacturing process and improved efficiency, while maintaining quality. The line of knives was also simplified to fewer patterns and sizes so that the number of tools, dies, and methods could be reduced. In addition, Camillus added more automation, such as multi-step dies, presses, and automatic polishers. Wallace believed it was management's job to improve quality, not the workers[19]: "Good working conditions pay the biggest dividend. Usually workers are blamed when it really is management that is at fault."

Wallace was also tasked with developing methods for producing stainless steel knives, which he began in 1919[20]. This coincides with the date that Wallace said Crucible had delivered their first cutlery steel to Camillus. They began to sell their stainless knives under a new brand, Stainless Cutlery Company, in 1924. They described the new knives in an article sent to the *American Cutler*:

In 1915 Brearley discovered the exact proportion of chromium necessary to make steel stainless, and in 1924 W.D. Wallace, of the Stainless Cutlery Company, discovered how to temper and harden this steel so that it could be sharpened and resharpened and always retain a fine cutting edge. It was left to the Stainless Cutlery Company, Inc., organized as a branch of one of the largest pocket-knife factories in the United States, to achieve what frequently had been declared by some manufacturers to be the impossible. Mr. Wallace, after spending five years of exhaustive study with some of the greatest metallurgists in England, Germany and the United States, finally developed the cutlery sensation of 1924. Not only did he throw by the board all previously known methods of heat-treating steel, but he also began at the bottom and developed a stainless steel adapted for the use of sharp-edged cutlery. But the problem of stainless steel was not finished when the company discovered a method of heat-treating stainless steel. A new and entirely different method of polishing had to be devised, and in this once more the Stainless Cutlery Company leads the field and is producing by a special method of its own a polish which is a real mirror polish, in every way superior in finish and color to the old crocus polish on carbon steel... It is therefore with just pride that the concern presents to the trade a small but complete line of stainless steel pocket-knives which "are sharp and stay sharp." They announce with the most complete confidence that their stainless steel pocket-knives will actually replace carbon steel knives, in the high-priced line at any rate; and so it can safely be said that the day of the dealer's losses on rusty knives is almost over. The Stainless Cutlery Company has ushered in "the dawn of a stainless tomorrow."[20]

Some readers of the *American Cutler* would send in criticism of this advertising, which was reported in the next month's issue in May 1925:

Several letters have come in criticising in a frank and friendly manner the article which was submitted to us by the Stainless Cutlery Company... The main exception to the article which the correspondents seem to take is its attempts to place W.D. Wallace in the same ranks with Harry Brearley in so far as the discovery and achievements in the production of stainless steel is concerned. The Stainless Cutlery Company assure us that that was far from their intention although what they do claim for Mr. Wallace is the discovery of an improved method of heat treating stainless steel.[21]

What this new heat treatment method would be is unknown; several logistics must be worked out to switch to the heat treatment of stainless steel, such as higher temperatures, better protection against scale and decarburization, and tempering by temperature rather than color. Perhaps it was the introduction of new temperature-controlled equipment or the working out of these logistics, which was being advertised as the "improved method of heat treating." Stainless Cutlery Company knives continued to be marketed until the 1930s. Not all problems with stainless were solved, however, as they continued to use carbon steel in the springs of the pocket knives[22]. However, the Stainless Cutlery Company was advertising stainless springs; eventually, they incorporated them into the knives[22]. Camillus dropped stainless steel for many years after the Stainless Cutlery Company was eliminated in the 1930s.

Early 420 Stainless Hunting Knives

Wade & Butcher

The first hunting knife released with stainless steel came from Wade & Butcher with their "Teddy" hunting knife[23], advertised by 1923[24]. Wade & Butcher started as a Sheffield razor company. Sigmund Kastor purchased Wade & Butcher in the USA in

1913, and then sold it to the Durham Duplex Razor Co. in Jersey City in 1918[23]. The hunting knives were first advertised in the US in 1919[23], which were in the style of Marble's. Durham reportedly remodeled the Sheffield plant[25], though the hunting knives were rumored to have been produced at the Durham factory in Jersey City[23]. Pocket knives and straight razors were produced in Sheffield, but stainless table cutlery was made in New Jersey[26], so it makes sense that the stainless hunting knives would also be made in New Jersey. Their manufacture in the USA is also likely confirmed by the fact that they advertised "Firth Sterling" stainless (located in the USA) rather than from the British Thomas Firth & Sons steel company.

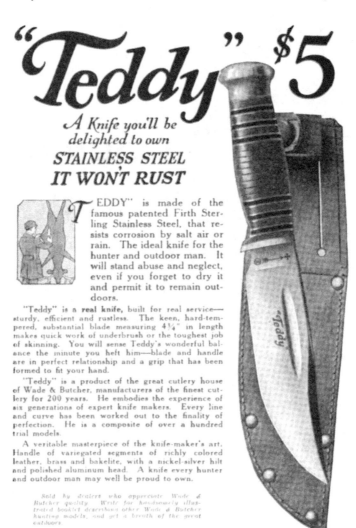

1923 Wade & Butcher ad.

Union Cutlery Company – KA-BAR

Other companies followed suit with stainless hunting knives, including Union Cutlery Company by 1925[27]. Union Cutlery started as a cutlery business owned by Wallace Brown, which began to manufacture its products in 1902 after purchasing the factory of the Tidioute Cutlery Company. Union Cutlery was famous for its KA-BAR brand, which was trademarked in 1924. The story of the name came years later from M.L. Brown:

> Years ago an Alaskan hunter had shot a bear who then attacked him and knocked his rifle from his hands. In order for the hunter to protect his life, he took a knife made by Union Cutlery company and successfully killed the bear, which was the Kodiak bear species. The hunter in appreciation of the knife having saved his life sent the bear skin to President

Wallace Brown. The thought then occurred to the management of the company that if the Kodiak bear is the strongest of the bear species, and the word bear is pronounced "bar", and further, if the cutlery produced by the Union Cutlery Company was the best and strongest of its kind, then it should be very significant that "Ka-bar" might truly represent the qualities of the company's products. Thus the Ka-Bar trademark was adopted.[28]

Another version of the story says that the hunter's writing was so difficult to read that instead of "kill a bear," all that could be seen was "k a bar."[29]

Union Cutlery primarily advertised the stain resistance of stainless steel rather than the superior edge performance. For example, in the 1925 advertisements for their stainless KA-BAR hunting knives, it said[27], "We make 12 styles of knives in finest Stainless Steel. While these knives require more frequent sharpening, their rust-proof quality makes them very practicable for the non-expert camper or hunter."

1925 Union Cutlery ad.

Remington

Remington, the famous gun manufacturer, began producing and selling knives in 1920 after purchasing a plant in 1919[13]. While their foray into knives was relatively short (they ceased operations in 1940), the high quality of knives, large number produced, investment into technology, and extensive advertising, meant that Remington became a dominant player in the knife market during those years. Remington started with pocket knives, and remained primarily a producer of pocket knives, but in 1925 began offering six fixed blade designs, reminiscent of Marble's style. In 1926 they began offering the RH134, a stainless hunting knife. They offered this knife until about 1930[30]. In 1931 they began producing stainless pocket knives as well[13], though by that time, they were likely using a 440-type stainless steel, as this steel is described in a 1937 internal sales manual[31].

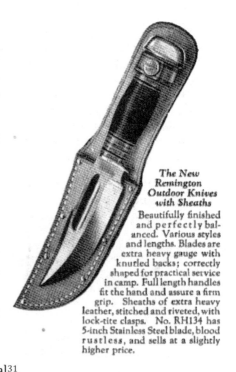

The New Remington Outdoor Knives with Sheaths

Beautifully finished and perfectly balanced. Various styles and lengths. Blades are extra heavy gauge with knurled backs; correctly shaped for practical service in camp. Full length handles fit the hand and assure a firm grip. Sheaths of extra heavy leather, stitched and riveted, with lock-tite clasps. No. RH134 has 5-inch Stainless Steel blade, blood rustless, and sells at a slightly higher price.

David Giles and 440 stainless steel

While some of the poor reputation of stainless knives was not necessarily the problem of the steel itself, as described above, the 420-type invented by Brearley with 0.3% carbon and 13% chromium still had limitations. The biggest was that with the 0.3% carbon it was limited to being heat treated to the low 50s Rc, while the best carbon steel pocket knives were in the 56-61 Rc range. Brearley was well aware of the reputation of his stainless steel[32]: "I have been playfully referred to as 'The man who invented knives that won't cut.' This amuses me because it is a witty saying, justified by a modicum of truth; though it sets me wondering why I should be regarded as a cutler, of whose craft I know next to nothing, instead of an analyst or a steel-maker." Simply raising the carbon content of the steel would increase the potential hardness, but would also reduce corrosion resistance. The solution was to increase the chromium content as well, to 16-19%, and those steels are now part of the 440 series.

The first of the 440-type stainless steels to be commercialized came from Percy Armstrong with the steel Delhi Hard which was being advertised by 1923[33]. This steel differed from later 440-type stainless steels in that it had a silicon addition of 1.5% along with very high carbon of 1.2%, with the typical 17% chromium of a 440 stainless[34]. The later 440C specification has a range of 0.95-1.2%, which would put Delhi Hard right at the edge of the specification. However, I wonder if the wide range of carbon allowable is due to the early manufacture of Delhi Hard. Armstrong did not recommend Delhi Hard for cutlery like he did Neva-Stain, but instead for high-temperature ball bearings. The Ludlum manual of 1924 said:

> Delhi Hard rustless steel is very similar to our Neva-Stain steel... The only difference between these two steels is the chromium content. Delhi Hard contains more chromium, about the same silicon, and a carbon content of over 1%... It can ... be made practically file hard, but a cutting edge cannot be made so rigid with Delhi Hard as it can be made with Neva-Stain. Delhi Hard is better suited for machine parts. Delhi Hard is particularly suitable for balls for valves and ball and roller bearings and super-heated steam parts, where maximum corrosion resistance is required with good hardness and wear-resisting properties.[35]

The Idea of increased chromium and carbon seems to have been gaining traction in research. For example, Jerome Strauss and J.W. Talley of the US Naval Gun Factory presented their studies on a range of stainless steels in 1924[36]. Among the steels tested were those with 0.65-1.13% carbon and 16.85-21.93% chromium, which were described as "very special materials, suitable for edge tools other than cutlery, and possessing also resistance to abrasion in varying degree." At least two of these were probably produced by Midvale Steel, which had a "Medium Carbon" stainless steel with 0.8% carbon and 18% chromium (similar to 440B), and a "High Carbon" with 1% carbon and 23% chromium[34]. However, I cannot find any marketing materials from Midvale on these grades. These products appear to have been short-lived, as they are not listed in a 1936 publication of Midvale steel products[37].

David Giles first patented 440A and 440B stainless steel for knives[38], which became a very common type in factory-produced knives. The 440A type was marketed as Latrobe Hy-glo in 1924[39]. Giles studied engineering at Carnegie Tech (now Carnegie-Mellon University) before working at Firth-Sterling Steel[40], presumably where he learned about stainless steel. In 1918 he joined Latrobe and set up the company's first metallurgy lab, which he headed. He also became the works manager in 1935 and vice president of operations in 1940[41]. The patent was for a steel with 0.5-0.8% carbon and 16-19% chromium, with the "preferred" steel being 0.6% carbon and 17% chromium, though the ultimate composition for Hy-glo ended up being 0.65% carbon and 17.5% chromium[37].

Interestingly, the patent says one of the primary goals was[38], "to provide a steel alloy of this type which is highly resistant to attack in its annealed as well as in its hardened state," as stainless steels only get their maximum corrosion resistance after heat treatment. The patent also mentions cutlery[38]: "In addition to this, the improved steel has a higher degree of hardening, retains a better cutting edge, and is easier and cheaper to manufacture than any known stainless steel on the market." Latrobe also introduced "Lusterite" by 1926[42], which had somewhat increased carbon, a 1934 source says 0.9-1.0% carbon[43], and a 1936 source says 0.8-0.9% carbon[37] (a 440B-type steel).

The 440-type stainless was adopted by the majority of other stainless steel companies, and by 1925 it was being written about[44] as the "hard cutlery type," said to be developed:

> In an effort to secure a greater degree of hardness and to find a cutlery steel outside the Brearley patent range... It includes the chrome-carbon compositions between 16 and 20 per cent chromium and 0.60 per cent and 1.10 per cent carbon... In the opinion of a few manufacturers this type of cutlery has the advantage of taking a high polish, giving a somewhat harder cutting edge and showing a greater resistance to wear than the standard type. This type has, however, the offsetting disadvantages of being considerably harder to work in forging, machining and grinding and is somewhat more difficult to heat treat than the regular quality ... at the present time it is not known to have any particular value outside of the limited cutlery field.[44]

The mention of it being patented to be outside the range of the Brearley patent is interesting since Latrobe was a member of the American Stainless Steel Co., paying for the rights to make stainless steels. And many companies began to make their versions, so it's unclear how the steel was licensed. The American Stainless Steel Co. called this grade "Type B" and the original Brearley/420 composition "Type A." They described Type B as being used in[45] "high-grade cutlery, particularly pocket knives, and in other hard heat-treated products." Firth-Sterling called their 440A with 0.7% carbon the "Modified Cutlery Type" or "Stainless B" and a 440C-type steel with 1.05% a "Ball Bearing Type."[45] Carpenter Steel introduced Stainless No. 2-B with 0.95% carbon and 17% chromium[45], which puts it at the border of 440B and 440C because of the carbon. Crucible introduced Rezistal Grade B (0.6% carbon), Grade B-100 (1.0% carbon), and Grade BM with 1% max carbon plus 0.5% molybdenum. I will explain the molybdenum addition later in this chapter. Crucible described all of their Grade B steels as suitable for cutlery and other applications. Interestingly, Ludlum modified the composition of Delhi Hard to match the other 440C-type bearing steels and by 1930 it had 1.05% carbon, 17% chromium, and 0.85% silicon[46].

Steel	C (%)	Cr (%)	Mo (%)
440A	0.65-0.75	16-18	0.75 max
440B	0.75-0.95	16-18	0.75 max
440C	0.95-1.20	16-18	0.75 max

Though 440A (and some 440B) became the most popular types for high-end pocket knives, pinpointing which companies first began using the steel is difficult since specific steel type was rarely advertised. Measuring the vintage knives is also challenging as the tang stamps on the knives that line up with the years produced often cover a relatively wide range. Stainless Cutlery Co. knives, released in 1924, could conceivably have used 440A, but measurements of a knife revealed a chromium content of around 13-14%, too low for a 440-type stainless and more likely it used the Brearley 420 type. A Schrade knife with a tang stamp Indicating 1920s-1946 measured with high chromium, indicating it was using a 440-type stainless steel, so Schrade was a company that switched to 440 relatively early on. The knife shown in the 1922 advertisement would have been made with the Brearley/420 type. Still, perhaps the stainless steel referred to in the 1926 Schrade catalog could be 440A, as it implied that they were using a more recent steel: "Stainless Steel represents the result of a great many years of experiment and research on the part of eminent metallurgists. During the past decade the fruits of this work have been put to practical use in the form of a Stainless Steel suitable for pocket knife blades and springs." A Remington 1937 sales manual said[31], "the stainless steel which Remington uses contains about 0.75% carbon and about 17% chromium... Stainless steels can be subdivided into two general classes: Type 'A' contains 12-14% chromium and about 0.35% carbon; Type 'B' contains 17-19% chromium and around 0.80 to 1.00% carbon. These two classes are known as Type 'A' and Type 'B' for purely historical reasons. Type 'A' is the original stainless steel, discovered by Brearley in England some twenty years ago. Type 'B' is a later development designed to give cutlery with somewhat better edge-holding properties. The type letters have no bearing on the quality of the material, Type 'B' being superior to type 'A' for many purposes. Both of these materials can be heat treated and are fully stainless. Laboratory edge tests show that the higher carbon material takes a better cutting edge and hold that edge longer than the low carbon." I also tested the composition of a W.R. Case knife that dates to the range of 1950-1964 that tested as a 440-type stainless. So, while W.R. Case is known in recent decades for working with 420HC, they also used 440A in their early stainless knives introduced in the late 1940s.

Queen Cutlery

Queen City Cutlery started in 1922 when several Schatt & Morgan supervisors began making knives on the side[13]. Reportedly they were fired in 1928 for producing some of their components at the Schatt & Morgan facilities. Schatt & Morgan went out of business, and in 1933 Queen City bought the plant and equipment. They changed their name to Queen Cutlery in 1945. In 1927 it was reported to the *Titusville Herald* that[47] "The company [Queen] has for some time been using what is known as stainless steel in the manufacturing of pocket knives and the output is eagerly sought in all parts of the country. Mr. Foresther stated yesterday that many orders recently have been received from Texas, where people have much trouble with pocket knives rusting. The stainless steel does not rust and remains bright even though used in cutting acid fruits and in other ways." It has been claimed that Queen Cutlery was the first to use a 440-type stainless in their knives[48], but looking into this claim, I am not sure if this is the case. In 1947 they trademarked the name "Queen Steel," and by 1956 this became their primary marketing name for their steel type[47]. In his book on Queen

Cutlery, David Krauss said that Queen first began using 440 stainless in 1946[47], but in a later article he wrote that they were using a 440-type stainless in the 1920s[49]. He also said that they started with 440C and later switched to 440A, but Cyclops Steel, their supplier, did not have a 440C-type stainless in the 1920s, but only a 440A-type. If Queen did start with 440C in 1946, that would make more sense. So, while it isn't impossible that Queen was using a 440-type stainless in the late 1920s, I have not found good information to confirm that they were doing so, or that they were the first.

Hardness of 440A and 440B Knives

One of the most significant benefits of the switch from 420 to 440 stainless steels was the increased potential hardness. 420 can be heat treated only to the low 50s Rc while 440A can be heat treated to the mid or even high 50s Rc, matching the carbon steel knives made by the best manufacturers. Unfortunately, the knife companies were not typically advertising the hardness of knives in the days of 440 steel's introduction to knives. So, we are limited to testing the hardness of vintage knives and the hardness advertised in later decades. A test of my W.R. Case Knife from 1950-1964 resulted in 53 Rc, which matches the 52-55 Rc range in a 1984 Blade Magazine article[50]. The same article said[50] that Ka-Bar's steels all targeted 58 Rc, including their 440A, Queen's 440A was 55-57 Rc, and Western's 440A was 56 Rc. These values are not exceptionally high, but were often in line with their carbon steel knives, such as Ka-Bar, which also targeted 58 Rc with their 1095, the W.R. Case knives in carbon steel were also in the 52-55 Rc range, and Western's chrome vanadium steel was 57 Rc, a point harder than their 440A knives. Camillus advertised their "Stainless Swordsteel" in their 1978 catalog as 56-59 Rc, and 440A as 57-58 Rc in the 1979 catalog, similar in hardness to their carbon steel knives. A 1983 advertisement for Schrade said their 440A was 57 Rc[51]. The better manufacturers heat treating their 440A to the 55-58 Rc range would have resulted in performance similar to carbon steels, though with somewhat better wear resistance and greatly improved corrosion resistance.

Molybdenum and Cobalt-Alloyed Stainless Knife Steels

Wilhelm Oertel, previously discussed concerning T15 high speed steel (Chapter 7), patented molybdenum-alloyed stainless steels, particularly for those with high carbon, in 1922[52,53]. The molybdenum was proposed to help with the corrosion resistance of high carbon stainless steels. Higher carbon is known to reduce corrosion resistance, and high carbon is desirable for knife steels. As noted in Chapter 9, the German Monnartz first found that molybdenum improved corrosion resistance, a fact that Oertel recognized in a 1927 article[54]. Oertel's patent resulted in several steels for Deutschen Edelstahlwerke[55], including Remanit 1540, a steel with 0.4% carbon, 15% chromium, and 0.25% molybdenum, which after some small evolutions, became the standard German stainless grade 1.4116, also called X50CrMoV15. Also, Remanit 1790, which is a 440B with 1% Mo, became the standard German knife steel 1.4112, also called X90CrMoV18. Crucible also released a 440B steel with 0.5% Mo by 1935, called Rezistal BM[56,57]; I'm not sure if it was inspired by the German steel or independently developed. Adding a small molybdenum addition to 440 stainless steels in the USA became common, which is why the specifications of 440 steels have a max Mo of 0.75%. The earliest I found a reference for either steel was in the mid-1930s.

Another interesting aspect of the Oertel patent is that he claimed that a nickel or cobalt addition would also improve the forgeability of high carbon stainless steels. The US version of the patent says[53]: "The alloy according to the present invention with a content of chromium exceeding 15%, a carbon content up to 0.5% and a content of molybdenum up to 3% may be easily forged. When the content of carbon and molybdenum is greater, the forgeability of the alloy decreases considerably. This disadvantage may be overcome by an addition of 0.5 to 2% Ni or 0.5 to 2% Co or by an addition of both elements up to 3% together. Nickel and cobalt are thus seen to be (as is common in the art of alloys) largely equivalents for each other." Nickel and cobalt are part of the same "group" in the periodic table, which explains why Oertel would think they may be interchangeable. Nickel was common in German stainless steels because of Krupp (Chapter 9), which also explains why Oertel would be looking at nickel additions in the first place. This resulted in Remanit 1790C, a modification of 1790 but with 1.5% Co, 0.5% Mo, and 0.3% V. Böhler N690 stainless steel, also given the designation 1.4528, is a higher carbon version of 1790C. Remanit 1790C was recommended for knives and shears, just like 1790.

Remanit 1790C is remarkably similar to what Kuehnrich ended up with in his "Crysteel" for cutlery (Chapter 8), and again it is difficult to know which direction inspiration came from. Kuehnrich had already independently made cobalt-chromium steels. However, his modification to a steel of a 440-type plus cobalt is more likely to have come from his knowledge of other 440-type stainless steels. The presence of 0.75% Mo and 0.25% V in Crysteel makes me think it was inspired by other steels

Steel	C (%)	Cr (%)	Ni (%)	Mo (%)	Co (%)	V (%)
Remanit 1540	0.4	15		0.25		
1.4116	0.45-0.55	14-15		0.5-0.8		0.1-0.2
Remanit 1740	0.4	17	0.5	1.5		
Remanit 1790	0.9	17		1		
1.4112	0.85-0.95	17-19		0.9-1.3		0.07-0.12
Remanit 1790C	0.9	17		0.5	1.5	0.3
1.4528/N690	0.95-1.2	16-18.5		0.8-1.5	1.3-1.8	0.25 max
Crysteel	1.1	18		0.75	3.3	0.25

-11-

WORLD WAR II CUSTOM KNIVES

World War II started in 1939 when Nazi Germany invaded Poland. However, the United States did not participate in the conflict until December 1941, when Japan attacked the Pearl Harbor naval base in Hawaii. At that point, the United States began wartime production of various ships, weapons, and other required materials, along with constructing many military training bases. Custom knifemakers, better-known and lesser-known, began making combat knives for the war. This included knifemakers like Scagel and Richtig, who made many fighting knives during World War II. Some knifemakers only made knives for sale during World War II, and some, like John Ek, started their businesses because of the War but then continued to make knives as a business afterward. Other knifemakers who made knives for the war effort include Ernest Warther, Floyd Nichols, A.C. Cornelison, Ben Rocklin, M.H. Cole, Henry Gill, A.G. Bimson, Bob Riggs, Donald Moore, and Taylor Huff. However, Bo Randall was one of the most famous knifemakers of World War II.

Randall Made Knives

Walter Doane "Bo" Randall, Jr. was visiting Walloon Lake in upper Michigan with his family when he found his fishing companion L.A. "Litch" Steinman working on a small boat. Randall said:

> He was there on the shore of the lake with a big, old rowboat turned over. He was cleaning it in preparation for painting and was the most beautiful knife I'd ever seen to scrape the boat with. I said, "Oh my God, Litch, let me see that knife!" Litch's knife truly was beautiful, with a stag handle and a blade a little over 6 inches long. It had the name "Scagel" on it. He told me how it was handmade down in Muskegon [Michigan] and that a friend knew this man Scagel quite well and had a lot of his knives. He had given Litch this one that he was scraping that boat with... It kind of played on my mind, the beauty of that knife, and I decided, by golly, I was going to make one.[1]

Randall went back home to Orlando, Florida, which was not then a major tourist destination. Randall wanted someone to forge steel for him to grind a knife out of, and he went to Al Marchand, who had recently opened a machine shop and blacksmith business. Part of Marchand's business was building or repairing automotive suspension springs, referred to as a "leaf spring." Leaf springs comprise several strips of steel 1.5-2 inches wide and 1/4" thick, so they were a good starting size for forging into a knife. Marchand had a book with specifications on the steel in the springs and how to heat treat them. Randall asked Marchand to rough forge two or three blades based on a sketch Bo made from his memory of the Scagel knife. Randall roughed out the knife using a 6-inch grinding wheel attached to a lathe head, then used a sanding drum. He returned the blade to Marchand for heat treating in a coal forge. They would heat the blade to a light cherry red and quench in fish oil. A bright spot was ground on the blade so the temper colors could be seen. They tempered by heating to a blue color and re-quenching in water. Marchand said[1], "On the blades that we did of the automobile leaf spring, you had to come to a light cherry temperature and then draw it to a blue to get quality. If you got it to a straw color, that wasn't right. We got blue, which Randall liked well."

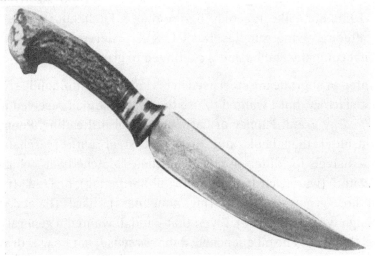

Randall's first knife made in 1937.

Another steel that they used on some occasional blades was Black Diamond Tool steel, which Marchand recalled using on some small ax heads. Black Diamond Tool started as a product of Park, Brother, and Company in Pittsburgh, though that company was one of the 13 that merged to form Crucible Steel in 1900. Black Diamond Tool was standard high carbon tool steel W1. A study in 1889 measured it at 0.86% carbon[2]. Marchand described it as "an old standard steel used for most mechanical tools, including hammers and wrenches... It was a very serviceable steel, of wearing quality. Most of your tools, all your old dies, and other tools were made out of this Black Diamond tool steel."[1] Marchand also liked it for its high toughness.

Randall's knifemaking went from a hobby to a part-time business in 1938. He added better equipment, so he wasn't using the lathe for everything, including a flat-belt sander, a drill press, and other tools. Randall began to do performance tests with his knives to ensure they were as good as they could be as Marchand recalled:

> We'd sharpen the knife so it would cut dry hairs on your arm. Then you get a piece of Florida hard pine, which is the hardest wood this side of maybe ebony and some of those maple probably - Florida hard pine is known to be awful hard on axes and things - and we'd take the knife that we'd sharpened to cut the hair on your skin, and then we'd cut through a broomstick thickness. And then we'd see if it'll still shave... Bo would take them and make those knives work for half a day, and just give them everything you could think about. I've seen him when he broke tips of the blades off; he'd done this and that. That was a time when he was trying to develop a quality with a temper that would last. He wouldn't quit till he got what he wanted. One (of my men) said, "He's a regular picky dicky." "No," I said, "he just wants something and to get it right." They said, "Hell, he'll get tired of this stuff real quick" But he didn't. I've known him for years, and he wasn't a quitter.[1]

Randall knives were 52-54 Rc, though around 1975, they had increased the hardness to 56-58 Rc[3].

Randall knives were selling well around Florida, and Bo slowly expanded. In September 1938, Randall contacted Abercrombie & Fitch, but they initially rejected the knives after receiving samples. But V.L. & A. ordered knives after Randall had contacted that company and business continued to grow.

Randall wanted to upgrade the steel used in his knives[1]: "Automobile spring steel was perfectly satisfactory, but I wanted to use the best." Randall received a copy of *Tool Steel Simplified* by Frank Palmer of Carpenter Steel in Reading, Pennsylvania, and Randall consulted that book and discussed steel with Marchand. Carpenter recommended steels for different uses with their "Matched Set" which categorized steels by "water" (water hardening), "oil" (oil hardening), or "red" (red hardness). And within those groups as "tough" (high toughness), "hard" (balanced properties), or "wear" (high wear resistance). Given that Randall wanted a general purpose high carbon steel that would be oil quenched, it makes sense that he was drawn to the "oil hard" steel, branded by Carpenter as Stentor, which is today called O2. However, Randall did not purchase Stentor from Carpenter, and instead purchased a Swedish O1-type steel similar to the original Halcomb/Crucible Ketos (Chapter 3). In his 1945 catalog Randall said: "I use the finest Imported Swedish Tool Steel, which may not necessarily be better than our best American steels, but has the reputation of being made from the purest of ores and of being the finest of cutlery steels." Randall told Robert Gaddis that he was also interested in Swedish steel because it was used in the Bofors 40mm antiaircraft gun, which was getting press in late 1939 as a defense against Nazi dive bombers. But as we also know, Scagel had also told Randall that he had been using "Swedish" steel (Chapter 6). Randall ordered steel from the Swedish American Steel Company in Brooklyn, NY, which may also solve the mystery of which

Swedish steel Scagel had been referring to. That company would have been importing from a Swedish steel company, and it was reported in a 1988 article that Randall Made Knives was using Uddeholm 01 steel (branded as Arne)[4]. Swedish American Steel Company branded the 01 as "Non-Pa-Reil" steel. Randall was critical of using recycled steel parts, such as in the 1948 catalog: "Some homecrafters try to make knives from files or automobile springs. This procedure might be successful if it were not for the guess work involved in knowing the exact amount of carbon in the steel."

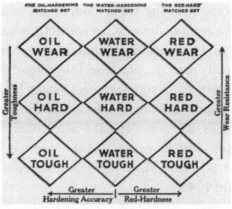

A selection of Randall knives in 1938.

In September 1938, Randall wrote to Carpenter Steel asking for information about "stainless steels for high-grade knives"[1]. They sent him technical data on grades that they called Carpenter Stainless No. 2, which was a standard 420 steel (0.3% carbon, 13% chromium), and Stainless No. 2-B, which was borderline between what is now 440B/440C with 0.95% carbon and 17% chromium. Randall requested samples of the two steels, and he and Al Marchand experimented with the two steels. They didn't like the high carbon grade, as Marchand said[1]: "When heated and hammered, it would crumble, go all to pieces. That one wasn't worth any of a hooploo." High carbon stainless steels can be more sensitive to overheating, which would have led to the crumbling; a lower forging temperature would have been necessary, but that would also mean greater difficulty in forging. Instead, they forged the No. 2 grade (420) from the 3/4" round bar stock to an all-stainless steel fishing knife, where the entire knife was one piece of steel, handle and all. He then added the all-stainless fishing knife to his standard line of knives, and a somewhat more refined version was added to his first published catalog in 1940. By early 1941 they were making fishing knives from No. 2-B, the 440-type stainless steel. These remained popular until they were removed from the line in 1945.

Randall's first stainless knife in 1938.

In 1939 Randall and Marchand also worked on stainless steel hunting knives[1], and in 1942 Randall began selling carving knives with forged stainless No. 2-B blades[5]. In Randall's 1945 catalog, he wrote about stainless steel: "Stainless-steel knives found on the market have the reputation of not taking or holding an edge. This is generally true because of the high-chrome, low-carbon content, the chrome being the element which makes them rust-proof. I have tried some lower-chrome, higher-carbon stainless steels and find they make excellent knife blades. This type of stainless steel, however, is far too difficult to fabricate into a knife to justify the extra cost." Also, in 1945 Randall contacted the Rustless Iron and Steel Corporation (later part of Armco, then AK Steel, then Cleveland-Cliffs), and he received samples in March. Randall requested 17-C-60 (440A), but the bars he received were 17-C-100 (440C). Randall again requested the 17-C-60[1]: "I intend to use all the flat stock for making knife blades of different sizes and thicknesses, and I feel the 17-C-60 material would be better than the 17-C-100, which I used before and feel is a little too hard." Randall introduced the

"Salt Fisherman and Household Utility" knife using the 17-C-100, made by stock removal from 3/4 x 1/8 inch sheet stock. In the 1948 catalog Randall said: "The high carbon stainless steel used for Randall Made Model #10 (Salt Fisherman) is not and cannot be used in the production of commercial knives." In the 1954 catalog, stainless steel blades were listed as an option for a range of models for an upcharge, and by that point, they had switched to 17-C-80 (440B) as the catalog says: "This is a special high carbon (0.8%) stainless steel not obtainable in ordinary knives. It is immune to the effects of salt water immersion." Randall was still using 440B steel from Armco by 1967[6], and reportedly the company continues to use 440B.

In 1943, Al Marchand closed his business to accept a commission as a warrant officer in the Air Force[1], which left Randall without a forging shop. An old master blacksmith in central Florida, Carl Christensen, was willing to come out of retirement to work with Randall because of the need for knives in World War II. Randall purchased a forge, anvil, hammers, and tongs and set up a small blacksmith shop in his backyard. Randall contacted the Cooley Electric Manufacturing Corporation requesting a heat treating furnace, but they denied him due to the war. Carl Christensen used the "baffle" method by using a steel pipe that was placed in the coal forge to create a region of uniform heating. By October 1943, Randall could finally obtain a heat treating furnace from Cooley Electric. So, at that time, Randall Made Knives was forging and heat treating its own knives.

Zacharias-Randall fighting knife. Photo from Ref. 7.

In 1942 Army Lieutenant James Zacharias, stationed at a nearby training camp, came to Randall to request a combat knife. Zacharias requested a "large Bowie-type fighting knife, " a relatively generic description, so Randall asked him to sketch it out for him. Randall said[1]: "It was a great big, huge thing with a blade close to 11 inches long. The first thing I said to him was that this would be entirely too heavy a knife for a man to carry with all of his other gear. Next, I told him it would also be much too long and would be getting in his way all the time." Randall modified the design by shortening the blade and changing the Bowie's sharpened clip point with a straight edge across the top. By the beginning of 1943, this design had evolved into the "All-Purpose Fighting Knife," later called the Model 1, perhaps Randall's most famous combat knife.

Early Model 1 Randall (1943-1944). Photo from SharpInvestments.net.

In 1943 many American soldiers that had been stationed with British soldiers wanted a knife like those soldiers were carrying, a Fairbairn-Sykes commando dagger. Randall remembered[1]: "The skinny, pointy thing was only good for assassination and sticking guys in the kidneys. Also, the handle on it was so small that it gave a lousy grip. I'll make the GIs one that's strong and has got enough of a rounded point that it'll be good for slashing as well as thrusting." This was Randall's stiletto combat knife, later called the Model 2.

In 1943 there were several newspaper articles about Randall's combat knives. And at the beginning of that year, several of his typical dealers, like sporting goods stores Abercrombie and Fitch and V.L. & A., were selling them. In 1944 Randall published a booklet called *The Fighting Knife* along with Rex Applegate with information about how to properly use the knives in combat. They began including these booklets along with each fighting knife. Randall sold about 4400 knives during World War II. In 1944 Scientific Films visited Randall and filmed a feature about his combat knives, released in 1945, which continued to grow the reputation of Randall Made Knives. Randall knives would continue to grow in popularity and influence by releasing an early drop point hunting knife, a knife for NASA astronauts, and other projects. The company continues to this day.

Bo Randall was the most influential knifemaker of his generation. A significant turning point was the start of World War II, which catapulted Randall's business, as he wrote in his 1988 catalog: "Then World War II began. A young sailor asked me to make him a knife for use in man-to-man combat. When his friends saw it, they placed orders, and their friends placed orders, and my knives were used in combat, and a reporter wrote a story and... All hell broke loose. Suddenly, unexpectedly, we were in the knife business."

Hoyt and Al Buck

In 1902 at 13 years old, Hoyt Heath Buck worked for the village blacksmith in Leavenworth, Kansas[8]. He learned everyday blacksmithing tasks like shoeing horses but also had to sharpen grub-hoes and reapers. He noticed that these tools seemed to require sharpening quite often, and Hoyt thought it might be possible to improve them. He began developing his heat treating methods with old files and rasps, and his "secret" was that the entire blade needed to be heated to a consistent temperature before quenching to avoid soft spots. In 1907 he moved to Washington state, and Hoyt worked various jobs, including insurance salesman and streetcar conductor. He married Daisy Green in 1909, and they had their first son in 1910, Alfred Charles "Al" Buck.

In 1937 Hoyt Buck moved to Mountain Home, Idaho, to work as a millwright at the Anderson Ranch Dam. In 1941 Hoyt became a pastor of Assembly of God Church, a small congregation. After Pearl Harbor and the U.S. Government's request to donate fixed-blade knives, Hoyt purchased a forge, anvil, and grinding mandrel and began making knives. Hoyt said[8]: "I didn't have any knives, but I sure knew how to make them." Hoyt first made knives for the men stationed at the Army Air Corps nearby, and more servicemen were traveling to Mountain Home to request one of his knives. He made knives using old files and used plexiglass for the handles. In 1946 Hoyt and his wife moved to San Diego to live next to their son Al, and he built a 10x12 foot lean-to on the side of Al's garage. He began knocking on doors of butcher shops, restaurants, and sporting goods stores offering sharpening services and handmade knives. Hoyt convinced Al to quit his job as a bus driver and join the knife business in 1947 and renamed the business "H.H. Buck & Son." Al recalled:

To tell the absolute truth, I didn't want to do it. The knife business was hot, dirty, noisy, time-consuming work. Besides, I liked being a bus driver, I really did. The pay was good. I liked the people. And since I had some seniority, I could choose any route I wanted. If Dad had known how much I had to learn about the knife business, he might have left me behind the wheel of that bus. He spent so much time teaching me what to do we produced fewer knives with two men than he'd been making by himself.[8]

Al learned how to laminate the Lucite plastic strips onto the handles and how to drill the handles. He also learned how to heat treat the blades. Al said[8]: "It sounded easy, but it wasn't easy at all. We heated the blades with torches, and if the torch was too close, the blade got too hot. If you didn't move the torch up and down the blade quickly enough, the heat wasn't uniform. It was tricky, but if you did it just right the entire blade would turn a light tan color just before turning red, and at just that moment you plunged it into oil. Our tempering process was a little bit different than the one used by other knifemakers, and it was one of the keys to our success." Al developed a demonstration of the superior heat treating of Buck knives that became a feature of their advertising: cutting through a bolt or heavy nail. One of the customers was very impressed by a knife and said, "Why, I'll bet you could cut a bolt in two with one of those!"[9] Al responded: "Let's try it and see."[9] Al put a bolt on an anvil and hammered it through the bolt without apparent damage to the blade. Al would then use this as a demonstration to potential dealers. As Al told the story:

Al Buck

> I used a sixteen penny nail. I'd go in and take along the nail and an old wrench, and to convince the dealer, I'd cut through that nail with the blade. That was the clincher. It was the backbone we put in our edge, the temper, that enabled me to cut through that nail... At the time, no other knife could cut steel. Most manufacturers intentionally made their knives soft, so they could easily be sharpened. We wanted to produce a knife that didn't need to be sharpened very often.[10]

It wasn't just the heat treatment that was necessary, however, also requiring skill and practice, as Al said, "You had to know what you were doing."[11] "There was a knack to cutting those bolts. Not everyone could do it. But after a while, people began to remember Buck knives as the ones that could actually cut steel! More than anything else, that's what made us famous."[8]

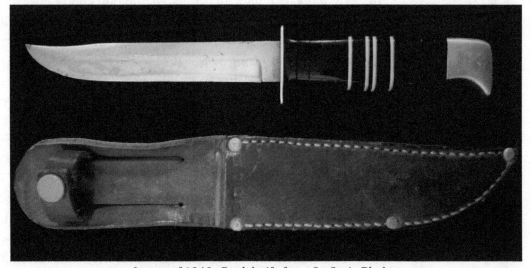

Image of 1940s Buck knife from St. Croix Blades.

Hoyt was still grinding all of the blades, but that changed when Hoyt's health problems were discovered. Al said:

> For the first couple of years we were together, Dad did all the grinding. The grinder was his pride and joy, and he wouldn't let anyone near it. In 1948, though, Dad discovered he

had cancer, and half of his stomach was removed. I think he realized then that his days with us were numbered. He also realized that I couldn't make Buck knives until I learned how to grind the blades. I knew it was important to him. So, for an hour or two every day, I worked at the grinder, trying to get the hang of it. I can't tell you how many blades I must have ruined in the process, but finally I thought I was ready. I remember calling Dad in and telling him to watch. I did one knife, and another, and another. Finally, his eyes started to sparkle, a smile came over his face, and he said, "Now I almost feel my life is complete - my son can make knives!"[8]

Shortly after, Hoyt moved back to Washington to be near his other children and passed away in 1949 at 59. Hoyt and then Al continued to use old files for their blades until 1961, which came from Consolidated Vultee, an aircraft manufacturer. Al said[8], "They were so cheap, we always took as many as he could bring us. We used to stack up those files like cordwood." Buck's transition to a true factory knife company is told in Chapter 18.

John Ek

John Ek owned a machine shop in Camden, Connecticut, and in 1939 he designed his first knives[12]. In 1941 he began making knives in volume. John Ek in early years sold his knives exclusively to military members and called them "John Ek Commando Knives." With the 69 employees he had during WWII, they were producing around 1,000 knives per week by the end of the war, with an estimated 100,000 knives over the course of the war. Ek's knives were relatively simple, with a full tang design and riveted wood handle. Because of the war, there were steel shortages, and Ek had to get approval from the U.S. Government War Production Board to purchase the 4340 "chrome-nickel-moly" steel used in his knives, which he heat treated to 56 Rc. Ek continued to make knives for future conflicts until he died in 1976, though the company continued without him.

Steel	C (%)	Cr (%)	Ni (%)	Mo (%)	Mn (%)	Si (%)
4340	0.4	0.8	1.8	0.25	0.8	0.25

David Z. Murphy (Murphy Combat and Gerber)

Murphy was making knives starting in 1938 in Portland, Oregon[13,14]. Joseph Gerber partnered with him to market Murphy's knives with Gerber branding. Murphy made carving knives with aluminum handles, and Gerber continued to produce knives of a similar style after the partnership with Murphy had ended. After Pearl Harbor, Murphy made about 90,000 knives under the Murphy Combat brand for World War II. Son Dave Murphy said[15] these were made from "old scrap hacksaw blades, made of high speed tungsten steel," which referred to T1 steel[16].

John Ek Commando Knife, photo from Ref. 17.   1944 John Ek Knives ad.

Image of Murphy Combat Knife from Rock Island Auction.

-12-

NEW DIE STEELS

Hot Work Die Steels

Steels designed for "hot work" require high toughness and some hot hardness to withstand the high temperature and stress on those types of dies. This type of work was previously done by steels like Peerless A (H21) and C.Y.W. Choice, written about in Chapters 5 and 7, as well as competitors' versions. However, these steels were not well optimized for the task, given Peerless A was simply T1 with its carbon and tungsten cut in half, and C.Y.W. Choice was initially designed for projectiles. With the growth of applications requiring hot work die steels, like the die casting of aluminum alloys, there was a need for optimized steels. The earliest of these new steels was Latrobe's Lumdie, which was introduced in 1923[1], possibly developed by David Giles[2]. This steel is somewhat of a modification of H21 with higher chromium (5%), lower tungsten (5%), and a small carbon increase (0.4%). This steel was later given the designation H14 and became the most commonly used steel for die casting by the early 1930s[3].

In 1932 a die steel was introduced by VASCO called Hotform[4], developed and patented by James P. Gill[5]. This steel was a modification of Lumdie/H14 with some of the tungsten replaced by molybdenum, and with Mo having double the effect of W, the "tungsten equivalent" was similar between the two steels. The critical difference, however, is that molybdenum makes steel much "deeper hardening," meaning that a larger die will fully harden through when cooling in air. In the patent, Gill stated[5]: "The uniformity in hardness and lack of distortion results largely from the inclusion of the molybdenum content. If the molybdenum were omitted and the tungsten content increased even to as high as 5.00%, the desired hardness and resistance to distortion could not be secured. The usefulness of this steel is largely dependent upon its ability to harden uniformly in air and show a negligible distortion on hardening." Gill also stated that the tungsten could be eliminated: "Furthermore, a heat of steel of similar composition but not containing tungsten showed practically the same characteristics as when the tungsten was present. The tungsten, however, has been added to obtain better wear resistance and slightly greater hardness at elevated temperatures." The original was later given the designation H12, and the Mo-only version ultimately became the more popular one and was given the designation H11. Research conducted in 1946 indicated an improvement in hot hardness by adding 0.5% vanadium which most of the steel companies incorporated into H11 and H12[6], and a 1% vanadium addition to H11 became a popular alternative[7] which was given the designation H13.

Steel	Year	C (%)	Cr (%)	W (%)	Mo (%)	V (%)	Co (%)
H21	1910	0.35	3.5	9.5		0.5	
Lumdie/H14	1923	0.4	5.25	5			0.5
Hotform/H12	1932	0.33	4.75	1.2	1.4		
Hotform 2/H11	1932-1945	0.33	4.75		1.3		
H13	1946	0.4	5.25		1.3	1	

A2 Cold Work Die Steel

The above die steels are too low in carbon to make suitable knife steels, but they are essential background leading to the development of A2 steel, which has been used in many knives. The 5% Cr plus Mo alloy design of H11/H13 became the basis for A2, though with 1% carbon instead. It looks like C.Y.W Choice but with higher Cr and Mo. This steel was developed by Universal-Cyclops steel and released as "Sparta" in 1938[8]. Norman Stotz of Universal-Cyclops wrote:

[T]he jobbing die industry has long been in need of a new steel in the lower cost brackets which would satisfactorily take its place in the gap between the oil hardening die steels (O1, O2) and the expensive high carbon-high chrome types (D2, D3), or the even more expensive high speed steels... Since speed of cooling is an important factor in controlling distortion, oil hardening types require detailed attention to the other factors to minimize its effect... The air hardening high carbon-high chrome types have a cost disadvantage at the outset. They also machine with some difficulty... Such a picture was so vigorously presented by the Tool Steel Committee of the Westinghouse Electric & Mfg. Co. at East Pittsburgh in 1935 that the Cyclops Company undertook to present a set of suggestions which might result in a grade of steel that would combine the distortion advantages of the high carbon-high chrome types with the obvious advantages of the manganese oil hardening types *without increasing* the first cost above the cost of this latter type... Three different compositions were suggested, and appropriations were made from the Westinghouse shop funds to purchase commercial quantities of all three suggestions for investigation ... the Westinghouse shops made up and put into service regular production dies, so that the possibilities predicted by the theoretical investigation were confirmed by actual service tests without long delays in debate and coordination. Two of the tentative suggestions fell by the wayside because they did not fully harden when cooled in still air but the third carried through to a thoroughly successful conclusion. As a result, there is a new grade of tool steel called Sparta... This new grade of material is said by Westinghouse to offer die economies of from 40 to 60 per cent over either of the better-known compositions available in the making of this particular lot of stampings.[8]

The three analyzed compositions he referred to had a base composition of 1% C and

5% Cr, with the second steel having 0.2% V and 0.5% Mo, and the third steel having the same composition but with Mo increased to 1.1%[9]. Increased Mo allowed the die to harden through, as Stotz said.

Steel	Year	C (%)	Cr (%)	Mo (%)	V (%)
C.Y.W. Choice	1892	1	3.5		
VASCO Choice		1	4		
VASCO Choice No. 2	~1934	0.9	3.9	0.5	
Sparta/A2	1938	1.05	5	1.1	0.25

In addition to good hardenability and dimensional stability after heat treatment, the new steel had a good intermediate level of wear resistance[8]: "It should be emphasized that this new material does not outwear high carbon-high chrome or high speed steel. It does, however, closely approach them... Abrasion comparisons with the manganese oil hardening steels have been shown to be from 2 to 4 times in favor of this new intermediate grade." Toughness of the steel was also good, being[8] "double those of manganese oil hardening types and triple the values for high carbon-high chrome steels." Perhaps because the composition was relatively simple, the grade wasn't patented[8]: "As so often is the case with an unpatentable material of merit, the value of this type of steel was immediately recognized by tool steel manufacturers generally, and several brands of the steel are already on the market in addition to the Sparta brand. This imitation is itself an indication of the merit of the steel, and is a promise for the future of this type of tool steel." This was proven true, as A2 became one of the most popular die steels. To this day, O1, A2, and D2 are the most commonly used die steels, which is likely why they have seen considerable use in knives.

A6 Steel and Manganese Air Hardening Steels

Shortly after the release of Sparta/A2, air hardening steels were developed using high manganese plus small additions of Cr and Mo. The relatively low Cr and Mo meant these steels were "easier" to heat treat because they could be austenitized at similar temperatures to simple carbon steels. The earliest of these steels was Bethlehem's "Bethlehem Air Hardening" released in 1940 with 1% C, 2% Mn, 2% Cr, and 1% Mo[10]. This steel was later given the designation A4. Carpenter released a modification of this steel with reduced carbon, 0.7%[11], called "Vega," and it was given the designation A6. A6 became the more popular version, and it has been used by a few knifemakers, such as Tai Goo[12]. Goo told me, "I wanted to use an air hardening steel because with warping it was a lot easier. You can straighten it as it is air hardening. Maybe in 1982 I started using it. I also used it in some Damascus steel with wrought iron or 203E. So I had air hardening layers mixed with layers that wouldn't harden. I actually had a Damascus cutting effect going."

Carpenter also developed a version of these steels with a 4% tungsten addition called Hi Wear 64, patented in 1960[13], sort of an air hardening version of earlier fast finishing steels like F2 and 1.2562, a point its inventor also made[13]. However, this steel does not seem to have succeeded, as I haven't found references to Hi Wear 64 after its initial announcement.

Steel	Year	C (%)	Mn (%)	Cr (%)	Mo (%)	W (%)
A4	1940	1	2	2	1	
A6	1942	0.7	2	1	1.3	
Hi Wear 64	1960	1.5	2	0.9	1	4

High Vanadium Die Steels

David Giles of Latrobe developed methods to make tool steel ingots more uniform, with less segregation and finer carbides and microstructure. This process was branded as "Desegatized" tool steel in 1946. Stewart Fletcher, a metallurgist at Latrobe, said in 1952[2]: "The 'Desegatized' process is the result of a long series of research experiments which were initiated almost 20 years ago by Mr. Giles. We at Latrobe feel that under the 'Desegatized' process, we are giving the customers the finest tool steels ever made... 'Desegatized' steel ... means that it is free from harmful carbide segregation throughout its entire length. This feat, which has never been matched by any other tool steel manufacturer, is the most recent and most dramatic product of our research program." I haven't found any reports of what their improved process entailed, but segregation is reduced by the rate of solidification, i.e. rapid solidification leads to a fine microstructure. Methods by which this is accomplished include a smaller cross-section of the ingot and controlling the pouring temperature, rate of pouring, and mold design. The use of copper and/or water-cooled molds can also accelerate cooling[14].

The refined microstructure from the improved Desegatized process was especially important for D2/D3 type steels and high speed steels due to their high alloy content and large volume fraction of carbides. It also allowed Latrobe to push the boundaries for alloy design, and Giles patented two cold work die steels with 4-5% vanadium for increased wear resistance[15,16]. One was a D2-type steel with 4% vanadium and the necessary higher carbon, branded as BR-4 and later given the designation D7. Interestingly this raised the carbon to levels equal to D3 steel, so it also looks similar to D3 steel with 4% vanadium added to it. Giles stated[15]: "Comparative tests of the abrasion resistance of my alloy as compared with alloys previously used have shown a marked superiority of alloys made according to this invention with respect to their resistance to abrasion... I believe that the extraordinary abrasive resistance of my alloy depends upon the formation of vanadium carbides which remain out of solution." The other high vanadium steel he patented was A2 steel with 4% vanadium and increased carbon to match, branded as BR-3 and later given the designation A7. Giles stated that this grade has better toughness and grindability than the D7 type[16]: "[T]he alloy of this invention is easier to form, easier to grind, and is able to withstand shock better than the alloys having a higher chromium content [like BR-4/D7]. This new alloy therefore serves a new field of usefulness in which resistance to shock or impact is necessary or desirable along with resistance to abrasion." However, this steel did not seem to have been released until 1960[17], and it had much higher carbon than in the patent (2.8 vs 1.8%). Also, in the early 1950s, Giles patented other high vanadium steels that did not seem to have developed as much popularity. One was a water hardening steel with 1.7% carbon and 4.5% vanadium, along with low manganese, silicon, and chromium, like other simple carbon steels[18]. Another was a stainless steel using 420 or 440A stainless as a base and adding 5% vanadium plus carbon[19]. As far as I can tell, neither of these was commercialized. However, there is a European water-hardening vanadium steel called 1.2838 with 1.45% carbon and 3.25% vanadium, which fits within the composition ranges given in Giles' patent.

Steel	Year	C (%)	Cr (%)	Mo (%)	V (%)
BR-4/D7	1950	2.4	12.75	1.1	4
BR-3/A7	1961	2.8	5.25	1.1	4.5

8% Chromium Die Steels - Vasco Wear

Vasco Die and 8% Cr die steels came from higher carbon (>0.45%) hot work die steels. In 1926 Cyclops steel introduced a steel for pocket knives and razors advertised as stainless with 0.6% carbon, 7.5% chromium, and 7.5% tungsten[20]. This steel was branded as "K-Rustless"[21] and seems to have been short-lived for knives but was then used as a hot work die steel, and versions with lower carbon were also introduced for that application. VASCO developed a modification of Hotform No. 1 (H12) with increased carbon (0.55%) and named it Hotform No. 3, which existed by 1945[22]. This steel was later given the designation A8. In the 1950s, there was a push for the development of hot work die steels for making jet aircraft and missiles, and a desire for products that were available in higher carbon contents (~0.5%) for certain applications[23]. One approach was Thermold J (A9) by Universal-Cyclops Steel, which was a modification of H13 with 0.5% C, and to maintain toughness they added 1.5% Ni[24].

VASCO developed a grade initially named Jet Forge, sometimes called X-8, which was similar to previous hot work die steels but with relatively high carbon (~0.5%) and 8% Cr like K-Rustless. VASCO also developed this grade as a flat rolled sheet for certain aircraft applications[23]. Their creative choice for maintaining toughness with the higher carbon was to increase vanadium to 1.4%[25]. VASCO metallurgists said, "The high vanadium content and the high carbon content combine to form a small, though noticeable, quantity of vanadium carbide, making the matrix carbon less than the other high-carbon types and the toughness somewhat equivalent to that of the lower-carbon types."[14] VASCO later released this steel under the name "Vasco Tuf"[26] as a high-toughness cold work die steel, or as an alternative to shock-resisting tool steels where greater wear resistance is necessary.

By 1960 Atlas Steels released "Beaver,"[27] a cold work die steel with 8% Cr like Jet Forge, and 1.5% Ni like Thermold J/A9, but even higher carbon (0.68%). How and if those two grades inspired Beaver steel is not clear. It may have also been partially inspired by chipper knife steel (see the end of this chapter). A powder metallurgy version of this steel was released much later as CD#1 (see Chapter 42).

Latrobe BR-4 (D7) 1950 ad.

In 1958 Latrobe patented "Koncor,"[28] a 4% vanadium die steel, also designed for hot work dies for manufacturing jet engine parts and gas turbine parts. Latrobe seems to have realized this steel had insufficient toughness, as shortly after they patented "Pyrovan," which had reduced carbon and vanadium[29]. Both were being marketed in 1962[30], and David Hughes and Stewart Fletcher developed both.

In 1968 VASCO released a steel called Vasco Die[31], a steel similar to Jet Forge and Vasco Tuf, but with higher carbon and vanadium, patented by Harry Johnstin in 1964[32]. A version with even higher carbon plus tungsten was released by 1973 called Vasco Wear[33]. The patent says Vasco Die was a hot work steel designed to manufacture jet engine parts. Johnstin started with a base composition of 0.85% C, 7.5% Cr, and 1.5% Mo, and looked at vanadium contents of 0, 2, and 2.6%, and also compared to Latrobe's Koncor with 4% vanadium, showing that Vasco Die has superior toughness. Johnstin found that the grindability was improved from a 2% vanadium addition because the vanadium carbides formed would refine the microstructure. Increasing vanadium to 2.6% resulted in the same grindability as 0% V, but with greatly improved wear resistance vs. the 0% V version due to the vanadium carbides. The composition of Vasco Die and Pyrovan are very similar apart from the chromium content, so I'm not sure if this was a coincidence based on both companies looking for a modification to Koncor, or if VASCO was aware of Latrobe working on Pyrovan and wanted an answer for it. Although the patent discusses hot work die steels, this new steel Vasco Die was advertised as a cold work die steel. It was promoted as having twice the toughness of D2 and ten times the wear resistance of A2. This steel would become much more popular for knives in a later powder metallurgy version called CPM-3V. Vasco Wear would see some use in custom knives and in factory knives from Gerber, and has seen popularity in recent years as a powder metallurgy version called CPM Cru-Wear (see Chapters 34 and 40).

Steel	Year	C (%)	Cr (%)	Mo (%)	W (%)	V (%)	Ni (%)
K-Rustless	1926	0.6	7.5		7.5		
Hotform No. 3 (A8)	~1945	0.55	5	1.4	1.2		
Thermold J (A9)	1954	0.5	5	1.4		1.0	1.5
Jet Forge	1956	0.47	7.75	1.35		1.4	
Koncor	1958	1.1	5.25	1.1		4	
Beaver	1960	0.68	8.25	1.4		1.0	1.5
Pyrovan	1962	0.75	5.25	1.1		2.5	
Vasco Die	1968	0.82	7.75	1.55		2.5	
Vasco Tuf	1969	0.53	7.75	1.35		1.4	
Vasco Wear	1973	1.12	7.75	1.6	1.1	2.4	

Other steel companies introduced 8% chromium die steels, typically with around 1% carbon and a relatively small vanadium addition (~0.5-1%) compared to Vasco Die. These included Daido DC53 in 1987[34], Hitachi SLD10 in 1989[35], Böhler K340 in 1989[36], and Uddeholm Sleipner in 1998[37].

Steel	Year	C (%)	Cr (%)	Mo (%)	V (%)
DC53	1987	1	8	2	1
SLD10	1989	1	7.5	2.8	0.4
K340	1989	1.1	8.3	2.1	0.5
Sleipner	1998	0.9	7.8	2.5	0.5

Chipper Knife Steel

An 8% Cr steel also became popular for "chipper knives," blades used for wood processing. The earliest of these I have found were Heppenstall T745 and T746, and these steels became known as "A8 mod" (it is A8 but with higher C and Cr) or simply as "chipper knife steel." I have found references to these Heppenstall steels by 1960[38], but Heppenstall was advertising "special analysis steel" for their chipper knives many years prior[37], so I don't know when they were developed. However, John Succop led the metallurgy department of Heppenstall for many years, so presumably he would have developed these steels. Chipper knife steel also became a topic of tariff discussions, and a specification was given for tariff purposes in 1979[39].

Steel	C (%)	Cr (%)	Mo (%)	W (%)	V (%)
T745	0.45	8.5	1.1	1.1	
T746	0.68	8.5	1.1	1.1	
Chipper Knife	0.45-0.55	7.25-8.75	1.25-1.75	1.75 max	0.2-0.55

Now let 1 die steel
do all your cold work

INTRODUCING

Vasco Die is a new cold work die steel of great versatility. Now, in this single general purpose die steel, you will find twice the toughness of D-2 and wear resistance up to 10 times that of A-2. It also has the machinability of A-2. ■ Costing less than high carbon, high chrome grades, Vasco Die is dimensionally stable during heat treatment. It tempers to maximum hardness at 950°F—a characteristic advantage over A-2 or D-2. ■ For long production runs or exceptionally tough applications, try Vasco Die! Ready Reference Data available on request.

VASCO
LATROBE, PA. 15650

A TELEDYNE COMPANY

1968 ad for Vasco Die.

-13-

BEARINGS AND SUPER HARD HIGH SPEED STEELS

Bearing Steels

In 1948 Latrobe released a series of steels branded as "Electrite MV" described as, "Intermediate Alloy High Speed Steels, the MV family is suitable for uses where higher alloys are not required."[1] These steels had about 4% Mo with no tungsten, so substantially lower "tungsten equivalent" than high speed steels, making them "semi-high speed steels." In the 1950s, with jet airplanes requiring higher flight speeds and greater engine power output, there was a desire for bearings that could withstand higher temperatures expected to exceed 500°F[2]. The standard bearing steels 52100 and 440C were unsuitable for the temperatures required because they have little hot hardness. The US Government set a goal of zero tungsten because of shortages created during WWII and the Korean War, so high speed steels alloyed primarily with Mo were studied. High speed steels like M1 and M10 were attempted, but the "fatigue life" was lower with high speed steel due to the coarse carbide structure in those steels. The lower vanadium Electrite MV-1 had a finer carbide structure and performed better in bearings tests, which began in 1954[2]. Electrite MV-1 became a standard bearing steel and was given the designation M50.

Steel	Year	C (%)	Cr (%)	Mo (%)	V (%)
MV-1/M50	1948	0.8	4.1	4.25	1.1
MV-2	1948	0.88	4.1	4.25	2
MV-3	1948	1.18	4.1	4.25	3.15
MV-4	1948	1.4	4.1	4.25	4.15

Birger Johnson of Latrobe steel then developed a series of stainless steels for bearing applications[3-5]. They were developed by modifying 440C, dropping its chromium content to ~14.5%, and increasing the molybdenum to 4%, similar to M50. The base version was called BG41, though referred to as "440C modified" before it was named, and was tested by 1957[6,7]. Shortly after, increased vanadium versions BG42 (1% V) and BG43 (2%) were released.

1960 ad for Latrobe BG41.

Johnson wrote about the development in the patent[3]: "There are many materials
which have been suggested for use as bearings and the like under various conditions;
however, no bearing material heretofore available has been entirely satisfactory for
high temperature use in corrosion atmospheres. In the past, Type 440-C stainless
steel has been used. This material, however, lacked many characteristics desirable for
this purpose. For example, in high-temperature bearings Type 440-C, although used,
did not have the desired hot hardness and wear resistance. Ordinary tool steels, on

the other hand, which did have the desired hot hardness and wear resistance did not have the necessary corrosion resistance... The alloy of the present invention provides a solution to the bearing problems of the high speed aircraft industry and similar industries where bearings having high hardness and resistance to corrosion at elevated temperatures are required." BG41 became common as a knife steel under Crucible's version, 154CM, introduced to custom knives by Bob Loveless (Chapter 16). BG42 saw limited use by 1974 in custom knives, but became more common in the 1990s (Chapter 41).

Steel	Year	C (%)	Cr (%)	Mo (%)	V (%)
BG41/154CM	1957	1.05	14.5	4	0.12
BG42	1961	1.15	14.5	4	1
BG42	1961	1.2	14.5	4	2

Matrix High Speed Steels

Understanding of the structure of high speed steels had been increasing over the years, with many reported studies. One of the methods developed for studying high speed steels was extracting carbides from the steel, which could then be analyzed separately. This technique was developed in the 1940s[8]. Once extracted, the weight fraction of the carbides could be measured, and their structure

Roberts   Hamaker

analyzed using x-ray diffraction. With those two pieces of information in hand, they could then calculate the "matrix" composition, or which elements are left in the steel after some portions are taken up in forming carbides. For example, some carbon would form with tungsten, molybdenum, chromium, or vanadium to form carbides. So, a steel with 0.8% C, 6% W, 5% Mo, and 2% V (M2 steel) does not have all of those elements "in solution" after heat treating, but might have something like 0.5% carbon, 4.5% Cr, 2.75% Mo, 2% W, and 1% V. The amount of those elements in solution controls things like hardness and "hot hardness" after heat treatment. The excess carbides contribute to wear resistance but are not necessarily needed to achieve a particular combination of hardness and hot hardness for a given application. In some applications, high wear resistance may not be needed, and in some cases, it may be desirable to eliminate the carbides because doing so improves toughness (resistance to fracture), machinability, and grindability. In studies on carbides and "matrix" compositions of high speed steel[9], VASCO metallurgists realized that they could take the measured matrix composition and make a steel with that same composition. This would create a high speed steel with little or no carbide content, thus having much greater toughness than the high speed steel it was based on but with similar hardness, and hot hardness, after heat treating.

In a patent application first initiated in 1958 by George Roberts and John Hamaker of VASCO, they said:

Under the microscope, hardened high speed steel is found to consist essentially of two phases; extremely hard excess alloy carbides and a matrix or background material... High speed steels while possessing high strength properties suffer from the disadvantage of being brittle. Ultra-high strength structural steels, on the other hand, are tough and ductile but are not as strong as the high speed steels. Steels with the strength of high speed steels and the toughness and ductility of ultra-high strength steels will therefore fill a long felt need in the art. In view of the foregoing an object of our invention is to provide a new family or class of steels of this nature, characterized by the fact that they have the same approximate hardness of high speed steel and the same approximate toughness and ductility of ultra-high strength steels... With the improved methods of determining the excess carbide analysis and content in high speed steels, the matrix compositions can be determined with acceptable accuracy, thus opening a new area for research and development. We have discovered that new steels having compositions which correspond to the compositions of the matrices of heat treated high speed steels may be heat treated to produce a new class of steels... The chemical analysis of our steel is such that essentially all of the carbides which are customarily present in excess quantities in the annealed or softened parent steel, will be dissolved in our steel when the latter is heated for hardening. The resulting steel throughout closely resembles the matrix of the parent steel in structure without the inclusion of excess carbides.[10]

The tested steels provided an excellent combination of hardness and toughness. Still, there was an unintended side effect vs. the original steel which resulted in reduced hardness[10]: "However, the absence of excess carbides permits grain boundary movement during heat treatment which results in grain coarsening. To obtain a fine grain size essential to maximum toughness and ductility in the new composition, the hardening temperature should be lowered about 175°F, or alternatively, to a point where a slight amount of excess carbide is produced for restricting grain growth."

The first steel they introduced was VASCO-MA in 1961[11], a matrix version of M2, probably because M2 was the most popular high speed steel then and still is today. The second matrix steel was Matrix II introduced in 1964[12], a matrix version of M42 (M42 will be discussed in the next section). VASCO-MA has seen some use in knives from a later powder metallurgy version called CPM-1V. There were not many more matrix high speed steels developed in the United States after VASCO's attempts, though this concept was extended in Japan. Hitachi developed the first Japanese matrix high speed steels[13] and introduced YXR3, YXR33, and YXR7. YXR7 has been sold in some Japanese factory knives. Nachi-Fujikoshi and Daido Steel have also introduced their own matrix steels. The concept of designing steel with a "matrix composition" and a target carbide structure gained traction from the introduction of matrix steels, beyond simply the steels classified as "matrix steel."

The earliest knifemaker I found to have used a Matrix high speed steel was Tim Wright, who talked to me in a 2022 interview:

I fell under the wing of Kuzan Oda. What he impressed on me, over 40 years ago, was that the steel with lower wear resistance, they had nice high quality edges but didn't have much life. Steels like D2 that had more durability didn't take a very good edge ... my dream was to find some steel that had better edge quality but also had higher durability... I was always playing around with steels. I heard about Matrix II and I thought, "Wow, this is the shit." I could only find it in round bar. I found a 2" round bar, and I paid $191, back in [1983]. I milled it so it was almost square, put it in my bandsaw, and held it together with wire and would slice pieces off it. That was pretty good steel, but it was an awful lot of work to get knives out of it. I made as many knives as I could out of that 18-inch bar.[14]

Steel	C (%)	Cr (%)	Mo (%)	W (%)	V (%)	Co (%)
VASCO-MA	0.5	4.5	2.75	2	1	
Matrix II	0.55	4	5	1	1.1	8
M2	0.85	4	5	6	2	
M42	1.1	3.75	9.5	1.5	1.15	8

Steel	C (%)	Cr (%)	Mo (%)	W (%)	V (%)	Co (%)
YXR7	0.8	5	5	1.1	1.1	
YXR3	0.6	4.3	2.9		1.8	
YXR33	0.5	4.2	2	1.6	1.2	0.8

High Hardness High Speed Steels

In the late 1950s, there was a desire for high speed steels capable of higher hardness; the current high speed steels were capable of about 65-66 Rc. T15 and M15 high speed steels could reach 67 and sometimes 68 Rc. New high speed steels were developed to achieve up to 70 Rc, called "Super Hard" or "Ultra Hard" high speed steels. They achieved this high hardness by taking high speed steel grades with high cobalt (high hot hardness) and increasing the carbon content. This increased carbon content was made possible by improvements in controlling segregation during solidification of high alloy steels[15], like the methods Latrobe developed for their "Desegatized" branding. Maximizing hardness means dancing on a tight rope in terms of composition, especially for carbon content, and improvements in closely controlling composition also led to the possibility of developing these new steels[15]. These steels were patented within a relatively short period, Braeburn's Braecut (M44) in 1960[16], Crucible Rex 49 (M41) in 1961[17], VASCO's Hypercut (M42) in 1963[18], and Latrobe's Dynacut (M43) steel in 1965[19]. However, they were all released within an even shorter window between 1960 and 1962[20-23]. M41 was a higher carbon version of M35, M42 from M33, M43 from M34 (bad for dyslexia), and M44 from M6. The new steels and the earlier high speed steel they were based on are shown in the table.

The metallurgists at VASCO[24] and Crucible[25] published articles on the design of their super hard high speed steels. VASCO reported that they created a series of experimental heats to find an optimal composition for a 70 Rc high speed steel. They used a constant "tungsten equivalent" of 20.5, and measured toughness at 65, 68, and

Steel	C	Cr	Mo	W	V	Co
M41	1.1	4.25	6.75	3.75	2	5
M35	0.8	4	6	5	2	5
M42	1.1	3.75	1.5	9.5	1.15	8
M33	0.9	4	1.5	9.5	1.15	8
M43	1.25	3.75	1.75	8.75	2	8.25
M34	0.9	4	2	8	2	8
M44	1.15	4.25	5.25	6.25	2.25	12
M6	0.8	4	4	5	1.5	12

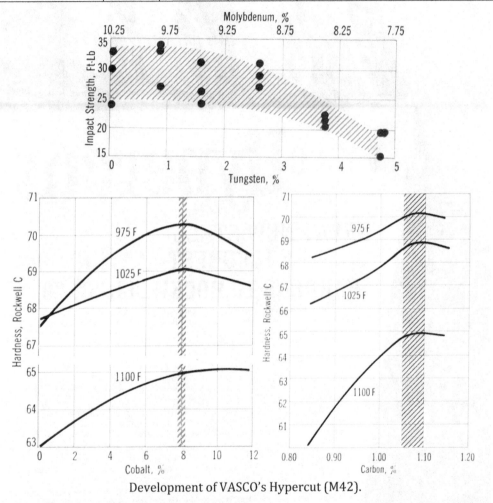

Development of VASCO's Hypercut (M42).

70 Rc (the band in the chart above represents the range of 65-70 Rc). The best toughness was found for compositions with primarily molybdenum, 9-10.25% Mo and 0-2.5% W, and increasing proportions of tungsten led to a significant reduction in toughness. Testing tungsten equivalents of 19, 20.5, and 22% also revealed that 20.5 had the best combination of hardening and toughness. They also found peak hardness with 1.05-1.10% carbon and 8% cobalt. Higher carbon or cobalt both resulted in reduced hardness. This was the process by which they ended up with a higher carbon version of M33, itself a version of M1 with 8% cobalt. Due to lower

vanadium, and thus better grindability, VASCO's Hypercut (M42), became by far the most popular of these M40-series super hard high speed steels[26]. The high hardness makes grinding difficult, so superior grindability was important for this new class of steel.

1962 VASCO Hypercut (M42) ad.

-14-

BILL MORAN AND TRADITIONAL KNIFEMAKING

William Frances "Bill" Moran, Jr. was born in Maryland in 1925. At ten or eleven years old, he made his first knife by stock removal from his father's crosscut saw. He used a hammer to break off the teeth from the saw and then used a grindstone to finish it. Finally, he used wood for a handle and wrapped it with cowhide. Moran said:

> There was a small smithy on the farm that was used to repair farm implements. It was there that I forged my first knife when I was twelve years of age. Believe it or not, by the time I was fourteen, I had sold several. I didn't know much about forging, so I would ask men in the area who had the reputation of being knowledgeable. But the only thing I ever

learned from another smith was how to temper springs, which didn't have a great deal to do with forging knife blades... My father had not been too pleased with the destruction of his cross-cut saw, so I began to use discarded harrow teeth as the primary source of my blade material.

One reason that I wanted to be a knifemaker was so that I could have all of the knives that I wanted and any kind of knife I craved. I thought it would be the most wonderful thing in the world to be a bladesmith.[1]

Moran was suspended from school at 14 years old due to his reputation as a troublemaker, and that was the end of Moran's formal schooling. From then on, he divided his time between the family dairy farm and making knives. Moran said[1], "I had terrible equipment, which mostly consisted of emery wheels, files, and a hand-held drill. There were no books on the subject, and I didn't know another person who made knives." According to Moran, his knives in the early 1940s were "rather crude" and unmarked[1]. He also sharpened knives and scissors and repaired farm implements to bring in extra money. After World War II, he began marking the blades "WFM," and in 1949 began using a stamp that said "Lime Kiln," named after the place where he lived, Lime Kiln, Maryland, near Frederick.

At the start of the Korean war in 1950, Moran started getting more orders for knives from servicemen, and many of them were for large knives like Bowies. In 1951 the fictional book *The Iron Mistress* about Jim Bowie was published by Paul I. Wellman, and in 1952 the movie adaptation starring Alan Ladd was released. This resulted in more orders, especially Bowies. Moran said that he "literally got tired of making Bowies."[1] Moran also found the book *Bowie Knife*, published by Raymond Thorp in 1948. Moran recalled[1]: "I must have read that book at least fifty times, and it certainly had a tremendous impact on me. It was during this same period that I began to attend gun shows in the Baltimore area. I was the only knifemaker who exhibited at these shows... People would stop at my table at a gun show back in the 50s and they would like my work, but when they discovered that I wanted $15 or so for a knife, they'd almost run away!" However, at these gun shows he met someone else that helped build Moran's business[1]: "[I]t was there that I first met Harold Peterson, whose book, *American Knives*, was to be published in 1958. Peterson and I became friends and when I received mention in his book, it resulted in another spurt of very welcome orders."

In 1960 Moran sold the dairy farm and the house, purchased a home, and built a new shop. He also added equipment like a 25-pound Champion power hammer and later a 100-pound Little Giant. In 1960 Moran also had his first brochure printed to advertise his knives.

Early unmarked Moran knives. Weyer photo.

Study of Historical Knives and Swords

Moran's 1960 catalog emphasized the ancient techniques used and the handmade, hand-forged nature of the knives:

> I am the only maker of classical ancient blades... I do not make any wild claims that these knives can be used for cutting bolts or metal as a test of the knife's quality. Moran Knives have long been acclaimed by many leading authorities as being the finest made today. They are probably the most expensive knives made because of the painstaking craftsmanship that goes into them. However, this results in a superior quality that one is unable to obtain in any other way. These knives will be collector items of future years... To my knowledge these knives are the only ones made today that are completely hand forged, hand tempered, and handmade in every respect... An immense amount of research and study of blades and forging techniques from ancient times until today has been required... The tempering process alone can only be learned by many years' experience. This technique is the same as that used by the blade smiths of the seventeenth century. It was this tempering that enabled those blades to withstand such severe strain in combat.[1]

Moran had developed an appreciation for historical knives and often used them as a basis for his designs:

> I am convinced that a fledgling knifemaker should study all knives, from all nations, and from all periods of history. You can thereby have a large amount of knowledge right at your fingertips. I spent a large portion of my income for many years buying antique knives, and after I had learned all that I could from a piece, I would sell it and buy another. Some of those were very difficult to part with, but it was the only way that I could expand my knowledge... On one occasion I went to an exhibition of world-class bird carvers in Washington, D.C., and I fully expected to see some of those wooden creations fly! Every feather, every muscle was correct, and it was obvious that the carvers had spent many hours studying the birds they were going to carve. If only aspiring knifemakers would study knives in the same fashion! If they would, we would see many more truly exceptional knives. I've taken a Persian knife on more than one occasion and studied that knife for hours - I've gone to sleep with knives in my hand. Then, and only then, would I attempt to make a Persian style knife.[1]

Moran believed that the key to making the best knives came from the study of historical pieces. As he wrote in his 1975 catalog:

> We must remember that the finest blades which have ever been made were developed at a time when one's life depended on his blade. Anyone who does not believe this is true should make a study of ancient blades. It soon would be apparent to them that these were the finest ever made. I have tried to follow the methods and techniques used by the ancient smiths. These methods have worked out quite well for me. In all the years I have been making knives I have never had one returned with a broken blade.[1]

Around 1975, Moran began forging a "distal taper" to his knives, where the blade was thickest at the guard and then tapered to the tip. Moran first discovered this technique when he was collecting naval cutlasses:

> One of these possessed a much better feel and balance than the others, and it didn't take much study to see why. This particular cutlass possessed obvious blade taper from the guard to the point, and the result was startling! I didn't think this would work on knives, but I had to see for myself, so I forged a moderately large Bowie with pronounced blade taper. I personally feel that most antique Bowies have a clumsy feel about them, and most are excessively blade heavy, but I couldn't believe how fine this particular Bowie felt. Since that discovery, I have used blade taper on all my knives, unless there was some specific reason not to do so.[1]

Steel Choices

Bill Moran's appreciation of the historical and ancient also extended to steel choices. He liked simple carbon steels and did not like high alloy tool steels or stainless steels. Moran had experimented with several steels for knives, including leaf springs, though he didn't have a particular brand of vehicle he liked, unlike Ruana with Studebaker. By the end of the Korean war (1953), Moran had settled on W2 tool steel as his favorite. Moran wrote about W2 steel in his 1960 catalog[1]: "The blades are made of the very best American high carbon, vanadium steel. There is no better steel than American carbon steel for holding a true sharp edge."

He wrote more about his choice of simple steel in his 1975 catalog:

> Today, we hear a great deal about new and super steels. In over 30 years that I have been engaged in knife making, I have of course tried most all of the steels available in search of a superior steel. It is true that the corrosion resistant steels of today are far superior to the old stainless steels, but, what most people do not realize is that these steels were designed to hold their hardness at a high degree of heat. This of course, is unnecessary in a knife. Most of the high alloy steels have 12 to 18 percent chrome. This is a very definite drawback to good edge-holding qualities. It is true that this steel does have some advantages, such as being corrosion resistant, and this can be important to some people who do not have the time to wipe and oil their blade... Any steel manufacturer will tell you not to use the complex high alloy steels unless you need tools that allow you to cut under high heat, etc. For most tools and especially knife blades, one should stick to the high grade carbon steels. I myself, have found after 30 years of experimentation that the steel with the best all around qualities for a blade is in the straight carbon range. The high alloy steels have far more drawbacks than they have advantages.[1]

Heat Treating

At the end of the 1940s, Moran had determined it was best to heat treat the blades so that the edge was hard, the spine relatively soft, and the tangs very soft. He had difficulties determining the right temperature to heat to before quenching[1]: "Someone would suggest that a blade should be heated to 'cherry red' color, but it was never specified how much light there should be. In a rather dark portion of the shop, cherry red might be the correct color, but when forging in relatively bright light, should you heat a piece of steel 'cherry red,' that would be much too hot. This is only one of the many problems I experienced in trying to follow directions that were not adequately explained." Moran would later use a magnet to find when the steel was ready to quench rather than interpreting the color of the steel[2]. When the steel transforms to the high-temperature phase "austenite," it is no longer magnetic.

Moran said his "tempering by hand" method was ancient[3]: "Of course this was a method used hundreds of years ago - especially in Europe - to make a better blade." Moran elaborated on this process in the 1960 catalog:

> [The blade] is then painstakingly hand tempered in the forge. This tempering gives the Moran Knife special qualities that which *no other* knife possesses. The blade is very hard from the center to the edge and the back is tempered to about the consistency of a spring. The point is slightly less hard than the edge and the tang is annealed for maximum strength. This unique method of tempering results in a blade that has a combination of strength and edge holding quality that is unattainable from production made blades. This can only be accomplished with this method of tempering by hand. It cannot be done in a furnace.[1]

Moran advertised his blades as 58-60 Rc, as he wrote in the 1975 catalog:

> Do not be misled by extremely high rockwells. This is comparable to the old adage that if a little bit is good, a whole lot is better. A knife should be tempered to usable hardness. If one wishes to go to very high hardness, then one turns to the true Damascus Blade, which has hundreds of alternate layers of iron and steel in it. I feel that 58 to 60 R.C. is the best. Also, remember that some day you are going to have to sharpen this blade. The very high alloys are exceptionally hard to sharpen, especially when they are very hard.[1]

Benefits of Forging

Moran felt that the forging process led to a better knife in line with his study of ancient blades and techniques. In his 1960 catalog he said[1], "The bar of steel is heated in a specially designed field stone forge ... and forged by hand on the anvil. Every blade is heated and reheated, then forged dozens of times before the forging operation is finished. This hand forging greatly improves the quality of the blade. The blade then goes through fourteen hand grinding and polishing procedures."

Moran also believed in a process called packing, though many bladesmiths now reject this idea (see Chapter 36). In his 1975 catalog, he wrote:

> I use a technique known as hammer hardening or packing. This is an important step that unfortunately is understood by only a few smith's. After the blade is forged almost to the finished shape, the blade is then hammered at a rather low heat with rapid, light blows. This causes the grain to be better aligned and the steel to be far more compact. This should only be attempted by a smith with a great deal of experience.[1]

In 1979 Jim Phillips summarized Moran's thoughts on packing:

> Moran readily admits that a very fine blade can be made from the stock removal method. However, he feels that a forged blade offers several advantages... During "packing," the blade is hammered at a rather low heat, which packs the molecules of the steel down and makes them much smaller and finer. Some people feel that once the blade is annealed or reheated, the steel will lose this, but it isn't true. In talking to a metallurgist, he found that the following action takes place. When the blade is reheated all of the molecules do expand, but, at the same rate. So, the packing process, along the edge of the blade, does add something to the steel.[4]

Moran's Contributions

The culmination of Moran's commitment to ancient steel techniques resulted in pattern-welded Damascus. Just as he felt that ancient blades had the best performance, so did the ancient practice of pattern-welded Damascus result in the highest-performance steel. But that is a story for Chapter 20. Moran also founded the American Bladesmith Society, covered in Chapter 21.

A Bowie knife of Moran's more refined style. Weyer photo.

-15-

440C, D2, AND A2 STEELS IN CUSTOM KNIVES

Daniel Edward (D.E.) Henry had a reputation from the early days of the Knifemaker's Guild. In the 1973 *Gun Digest Book of Knives,* they wrote[1]: "Perhaps the best picture of the image Henry projects at such events (gun shows, collector meetings) can be obtained by asking those who have met him at such affairs their impression of the man: 'cold,' 'snobbish,' 'arrogant,' 'intolerant' and a few other terms not accepted in mixed company are frequently employed. None, however, deny or even question the man's ability to produce fine knives." Henry was known for his Bowie knives and was a student of historical pieces. He had a collection of 19th-century Bowies, and eventually wrote a book on *Collins Machetes and Bowies 1845-1965*, demonstrating the depth that he was studying these subjects. He studied historical Bowie knives, not simply the famous Jim Bowie knives, but many different historical models he patterned his knives after, though not religiously. "I adopted the good points and rejected the bad points," he said later[2]. And when asked whether he learned from other knifemakers, he reported[1]: "By studying their work, I learned to avoid their mistakes." However, he did give early knifemakers credit[2]: "They were important, those pioneer knifemakers, for their pure existence, for being there. They popularized the custom knife movement and created a market for the many that would follow."

Henry talked about his early knifemaking in a 1993 interview:

[I made my first knife in]1943. I was a marine stationed aboard the Navy Transport USS Rochambeau, somewhere between San Francisco and New Caledonia. I just about majored in history at the University of California. I was mostly interested in Western history and its characters, including (James) Bowie. When Raymond Thorp's book, "Bowie Knife," came out, this further stimulated my interest. Thorp got a lot of bum information, like Harold Peterson did in his book, "American Knives." However, they were pioneering efforts and should be given credit for that. Thorp's biggest mistake was he didn't study the old Bowies. They weren't easy to find in those days. I was lucky. I had access to Bob

Abel's collection. I met him at a gun show and later I visited his shop in New York. I bought a Bowie from him. He practically gave it to me.[3]

Many of the historical Bowie knives in the USA were made in Sheffield, England, rather than America, and Henry was reportedly the first modern knifemaker to copy these designs[2]. Henry enjoyed designing knives and criticized makers who made[2] "degenerate, overly ornate, non-functional knives." Instead, Henry said[2], "I get my real kicks out of solving the problems involved in new designs." Henry would release a new "generation" of Bowie knives every few years or so as he would begin a new historical style of Bowie design.

Henry contributed more to modern custom knives than just Bowie designs, however. He introduced the hand-rubbed satin finish in the 1950s, which became a signature of his work, and has been widely used by custom knifemakers ever since. When asked about the impetus behind this finishing method, Henry said[3]: "I taught myself all my techniques of knifemaking, including finishing. I have always felt that knives should be polished by hand - not buffed. I never had any buffing equipment. There was no reason for that finish. That was just the way I did it."

He was greatly interested in knife steel and tried several. Henry used 52100 in his early knifemaking years[3]: "A friend of mine was forging out blades from big ball bearing races of E 52100 in the early 1950s. I spotted some 52100 1/4-inch flat bar stock, made by Timken, on a listing from Allen-Fry Steel Co. in L.A. and obtained a couple of bars in 1952... I pity those poor boobs who expend such great effort in forging out round ball bearings. I guess they don't know that E 52100 is available (in flat stock)." Henry was not impressed with forging as a knifemaking process:

> I have no bones to pick with knifesmiths - just with some of the fantastic claims they make. I doubt that many knifemakers who forge their blades have sufficient skill, experience or equipment to achieve optimum results... The reasons for forging in the Bowie period are historical - it was the cheapest way of forming steel. Good steel was in short supply... To me, forging is a stupid way of making a knife. Like handforging a car. Who wants a car with Damascus bumpers?[3]

D.E. Henry 1st Generation Bowie.

Henry was the first to use L3 tool steel[3]: "In 1960 on a trip, I went to VASCO-Pacific Steel Co. in L.A. and talked with the proprietor, a well-known metallurgist. I told him I was using 52100. He told me about his Vasco BB (L3 tool steel, Chapter 3), also an electric furnace steel, but containing vanadium and other elements which made it superior to E 52100. I bought some bars and used it until five or six years ago (~1985-

1987) when I started using 440C exclusively in Bowies." Speaking of 440C, Henry claimed to have been the first to use it: "I was the first to use 440C in knife blades - that was in 1960 when it was only available in round bars and I had to have it flattened out in a forge shop. And in 1965 I was the first to use high-carbon, high-chrome steels - D2 and D3."[3] He also added[3], "since 1965 I have been using D3 and D2 for hunting knives." It is unclear how much D2 he used as he stated, "I prefer D3 for edge holding - abrasion resistance - in hunting knives. I prefer 440C for its corrosion resistance properties. Vasco BB was a very good compromise, but many customers did not take care of their BB knives and some of (the knives) rusted slightly." In the 1973 *Gun Digest Book of Knives,* the authors stated that they only knew of Chubby Hueske and Ted Dowell using D2 early on, despite also writing an article on Henry. He was also not particularly known for introducing 440C to custom knives, perhaps explained by a 1973 interview where he said he only used it "In a few models where corrosion resistance is a real design criterion."[4] And in another 1973 interview[5]: "I have only used it a few times and have found its negative aspects far outweighs its anti-corrosion properties." In early editions of *Knives Annual,* he listed only that he "grinds all steels," though in *Knives '86,* he listed D2, D3, and 440C as his preferred steels.

Henry's poor reputation for politeness was discussed in a 2014 *Blade Magazine* article[6]. A.G. Russell said[6], "For years I was one of Ed's (D.E. also went by Ed) very few friends. At one time he asked me, 'A.G., what do you tell people when they ask about me?' I replied, 'Ed, I tell them that you are the most arrogant SOB I have ever met.' For years, he signed his notes to me, 'Your arrogant friend, Ed.'" A.G. said that their relationship could be tumultuous[6]: "Our friendship ended when a writer quoted me as saying that someday Bob Dozier would be as fine a maker as Ed Henry. I got a note that read, 'Never in 10 billion years. D.E. Henry.' He was wrong, of course, but it was at least 15 years before I heard from him again." Some defended Henry's bad behavior. Loveless, in 2001, said[7], "D.E. had a very harsh and abrasive kind of personality. He had no patience with most people but that was largely the effect of an automobile accident. It was like his head had been broken open and (the resulting surgery was botched) but, anyway, his wife told me his personality changed radically after that." Jim Sornberger told me that it was Henry's diabetes that made him cranky and short-tempered. A couple of writers that had interviewed Henry reported that he was courteous and patient with them, including Jack Edmondson and Kathie Miloné. Edmondson reported[6]: "If you got Henry away from a knife show, he was fabulous... But I witnessed him in a show environment... He became haughty, rude and abrupt. I would sit at the table with him and cringe at the way he would treat people." Miloné said[5], "I had been told that this was a recluse; an arrogant, ill-tempered snob. Nothing could be further from the truth." She also asked Henry why he had developed this reputation:

This is because of encounters with me at gun shows. People who have read a book and think they know all about knives, are inclined to ask questions that have no bearing on advanced knife making. When I question them about their knowledge, they become insulted and then become insulting in response. If a person is courteous and friendly to me, I'll talk to them in the same way. I have had to walk away from many situations where I came close to punching someone in the mouth. I get really steamed.[5]

Gil Hibben

 Gil Hibben taught many influential knifemakers, including Steve Johnson, Harvey Draper, and Rod Chappel. He also influenced makers like Don (Zack) Zaccagnino and Buster Waresnki. A.G. Russell said, "He's taught the best."[8] Hibben became a full-time knife maker in 1964 but has been making knives since the 1950s, first becoming interested from the film *The Iron Mistress*. He made a Bowie knife in 1956 and sold his first Bowie in 1957 while working for Boeing as a machinist. Hibben recalled[8]: "In the beginning, I experimented with different steels. At this time there were about three other makers besides myself (Ruana, Moran, and Randall). I started with Boeing. They had some great metals. In 1963 I ran across 440C." 440C was not available in flat stock[8]: "We had to hand forge everything we did. I used round bars. Had a fifty pound trip hammer, a big anvil, and wore wrist supports while pounding my way through all those knives. I pioneered 440C specifically for custom knives." While both Randall and Henry had previously used 440C, it may indeed be true that Hibben was the person who popularized it. For example, Hibben-apprentice Harvey Draper was advertising knives in 440C in the beginning of 1967[9]. Hibben designed all of Browning's knives in 1968. Hibben prided himself on doing his own heat treating, at least by 1980: "I freeze treat my pieces also for maximum martensite transformation" to 56-58 Rc[8]. Cold treatments are explained in Chapter 19.

Already by 1980, Hibben had made over 10,000 knives[8]. Hibben is also known for being the first knifemaker to do a mirror polish on his knives which became very popular. In 1987 Sylvester Stallone commissioned Hibben to make the knife for Rambo III, and he began designing knives for United Cutlery[10]. This made him one of the most well-known knifemakers in the world. Hibben was inducted into the Blade Cutlery Hall of Fame in 1990[11]. While Hibben is perhaps best known now for Rambo movie knives and Paul Ehlers-designed fantasy knives, he has made many different types of knives over the years and considers himself a versatile knifemaker. He has also made many working knives: "I try to make a knife for everybody: Collector, hunter, black powder guy, or a guy who likes to throw knives, a spear, tomahawk, or whatever. I appreciate the challenge of trying to satisfy the customer."[12]

Hibben fantasy knife. Weyer photo.

G.W. Stone

George W. Stone became a full-time knifemaker in 1962. He first made knives as a hobby in the early 1940s, making a few knives for servicemen in WWII. Before going full-time, he worked for an aircraft company where he learned about metallurgy, machining, casting, and forging. Early knives looked like Randall's, but he developed his own style by the early 1970s. Stone would demonstrate the toughness of his knives by hacking at blocks of ebony or steel columns. Or to bend his very thin fillet knife in a vise without it breaking. Stone had a reputation for superior heat treating of 440C in the early 1970s[13], a steel he learned about through his previous experience with aircraft. His heat treating method was a secret, though he said it was "a process between the first and second draws in the tempering, which leaves the blades rather hard at RC 59-61, but not brittle."[14] This presumably refers to some type of cold or cryogenic treatment, one of the earliest references to such a treatment of custom knives I have found, especially in connection with a claim of improved performance (see Chapter 19 for more about cold treatments).

G.W. Stone folder in 440C. Photo from Friendly Brothers Knifeworks.

Chubby Hueske

E.F. "Chubby" Hueske made his first knife in 1936 and advertised his knives in 1968. In a 1973 publication Hueske said[15]: "[I] have always hunted and liked knives and guns. Wanted a better knife than I could buy. Did not know about custom knife makers, so I made my own. Went the whole route - planer blades from saw mills - files - hacksaw blades - etc. Then tool steel and for 4 years the special steel I now use." Chubby Hueske is said to have been the first knifemaker to offer D2 as a "standard blade material" in 1968[16]. Hueske sought out a metallurgist in Houston who told him that D2 would work well in knife blades[16]. Hueske was asked what his favorite steel was in 1991[17]: "D2, because it holds an edge, is relatively easy to resharpen, is stain resistant and, although I find it more difficult to grind and polish than some of the others, I feel the end result is worth. I like D2!" Hueske heat treated his knives to 57 Rc[17]: "At Rockwell C 57, D2 will hold its edge as it will at 59-60, yet at 59, D2 is much harder to sharpen than at 57. Going over 56-58 adds nothing to edge holding, and it detracts in other ways from the overall quality of the blade."

Image of Hueske knife from Arizona Custom Knives.

Jimmy Lile

James B. "Jimmy" Lile was born in 1933 and carved his first knife out of wood at eight years old and was making them out of steel by eleven. He made and sold knives to friends throughout childhood and even said[18], "knifemaking put me through college." In 1955 Lile started as a school teacher, then began a construction business in 1959. Unfortunately, an issue with unreliable subcontractors left him $200,000 in debt, so he sold the business in 1965 and worked as a construction superintendent. During that time, his wife Marilyn would grind knife blades on a bench grinder during the day and then Jimmy would finish them at night. In 1969 Lile decided to work towards becoming a full-time knifemaker, but they immediately fired him when he mentioned his plans at work. So Lile told his wife[18], "It looks like we're in the full-time knife business a little sooner than planned."

Lile's first knives were made with recycled materials before switching to O1. Lile started making knives in D2 in 1968[18], made by Carpenter Steel (branded by them as "No. 610"). Butch Winter said[19], "Jimmy and Chubby Hueske carried on a continual, friendly argument as to who had started using D2 first." In his 1969 catalog, Lile said[18], "I use high carbon tool steel in all my blades, striving to achieve a degree of toughness and edge-holding ability which will give the professional and amateur alike the satisfaction of knowing that when they own a Lile knife, they can depend upon it to perform the task for which it was purchased... I myself prefer high carbon tool steel, but if so desired and specified, the customer may select 440C stainless steel at an additional cost of $10.00 per knife." The 440C he used came from Cyclops Steel[20]. Lile forged some blades, but stock removal became his preferred knife-making method. D2 was his favorite steel throughout his career. In an interview conducted shortly before his death, Lile said his favorite steel was[17]: "D2. I feel that this steel has the best edge ability of any steel. It is tough and resists staining. This combination makes it my choice." He heat treated his knives to 59-60 Rc as he felt that[17] "is the proper amount of hardness for a working knife."

Lile is perhaps most famous for designing and making the knives used in *First Blood* and *Rambo: First Blood Part II*. Joe Ellithorpe of Pony Express Sports Shop became a dealer for Jimmy Lile Knives. When Sylvester Stallone asked Ellithorpe if he knew a knifemaker to work on a special knife, Ellithorpe told him about Jimmy Lile. Stallone called Lile and said[18], "I want one like everybody ain't got one like." Lile said, "It was not my intent to create just another attractive movie prop, but rather to make ONE basic tool to do a variety of jobs."[18] Lile made a series of prototypes in D2 steel for Stallone, though the final knives were made with 440C. Six were given to Stallone, and seven more were kept by Lile.

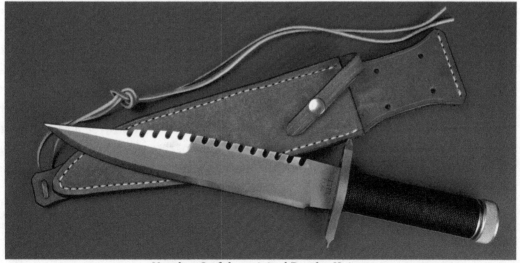

Number 8 of the original Rambo Knives.

Ted Dowell

Theodore M. "Ted" Dowell had a master's degree in Mathematics and taught high school, and then college math at Central Oregon Community College. Dowell began

Ted Dowell

making some knives from recycled materials like files, hacksaw blades, and planer bits. He made his first knife in high school out of a file[21]: "It must have taken me 20 hours to grind that blade, but it required only a second to break it... The first good whack I took at a handy sapling left me holding the handle and about an inch of blade." In 1966 he had a one-year sabbatical at Florida State University, and while there he traveled to Orlando to visit Randall's knife shop. There he bought two heat treated blades which he put his own handles on, and then made sheaths for them. Dowell found that the Randall knives did not hold an edge well after using them during deer hunting season and wanted something better. After making many more knives, Dowell first advertised his knives in 1968 in the *American Rifleman*.

Dowell is perhaps best known for introducing integral knives to the custom world in 1971 with his Integral Cap and Hilt model. He also coined the term "integral," which became the standard term for this type of construction. This type of knife required a lot of material to be ground away using the stock removal method. Dowell developed this style after a screwdriver he saw in his youth with a "hilt and cap" handle design machined from a single piece. Integral knives had existed in the past but not in recent custom knives. He also tapered the tangs of these integral knives[21]: "Grinding the tapered tang on a regular blade is quick and easy on the 4-inch wide belt grinder, but on the integral hilt/cap model, you have to sneak in from the side on a narrow belt and grind by eyeball, to keep from cutting off the humps. A slip in either direction ruins the knife!" Dowell preferred stock removal to forging knives, especially early on[22]: "I use stock removal, I can't honestly say I prefer it because I haven't tried forging. I have always thought that the steel used was more important than the method in producing the blade I was after. I do know that if you are going to forge you'd better do it right so as not to wind up with a stress loaded blade or a blade with minute cracks." Dowell became a full-time knifemaker in 1974. Dowell also branched out into forged knives after being inspired by Moran, though only about 40 knives from simple steels were sold starting in about 1979. He made about 20 knives in his own Damascus, first mentioned in his 1986 newsletter.

Dowell experimented with many steels and was often interviewed about his choices. He would thoroughly test them through his hunting[23]: "The best test of a knife's steel is to use the knife over a period of time in the field. In fact, this is the *only* meaningful test I can think of." Dowell had another trick for testing knives, as he said in a 1999 article[24]: "I've played with a lot of different steels and selected only the ones that work best. I test any new steel by making a paring knife for my wife (Betty) to use in the kitchen. If it doesn't hold up, we find out." In 1973 Dowell said[25]: "At the core of every serious knife maker's work must be endless experimentation directed at finding the best steel and heat treatment for achieving a knife of ultimate edge-holding ability and toughness. And I'm still looking!" His early favorite steel was F8, a relatively rare high speed steel with 1.3% carbon and 8% tungsten, and M2, a common high speed steel. Dowell was attracted to F8 steel for its high wear resistance. However, he then learned about D2 from Chubby Hueske and began using it shortly after, remaining his top choice throughout his career. In 1973 he compared F8 and D2[26]: "Right now it's a toss-up as to which of these holds an edge better. The D2 is much tougher." The greater toughness of D2 probably explains why he largely replaced the F8 with D2. In his 1973 newsletter, Dowell compared D2 to M2 and M3 high speed steels but had "not gotten as good results" with the high speed steels. He dropped A2 steel as an option in the February 1972 newsletter and, in January 1973, said: "A2 is tough even at 62 Rc, but in my experience it does not hold an edge particularly well at any hardness." He was also testing 154CM at that time.

Images of Dowell "Cap and Hilt Integral" from Nordic Knives.

Dowell had his blades heat treated in California to 58-59 Rc, though later would perform his own heat treating. He said about the hardness of knives that[25] "I think that ultra hardness in a blade contributes more to advertising value than to edge holding. I would say that the chemistry of the steel is much more important than the hardness, assuming we are talking about hardnesses in excess of about 56 Rc." Dowell offered 440C relatively early on, but at the time only offered it to have a stainless option for certain knives. But in 1973 when he tried 440C with a -325°F cryo treatment rather than -100°F he said[21] that it went from being a "one deer knife" to a "four deer knife" which he offered by December of that year. In January 1974, he said that he could use 154CM if the customer wanted it, "but I don't recommend it for edge holding ahead of regular 440C and certainly not ahead of cold-treated 440C. I haven't cold treated any 154CM as yet – this will be done as time allows. Bear in mind that the foregoing on steels is my opinion. True, it's based on evidence and test, but it's subjective and certainly subject to error nonetheless." He still preferred D2 for hunting knives but said 440C is superior for humid climates or near salt water. In an August 1974 newsletter, he reported the results of Don Wenger on an African safari where he tested 440C and 154CM knives, which were both given the -325°F treatment, and he said both performed well, but the 440C blade was slightly superior.

Other steels Dowell tried by 1974 were D3, S5, Loveless "Special Melt," Vasco BB, Versa Steel, and Staminal[27]. 440C and D2 became his standard steels by 1973, though later he began using Vasco Wear[28], BG-42[24], and CPM-440V[29], which will be discussed in later chapters.

Some Thoughts on D2 and 440C

D2 has been a mainstay in knives, or at least has had several resurgences over the years, despite criticism from respected people like A.G. Russell[30] and Bob Loveless[31]. 440C has seemingly been known as a "good, not great" steel since its first use in knives. Both 154CM and D2 were said to have better edge retention than 440C. It continued in use in decent volume through the 80s and 90s, but not always because it was a "hot" or "super" steel. Butch Winter, writing in 1984, said[32], "In all my years of associating with custom knives and custom knifemakers I have heard more disparaging remarks made about 440C than all the other steels combined. Yet 440C continues to be one of the most used steels in the custom knifemaking industry. Why? Because it's a good steel, that's why." Makers like Dowell increased its performance through cryogenic treating, and even Loveless believed that it was competitive with 154CM with such a treatment[31]. 440C has had its defenders over the years, such as articles in the June 1984 *Blade Magazine* by Robert Campbell[33], or in *Knives '87* by Jim Sornberger[34]. Once Loveless introduced 154CM, the new steel had a better story and more hype. 440C has never shaken its reputation as a has-been ever since.

Ron Lake, A2, and Custom Folding Knives

Ron Lake began making knives in 1965 after he built a small workshop in his garage. A friend brought him a knife for repair which inspired him to make a Bowie knife:

> A fellow worker and good friend at Stewart-Warner Corporation, John Brown, asked me to help him refurbish a German SS dagger which he had brought back from World War II. He had been an Army Ranger during the war and scaled the cliffs at Normandy. Then, I made that Bowie knife and went from there. There wasn't any [knifemaking] reading material in those days, but I had a good machinery background. The real ingredients are desire and determination.[35]

In 1971, he joined the Knifemakers' Guild and went full-time as a knifemaker. Lake's previous machining experience helped him with his knifemaking and steel selection[36]: "As a vocational machine-tool graduate and a model maker (prototyping) by trade, I have had extensive experiences in die and mold making and have worked with a wide variety of tool steels, all with different metallurgical compositions."

Lake's first choice for a steel, and his standard for many years, was A2 tool steel heat treated to 62 Rc[37]: "I feel it is a good balance between edge-holding qualities and toughness. True, the steel will tarnish and rust even with five percent chrome, but so will a gun if it isn't cared for in a proper way." Lake said in 1973[36]: "Blade steel selection and proper heat-treating is the most important (factor of knifemaking). This is the nucleus - other phases well executed make a good custom knife." Lake was the first knifemaker to use A2 regularly as a knife steel. Lake told me in 2019[38]: "At the first Knifemakers' Guild show in 1971, I didn't find any other makers using it; almost no one had heard of it apart from Bob Loveless." Ron Lake also told me that George Herron began using A2 in the early 1970s, though he doesn't know if Herron used it due to Ron's influence or began using it independently. In the early years, Lake didn't want to use stainless because he didn't think it held an edge as well as A2[39], though he was using 154CM and ATS-34 as his primary steels by the mid-1980s[40]. Lake made his knives by the stock removal method[36]: "Although a few of the cutlers use the forging process and think highly of it, I feel just as strongly about the stock removal process. I feel the controlled rolling at the steel mill is the best way of aligning the grain structure and it cannot be surpassed by eyeballing the temperature and hammering the steel."

Lake is best known for the "interframe" folding knife, which had a metal handle with an inlaid handle material, allowing delicate materials to be protected by the overall handle. He also made these knives with a "tab lock," which was a modification of a lockback. A friend had brought over a factory folding knife for repair, and Lake looked at the good and bad points of the design, which gave him ideas for his own folding knives[41]. He wanted it to be easy to open and close, so he put a hardened steel bushing at the pivot and a nail nick on either side for opening with either hand. He noticed that his friend's knife had chips and cracks around the edge of the handle, which gave him the idea to protect the handle material with the metal frame. Lake described how he ended up with the tab lock in a 2011 interview:

> I had only been making knives for six years. I didn't even know another knifemaker and didn't know what knives were supposed to look like. I came up with the interframe design and then thought about where I could put a release mechanism. I still make interframes and that's about all I do make, and they are with the tab... The tab lock is really not a lock mechanism. It is just a tab on the end of a lockback that offers a mechanical advantage. If it isn't there, I just have a bar to release the lock, and depressing that bar can be hard to do when the hands are soft, cold or wet. The tab gives you a larger surface to push on.[42]

Lake brought several interframe folders to the first Knifemakers' Guild Show in 1971. He was not confident in how they would be received[43]: "I was the first in the Guild to make folding knives when at that time people thought that they should cost less than a fixed-blade knife. And we're talking about a time when you bought a Buck folder with $5, and a Loveless fighter cost $115. So, I sold my folding knives for 115 dollars as well." Lake was the first member of the Knifemakers' Guild to have a milling machine, and with that and his machining experience, other knifemakers were shocked at the high quality of his folding knives. Loveless was particularly impressed[44]: "He looked at my knives and said, 'Where the hell are you from?' He kept my knife overnight in his room. If I hadn't gone to that first show and had Bob laud my knife in front of those other makers, it would've taken me years to achieve the notoriety his praise gave me. When Loveless made positive comments, everybody perked up and listened." Lake would patent the interframe design[35]: "What was significant about the interframe was that manufacturers and individual makers realized you could make a folding knife out of a solid piece of metal rather than liners and bolsters. It's an evolution and people got the idea that you could use a solid billet. In the patent language I called it a frame, and I believe it was in conversation later with A.G. Russell that I first used the word interframe." In 1987 Lake coauthored the book *How to Make Folding Knives* with Frank Centofante and Wayne Clay.

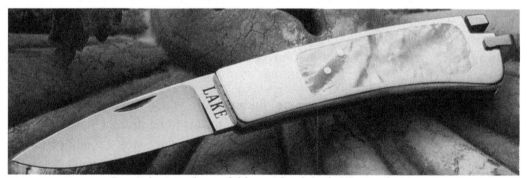

Ron Lake interframe folder with tab lock. Weyer photo.

-16-

BOB LOVELESS AND 154CM

Robert Waldorf "Bob" Loveless was born in 1929 in Warren, Ohio. During World War II, Loveless wanted to help fight the Axis powers, but he was too young. At 15 years old, he lied about his age and joined the Merchant Marines until he was old enough for the military. The war ended, but at 17 years old he joined the Army Air Corps, the forerunner of the Air Force. During a two-year tour in Guam and Iwo Jima, he had a lot of time on his hands and became interested in design and drawing. As Loveless remembered[1]: "There was an old house trailer on Iwo Jima. I was fascinated by it and started drawing variations of it." Because it was difficult to increase in rank after the war, he retired from the military at the end of 1948. He enrolled in the Institute of Design in Chicago and was accepted despite not having a high school diploma, based on drawings he had done in the Air Force. Loveless didn't appreciate the training there:

> The Institute of Design was an outgrowth of the Bauhaus school: Avant Garde ... form follows function and all that... I began to realize what a phony bunch they were. It was intellectualized. Everybody at the Institute of Design was interested in shopping centers and office buildings. I was interested in practical architecture - things like family dwellings. I was intrigued with the possibility of making mobile homes on a production line but I got no encouragement from the teachers. I finally just got mad one day and quit. I didn't even check out.[1]

Later he did start to appreciate things he learned, however: "They taught me to think in functional terms, especially with materials"[1].

Loveless was working on an oil tanker in 1953 when he read an article in *True* Magazine about Randall knives[2]: "Knives were used on a daily basis aboard ship. After reading about Randall's knives and how well they stood up to abuse, I decided I wanted one." Loveless visited an Abercrombie & Fitch sporting goods store but found the salesperson unpleasant. "He was a kind of snippy dude," Loveless recalled[3]. "I am certain that my oil-stained denims, pee coat and watch cap didn't really make me the typical Abercrombie & Fitch customer"[2]. After asking for a Randall, the clerk answered[2], "We simply don't have any. And it will take at least nine months to get one." Loveless recalled[2]: "The clerk didn't even ask if I wanted to put my name on the waiting list for a Randall knife. It was evident that he didn't want to be bothered with a young guy wearing the working clothes of a merchant seaman." Loveless decided to make his own knife instead:

> I told the taxi driver to let me off [at the junkyard] to get some pieces of automobile springs. I don't know exactly why, but I thought automotive springs would make good knives. The fellow at the junk yard told me to "Go see the guy out there with the burning torch." There, I found the biggest pile of auto springs you've ever seen - maybe 40 feet high and 40 to 50 feet wide at the base. I told him I wanted some good metal to make a hunting knife. He got a 1937 Packard spring, and said, "This is the best spring in the yard." I got four or five pieces, each about a foot long. I think I paid the fellow in the office a dollar and gave the yard man a couple of dollars for his trouble. I took the pieces of spring back to the ship and roughly ground them - with a snag grinder, with a hard-clay wheel - to the outline I wanted. It took me over a week to get the knife done and it looked pretty good.[1]

Loveless then heat treated his knife:

> I fired up the galley stove which was a fuel oil burning model with a pressure blower on it. We could get some pretty heavy heat on that outfit. After I got the blade nice and red I stuck it down into a five gallon can of refrigeration oil and I happened to think at that time how beautiful it was to see that red hot steel go down into that heavy oil. When it came out the carbon layer broke loose freely and my blade was a nice pearly gray. I went back to the grinder for a little touch up work and then finished up with sand paper. Well, I started using my knife. I put a nice handle on it and since there are a lot of lines on a boat like the one I was on I got to use that knife a right good bit. After a lot of cutting the edge would go dull but in a funny way. The first mate of that boat collected stamps and that dude had a twenty power magnifying eye piece. I borrowed it and started looking over the edge and it seemed that it was actually chipping away. I talked to the chief engineer about my problem and he inquired as to whether or not I had drawn the blade. I didn't know what he was talking about. He explained the basic principles of drawing a blade after hardening and I fired up the galley stove again to finish the job I had started. I got that blade to a nice straw blue color and when I had put my knife back together I found out that it was sure enough a darn good knife and would hold an edge real well.[3]

Loveless was very proud of the knife[1]: "About five or six weeks later, I took it to Abercrombie & Fitch. I was going to show that young punk my knife." However, Abercrombie had fired that clerk, and instead he talked to the floor manager:

> I ... talked to Pat Devlin who was the buyer at the time. He liked the knife well enough to order a couple of them. It took me about a month and a half to make three more knives... I took them back in there and he bought all three. He gave me the money right there on the spot... I went back to the ship and was reassigned to a big tanker and so I lost track of

things at Abercrombie's for a while. To me, my mission was accomplished. I had made a knife for myself and three for them. Some time later I happened to be back in town and I dropped by to see how things were going. I got an order from Pat Devlin for a gross of knives and he wanted them real quick. He said he had tried every way to get in touch with me after the first three knives had sold almost immediately. I wouldn't take the order for a gross but I did accept an order for six dozen. He gave me a purchase order for them and I took it to the bank and borrowed $500.00 on a ninety day note. I bought a grinder and a few hand tools and went to work making knives for Abercrombie's along with being in the Merchant Marines.[3]

When asked what gave Loveless the confidence to make knives without prior training, he said the following in a 1973 interview:

When I was a little kid we lived out on a farm. It was during the depression and we didn't have much money. If you didn't have enough money to buy some fancy toys and you wanted to play cowboys and Indians, you picked up the end of an orange crate and whittled your own gun. My guns were pretty good and I even sold one or two for a nickel or a dime to some of the other kids. I never once doubted that I could make a good knife. I just went ahead and did it.[3]

Loveless Designs

Loveless developed his own style over time:

[My first knives] looked a lot like knives that Bo Randall produced. The knives all had narrow tangs, single guards, leather washer handles with finger grooves and an aluminum pommel. Looking back, there is really no comparison between what we produce now and those early knives. They might have been good, solid edged tools, but they were lacking in aesthetics. Thankfully, over the years I've been able to break out of that early mold and create knives that not only are functional, but also look good.[2]

Loveless felt that both aesthetics and function were important[4]: "Eye appeal is the very first thing - it has a bearing on whether or not the knife will be accepted... If it is not appealing to the eye, then the prospect won't pick it up. If he doesn't pick it up - you're lost. But for the intelligent outdoorsman, you have got to get him to pick up the knife." Then, after the prospect picks up the knife, feel is next in importance:

A knife must feel good in a man's hand. Whether we realize it or not, our sense of touch is very critical to us. We get a lot of our input from our fingertips ... it has to feel good ... since we are directing our effort to the user's hand primarily ... it's important to the knifemaker to understand the function of the hand... The knife designer, if he's good, realizes that the human hand defines everything from the guard on back, in his work.[4]

Loveless felt his success came from this superior design[4]: "I like to think that it's been so successful because we've thought out very thoroughly and carefully how the knife should be, and we use good design principles, and stayed within the discipline of the good product designer in making these knives."

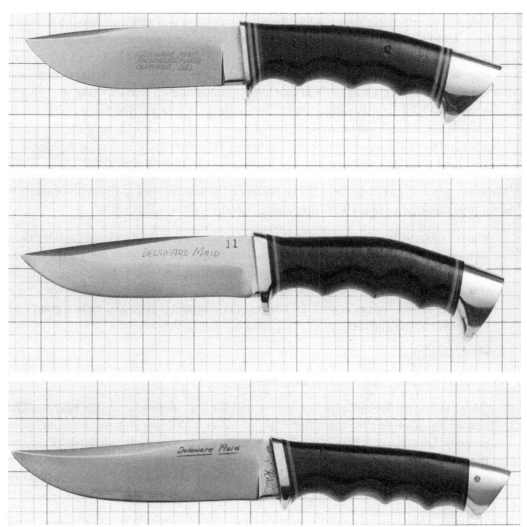

Early Loveless knives from Abercrombie & Fitch. Weyer photos.

One of the changes to his knives was to use a full-tang construction, as he wrote in his 1972 catalog: "After making many hundreds of such knives (narrow tang), I began to realize that this method of construction suited the convenience of the knifemaker, but did not benefit the customer. Beginning in the early 1960s, I made a few full-tanged knives. As I got deeper into this kind of knife, I began to feel that this was the only logical way to make a knife..." Around 1966, he started tapering the tangs of his knives:

> Before the tangs on my blades were tapered, the knives we made always seemed a bit handle heavy. Just a couple of years into knife making, a friend gave me an old Lamson and Goodnow knife that had a fully tapered tang. What was interesting about that knife was the design dated back to the mid-1800s. Later on, when attempting to salvage some blades that had warped during heat-treatment. I decided to taper the tangs as a method of restoring straightness to the blade blank. After straightening the blank and assembling the knives, I realized that the tapered tang was the way to go. Not only had the tapering

saved a number of warped blanks, but it also changed the knife balance, moving it forward to the mid-point of the knife near the guard. Some makers drill holes in their tangs to accomplish the same thing. I just think a tapered tang gives a knife that very special look.[2]

The drop point had existed for centuries, and Randall had released his drop point hunter in the early 50s, which he advertised for its benefits in skinning. However, Loveless was known for popularizing the drop point design in hunting knives:

> It just made sense. When an animal is field-dressed, a hunter must open the abdominal cavity. Unless you're very careful, a flip- or trailing-point blade can easily slice into the underlying viscera. When that happens, you end up with stomach contents, including caustic digestive fluids, all over the inside of the animal. Unlike both the clip- and the trailing-point blade patterns, which are created by taking a concave "clip" out of the back of the blade, the dropped hunter uses a convex curve to shift the position of the blade tip down and away from the direct line of the back of the blade. Configured in this manner, when you're field-dressing the tip of the blade is less likely to come into contact with the abdominal content. The same thing is true when skinning. The dropped blade tip is positioned away from the muscle tissue beneath the hide.[2]

Typical Loveless Drop Point Hunter with engraving by Ray Viramontez. Weyer photo.

Loveless thought that actually using the knives you designed was integral to the process:

> Unless you've had field experience, you can't make hunting knives. I see knives with cutting edges that are too thick. With blades that are too long... With awkward and ill-shaped handles. Too many knife makers don't understand their own hand. Their knives are not user friendly. You can't teach a guy to make a good hunting knife if he's never been in the woods, shot a deer, and been up to his elbows in deer guts. It's like giving a typewriter to a man who is illiterate.[1]

Despite being known for his contributions to knife design, Loveless did not see himself as an artist:

> In primitive societies everybody works to survive. There are no artists. People come up to me and say, "Boy, Mr. Loveless, you're a great artist!" I say, "No, I'm an artisan." I deliberately took a common hunting tool and refined it to its ultimate degree. I abhor the term art knife. The term is an oxymoron. We make working knives. A Loveless knife is the antithesis of an art knife.[1]

Loveless was very proud of his designs, as they served customers well and were copied by many other knifemakers[1]: "I arrogantly feel I'm the best damn knife designer in the world. I hardly ever make a knife I didn't design. I make knives to please me. Those who do it just for money hate their wives and kick their dogs."

Steels Used by Loveless

Steel	C (%)	Cr (%)	Mo (%)	Ni (%)	Si (%)	V (%)
Jessop 139B	0.8	0.25		2.6		
S5	0.6	0.25	0.45		2.0	0.2
154CM	1.05	14	4			

After starting with automotive leaf springs, Loveless moved to a new steel in 1954:

I used Jessop 139-B which is a nickel steel alloy. It was very tough and there are still several hundred of those knives out there kicking around and doing an honest job for people... It has better than 2-1/2% nickel in it and nickel is generally about twice as effective as chrome for corrosion resistance. We used to run the 139-B at about 58 Rockwell and it was good and tough and held an edge well. It didn't want to rust hardly at all. I'm still surprised at how good it really is for knife blades. It became hard to get because they manufactured it primarily for lawn mower blades.[3]

Initially, Loveless was forging the knives from 139B round bar:

I had no idea that flat stock in the steel of my choice (Jessop 139B) was available. When I ran out of steel, I contacted the manufacturer. The woman who answered the telephone suggested that I try their flat stock. I realized right then and there that I'd forged my last knife. Her suggestion was what you might call "a light bulb moment." When that light came on in my head, I realized that there was no point in beating a piece of round steel stock flat, when you can buy it flat already. No matter what the forging community believes, you can't improve a piece of steel by heating and beating on it.[2]

Around 1966 Loveless switched steels, first looking at S5 "silico-manganese" steel and then a modified version:

Speaking for myself, I "got an itch I couldn't scratch," and went on to working with the Shock-Resisting steels, with a grade called S5. I did then, and do still, consider the matter of toughness to be of vital importance in a working knife. So after reading everything I could get my hands on, I got an order of S5 from an Eastern mill (in the United States). After working with it for a year or so, it had gotten to the point where I wasn't sleeping very well, and "there-in hangs a tale": S5 has almost 2% Si and 1% Mn, and when everything went just right, we got some very fine working knives out of it. Very high toughness of course, but also very good edge holding. But it was a Brine-quenching steel, and with S5 two problems were never licked: sudden, brittle fracture during testing, and rusting that seemed to pop out of the blade before my eyes. The brittle fracture just shouldn't have happened, and two out of three times it didn't. But the third time, I would put knife after knife in the vise for a simple bending test, and watch the damned things bust with very small bends. I hollered at metallurgists, kicked the dog, my kids wouldn't bring me their report cards, my internist was pulling his hair, and my Old Lady was making me eat out in the shop.[5]

It isn't clear what was causing fracture in Loveless's knives, though the fact he mentions it being a "brine quenching" steel may mean he was quenching the steel too fast and seeing quench cracking. The steel can undoubtedly be oil quenched, especially in knife thicknesses. Loveless continued:

> And still I wouldn't quit. I went East (within the United States), talked hard, and found a mill that would make me a special modified melt of slightly different Silico-Manganese Shock steel. That stuff made up into knives that were so good I didn't believe it myself, and started thinking my customers were kidding me. But... you guessed it... that old problem of two-out-of-three! Again, that third time, when the knife would break I'd end up mumbling to myself, watching the moon go down and the sun come up trying to figure out what was going wrong. Never did, either. So I have a bad taste in my mouth over this kind of steel, which has been recognized for years as the best stuff to make chisels out of...the properties of the Shock steels intrigued me, and still do. But there was the rusting, and I got tired of having to hone my knife everytime I started out hunting, so I began wondering about things like that.[5]

Loveless wanted to use a stainless steel instead to fit the trends of customers. He gave his thoughts about why customers wanted stainless:

> In a word: laziness. When I was a child during the Great Depression, we didn't have much of anything. What we had, we took care of... Today, nobody seems to take care of anything. We have become a disposable society... It takes far less care than a comparable carbon steel knife. And that's the primary reason carbon steel knives have fallen out of favor.[2]

Loveless described the process of looking for his perfect steel in the 1972 catalog:

> So all during those years I was building in my mind a picture of what the perfect Loveless Knife steel would be, and what it would offer: working hardness exceeding Rc 62, working ductility and toughness at that hardness, and finally rust-resistance equal to the usual Knifemaker's 440C, which I had tried but rejected due to poor edge-holding. During the summer of 1971, I learned of a steel made by a leading Tool Steel producer in the East (Crucible). They had developed it for service in the high-temperature regions of the fan-jet engines on the 747 aircraft, and it was alloyed to hold strength at 700 deg. F. My examination of the alloying, and a study of the heat-treating procedures suggested by the maker, led me to think it might be a candidate for fine hunting knives.

Loveless wrote in 1975[5]: "I managed to get a sample, and ran initial heat-treat tests at Downey Steel Treating Company, with my good friend Art Jaekel. Together we were able to develop hardness (after tempering) of Rc 62-63, and impact testing showed a very useful ductility at this hardness." The heat treating method Loveless developed was also shared in 1975, though before providing it, he poked fun at makers with secret heat treatments[5]: "[I]f you've ever wondered what goes on behind closed doors at the dark of the moon get ready!" He used a 2000°F austenitize in a salt pot, quench in a 1000°F salt pot, then quenched in oil, and then a -120°F cold treatment, followed by multiple tempers at 975°F. He also added[5]: "If there's any 'secret' to making knives like this, I've given it above, and I encourage you to try for the same results."

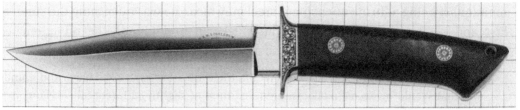

Loveless fighter with engraving by Lynton McKenzie. Weyer photo.

Loveless continued the story of 154CM in his catalog:

> I ordered a quantity of 500-lbs, which was delivered in August. Test knives were made immediately, and placed in the hands of friends here in the West and in Canada, and I began hearing from these men, all critical knife users, within two weeks, with their comments. Previously knife buyers had to make a choice when they ordered a knife. The knifemakers, this one included, could offer edge-holding, or rust-resistance, but not both qualities, in the same knife. Typically, those of us who valued field performance preferred to make our knives from one of the better medium or high-carbon Tool Steels, in the judgment that edge-holding was the main reason for the existence of the benchmade knife. But two things were immediately apparent, in the letters commenting on the new knives. First, they did hold their edge, better than did my previous alloy. And second, they didn't rust, in normal service, even if not cleaned up right away. We had to unlearn some things, and learn some new things, in the way of Shop practice, in making these new knives. But it was soon quite apparent that we had found a new kind of steel, and that this new steel was yielding the results I had been looking for. Knives made from Grade 154CM are indeed fine working knives... Exposure to blood acids in big game stains the blade slightly, if anything, but badly-pitted blades just don't happen, and even the staining is rare. And the cutting edge holds up better than did the older Silico-Manganese Alloy knives. Blades are put in service at a working hardness of Rc 62-64, and are quite ductile and tough at that hardness... All Loveless Knives made since September of 1971 are of this fine new steel. We have almost 300 knives in the field now, and the results have exceeded my early hopes. The 154CM Alloy has brought new and better quality to Loveless Knives, and it's the kind of steel I had hoped to find for years.

Loveless's statements on steel were part of a more extensive article he wrote on the subject[5], and in that article he recommended the book *Tool Steels* by Roberts, Hamaker, and Johnson[5]: "This book is one of very few 'bibles' in the tool steel business, and my own copy is well-thumbed and beloved." Loveless summarized his article:

> Eight thousand words or so ago, I set out to tell you something about steel, and knifemaking. Whether I succeeded or not is up to you, and how hard you want to work at knowing the world of fine knives. You won't learn it all from an article, and may never learn it all in one lifetime for all I can see. But that's what makes it interesting, even challenging, and yes ... fun. There's a little more to it than you thought there was. Right?[5]

Loveless seems to have dropped hardness over the years, or at least the advertised range of hardness. In a *Blade Magazine* article in 1977, Loveless stated that his 154CM was 61-62 Rc[6]. And in a 1986 article in *Knives '87* on Loveless, it said his 154CM was 60-61 Rc[7]. His thoughts on 440C also evolved. In that same 1977 *Blade Magazine* article[6] he stated that 440C could be "field competitive" with 154CM if given a cryogenic treatment at -325°F, but not a "sub-zero treatment of -120 Fahrenheit."

When there were difficulties with procuring 154CM from Crucible Steel, and issues with the cleanliness of the steel they were making, Loveless began using the same steel made in Japan by Hitachi called ATS-34 in the 1980s. Loveless also tried the somewhat more wear-resistant, vanadium-alloyed version BG-42, first in 1975 and then in somewhat greater quantities in the late 90s, but ATS-34 remained his base steel. As Loveless said later[2]: "Most Loveless knife customers seem to be happy with the performance of ATS-34. If you're going to use one of our knives for a lot of field-dressing, skinning and other field care activities, then we suggest using BG-42. That particular steel provides about 10% more edge retention over ATS-34. That said, BG-42 also demands about 10% more effort to make blades."

Loveless Thoughts on Knives

Loveless was known for his writing and interviews on knives and knifemaking. Below are a few examples of his thoughts:

> Knives are, by far, the oldest created tool. The first tools were the club and the stone. Then, man lashed the stone to the club to make it more effective. Then he honed it to a sharp edge. Almost everyone has a feeling about knives. There are very few men who can regard knives neutrally. There's something downright atavistic to a man when he picks up a knife or regards a knife. It may be a throwback to the hunter in man.[1]

> Knife making can be thought of as an endless chase after perfection in design and execution - an ongoing pursuit of better steels and materials, and, of course, of working hard enough and being lucky enough to realize the measure of success that will give us a decent living. Some of us keep at it, often working long hours into the night after the demands of a regular day job. Sometimes, it becomes too much and we drop out. Yet it keeps calling us back - perhaps later when we can afford a better shop - to try again.[1]

> Years ago I did what I set out to do in terms of design, metallurgy and working techniques. Some people would say we were riding the crest of our success. Other people would say Loveless is lazy, doesn't bring out anything new and hasn't for years. Well, I don't really care... Knives, after all, are a rather (mundane) kind of thing. Of course, you can't say that in the magazines because you've got to make (bleeping) heroes of us all.[8]

Loveless had a significant impact on knife design and steel selection which has continued ever since; many knifemakers discussed in this book spoke of Loveless's influence on them. He was instrumental in creating the Knifemakers' Guild (Chapter 17), was the first knifemaker to have a "collaboration" knife with a major knife manufacturer (Chapter 18), and was even the first president of the Japan Knife Guild (Chapter 35).

-17-

A.G. RUSSELL AND THE KNIFEMAKERS' GUILD

Inspiring a Generation of Knifemakers

In the late 1940s and 1950s, there was a surge of interest in James Bowie, James Black, and the Bowie knife. Raymond Thorp published the nonfiction book *Bowie Knife* in 1948, and Paul Wellman published *The Iron Mistress* in 1951, a fictionalized story of James Bowie's life. *The Iron Mistress* was adapted into an Alan Ladd-starring film in 1952, and several more James Bowie movies would follow. *The Iron Mistress* would inspire many custom knifemakers and collectors. Many knifemakers that began in the 1950s and 1960s started with Bowie knives. While the legend of James Bowie and his knife were already known then, *The Iron Mistress* significantly increased interest in Bowie knives. Bo Randall said[1], "Up until the movie, *The Iron Mistress*, I never had an order for a Bowie. Then, almost overnight, I received a number of orders and I had to do some research to find out just exactly what an authentic Bowie looked like."

Despite claiming to be nonfiction, the Thorp book portrayed a fantastical version of history and has since been criticized for bad citations, altered quotations, and even invented material[2]. For example, Thorp shared a story from Governor Daniel Webster Jones about James Black that feels pulled right out of the fictional *The Iron Mistress*:

> Time and again, when I was a boy, he would say to me that ... when I had reached maturity, he would disclose to me his secret of tempering steel... On May 1, 1870, which was his seventieth birthday Mr. Black told me that, since in the ordinary course of nature he could not expect to live much longer, he had decided that the time had arrived. He stated that I was old enough and sufficiently well acquainted with the affairs of the world to properly utilize the secret, and that if I would procure pen, ink and paper, he would communicate his knowledge to me. I lost no time in bringing the materials to him.[3]

Three times Black would tell him to leave and come back in an hour, rubbing his brow "as if trying to reconstruct something in his mind."[3]

When I came to him at the end of the third hour Mr. Black burst into tears, saying: "My God! It is all gone from me! All these years I have accepted the kindness of these good people in the belief that I could partly repay it with this, my only legacy. Daniel, there are ten or twelve processes through which I put my knives - but I cannot now remember even one of them. A few hours ago, when I told you to get the writing materials, everything was fresh in my mind. Now it has flown. I have put it off too long!" I looked at Mr. Black in awe and wonder. His forehead was raw and bleeding, where the skin and flesh had been rubbed off by his fingers. His sightless eyes were filled with tears, and his face expressed utter grief and despair. I could only say: "Never mind, Mr. Black. It is all in the wisdom of God. He knows best; and undoubtedly, He

had His reasons for allowing the secret of the Bowie-Knife to remain with you." The inventor of the Bowie-Knife lived with me slightly more than two years following this scene - but from that moment he was a hopeless imbecile. The struggle to impart the secret had destroyed his mind. God gave him the secret for His own purposes, but was unwilling for him to impart it to others.[3]

Thorp also described some of the qualities of ancient Damascus steel[3]: "The introduction of the Bowie-Knife revived the ancient tales of the famous blades of Damascus. The original Bowies, as made by James Black, owned all the qualities of the Damascus swords, including resiliency, temper, and cutting edge. Since no one can provide the formula for the Bowie-Knife, we should investigate what is known for the Syrian blades." He then described a few historical attempts to recreate Damascus steel. Thorp's book inspired knifemakers like D.E. Henry and Bill Moran. Moran said[4], "I must have read that book at least fifty times, and it certainly had a tremendous impact on me." Thorp's book was likely one of Moran's inspirations to make Damascus steel. Randall released a Raymond Thorp-designed Bowie in December 1955, which Randall described as "Probably the most nearly authentic Bowie type."[5]

The Iron Mistress, both the book and the movie, also described special steels and secret heat treatments, and James Black incorporated a piece of meteorite into the steel used in the Bowie knife. In the book, the fictional James Black described the meteorite:

[A]n examination of the smaller fragments by a noted scientist in the University of Pennsylvania revealed one possible answer to this superior steel. It contains alloys of some kind. Just what these are or how they are combined is impossible to tell. Under our methods it is not possible to combine alloys to make this transcendent steel - at least to my knowledge. But in this small piece is some combining factor. Find its secret for mating alloy and steel, and we will have made a truly great upward step in metallurgy.[6]

In the movie, James Black said, "This is steel, pure steel from another world. Tougher and harder than anything on this earth. I've tried to match it, but of course I never can. Yet I think I've come as close as any man." In both versions, Black's heat treatment was also described as exceptional. In the movie, Bowie told Black that he had heard he had "a different method of tempering" and Black responded that it was a "trade secret." Both versions said that he quenched the knife seven times. However, the book said he quenched in "panther oil," while the movie said, "bear oil." While these are fictional accounts, these stories helped feed the desire of new knifemakers to try different steels and heat treating methods. Several knifemakers have even made knives using meteorite, including Bill Moran in 1980, and he got the inspiration from *The Iron Mistress*[4]. He forged and folded the meteorite ten times and then forge-welded it to a piece of W2 to be the core steel.

Moran San-Mai knife with meteorite sides and W2 core steel. Weyer photo.

Randall also helped inspire new custom knifemakers, as he was the most publicized maker of handmade knives. Such as the *True* magazine article that Loveless read before going to Abercrombie & Fitch to purchase one. When more articles were published about handmade knives, beginning in 1966, this led to a significant increase in the number of knifemakers. Ken Warner wrote the first of these articles for the 1966 *Gun Digest*[7], called "The Best Knives Made." Warner wrote about Bo Randall, Bill

Moran, Merle Seguine, Rudy Ruana, Lee Olsen, Jr., and Harry Morseth, along with the companies Buck and Gerber. Then he briefly discussed the "Commercial Knives" by Ka-Bar, Case, Western, Schrade-Walden, and Marble. Warner did talk about steel, but he did not mention any specific alloys, and he somewhat downplayed the importance of the steel choice[7]: "You can forget all the jazzy ad copy about secret alloys and new ingredients and miracle stainless and similar claims. That goes for any kind of stainless, incidentally. High-carbon stainless is almost as good as high-carbon plain, but it's tough to work... The alloys do have purposes, of course, but these purposes are not concerned with the basic business of a knife." He did say that the steel is an area where you have to pay more, however[7]: "This is why the custom guys mostly beat the factories. They are willing to put up with working the steel as hard as they want it to be. The big factory can't do this on a knife that sells for $5 or so, though they are proving that they can do it on knives that go at $12 to $20." Warner also wrote about the early forged vs. stock removal debate. Seguine and Olsen made knives by stock removal; they told Warner, "forging is completely unnecessary ... proper heat-treating is all it takes." However, Moran, Randall, and Ruana argued that forging made for a tougher knife.

After Warner's article B.R. Hughes said[1], "Almost overnight dozens of other knifemakers emerged, practically from nowhere. Between the years 1966 and 1971, we examined the work of approximately seventy-five knifemakers, less than ten of whom we had heard of in 1965. Other articles followed, in periodicals such as *Gun World*, *Guns & Ammo*, and *Guns*. Each new article appeared to inspire others to enter the field of knifemaking." Many of these makers had been making knives as a hobby, but this newfound recognition of custom knifemaking led to selling them. In addition to inspiring some to try knifemaking for the first time, of course.

A.G. Russell and the Knifemakers' Guild

Andrew Garrett Russell III went by Andy or A.G. In the early 1960s, he was trying to buy an Arkansas stone to sharpen his knives but could hardly find any stores selling them. When he did find a store with some in stock, he was told he had to buy three or more. "Here we were less than 120 miles from the world's only deposit of Arkansas Stones, and they were practically unknown and almost impossible to obtain."[8] Russell talked about the start of this business in 2010[9]: "In 1964, I started a small mail order business selling Arkansas sharpening stones through tiny ads in gun magazines like *American Rifleman* and *Guns & Ammo*. Very quickly, I began to sell knives in those ads as well. Sometime in 1967, I made a trip to sit down face to face with the handmade knifemakers I had come to know as a result of this new business." Russell visited California, Washington, and Alaska, meeting with custom knifemakers about endorsing the stones. He would purchase magazine ads for his stones, featuring a picture of a custom knife with the maker's name and address. "Surprising[ly], few of the top knifemakers had even heard of these stones, but they were willing to learn."[8]

Early in his trip, Russell met with D.E. Henry as he reported in 2010:

> Ed had a small knife collection and on the evening of my visit, I was thoroughly enjoying the opportunity to examine those knives up close and personal. After looking at several nice pieces, I picked up a knife that I did not recognize, and I kid you not, the hair at the back of my neck lifted as I looked at the most stunning fighting knife I had ever seen. The mark was "R.W. Loveless, Lawndale, CA." I asked Ed, "Who is R.W. Loveless?" He refused to tell me anything unless I agreed that I would not provide any promotion for Loveless. I told Ed I could not make such a promise and he finally gave in, telling me that he and Loveless had traded knives. During that time most knifemakers were envious of each other and Ed Henry was no exception. He did not want me to use my little company and my contact with gun writers to promote Loveless. After my visit with Ed, I went on to the San Francisco airport and called R.W. Loveless. I was surprised to learn that he knew who I was and that he was quite excited to hear from me. He insisted on picking me up at the Los Angeles airport that afternoon. Bob and I spent hours talking knives and the knife business. We formed a friendship that, even after more than 40 years and being separated by half the county, endures to this day. Bob and I talked endlessly on the phone, working to promote his business and knifemakers in general. One of the results of this was the formation of the Knifemakers' Guild in 1970. In the beginning, the only support we could find came from Dan Dennehy. Most makers were afraid that others would steal their customers.[9]

THE WORLDS BEST SHARPENING STONES

R. W. Loveless, knife maker, Box 837-A, Lawndale, Calif. 90260. Says "Russell's WASHITA/SOFT ARKANSAS is the only stone for a really fine knife.

WASHITA/SOFT ARKANSAS

WP—0515	3¾ X 1⅜ X ¾	postpaid	$2.50
WP—0513	3½ X 2½ X ⅜	postpaid	$4.20

ABOVE IN PLASTIC SHEATH

WB—04	4 X 2 X 1	postpaid	$6.50
WB—05	5 X 2 X 1	postpaid	$8.00
WB—06	6 X 2 X 1	postpaid	$9.50
WB—08	8 X 2 X 1	postpaid	$13.00
WB—011	11½ X 2½ X 1	postpaid	$26.00
WB—0116	16 X 3½-4 X 1¼	postpaid	$53.00

Use one of these and if you dont agree that it is the best sharpening stone you have ever used, return it for a full refund. Instructions with each stone. Dealer inquiries invited.

RUSSEL'S ARKANSAS O:/STONES
Box 474-H, Fayetteville, Arkansas 72701

1970 ad for A.G. Russell Arkansas Stones

Russell said in 1986[10]: "It occurred to me that if we could get all or most of them together (knifemakers), they would get to know one another and they could exchange information concerning procedures, materials and the like. Moreover, working through this proposed organization, it would be easier for the various makers to receive favorable publicity."

In February 1970, Russell purchased 20 tables at the Sahara Gun Show in Las Vegas and invited the knifemakers he knew to sell there. Russell said[11]: "At the end of the show, Bob Loveless, Dan Dennehy, and I talked to some of the other makers about forming a

Knifemakers Guild. I rented a block of tables for the Gun Report Show in Tulsa, Oklahoma, for the summer of 1970. That was where eleven knifemakers and I met to form the Knifemakers Guild." The original Guild knifemakers were John Applebaugh, Blackie Collins, John Nelson Cooper, Dan Dennehy, Ted Dowell, Chubby Hueske, Jon Kirk, Bob Loveless, John Owens, Jim Pugh, and G.W. Stone.

Left to right: Jimmy Lile, Bill Moran, Clyde Fischer, Chubby Hueske, Bob Loveless, and Bernard Sparks in 1972.

The leading position in the Guild was initially called the "Secretary." The first Secretary was Bob Loveless from 1971 to 1972, with A.G. Russell as the Honorary President. Later the leading position was renamed "President." Bill Moran was the President from 1972-1973, Loveless again from 1973 to 1974, and Loveless was again elected in 1974, but he resigned to allow Ted Dowell to be President. Other Presidents have included Jim Nolen, Buster Warenski, Jimmy Lile, George Herron, Frank Centofante, D'Alton Holder, and Gil Hibben. A Knifemakers' Guild Show was first held in Kansas City, Missouri, in 1972, in conjunction with a gun show. Shows were also held in Houston, Texas, and Overland Park, Kansas. Bylaws for the Guild were crafted in 1971-1972 by Bob Loveless and Ted Dowell. The bylaws were approved by the general membership at the 1972 Guild Show[12]. However, things were not always smooth, and there were arguments between Loveless and Moran in the early years. B.R. Hughes wrote in 1986:

> Future knife historians should be aware that Loveless and Moran had little in common. Loveless was an advocate of the stock removal method of making knives using stain-resistant steels, while Moran was a champion of the traditional forged blade. There were

other, deeper differences. Loveless felt that quality should not be a condition of membership and that requirements should be as loose as possible to encourage new members. Moran was of the opinion that the Guild should have moderately strict guidelines and that members should be responsible to the Guild for unethical behavior... Be that as it may, the two, not friends by any means, 'buried the hatchet,' at least publicly, for the welfare of the Guild.[10]

For his part, Loveless said in 1992 that ethics were a part from the beginning[13]: "Our goals were to promote knives and knife makers ... to encourage more ethical practices and help others make better knives." And a code of ethics was included in the bylaws that Dowell and Loveless wrote.

Discussions between the Guild remained challenging as the organization became more structured at the 1974 Overland Park show where Lile said, "It took a stormy two-day session to accomplish what could have been done in thirty minutes."[14] They established a five-member board of directors with a president, vice-president, secretary-treasurer, and two directors. In 1975, they had the first independent Knifemakers' Guild Show which was not affiliated with a gun show, and the standard site became Kansas City in 1977. Dowell was the President at the time of the first all-custom-knife Guild Show[15]: "I had a good bit of opposition to that first show. People said it wouldn't work, but people found out we could do it." Dowell had a vision for the Guild that he felt wasn't being accomplished:

> I tried years ago to institute ... "A (system) of laws and not of men." That is to say, that everyone was treated the same and in a consistent, rather than an inconsistent, fashion. The Guild wasn't ready for that hence my resignation in my second term. Billy and Beverly Mace Imel were of a like mind later and, as secretary/treasurer, they have done a wonderful job of establishing that concept, and it is now pretty much in existence in the Guild, for which I am thankful.[15]

There were growing calls for stricter guidelines on what to do when a knifemaker participates in unethical business practices. For example, Jimmy Lile in 1976 said:

> Much to the dissatisfaction of the President and Board of Directors, past and present, of the American Knifemakers Guild, and of almost all its growing membership, its ranks contain knifemakers who engage in unethical practices and treatment of their customers. These practices are not sanctioned by the Guild and at the Guild Show and Meeting in Dallas, I will submit proposals for changing the bylaws to provide strong remedies for these situations created by makers who spend more time tooting their horn that they do at honoring their commitments and obligations... Not one of us can lay claim to never having been late on a delivery, always doing a perfect job, or never having a dissatisfied customer. We can hold these problems to a minimum by communicating with the customer, but the BEST way to handle them is to prevent them. This can best be done by the Guild policing its own ranks and screening its prospective members more closely and this can be done during the two year probationary period of the prospective member.[16]

B.R. Hughes, in 1986, credited Lile with[10] "turn[ing] what had been a loose-knit organization into an association operated on sound business and management principles. He was also instrumental in the Guild's requiring members to take a more ethical attitude toward its customers."

A.G. Russell Mail Order Knives and Morseth Knives

A.G. Russell began selling knives in 1968 and purchased Morseth Knives in December 1971[17]. Gordon Morseth and grandson Steve Morseth were running the company, but they were struggling to operate it, were low on cash, and had promised delivery of hundreds of knives to customers that had yet to be delivered. They contacted Russell for help, and he offered to purchase the company. Russell hired knifemaker Bob Dozier to make the Morseth knives, and they fulfilled the back orders and continued to sell new knives produced by Dozier. Steve Morseth taught Dozier how to make Morseth knives. They continued to make the knives with the laminated steel hardened to 64 Rc, though changes were made to the knives to fit Dozier's knifemaking style, and Russell refined the designs. A.G. Russell would design several knives for his company, including the boot knife Sting in 1975.

More Publications about Knives

A.G. Russell also developed relationships with several magazine writers, including Tommy Bish, Sid Latham, John Lachuk, Ken Warner, John Wootters, and B.R. Hughes[10]. Russell would give the writers the names and addresses of top makers, increasing the knifemakers' publicity. As a result, more articles were coming out about custom knives, and a broader range of makers and subjects related to knives were being covered.

Warner again wrote about knives in the 1971 *Gun Digest*[18] and covered his favorite knifemakers Randall, D.E. Henry, Harvey Draper, Moran, Bob Loveless, and then 11 more knifemakers Ralph Bone, John Nelson Cooper, Ted Dowell, Gil Hibben, Clyde Fischer, Walter Kneubuhler, Jimmy Lile, Harry Morseth, Rudy Ruana, Merle Seguine, and G.W. Stone. At the end of the article was "A.G. Russell's List of Knifemakers" which totaled 44 makers. Warner had an expanded section on knife steels, talking about Loveless getting 660 pounds of a new steel (presumably 154CM). He said that since 1966, "About the only material change is that stainless steel makes more sense now than it did then. There is evidence ... that the absolute top edge-holding ability still goes to non-stainless steels, but 440C and some other 400-series steels that are stainless have proven over the past 5 years to be very, very good indeed."[18]

By 1971 it seemed that the knife enthusiast's view of steel was changing, and more attention was paid to specific steel types. For example, in the December 1971 *Shooting Times* magazine, John Wootters wrote:

> Fundamentally, steel is the justification for a handcrafted working knife. Mass-produced blades must be stamped to shape, and stamping dies are expensive. If the factories used steel like these bench-cutlers', those dies would fail so quickly that the manufacturers couldn't compete.

[N]o one alloy has all the characteristics desired for even one kind of knife yet, much less for all the varied special-purpose blades being made today. So the knifemakers are in a continuous race to find super-pure, super-performance steels, and then to discover new ways to push these metals to new limits in heat-treatment and tempering. Most have their own secrets in heating, quenching, and drawing. Some do it by eyeball judgment of subtle colorings of the steel in the charcoal, while others use ultrascientific methods, vacuum ovens, supercooling, etc. Still others don't say. The result is the Superblade, a knife that will test as high as Rockwell C64 in hardness and still not be too hard to sharpen or too brittle to hold its edge.[19]

In 1973 there were entire books on handmade knives, such as *The Gun Digest Book of Knives* by B.R. Hughes and Jack Lewis, and *The Custom Knife* by John Davis Bates, Jr., and James Henry Schippers, Jr. These books featured profiles on knifemakers and many topics such as knife steel, different knife types, knife pricing, whether custom knives are "worth it," and Rockwell hardness testing. Hughes and Lewis polled twenty knifemakers for who they thought the best knifemakers were[20]. Loveless was voted number one and Moran as number two, followed by an "appreciable span" to a tie for third place between D.E. Henry and Corbet Sigman. Ted Dowell, Lloyd Hale, and Ron Lake shared sixth place, followed by Clyde Fischer in eighth. Blackie Collins and Bernard Sparks tied for ninth, and Gil Hibben was number eleven.

In 1973 the most popular knife steels were O1, A2, W2, D2, O6, 440C, and 154CM[21]. Most knives then were hardened in the 56-60 Rc range depending on the steel and the knifemaker[21,22]. Most knifemakers were advertising the steel they were using and the hardness range they were heat treated to[23].

Bob Schrimsher, O6, and Knifemaking Supply

One puzzling inclusion in the list of most popular steels in the previous section was O6 tool steel. This steel was patented in 1933 by Frederick Bonte of Timken[24] and marketed as Graph-Mo, the "Graph" being short for "graphitic" and the "Mo" referring to molybdenum. The steel is high in carbon (~1.5%) and silicon (~1.0%); the silicon destabilizes the iron carbide leading to the formation of graphite instead. The free graphite in the steel makes it easier to machine. The molybdenum addition in Graph-Mo gave it higher hardenability so the steel could be oil quenched in large sections. However, the free graphite isn't always a positive, as Bob Loveless reported that Graph-Mo could look pretty good if finished to 600 grit, but was not suitable for a polished knife[25]. Moran made several knives using O6 and liked it for its cutting ability, but he didn't like that it would break when bending[26]. Graphite is a benefit for machining and grinding but not for final performance or finish, so it is an odd choice to have been so common in the early custom knife industry.

Looking at the profiles of the approximately 89 makers listed in the 1973 *The Custom*

Knife by Bates and Schippers, at least nine knifemakers reported 06 as one of their primary choices[23], more than 10% of the knifemakers. Still, I haven't found any claims from knifemakers that they were the first to introduce it. How did it become so popular? The answer comes in a 1973 *Blade Magazine* advertisement for Bob Schrimsher's Knifemakers Supply, which advertised the steels 440A, 440C, 154CM, and 06. The influence of knifemaker supply companies is sometimes underestimated regarding what materials they use. It isn't just the tale of the knifemaker searching for and discovering a new steel for knives. Eventually, he (or she) has to buy that steel. Many steels are standard in custom knives simply because they are easily accessible from knifemaking suppliers. A.G. Russell said[27], "(The suppliers) are important because they buy material in quantity. (In the old days) a knifemaker couldn't find certain things in small quantity. In the 1960s, I had to buy 2,000 pounds of 440C steel just to get it. There were no hardware stores carrying steel like that, so I bought it and solid it to knifemakers." Russell continued[27]: "I was supplying steel and handle materials to knifemakers simply as a service to them and literally at my own expense. Bob Schrimsher called one day (in 1969) and told me he going into the knifemaker supply business. I said, 'Wonderful! Here's my list of suppliers and customers. You do it.' Bob was simply too good to everybody."

Many knifemakers reported that Schrimsher was very generous. Bob Dozier said[27], "Without Bob, a bunch of makers would never have gotten started. I've been into Bob for as much as $2,000 back then, and you would pay him as you could. He went out of his way to do lots of things for knifemakers. He was the epitome of this. He was the best and did more for makers than anybody else." Blackie Collins discussed meeting Schrimsher around 1970:

> When I first met Bob I called him and told him about some of the supplies I needed. He invited me to come to his place in Emory. I thought I was in a knifemaker's mecca because in the back of the warehouse were stacks and stack of things that, to me, were like gold - Micarta, ivory, stag, every kind of steel, every kind of brass, everything [a knifemaker] could need... [O]ne thing that all knifemakers had in common in the '60s and '70s was a lack of money. All of us were sorta [broke] - with a few notable exceptions. [But] Bob said just take what you need and I said, "I'd like to but I can't afford what I need." He said, "Pay me when you get the money," and I was startled; he didn't know me. But I came to find out that Bob was like that to just about anybody. He would let you go into his warehouse ... and pick out what you needed, and make a list of what you wanted, tell him what you

were getting, pay him what you could and pay him when you could. I probably owed Bob a lot of money for a long time, as did ... a lot of [knifemakers].[28]

With Schrimsher's success, there were soon many competitors, of course. As a result, by 1986 there were several knifemaking supply companies including Sheffield, Koval, Holt-Sornberger, Atlanta Cutlery, Knife & Gun Finishing, and Texas Knifemakers Supply[29].

Decline of the Guild

The Knifemakers' Guild was the most influential knife organization throughout the 1970s, 80s, and 90s. Nearly all the top custom knifemakers were "voting members" of the Guild. Membership peaked around 1999-2002 at about 500 members, but had dropped to less than 150 members by 2018. However, the current Board is attempting to revitalize the organization. The Guild had the only knife show through most of the 1970s, and the show remained the biggest available for years after. Guild membership was required to attend. However, even the 2000 Guild Show was significantly slower than in years past, as Steve Shackleford said[30], "Putting it kindly, the most recent Knifemakers' Guild Show in New Orleans was not one of the organization's more stellar events in terms of overall knife sales." And Shackleford wrote an article that year called "5 Ways to Recharge the Guild"[30] with suggestions on strengthening the organization. I asked knifemaker Bob Terzuola what happened with the Guild:

Basically, it can be summed up in one word: technology. One [rule of the Guild was] you had to have a catalog... The reason you had to have a catalog, at that time there was no internet, no shopping online. Mid 80s there really wasn't any way of buying knives apart from the mail, magazines, and going to shows. That was basically the only way to buy a knife... It was still mail order, telephone, and you had to have a catalog... The Guild Show suffered diminished attendance. Shows have become expensive, and you no longer need Shows to sell knives with Instagram, Facebook, and the Internet. Things like the catalog started to disappear. There was no point. Knifemakers had sites for people that wanted a knife and two days later you get the knife, don't even have to go to a show.[31]

An essential element of the Guild was that the knives were required to be "handmade" for the knifemaker to be a member, and an increasing number of knifemakers did not fit the Guild's definition. Terzuola said:

What really caused the decline of the Guild, there were CNC machines that were fairly inexpensive. You could get one of these machines in your garage, not much bigger than a couple of refrigerators. The advantage of having CNC is you can make 100 of the same knives and they're all exactly the same. Keep your costs down. Don't have to be locked into one design because a stamping die can only make one thing. The younger guys coming into knives started doing this. There was a big controversy. If you are running the CNC machine, are you really doing your own blade? It became too much of a hassle.[31]

-18-

A NEW GENERATION OF FACTORY KNIVES

Gerber

Joseph R. Gerber wrote about Gerber's origins in 1960:

> One day during the Pendleton Oregon annual Roundup in September 1939, in the saddle shop of Hamley and Co., I was idling away a little time, waiting for Lester Hamley to take me to lunch. My mother used to remind me that "an idle mind is the devil's workshop" and I should have remembered. But instead I idly picked up a workman's knife shaped like a single bladed cabbage cutter. Scraps of beautiful saddle leather were lying around, I suppose it was the whittler's instinct that set me to slicing leather, but I grew so enthusiastic that I decided to have a carving knife for myself that would cut like that. The man who made Hamley's leather knives lived in Salem. I found an excuse to go there, but he claimed to be overworked and under-interested. I kept looking and found a man who would grind a carving knife for me (David Z. Murphy). The knife proved an excellent cutter, being of such terrific alloy quality, but its design smacked of a horned toad. Our Art Director (Dean Pollock) and I set out to make a design that was artistic as well as sound and functional. The design we settled on was accepted with enthusiasm. At Christmas in 1939 we sent out 24 knives as gifts to friends. The day after Christmas these recipients began asking "How can I get these knives for my friends?" I hadn't thought of selling them commercially, but one of my friends, Bill Haskell, then of the New York Herald Tribune, took a packaged knife to Abercrombie & Fitch and came back with an order for "all you can make until further notice." Abercrombie and Fitch have always been our biggest outlet and still are after 20 years.[1]

David Murphy said[2]: "That crazy S.O.B. says he can take three of my knives and put them in a wooden box and sell them for $25. If he's that crazy, I'm just crazy enough to make them for him." Murphy became unhappy with the relationship at the end of 1940, and then began making his Murphy Combat Knives at the end of 1941 (see Chapter 11). After Murphy left, Joseph Gerber restarted the knife company in 1945 with his son Francis "Ham" Gerber as president. Ham named the company Gerber Legendary Blades and named the various knife models after legendary swords like

Excalibur, Balmung, and Joyeuse. Ham said[3]: "Gerber Hand Made Blades, with the help of a manysided craftsman named Otto Nedvidek, got back into the operation in late 1945, doing most of the physical work upstairs over a plating shop at S.E. 10th and Lincoln, and shipping from S.W. 12th and Jefferson, the address of the Gerber Advertising Agency. From 1947 to 1951 the factory was on N.W. 29th Avenue in Guild's Lake, in a defense housing duplex and then also in a built-on annex. From the fall of 1951 through January, 1966, the Willamette district of West Linn was the home of Gerber Legendary Blades." In the early days, Gerber continued to use recycled T1 steel blades like Murphy[3], making the knives "out of old Power Hack Saw Blades, made of an apparently magical stuff called high speed tool steel." They used them in the hardened state, grinding away the teeth and the remaining steel into the final blades. This lasted until roughly 1948, when they had to move on because newer hacksaw blades were made with hardened steel welded to a mild steel body[2].

Ham, Joe, and Pete Gerber.

Gerber Legendary Blades then switched to M2 high speed tool steel supplied by Crucible Steel[1]. Joseph Gerber said[1], "The basis of these knives is the steel they're made of. We have never changed the formula. It is the accepted standard for more uses than any other tool steel. We have learned to plate and polish it until it is bright, and stays that way. Better than stainless, it takes and holds the finest cutting edge - a quality badly lacking in 'stainless' steel." The early advertisements from Gerber almost always promoted that their carving knives were made with superior steel to any other available knives. They began chrome-plating their blades in 1950[4]. The main knives that Gerber sold early on were carving knives with cast aluminum handles with fancy wooden boxes to hold them in. However, sporting knives were also released early on, including a hunting knife with a modified carving blade with a handle designed by Thomas Lamb in 1947, and another hunting knife in 1948 designed by outdoorsman Hale Woolf of Montana. When younger brother Joseph R. "Pete" Gerber, Jr. graduated from college in 1951, Ham wanted to move into the company's research side, leaving Pete as the company's president.

"MIMING"
Gerber Steak Blades

MIMING, the Gerber individual steak blade, is beautifully designed and hand made from the greatest cutting steel ever known, anywhere, anytime, by anybody. And how it cuts meat! A Miming blade at your guest's plate makes meat, fish, or fowl seem more tender and tasty. Gerber blades' functional grace and clean design enhance any silver pattern.

In sets of 4, 6, 8 and 12, in solid hardwood cases. Be sure you get Gerber blades—the originals. At famous stores. Write for selection sheet showing all styles of Gerber carver and steak blades and Ron, the holding fork. Address Gerber, 1311 S.W. 12th Ave., Portland, Ore.
M8 set $29.50

1950 ad

A good man deserves a

GERBER
Legendary
BLADE

ERD set $25
Includes Excalibur, 11" carver; Ron, the holding fork; and Durendal, 6" boning blade.

Carving a good roast is one of life's high spots —IF you have a really fine blade. Gerber Blades are bringing back the nearly lost art of home carving. Hand made and individually hand-finished from the hardest, toughest, most expensive steel ever used in cutlery, they take a better edge and hold it far longer. Their perfect balance and exquisite design make them the choice of discriminating people everywhere.

There are 6 great blades and a holding fork in 16 set assortments, in hardwood chests from $8.50 to $80. At leading stores, or write for free catalog to Gerber, 1307 Southwest 12th Avenue, Portland, Oregon.

1950 ad

Balmung Lives Again!

the blade that slew the dragon . . . now carves your roast

The fabulous steel of Siegfried's blade "Balmung" could not have been the equal of the steel from which Gerber Legendary Blades are made. This modern steel, the product of metallurgical science, is the hardest, toughest and costliest steel ever used in cutlery. Hand-made Gerber blades take and hold a sharper edge, give a lifetime of superior service. A real satisfaction to use—and to own.

For carving medium size roasts or hams, Balmung, 8½" blade is supreme. Other sizes are Excalibur 11" and Durendal 6"; Snickersnee, the sensational server-carver; Ron, the holding fork; and Miming, the individual steak blade. At famous stores, in assorted sets from $8.50 to $80. Write for selection sheet showing full size illustrations of each blade and 16 priced sets in hardwood cases. Address Gerber, 1309 S.W. 12th Avenue, Portland 1, Oregon.

BD set $16

GERBER
Legendary
BLADES

1950 ad

A knife is only as good as the steel in its blade...

The steel from which Gerber knives are made is considered far too good to use in cutlery . . . so we use it anyway. We like the way it cuts and holds an edge that won't feather; and the way it wears and wears. Especially we like the way it makes all meat seem more tender.

Nearly one fifth of the weight of our steel is tungsten, molybdenum, vanadium, chromium and carbon. These costly alloys explain in large measure the outstanding characteristics and performance of Gerber knives.

With an occasional touch-up on "Gungnir," our sharpening steel, or on an electric sharpener, you can have really sharp blades all the time. And our Siegfried finish insures permanent brightness.

There are 19 sizes and shapes of Gerber carvers, dining knives, kitchen and hunting knives in 15 set assortments and in individual walnut scabbards or cardboard gift boxes.

It is time to usher in a renaissance and to restore cutlery that cuts like Siegfried's blade of legend. Won't you join our crusade for better knives? See your favorite store or ask for free copy of our Selection Catalog.

Write Gerber, 1305 S.W. 12th Ave., Portland 1, Ore.

"EXCALIBUR"—11½" Carving Blade in walnut scabbard, #14

1960 ad

New from Gerber
FOLDING HUNTER

The most advanced folding hunting knife for outdoorsmen who care about quality. Safety-featured to do tough jobs.

SAFETY CLOSING
Blade will not spring or snap shut. Safety pause at 45°.

POWER-GRIP HANDLES
Shock-proof duPont Delrin® or solid black walnut. Slip-free grip.

BLADE CHOICE
Ultra-sharp high alloy tool steel or corrosion-resistant 440C stainless.

as illustrated (hand checkered black walnut handle and cowhide scabbard)

GERBER
Portland, Oregon 97223
See your sporting goods dealer

1968 ad

GERBER

TROUT & BIRD

NEW for fishermen and bird hunters

Specially designed for cleaning trout, bass and other game fish. Equally effective for game birds. Narrow 3¼" blade has tip sharpened to double-edged point for making the initial incision and cutting either direction. Spoon on handle strips out blood sac along backbone. Extremely hard 440C stainless blade (Rockwell C57-59). The only blade that's better is Gerber high speed tool steel—but 440C is rustproof. Long, solid-cast, non-slip Armorhide® handle gives sure grip, even when your hands are wet. With cowhide belt scabbard **$8.50.**

See the complete line of Gerber Outdoor Knives at sporting goods stores.

GERBER • PORTLAND, OREGON 97223

1970 ad

When they built a new factory in 1966, they were able to start expanding their product offerings in new directions. The first new knife after the opening of the new factory was the Mark II[4,5], which was a combat knife based on a sketch from Army Captain C.A. "Bud" Holzman, an evolution of the Fairbairn-Sykes and the Case V-42, using a drop-forged blade. The steel in these knives was L6 tool steel, Carpenter's Reading Double Special steel[6]. They also added a pocket knife, the Fh or Folding Hunter, in 1968, and a series of fishing knives in 1968-1969. These knives were also advertised with 440C blades. While they were not the first knife company to use 440C, this is the earliest time a knife company advertised the steel by its designation: 440C. In addition, this may be the earliest example of when a specific steel type was named in advertising by a significant knife manufacturer. Before this knife, companies would refer to steel under generic terms like "highest quality" or perhaps as specific as "English Crucible steel." While advertising specific steel types is commonplace now, it makes sense that the knife companies would assume that consumers would not know the names and compositions of various steels. Perhaps the advertisement of 440C by name was an influence of the growing custom knife market.

Pete Gerber talked about how the custom knife market affected their business in 1976[7]: "The custom craftsmen have made people realize fine knives aren't cheap and they have also made them aware of quality in knives." Another key person in the Gerber team was Al Mar, who was hired as a "design chief" in 1967 after his prior experience working for an industrial design firm in Los Angeles[7]. Mar was a hunter, and he also visited knife shows, talked to custom knifemakers, and exchanged ideas with them. In 1974 Mar spoke about his ideas behind product lines he was developing for Gerber[8]: "Clean, sharp, functional lines work best with a knife. After three years of research we've come up with a line of knives we feel is designed with the times, easy to manufacture and aesthetically pleasing to the eye." Gerber also introduced knives with designs from custom knifemakers, such as the Paul knife by Paul Poehlmann in the late 1970s or the LST (light, smooth, tough) by Blackie Collins in 1981[5].

Kershaw, Silver Knight, and Al Mar Knives

Pete Kershaw, a Gerber salesman, decided to start his own knife company in 1974. Kershaw said[9], "What I had in mind, was to make and sell the Nikon of pocketknives." Kershaw began a partnership with Kai Cutlery and was purchased by Kai in 1977. Kershaw knives were initially made by Sakurai Cutlery (later called Moki), and in 1982, 75% of Sakurai production was for Kershaw knives[10]. Kershaw knives were initially marketed in 1976[11] as being made with "Kai Stainless steel tempered by an advanced sub-zero quench treating process." In 1978 they advertised[12]: "High carbon Kai Stainless Steel blade from the AUS series is an improvement over famous 440-C." These steels were AUS-6 for folding knives and

AUS-8A (see Chapter 35) for fixed blades at a hardness of 57-59 Rc[13]. Kai later built a Kershaw factory in the USA in 1996[14].

Gerber introduced a line of knives called the Silver Eagle in 1977, though Trailways Bus Co. sued over the use of the name, and Gerber changed to Silver Knight[5]. Pete Gerber wanted to exploit the favorable exchange rate between Japan and the USA. Pete Gerber contacted Gerber's Japanese distributor, Sake "Stanley" Wada of Osaka. Wada contacted Susumo Sakai of Seki City. They agreed, and Sakai and his five sons built a new factory in Seki called Gerber-Sakai. The name was shortened to G. Sakai in 1982.

Al Mar left Gerber knives in 1979 to start his own imported knife company. Many of his designs were too radical for Gerber, leading to his decision to start on his own[4]. Of his early knives, the "better" ones were made with RS-30 steel which Mar compared with ATS-34[15]. This steel had 1.1% carbon, 14% chromium, 0.6% molybdenum, and 0.2% vanadium. By 1984 he had knives made in ATS-34, the first factory knives available in the USA with the steel[16,17]. Mar designed many knives and worked with custom knifemakers[18]: "Custom makers - the really good ones - are always experimenting, always taking a risk or two. There's a bridge between production and handmade knives; you just need to find the right people to cross it with." Mar had his knives produced in Japan out of necessity[18]: "There simply isn't anyone well funded enough to take on the challenge of building good product here (USA). Or, they're satisfied with being average, with making just what they've made for the last half a century." G. Sakai made most of Mar's knives, with whom he had established a relationship while at Gerber.

Buck

Before the 1960s, H.H. Buck & Son was still a handmade knife company, producing about 25 knives per week. Al Buck said[19], "Looking back, it's hard to believe we survived. It was really a Ma and Pa operation. I made the knives. Ida (his wife) handled the book. When he got older, Chuck (their oldest son) chipped in whenever he could. We sure weren't General Motors, and that's for sure." The company struggled to make money, and the Bucks fell into debt. Al was discouraged[19]: "I came within a gnat's eyebrow of quitting. I was working day and night, and for what? It was all very discouraging, believe me." In the fall of 1960, Al's pastor, Robert Wilson, noticed that Al was struggling and offered to find help. Wilson spoke to Howard Craig, a quality control manager at Ryan Aeronautics, and also part-owner of Precision Metals, a small company that

Chuck Buck

performed custom welding of airplane parts. The three met and decided that Buck Knives should incorporate and sell stock to build a factory to mass-produce knives. They filed for incorporation in April of 1961 and created a board of directors with Al Buck as the president. In the beginning, they managed to sell shares to only five investors. Al said[19], "As you can tell, we didn't exactly have people lining up to buy shares of stock in our company." The board consisted of son Chuck Buck, Howard Craig, Don Ham, chief accountant for Precision Metals, and Bill Kupilik, another manager from Ryan Aeronautics. Howard Craig became director of manufacturing and engineering, Chuck Buck was in charge of production and personnel, and Don Ham served as CFO and later the director of sales and marketing. Kupilik said[19]: "The key, from the very beginning, was the cohesiveness of the board. We all got along well. We all cared for each other. Most importantly, we all cared for the company, and we never lost sight of our goals."

BUCK KNIVES

·FAMOUS FOR HOLDING AN EDGE·

An honor product of three generations of knife making and tempering.

H. H. Buck developed the Buck blade at the turn of the century. The process for tempering it has been handed down as a family secret for three generations. The amazing ability of the knife to hold an edge is due to the high carbon content of the steel and the tempering process.

Knives are equipped with phenolic handles in ebony and are specially designed to fit the hand. This material is impervious to heat, cold and shock and is practically indestructible. Guard and trim are made of tempered dural and hard fiber. A dural butt is used for lightness and balance. All knives are equipped with a sheath of genuine saddle leather.

A superior knife, like a fine gun, rod or reel, is a once in a lifetime investment. The initial price of a Buck Knife is small when you realize that this knife is guaranteed for life and is a masterpiece of craftsmanship. Buck Knives are made of steel of the highest quality. This steel has stain resistant qualities, an amazing temper that only a quality-priced knife can be crafted of it. The hand work and special detailing that goes into a Buck Knife makes each one a masterpiece of skilled workmanship. Buck Knives bring to their owner the pride of ownership and prestige that comes from possessing only the finest.

1965 Catalog

Their new headquarters was a prefabricated steel building in San Diego. They were producing six models of hunting knives with black handles. With increased production, they could no longer make the knives from used files and instead began using 440C stainless steel. The steel name was not mentioned in the advertising, but the promotion material said, "Buck knives are made of steel of the highest quality. This steel has stain resistant qualities, an amazing temper that only a quality-priced knife can be crafted of it." Buck's knives were more expensive, $12-20, when the competition was selling knives for $2. Al recalled[19]: "For a while there, we were the laughing stock of the industry. Our competitors were taking bets on how long we'd last. We were convinced, though, that people would pay additional money to get additional quality. And eventually, that proved to be true." The company lost money for the first year, and Al and Ida Buck spent two months in 1962 visiting potential dealers, 250 total, and 100 of them submitted orders. Kupilik said[19]: "That sales swing by Al really set us apart from the competition. Here is the president of the Buck knife company, Mr. Al Buck himself, walking into the store to chat. On that one long trip, Al showed retailers that we were going to be different from the other major knife companies."

Buck Knives also began to advertise, using an illustration of a knife cutting through a bolt with a hammer. Al said[19]: "The ad wasn't fancy, not by a long shot, but it really made an impact on the general public. At the time, no other knife could cut steel. Most manufacturers intentionally made their knives soft, so they could easily be sharpened. We wanted to produce a knife that didn't need to be sharpened very often." Buck also offered a Lifetime Guarantee to repair or replace any Buck knife that failed due to manufacturing issues. Al said[19]: "It was risky, and I think all of us were concerned that people would try to take advantage of us, but the Lifetime Guarantee turned out to be a very good thing. We were telling people how good our knives were. The guarantee proved we were willing to put our money where our mouths were. Even more important, it kept us on our toes. With that guarantee in effect, we knew we couldn't afford to produce any second-rate knives. It was the best tool our quality control guys ever had!"

Earliest known Buck Model 110.

Buck Knives developed several new products in 1963, which went into production in 1964, including a Buck Bowie, a fishing knife, and a line of kitchen knives. The most significant new knife was the Folding Hunter, named the Model 110. Al Buck[19]: "We went out and bought three of the rival knives. We handled them, used them, tried to get a feel for them. Then we took them apart to see how they worked. In the end, we agreed that all three had features we liked and some features we didn't like. And we didn't like the looks of any of them. So we asked one of our engineers (Guy Hooser) if there was a way to get all of the good things into one good-looking knife." Sales of the 110 grew significantly during 1965-1967, Al Buck said[19]: "We introduced other new products during that period, but the best seller, by far, was the 110, and the spinoff models of the 110, such as the Model 112 Ranger." While large lockback folders had existed historically, Buck knives popularized the lockback in modern times. This knife became a standard in the industry, and competitors widely copied it.

From the introduction of 440C steel in 1961, the blades were forged, but in 1967 they began blanking the blades[20]. To blank the 440C blades, they heated the steel sheets to 700-750°F before punching them out, which was necessary as the relatively coarse microstructure of 440C would otherwise cause cracking during blanking[21]. Universal-Cyclops made the steel they were blanking in Pittsburgh[21]. In 1981 they switched to Allegheny Ludlum 425 Modified stainless steel, with 0.52% carbon, 13.5% chromium, and 1% molybdenum. Chuck's son, CJ Buck, said in 2003[22]: "440C held a great edge

but was prone to rust, pitting and breaking half moons out of the blades of users. In 1981 we switched to 425M. It was a hard choice as our brand was built on the edge performance. The little we lost in edge performance we more than made up for in ductility and corrosion resistance. And we gained the ability to Fineblank which turned out very precise parts and better action in our folders then we had ever gotten." The lower carbon and chromium content of 425M gives it a much finer microstructure which allows the more accurate process of "fine blanking"

CJ Buck

without cracking of blades and less die wear than with a steel like 440C. This switch in steel occurred shortly after Buck opened a new factory in El Cajon, California, in August 1980. Another new part of the factory was a heat treating facility built for Paul Bos, who had previously heat treated Buck knives from the commercial heat treating company where he worked. Bos had grown up with Chuck's brother, Frank, and Bos first started heat treating knives that Frank made from O1 steel[23]. Bos purchased Star Heat Treat in 1969 and was a major heat treater for companies like Boeing and Lockheed in addition to Buck Knives and custom knifemakers. Bos helped Buck set up their own heat treating facilities in the late 70s, and Buck Knives and Bos agreed that Bos's custom heat treating could be performed at the Buck facilities. This allowed Bos to focus on knives and sell his previous business (see Chapter 19 for more on Bos). In 1994 Buck switched from 425M to 420HC, which has 0.44% carbon and 13% chromium. With the high-quality heat treating of Bos, they could maintain a 58 Rc hardness target through the switch to 425M, and finally to 420HC.

Schrade-Loveless, Factory Collaborations, and 154CM

In 1972 Henry Baer (Uncle Henry) of Schrade Cutlery talked to A.G. Russell about a knife for the Knife Collectors Club. Russell suggested that Baer call Bob Loveless, and Schrade decided to collaborate with Loveless on a design[24]. This was the first of the custom knifemaker-factory collaboration knives. Loveless visited Schrade, toured the facilities, and met with Dave Swinden, Schrade Vice President and general manager, and Bob Ernst of sales. They came up with a design that could be produced at the factory. Loveless later said that factory collaborations were a win-win for the consumer[25]: "There is still far more demand for one my knives made right here in Riverside than both my business partner and I have the time and energy to build. I've designed knives in collaboration with other cutlery firms... The work was primarily so that the average guy can own a Loveless-designed knife at an affordable price. That's a 'win, win' for both the customer and myself."

They decided to use 154CM steel, which Loveless had recently introduced into the custom knife market, and this would be the first factory-produced knife in the steel. It took three years for the knife to be released in 1975. Dave Swinden said:

> Technical problems were the prime factor in production. Any new item will give problems, but in this case we were working with an exotic steel, 154CM, and it meant re-educating our people and learning different methods of assembly. A one or two man custom shop can do things it's impossible for a large factory to do. Heat treating is one example. We didn't have the capability of heat treating 154CM and went to Benedict Miller of Lyndhurst, New Jersey. They do excellent work and have a staff of highly skilled metallurgists we could turn to as problems arose. We didn't feel awkward in going to an outside company. After all Bob Loveless has his heat treating done by specialists and 154CM is a highly complex steel.[24]

Ultimately they found the 61-62 Rc blades to be brittle, and after discussions with Loveless they dropped the hardness to 58-59 Rc. They also had difficulties in blanking, drilling, and machining. The steel had to be heated before blanking, or it fractured. They had to use cobalt-alloyed high speed steel drill bits (see Chapter 7). Dave Swinden said[24], "If we were only making a couple of hundred knives we wouldn't go to that expense, but we're talking of many thousands. Everything had to be changed; drill speeds, grinding wheel motors had to be setup to gain a few more hundred RPMs and we even made a special soldering stand to do the delicate job of soldering the brass guard to the blade." Schrade could not work with Micarta, one of

Loveless's favorite handle materials, because it needed to be cut and shaped, so they used Delrin because it could be molded. Schrade usually did all of the blade grinding after heat treating, but because of the high wear resistance of the material, they did some of the machining before heat treating. They called Loveless several times to discuss production issues, as Swinden reported[24]: "Loveless is a knowledgeable metallurgist and, after all he was the knife's designer, and we felt no hesitancy in calling him for suggestions."

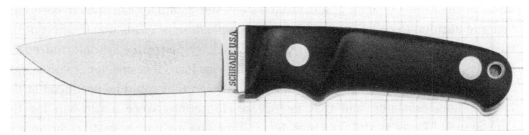

Schrade-Loveless hunter. Weyer photo.

Ultimately Swinden wasn't sure the knife would make money:

> There is no way this venture can be profitable from a strictly dollars and cents point of view. It was an exercise in education. It was also an exercise in finding out if a good so-called custom knife can be produced in a factory. We found this out, we can do it and our education will eventually pay off in upgrading all of our knife products. The workmen now know what is expected and is demanded of a knife that can command a high retail price in the market. For future business and future knives it will be something that could be very profitable for us.[24]

Initially, they wanted to sell the knife for $50, but after the many difficulties, the knife sold for $100 ($550 in 2023), and 3,942 produced[26]. Swinden said that they also learned lessons about how to do collaboration knives in the future[24]: "We'd have to work more closely with the artist right at the very beginning. We'd have to work at the design stages, look at the methods we now have that can be adapted to designs and methods of construction and manufacture. It would be less painful for everyone." Loveless said[24]: "I'm reasonably happy about the knife. They made some minor changes, and I suppose every factory would have to do the same, but in general I'm delighted and I think it will make one helluva deer hunter's knife." Future Schrade knives did not use 154CM; even the future Loveless collaborations were in 440C[27]. Crucible told Schrade that they would no longer make 154CM (see Chapter 39 to learn about the return of 154CM). The Schrade-Loveless knife, however, was the beginning of many collaborations between custom knifemakers and knife factories, both from Loveless and other custom makers.

-19-

FURNACE HEAT TREATING

Cryogenic Processing and Cold Treatments

Metallurgists as early as 1890 discovered that high nickel steels were non-magnetic at room temperature[1]. French metallurgist Floris Osmond reported in 1899 that if these high nickel steels were immersed in cryogenic "liquid air," they would become magnetic again[1]. Typically, steel transforms to non-magnetic austenite at high temperatures, such as 1350°F/730°C for simple carbon steels. If slowly cooled, the steel transforms to magnetic, soft, ferrite, and if rapidly quenched, it transforms to magnetic, hard, martensite. Nickel lowers the temperature at which martensite forms so that at room temperature, there are still large amounts of non-magnetic austenite, called "retained austenite." Osmund discovered that these steels had low martensite transformation temperatures, so the low-temperature liquid gases could cool the steel sufficiently to transform to magnetic martensite.

Although knife steels do not contain large amounts of nickel, other alloying elements can also stabilize austenite and reduce the transformation temperatures of martensite. The most significant element is carbon, and most knife steels are high in carbon. Chromium does not have the strongest effect, but when present in large amounts in stainless steel the effect adds up, especially in combination with high carbon for high hardness. This can make it more challenging to avoid retained austenite without precise control over the high-temperature austenitizing temperature used before quenching. One way to prevent excessive retained austenite is to use lower austenitizing temperatures so that less carbon and chromium are put "in solution," from dissolving chromium carbides. But higher carbon and chromium in solution are desirable because that increases corrosion resistance (Cr) and hardness (C). There is a maximum usable austenitizing temperature because the hardness will go down from excessive retained austenite above that point. An example chart from D2 steel is shown on the following page. The peak hardness of 65 Rc was reached from 1885°F/1030°C when the retained austenite was about 15%.

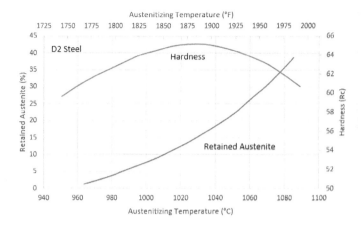

Data from Ref. 2.

The first company to use cold treatments in knives was Henckels, which patented a rather creative method in 1939[3]. They heat treated the stainless steel at a high temperature, 1100-1200°C (2000-2200°F), so that after quenching, the steel would contain a high content of retained austenite and thus have low hardness. The steel would then be machined in this soft condition. The steel was then cold treated at -50 to -70°C (-60 to -95°F) to convert retained austenite to martensite. Henckels also stated in their patent that the increased carbon and chromium in solution would improve edge retention and corrosion resistance. Henckels began to use "Friodur" branding for this process in the 1950s.

Robeson cutlery began advertising their "frozen heat" knives in 1949, which were given a cold treatment at -100°F/-70°C. This procedure was introduced by Emerson Case of the W.R. Case family, who was hired to save the floundering company in 1940. These steels were also austenitized at a relatively high temperature of 1100°C/2010°F before quenching in oil and cold treating for increased hardness vs. conventionally heat treated knives[4]. The knives were even guaranteed to stay sharp for three years.

Commercial Furnaces

Knives have been heat treated in a wide range of different furnace types. The most common method has been standard box furnace heat treating, often in batches. The knives are placed in a basket, heated to the target temperature, and then dropped into an oil quench bath. Another common type is the continuous belt conveyor furnace, where knife blanks are placed on a mesh conveyor belt that slowly takes the knives through the furnace and then are discharged into an oil quench. These furnaces are best for knives of similar size as the time at high temperature needs to be closely controlled. The earliest reference to a belt furnace I found being used for knives was from 1920[5], though they probably existed earlier. These furnaces could be heated by oil, gas, or electricity, and used with or without a protective atmosphere.

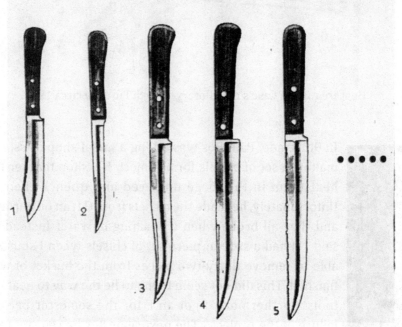

1949 Robeson Cutlery ad.

Some companies were faster than others to incorporate new technologies. Case Knives used lead bath furnaces to heat treat their knives until the late 1940s when they started using a General Electric "rotary-hearth" furnace which has a circular rotating table on which knives are placed. This furnace was atmosphere controlled for reducing scale and decarburization. The batches of blades would be manually removed after traveling around the rotary furnace and then quenched in oil[6]. They also began using induction heating to soften the tangs of their knives, which was needed for their stainless knife production.

Heat treating in Case's new Rotary-Hearth Furnace circa 1948.

Paul Bos

In 9th grade, Paul Bos was taking a metal shop class, and he was making a set of chisels for a project. His shop teacher told him to heat them in the forge until red and quench them in water. Unfortunately, he made the chisels from O1, an oil hardening steel, and they all broke when quenching in water instead of oil. Bos said[7], "I had a sixteen-piece set of chisels when I started and was able to remove thirty-two pieces from the bucket of water when finished. This did not seem to me to be the way to heat treat metal tools. Furthermore, I got an F for the semester because of the failure of the project." The next semester he repeated the project

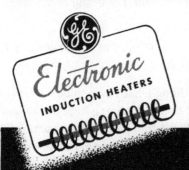
1949 General Electric ad about W.R. Case furnaces.

but went to Star Heat Treat and Fishing Supplies in San Diego for advice. There he learned that he should quench the O1 in oil to avoid cracking. Bos enjoyed this experience of learning to heat treat so much that he went to work for the same commercial heat treating shop while in high school for no pay.

Bos is an enthusiast of drag racing, and one of his racing buddies was Frank Buck, son of Al. Frank was making some custom Bowie knives in O1 steel and asked Bos to heat treat them for him, and Bos did so at night after he finished the regular commercial work. In the late 1960s, Bos met Al Buck, and he began heat treating for Buck Knives. Bos also met locals Bill Duff and Don Collum of Rigid Knives, and he would trade heat treating for knives. In 1969, Bos and a partner started Certified Metal Craft, and they began heat treating for companies like Boeing and Lockheed.

Bos was heat treating for Rigid and Buck and had heard of some custom knifemakers in Arizona and California. However, he did not start heat treating for custom makers until he visited a knife show with Bill Duff, where he also met Jerry Poletis and Jody Samson. In a short time, dozens of makers were sending Bos knives for heat treating. Bos said[7], "I even had a couple of makers send me their blades that had been treated by somebody else but had been warped in the process. This kind of blew my mind; to think that they had enough faith in my abilities to trust me to straighten out the blades after they had warped. I began to realize that there may be a full-time occupation in heat treating custom or handmade knives only. My hunch soon proved correct."

Bos would send letters to several knifemakers introducing himself and his heat-treating services. In 1982 he said:

> The word of my work seems to be getting around rapidly. I had originally sent out no more than ten letters of introduction to those makers I admired so. It wasn't long until I was receiving five or six packages of knife blades per week, many of them from people of whom I had never heard. I guess word of mouth was the answer. They all told me they had seen somebody else's knife that I had treated and wanted to give me a try. One maker, for instance, stated that he would give my work a try but if he didn't like the work, I would never hear from him again. It wasn't but a week after I sent back the blades that I received the second package and have been doing his blades ever since. I'm proud of the long list of famous makers who rely on my work.[7]

Another story Bos has told was about Bob Loveless, who he met at the original 1972 Guild Show. Bos told a memorable story about Loveless:

> [Loveless] made it possible for me to make a living doing what I love to do - heat treat custom knife blades. I introduced myself to Bob and told him I was a heat treater and I would like to heat treat his blades, and that I would do the first batch for free, which I did for all the other makers. Bob said, "So you think you're a good heat treater - ha!" I replied, "Yes sir, I do." A few months later Bob called and said he wanted to bring a few blades down to me for heat treating. He knew I was located across the street from Gillespie Field [an airport near El Cajon, California]. He said to meet him there in an hour. I went over

and waited for him. It wasn't long before I saw this vintage plane coming in for a landing. It was Bob. When he taxied up he was wearing an old leather flight helmet and goggles. He looked like the first picture I ever saw of World War II fighter ace Greg "Pappy" Boyington of Black Sheep Squadron fame, another one of my heroes. Bob and I went into the airport café and had a cup of coffee. He gave me the blades, told me how he wanted them heat treated, got back in his plane and took off. It's been a friendship ever since with both him and Jim Merritt [who worked with Loveless]. I will never forget that year and all the Guild shows that followed, and all the great friendships that started that day.[8]

In 1982 Bos was heat treating for about 80% of the Knifemakers' Guild, and more than 300 total makers[7]. In 2007 it was estimated that he was heat treating for about 70% of U.S. knifemakers[9].

Buck conveyer belt furnace.

In the late 1970s, Buck Knives and Bos developed an agreement where Bos would set up a heat treating shop for them, but Bos would be able to do work for his custom knife heat treating business out of the same shop. As a result, Bos gave up significant income but he loved heat treating knives, as he said in 1982:

> My commitment to heat treating for custom knife makers is total. I still own part of a commercial treating business downtown - a business that was my life up until a year or so ago. I gave up a large income - my income last year was about one-fourth what it was the previous year - and long days and nights as the boss, to devote myself to this work. I am happy, I love the kind of people I now work for. Knifemakers are simply wonderful people, as far as I'm concerned. I was working fifteen hours a day, seven days a week. I had no privacy. I make less money now but I'm much happier.[7]

Bos said of his heat treating[7]: "I'm confident of my abilities. I believe that my work is the best that anybody can do. I eat, sleep and live knives. Even the Bowie knife or the Arkansas toothpicks I treat, which I know are never intended to be used - they will be placed in the owners' collections - are properly heat treated, and if taken out and used, would stand up to heavy usage." For high alloy steels and stainless steels Bos used a furnace with blades hanging from wires; they were then quenched with pressurized argon[7], given a cold treatment, and tempered 2-3 times. Later Buck Knives added a conveyer belt furnace[9]. Bos retired from Buck Knives in 2010, and Paul Farner took over Bos heat treating.

Pacific Heat Treating and Holt-Sornberger

Pacific Heat Treating developed a reputation for heat treating knives under Bill Holt. Pacific Heat Treating was founded in 1961 by Leonard Bazzani in Sunnyvale, California[10]. Bill Holt began working for the company in 1962, and within 12 years, Bill Holt was the owner and manager[10]. In the 1977 *How to Make Knives* by Barney and Loveless[11], they said that Holt had "been handling the famous D.E. Henry Bowie blades for ten years or more, little more needs be added... Billy has the technical background to work with the newer steels, and his facilities include a fine and up-to-date test lab." Bill Holt and Jim Sornberger started Holt-Sornberger Knife Supply, a subsidiary of Pacific Heat Treating. They were advertising in Blade Magazine by 1983[12].

The Holt-Sornberger 1985-1986 catalog[13] described Bill Holt as "a well known and highly respected heat-treater with many years of experience in heat-treating knives for some of the finest knifemakers in the country. Bill owns a modern commercial heat-treating plant housing some of the finest heat-treating equipment available. Bill has a sincere interest in the knifemaking field, over the years he has worked with knifemakers in researching and testing various steels and developed procedures and techniques for properly heat-treating the highly specialized steels we use in knifemaking today." Sornberger is a well-known knifemaker and engraver who has since been added to the Blade Cutlery Hall of Fame[14].

Pacific Heat Treating uses (and used then) vacuum furnaces and a pressurized gas quench along with liquid nitrogen cryogenic processing. The catalog said[13], "Over the years, we have heat treated thousands of blades and run many tests on materials commonly used by knifemakers. These tests were performed by people who use knives in their daily work: Shipping and Receiving clerks use knives day in and day [out] for cutting cardboard, twine and - OOPS! - an occasional staple; professional guides and cattle ranchers (for any chore at hand). The test results were all the same: the knives outperformed (many times over) any production knife they had ever used. Based on these test results, we have established hardness ranges for several of these materials which we feel will produce a good combination of ductility and edge holding qualities." Loveless and Jack Wolf brought in about 1000 lbs. of ATS-34 to sell to some knifemakers, but Holt-Sornberger Supply brought in the first big batches, selling it in significant quantities to knifemakers[15]. They advertised the ATS-34 as "cleaner and a little easier to work than 154CM, as well as being available in thinner sizes."[13] 154CM was only available down to 3/16" while the ATS-34 was available down to 1/8", unless the steel was purchased as Blanchard ground.

Brad Stallsmith and Peters' Heat Treating

For the past 10-15 years, Peters' Heat Treating has been the most well-known heat treater of high-end knives. The Peters family formed the company in 1979 in Meadville, PA[16]. They had done some heat treating for factory knives such as D2 steel for Queen Cutlery in nearby Titusville, PA. Bill Howard of Queen Cutlery in 2004 said that they used Peters' because of their familiarity with D2 from their work in the tool and die industry[17]. Brad Stallsmith of Peters' had always been interested in knives, making the knife industry appealing[18]: "I have always been a knife enthusiast my entire life. Anything with a sharp edge on it I was into it. I was a collector before we began to pursue [the knife] industry." This is a significant part of why Brad wanted to build their knife heat treating at Peters':

> Going back to the beginning, we had Queen Cutlery as a customer, so we had the factory [knife] connection early on. Then we had a single custom maker, local, that started bringing his knives in. His name was Frank Parise, our first custom maker. I was at a point in my career that I had come to a crossroads as to whether I wanted to continue with just the commercial heat treat side or pursue the knife industry. I spoke with the boss (Doug Peters) and I told him that I thought there was an opportunity for us to dive into this because Paul Bos was the only guy that was doing any of the custom stuff. I did a little research on Paul and found out that he was getting up there in years and was going to soon retire. That prompted the boss and I to start to dip our toes in the water to see if there was a market there for us.[18]

By 2005 Peters' was advertising in *Blade Magazine* for heat treating of custom knives[19]. "We joined the Knifemakers' Guild and my wife (Linda) and I began to go to some shows. [Soon after] Paul Bos retired, and that really opened the door for us to get into that market. The rest is just word of mouth and turning out good product and it just continued to grow from there. Getting into the Guild got us into the custom side of things, and the custom thing really took off from that point. The production side of things followed behind that."[18] The first show they attended was in 2006. Peters' has been involved in the BladeSports cutting competitions as a sponsor, but Stallsmith was also a competitor. He said:

> I believe I was talking to Gayle Bradley and he mentioned the BladeSports thing, and that first year that we went to the Guild show in Orlando they were already having competitions. I watched the whole competition, and after the competition I guess I was standing there looking intrigued by that and Gayle Bradley asked if I would like to try to cut a 2x4. I borrowed a knife, signed a waiver, and chopped that one single time. And then

at that point I was also hooked up with Dan Keffeler, helping him to produce his samurai swords, and I got to talking to him about that and he was very intrigued by that. Dan joined and we picked him up as a sponsored cutter. I talked to the boss, and I said this young fella has a potential to be really good at this sport and we should pursue sponsoring him, so we did. Dan actually competed for a couple years before I then got involved as a competitor. We have been involved ever since. Peters' Heat Treating and Spyderco Knives are the oldest continuous sponsors of BladeSports.[18]

Peter's has continued to be the most popular heat treater for custom knives and also heat treats many semi-custom and factory knives.

Furnaces at Home

Heat treating services were very popular with stock removal knifemakers, especially those working with stainless steel. When furnaces were marketed to knifemakers, the number of makers heat treating their knives steadily increased. The first of these furnaces was the Paragon KM-14 in 1986[20]. Paragon is a furnace company that dates back to 1948[21]. Their advertising promoted the benefits of performing your own heat treating[20]: "Will your beautifully finished knife really last a lifetime? ... Will you really know for sure without doing your own heat treating? ... Your heat treater may well be worthy of your trust. Yet when you let someone else do your heat treating, you also miss an opportunity to deepen your understanding of knives... Knifemakers who heat treat learn a great deal about the characteristics of steels at different hardnesses, because they're free to experiment." Robert Smith designed the KM-14[22] and wrote an article on how to heat treat knives for the February 1987 *Blade Magazine*[23]. Evenheat is another furnace company that started in 1948[24]. Evenheat began selling knifemaker models KF 13.5 and KF22.5, and in 1991 Koval Supply began marketing these furnaces[25].

2005 Peters' ad

1986 Paragon furnace ad.

-20-

BILL MORAN'S REINTRODUCTION OF DAMASCUS

Recreating Damascus

Bill Moran was intrigued by the legendary status of the ancient Damascus blades, both for their beauty and reported properties. However, he also felt that Damascus was the key to selling "art knives," a category that hadn't sold much before; most of the market wanted knives to use, not display.

"When I first began trying in 1969," Moran recalled in 1990, "I did so primarily because while there was no market at that time for art blades, I felt that such a market was developing and, if my belief was correct, Damascus would be the ultimate material for the blade of an art knife"[1]. "Ever since I can remember, I, like all who love fine blades, have been fascinated with the beautiful Damascus blade. Many legends abound throughout history pertaining to the wonderful quality of these remarkable blades. After years of study and research, I felt I would like to try making them"[2].

While there were books with pictures and descriptions of Damascus blades, he could not find any information on the process of making them. He even had friends research at the Library of Congress, but they could not find anything helpful either[3]. Everything written was by collectors, not the craftsmen who made the steel. And when collectors gave their ideas about how Damascus was produced, they were often wrong. Moran also knew that Paul Müller of Germany was known for being a Damascus smith and sent him a letter but never received a response[4].

Moran talked about the perseverance required to make Damascus work:

> [M]any metallurgists I talked to told me that it could not be done, working with modern steel. This was because the slag in the old iron and steel made it much easier to weld the layers together. By then, of course, I had had many years of experience in forge welding, so that when I started my project I did not have to learn how to weld. The fact that I had those years of experience in welding iron to steel was a great help[5].

> I gave up half a dozen times, but I always went back to it. There were so many problems, and no one had any answers. I had to proceed through trial and error, and I made more than my share of errors. But, I did have an advantage. I had been forging iron to steel to

make throwing tomahawks for buckskinners and, thus, whenever someone would assure me that you couldn't hammer-weld iron to steel, I knew better.[1]

I soon learned that the heat range on forging Damascus steel was very critical. If the heat was a bit too high the steel would turn out to be a poor quality. If it was the slightest bit too low the result would be poor welds. After learning this I started experimenting with various numbers of layers. After going to over 2,000, I found that the bar then becomes carbonized all the way through and loses its strength. So, after making many blades with various number soft layers I came to the happy conclusion that around 512 layers was the best.[5]

On another occasion, Moran described the process a bit differently:

I have seen Damascus blades from the area of the Saracen Empire and Northern India, but when I first started to work out this process I didn't have the vaguest idea whether there were five layers or five thousand layers. I didn't really know where to start, so I began welding blades with various numbers of layers, from sixteen to over two thousand, and I soon came to the conclusion that somewhere around five hundred was the best number. Such blades closely resembled the originals I'd studied.[6]

B.R. Hughes wrote about other difficulties in recreating Damascus, including finding an appropriate flux for forge welding. The flux acts to dissolve oxides and prevent further oxidation so that the two metals can diffusion bond:

When it came to flux, Moran began by using a commercial preparation, but when his efforts were frustrated, he turned to that old household standby, borax. To this day he still uses borax, the bonding agent used in the teaching of Damascus at the ABS/Texarkana College Bill Moran School of Bladesmithing that bears his name. One of the more aggravating aspects of Bill's quest was the fact that when he did succeed in getting a number of alternating layers of iron or mild steel to bond with a cutlery steel, there was no visible pattern, which seemed to nullify any degree of success. Finally, he hit upon the idea of treating the layered steel with acid, which reacted differently to the two metals.[7]

But even etching with acid did not solve all of Moran's problems, as the combination of steels also affects the response to etching. So he began to use O1 and mild steel:

One thing that really frustrated me was that when I did manage to weld several layers together, it was virtually impossible to see any pattern. I mulled this over and decided that since O-1 steel had a mild chrome content, it just might react to the acid treatment differently than W-2, and so it proved. When I acid-etched the blade made from O-1 and a milder steel, I got a beautiful pattern. It took me a while to come up with what I deemed to be the correct blend of two parts of iron to one part of steel, but when I did I knew that I had something.[8]

"Over a period of two years, I must have thrown up my hands and quit at least three or four times, but I always went back to the project because I found it so engrossing"[8]. Finally, Moran gave a final long push to make Damascus: "I cannot even estimate how many blades I made and discarded before I made one that had all the qualities of real Damascus steel. I worked on them for about nine months before I was satisfied that they were good enough to sell"[3]. "I was satisfied that I had really recreated this wonderful steel"[2].

Damascus Patterns

Moran's first Damascus steel was "random" pattern. Meaning he forged the knife using the "straight" laminate of alternating layers of steel. The pattern comes from where the different layers intersect the surface, and the exact pattern comes from natural variations in forging. However, Moran was not satisfied with random pattern: "Then I tried twisting my Damascus billet, and the results were even more pleasing"[8]. One of Moran's earliest patterns was "maiden's hair," which was a "slow" twist, meaning the billet was not twisted very many times. This results in flowing lines that can look like strands of hair.

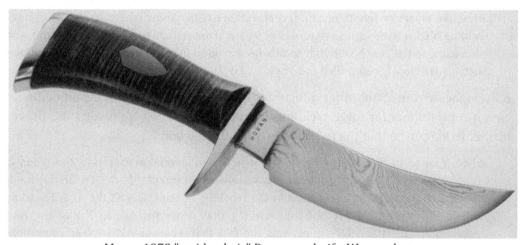

Moran 1972 "maiden hair" Damascus knife. Weyer photo.

Another of Moran's patterns was "Mohammed's Ladder," more commonly now called only ladder pattern. This pattern originated with ancient Wootz Damascus blades under different names, including the aforementioned Mohammed's Ladder, 40 Steps, and Kirk Narduban. Moran used a frequently published name for the pattern, Mohammed's Ladder, which means he was likely familiar with the ancient blades with the pattern. For example, Cyril Stanley Smith's *History of Metallography* called it Mohammed's Ladder and also had Smith's proposal for producing the pattern: cutting grooves into the steel followed by forging flat. And indeed, this was how Moran chose to make the pattern.

Kirk Narduban (ladder) pattern blade in 17th century Persian Wootz sword[9].

It is Difficult to Make Damascus

Moran described the difficulty of making Damascus steel in almost fantastical terms, either in his brochures that he handed out or in interviews for articles. Moran said, "The making of a Damascus blade is difficult beyond belief. So much so, that forging any other blade is simplicity itself. For instance, it takes over 150 pounds of high grade coal to forge one Damascus blade, whereas an ordinary blade can be forged with about 5 pounds of coal"[10].

B.R. Hughes reported about Moran's difficulties:

> The heat from the forge gets up to around 3,000 degrees Fahrenheit during this process. This heat makes it almost unbearable to forge these blades in the summer. In the dead of winter, Bill Moran can stand outside his shop and look like he just ran the Boston Marathon after making a piece of Damascus. He wears a heavy leather sleeve during the forging process, because the heat in the forge has blistered his skin from three feet away. It takes several hours to weld one of these pieces of steel.[3]

Moran told Sid Latham, "Caution, don't try to grind the Damascus blade on a belt grinder. The steel is too hard and will tear the belt." So instead, Moran used a 2-hp 36-grit emery grinder to do the rough grinding first with the wheel. Then he would clean up the grind with an 80 grit belt, though Moran added that you must "always us[e] a new belt for a Damascus blade"[11].

Properties of Damascus

Another aspect of advertising the new Damascus knives he made was the reported properties of the Damascus. Moran said this was not simply a steel produced for its beauty but also because it would make the highest-performance knives available.

> No one was more surprised than I, when I found this steel did have most unusual qualities. I found these blades could be left as-quenched and still have good flexibility, they also would hold a very good edge. Indeed I am now convinced that many of the old legends concerning these blades are true. These blades were far superior to any other blade because of the hundreds of alternating layers of iron and steel. The principle is the same as plywood or the laminated bow. I am now convinced that Damascus blades stopped being made so long ago for the same reason the wonderful composite bows of the Turks died out; that reason being they were so terribly difficult to make and after the fall of the Saracen empire, man went back to the self-bow which was made from a single piece of wood. The same being true for the Damascus blade; it was so much easier to make a blade from a single piece of steel.[10]

B.R. Hughes expanded on Moran's thoughts about Damascus steel's unique properties:

> A strange phenomenon that had been told to Moran years before was that a Damascus blade would cut a person quicker than a standard steel blade. As he researched, references were found that claimed this blade made wounds that were more difficult to heal. Moran found out for himself just how quickly Damascus steel cuts, as the many scars on his hands

attest. Moran feels this may be a result of the grain flowing in a wave pattern toward the edge of the blade. The microscopically fine teeth are more jagged, and the blade seems to catch and hold the flesh. Field tests have proven that a Damascus blade will last through a long day of dressing even large game... In a Damascus blade, the laminations, layers and patterns are handmade. This is what makes the Japanese blade so desirable. This process makes a better weapon. A Damascus blade tends to break in layers, rather than clean through – a very important factor when your life depends on a blade. The steel in the Damascus is based on the same principle as a piece of plywood. There are alternating layers of hard and soft steels, which make the blade hard, yet flexible and strong.[3]

In recent years, bladesmiths do not commonly claim improved performance from forge welding Damascus. These ideas have evolved over time; see Chapter 23.

Response to Damascus

Moran introduced several Damascus knives at the 1973 Knifemakers' Guild Show in Kansas City. The new knives created quite a stir with knifemakers and collectors alike. Knifemaker Ted Dowell said, "There was considerable interest on the part of the knifemakers in what Bill was doing. Bill promoted it and got it introduced and he has gotten largely the credit for it, and it's well deserved."[7] Knife collector Butch Winter reported, "In those days the Guild Show crowds weren't that big. There were a few people looking at Moran's knives but they more or less were agog and didn't understand what he'd done. The other makers were saying, 'I wish I could do that,' or 'How'd he do that? "I sidled up to Miss Margaret (Mrs. Bill Moran) and put my arm around her neck and said, 'Margaret, would you somehow convince Bill to sell us one of those knives?'" The Winters paid $1500 for their new Bill Moran Damascus knife[12]. "It was more than we'd ever paid for a knife before"[7]. $1500 was practically an unheard-of sum in those days when even $200 was a lot for a custom knife.

Moran ladder pattern dagger purchased by the Winters in 1973[12].

Some were not believers in the Damascus knives, however. Rita Winter said, "I remember some of the makers saying, 'I don't know what's great about (Damascus), it's not going last.'"[7] And knifemaker Steve Johnson said, "It was interesting to have something that old to be reintroduced, and it certainly was newsworthy. There was the usual skepticism and excitement among the knifemakers, and (even those who said) if (they) hadn't made it then it couldn't be any big deal."[7]

Early Bill Moran ladder pattern dagger. Weyer photo.

Moran's first Damascus steel knife. Francesco Pachi photo.

However, the knives sold, and Damascus was not a quickly dying fad. "Not only did Damascus help to bring back forging," Moran said in 1979, "the first year I showed my Damascus blades in Kansas City, I took in the most money I had ever made at any one show. For the first time, it occurred to me that it just might be possible to earn a decent living making knives. However, making Damascus cut my production by about two-thirds, and as I get older, it seems that I can't make knives as quickly as I once did, and I was never a very fast bladesmith"[3]. "I think this really gave me the most pleasure of anything I have ever done. I certainly felt more sense of achievement from that than everything else put together!"[6].

-21-

AMERICAN BLADESMITH SOCIETY

Bill Moran started discussing the idea for a new organization in 1972 to "encourage and promote activities involving the art and science of forging" blades. Moran had led the Knifemakers' Guild in 1972. However, he was interested in an organization that would promote forging, as the number of forging bladesmiths was only about a dozen, and the number of stock removal makers was growing rapidly. Moran had also been frustrated with the bureaucracy and politics of the Knifemakers' Guild, which had given him ideas about how his own organization would be led. However, the goal of the ABS was not to compete with the Guild but to serve a specific purpose when it came to forged blades.

Also, as we have covered already, Moran re-introduced Damascus blades at the 1973 Guild Show. Damascus steel led to many knifemakers becoming interested in forging as it was the only method for making Damascus. B.R. Hughes has said, "It would have been virtually impossible for the ABS to have succeeded had not Damascus steel reentered the picture."[1] Bill Bagwell was the first to follow Moran in making a Damascus knife, and Don Hastings after. The 1976 Guild Show was in Dallas, and after the show, Bagwell invited Moran, Hastings, Hughes, and gun writer Bill Jordan to visit his shop in Louisiana. Moran, Hastings, and Bagwell forged blades and tested them. Jordan would write an article on forged blades for the December 1976 issue of *Guns* magazine. After Jordan left, Moran proposed the idea of a bladesmith organization. Moran said to Hughes, "Bill, if we play our cards right, one of these days we could have 25 members!"[1] Hughes later said, "Privately, I felt that Bill was being wildly optimistic, because there seemed no way that the number of bladesmiths in America would ever reach such a lofty number!"[2] At the time there was a feeling that forging was being phased out, as Bagwell said in 1978[3]: "It's easy to find where some well-known knifemaker criticizes smiths and forged blades. Because forging is an ancient art, the critics seem to feel that it is an outdated procedure that results in inferior blades. Nothing, absolutely nothing, could be further from the truth. To a great extent, I can almost excuse them for their misconceptions, because very few people have much, if any, knowledge of proper forging techniques."

Don Hastings, Bill Bagwell, and Bill Moran during the ABS organizational meeting. Photo by the other founder Bill Hughes.

Over the next few months, they would exchange letters and have conference calls to decide on the structure of the ABS and to write the bylaws. On December 4, 1976, the four met again and signed the bylaws, electing Moran president, Hastings treasurer, Bagwell secretary, and Hughes as a director. Jimmy Lile was added to the board as a director in 1977, who had been a President of the Guild. By April 1978, the organization had 16 bladesmiths and 12 associate members, primarily collectors and writers. The first Master Smith stamps were not awarded until 1981, given to Bill Bagwell, Jimmy Fikes, Don Fogg, Don Hastings, Bill Moran, and James Schmidt. At that time, it was relatively informal; the best forging bladesmiths were awarded the title, and there was no explicit judging of blades or performance tests. In 1984 Bagwell resigned from the ABS due to "philosophical differences"[2], and Lile resigned for health reasons. In 1986, Hastings passed away from leukemia. Hastings' cancer had been in remission for several years but had taken a turn for the worst. He had even recently been featured in *Blade Magazine,* which created many new orders[4]. The ABS created the Don Hastings Memorial Award in his honor. Hughes said that Hastings was "a gifted teacher, and a man who literally had not an enemy in the world."[2]

In 1988 the Board put the requirements for achieving Journeyman and Master Smith stamps into effect. The Board of Directors, of which Moran and Jimmy Fikes were a part, created the procedures. The first to attain a Master Smith under the new guidelines was Wayne Goddard. The requirements for a Journeyman Smith are to cut a free hanging 1" manila rope, chop two pine 2x4's without damage to the cutting edge, shave arm hair afterward, then bend the knife 90 degrees in a vise. The edge can crack, but the spine cannot. The applicant then presents five finished knives to a panel, which decides if the quality of the knives is worthy of the JS title. The requirements for Master Smith are similar, but the performance test must be performed with a Damascus knife, and the five submitted knives are given greater scrutiny. One of the five knives must be a Damascus "quillon" dagger.

The ABS also created a bladesmithing school in May of 1988 with Texarkana College. Moran said of the school[2]: "This school will be vital to the future of bladesmithing. It will make it unnecessary for each generation of smiths to reinvent the wheel, so to speak, and I think I can truthfully state without fear of contradiction that a student will be able to learn more in two weeks at this school than he could in ten years of trial and error." Moran stepped down as president in 1991.

Carbon vs. Stainless Steel

Moran was not only a fan of the forging method, but also believed it created superior-performing blades. He believed in "packing," which supposedly led to tighter "molecules" in the steel for better performance. And while stock removal makers were promoting stainless steel and cryo heat treatments, Moran promoted heat treating by eye in a forge and tempering with a torch. The preference for simple steels has continued with the ABS ever since, with most knives produced by ABS Journeyman and Master Smiths being made with carbon steels and low alloy steels. The guidelines for the Journeyman performance tests require making the knife with "carbon steel." There was also a belief by Moran, and other founders of the ABS, that the secrets to the highest performing blades were found in historical and traditional methods. As Bagwell said in 1978[3]: "We hear and read a lot about today's knives being the finest that the world has ever seen, but that is simply not true! The finest blades ever made were produced during the Middle Ages when knives and swords were the principal means of self-defense. During that period skilled smiths made blades that are far superior to anything being offered today, although a few of today's skilled smiths are coming close." He also added: "[Y]ou must remember that the primary function of a knife is to cut, and there is no getting around the fact that a properly forged blade will take and hold an edge for a longer period of time than will any other type of blade, and in addition the forged blade will be stronger."

Moran promoted traditional methods and materials, which has largely remained the preference for members of the organization that he created. Despite the increasing number of ABS bladesmiths, the number of bladesmiths who forge stainless steels and high alloy steels has remained small. This is in part due to the difficulty of forging these steels. Another challenge is that heat treating high alloy steel in a forge is very difficult, typically requiring a furnace to heat treat properly. But the culture around forging bladesmiths also promotes simple carbon steels. Bagwell said in an interview[3]: "Frequently, I note in some bit of literature that so-and-so is using a fantastic jet-age steel for his blades. Frankly, just because a given steel is great when used for the turbines of jet aircraft engines doesn't necessarily mean that it is the ultimate cutlery steel. A blade that is forged, however, is made from steel that the smith has in effect manufactured expressly for a knife, and this makes possible the toughest, strongest, most effective cutting edge possible, but at the same time this blade will be relatively easy to sharpen once it gets dull."

The performance tests required for Journeyman or Master Smith titles create a bar over which the bladesmith must clear and are designed to show qualities made by using simple carbon steels. The 90-degree bending test requires that the knife be given a "differential temper" which is difficult to achieve with high alloy and stainless steels. The primary methods for achieving this include: 1) "edge quench," leaving the spine soft, or 2) tempering the spine with a torch. A high alloy steel or stainless steel, if "edge quenched," will still have a hard spine because the steel hardens in air. Stainless steels and high alloy steels also temper at much higher temperatures, so having a torch-softened spine is extremely difficult to achieve. A stainless steel laminated with softer stainless steel sides would fit this required performance, but this is not allowed for the test. The ABS bylaws explicitly state that stainless is discouraged for the following reasons[5]: "[T]he Society discourages the continuous use of stainless steel for forging blades unless and until forged stainless steel blades have been proven, to the satisfaction of the Board, capable of being forged into quality blades that will consistently pass the standard ABS cutting and bending tests." This statement is applied to all ABS blades, not only those used in the performance tests.

Despite the belief among many ABS bladesmiths that forging leads to superior carbon steel blades, a popular opinion over the past several decades has been that stainless steel does not benefit from this process[6-9]. B.R. Hughes, in 1977, wrote[9]: "440C, D2, and 154CM are all ... stain resistant steels, and all three will take and retain an edge for long periods of time even under heavy usage. It was not, incidentally, feasible to attempt to forge such sophisticated steels as those mentioned in the last two paragraphs, and the [common] view that many modern cutlers take of forging may in large measure be due to the fact that a few misguided souls even tried to forge 440C and even D2 by guess and by gosh, and the results were mostly awful." In a 1985 *Blade Magazine* article, Dan Petersen wrote[10], "Tests recently performed by members of the American Bladesmith Society would attest to the superior performance of forged high carbon steels over stainless steels in both edge holding and toughness. The simple high carbon steel is the superior blade material and will produce the better knife particularly when the knife is forged... Chromium steels do not generally produce tougher and better edge holding knife blades. Simple high carbon steels will stain or rust more easily than the chromium steels but they make better knives."

Knifemakers Forging Stainless Steel

There have been those that have forged stainless steels, going all the way back to Marchand forging stainless for Randall (Chapter 11). And both Hibben and Henry had to forge 440C to use it early on (Chapter 15). Some have even promoted the forging of stainless and high alloy steels. Karl Schroen published the book *The Hand Forged Knife: An Introduction to the Working of Modern Tool Steels* in 1985 which included instructions on how to work with high alloy and stainless steels. Schroen wrote in his

Karl Schroen

book[11]: "The first high alloy tool steel that I began to experiment with was D-2. As it turned out, it was also the hardest to master. Hammering W-1 was like hitting a soft sponge in comparison to D-2 which felt like a stiff pine board. All the other steels tested arranged themselves somewhere in between these two extremes. A close runner up to D-2 is 154CM which feels about the same." In his book, Schroen described the properties, including how easily they forged, of W1, S5, O1, S1, L6, A2, Vasco Wear, M2, D2, 440C, 154CM, ATS34, BG42, CPM-440V, and CPM-10V. Though he said that L6 was his "favorite steel," he also said he liked the high alloy and stainless steels despite finding the forging, though especially the finishing, difficult. About CPM-10V, for example, he said, "the advantages to me far outweigh the disadvantages" despite the difficulties in finishing from the very high wear resistance and vanadium. He added that "Modern tool steels are valuable materials for the traditional bladesmith who wants to take the time to understand them."[11]

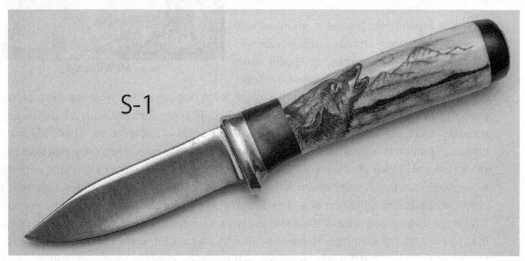

S-1

Karl Schroen knife. Photo from Schroen.

Schroen also defended the use of stainless steel[11]: "The classic, universal complaint that I have heard about stainless steel is that knives made of it do not hold an edge well and do not sharpen easily. I would certainly agree with these complaints in general. But I have found that the fault does not seem to be caused by the material but rather by the techniques or process used in making the knives. The issue is further complicated because there are a number of different types of martensitic stainless steels manufactured and it would be hard to sort out all of the many variables connected with the problem." Schroen found that 440C has a "marked tendency to crack," though he "solved this problem with lighter hammering at a reduced temperature." He did find that 154CM "does not crack under the hammer as much (as 440C)" and further that ATS34 "seems to forge more easily than 154CM and ultimately produces a tougher, longer lasting edge than 154CM." Schroen concluded his book by saying, "The combination of modern tool steel with traditional blacksmithing techniques has just begun. I hope to stimulate an understanding of this material since I believe that only through a thorough knowledge of this material can a bladesmith take full advantage of it as a medium. I will continue to work with these materials and explore new ways of adapting them to the ancient techniques of blacksmithing."

Bladesmith Sean McWilliams said that Schroen was one of the knifemakers who encouraged him to forge stainless when experimenting in 1983[12]. Dan Dennehy and Jim Schmidt also encouraged him. McWilliams was advertising forged 440V knives in 1986[13]. McWilliams wrote about forging stainless in 1988[14] and 1989[15] *Knives Illustrated* articles:

Sean McWilliams

> From the knifemaker side of it, stainless has generally been considered a stock-removal-only material. Bladesmiths generally look down on stainless steels. Some who have tried to forge it not only look down on it - they glare. So as one who does forge stainless, I have been sort of avoiding other bladesmiths until I could come forward with a hand forged stainless knife of superior strength and cutting ability. Superior to what? My own reference point has been to make them stronger, sharper, and easier to sharpen than the stock removal blades of either 440-C or O-1 that I used to make before switching entirely to forging about three years ago. If forged stainless seems a well kept secret, I guess it's partly my fault, since I'm one of so few makers who forge stainless steel. Some makers have tried it and given up, and their bad experiences have augmented the common belief that you can't forge stainless. There are some tricks to it for sure, but the real secret to forging stainless is "hard work." And that just might be the best kept secret in America today. Stainless steels are forgeable. After all, they are at least 80 percent iron.[14]

McWilliams also did performance comparisons:

> How does forged stainless steel compare to forged carbon steel? To answer this, I forged and heat treated three test blades: one each of O-1, 440-C and T-440-V. These were tested for edge holding and toughness. For the edge holding tests, I made cuts on one inch hemp rope until the blade lost its bite on the rope. The O-1 blade made 13 cuts; the 440-C made 10 cuts; the T-440-V made 25 cuts. Toughness was tested by repeated bending in a vise. Blades were bent about 30 degrees, and they all sprung back straight. The forged stainless performed exactly the same as the O-1. But there's one test that the O-1 failed: cut a lemon and lick the juice off the blade. The stainless blades, no taste. The O-1, yuck! My experiences show that the old either/or - sharpenability or corrosion resistance - is not a choice that the knife user has to make any more. Neither is the choice between a forged blade or one that won't rust.[14]

Sean McWilliams "Ranger 7" in forged CPM-440V/S60V.

These bladesmiths created a minor surge in the forging of stainless blades, but it was relatively short-lived. Master Smith Wayne Goddard, who worked with both stainless and carbon steels, forged carbon steels, but used stock removal when working with stainless. He said in 1989[6]: "There is a trend to forge stainless steel, which I feel is a waste of time and effort. My opinion is based on theory of metallurgy and actual testing of other makers' forged stainless. I have not forged stainless blades and tested them simply because I haven't felt that it would prove anything. Steel bar stock as it comes to the maker may have a good enough grain structure so that the work that a bladesmith performs on it may not improve it enough to gain anything in edge holding or strength." In the same year, Jim Fister said[6]: "My specialty is forging. Most of my knives are forged. I have customers who prefer stainless over carbon and those who don't like rust. Since I don't forge stainless, I use the stock removal method for (stainless) knives. That's why I do both, although I prefer forging." Tai Goo added:

> The only advantage to stainless is corrosion resistance. It is difficult to work with. It is hot hard - in other words, it is hard even when heated to its proper forging temperature. The hot hardness makes it difficult to forge. A high degree of shock from hammering is transferred to the smith. This shock will eventually result in tendonitis, arthritis, and pain in the wrists, elbows and shoulders. This is the main reason I do not forge a lot of stainless. Also, I do not like the way stainless looks - cold, sterile, unchanging. If stainless is properly forged, it will improve its structural integrity. However, a properly forged stainless blade will not be as hard, tough and easy to sharpen as a properly forged piece of plain high-carbon steel. Stainless is difficult to forge, therefore it is difficult to do a good job (with it). The risk factor is high.[7]

Many myths about forging stainless have continued over the years. For example, Master Smith Joe Szilaski wrote in 2007 in response to a question about forging stainless:

> I am not surprised that your friend heard that stainless steel could not be forged. I have heard this said many times and I disagree. I have also heard of many smiths who feel that though you can forge stainless steel, it is not improved by forging, so it is a waste of time and effort... I have forged stainless steel fillet knives out of 154CM and 440C, with great

success, for salt-water fishermen. Both steels worked equally well under the hammer. Hunting and camp knives are also good candidates for a forged stainless blade. I tested the resulting forged stainless steel blades by cutting some manila rope, and, in my opinion, they held an edge better than a non-forged stainless blade.[8]

Szilaski added that the traditional nature of the ABS is what has maintained some of these beliefs[8]: "Maybe what your friend heard was that forging stainless steel is not as traditional as forging the non-stainless carbon steels that most smiths used hundreds of years ago. The American Bladesmith Society bylaws allow a forged knife to be stamped with a journeyman- or master-smith stamp only if the blade is forged of high-carbon steel. The ABS is trying to maintain a tradition as close as possible to what once was."

To this day, the ABS bylaws disallow stainless steel to be stamped with a JS or MS[5]: "Stain resistant steels or blades that are not forged by the bladesmiths shall NOT bear the MS or JS stamp or any hallmark of the Society." Perhaps in the future this outdated rule will be removed.

-22-

WAS PATTERN-WELDED DAMASCUS EVER LOST?

German Damascus

Pattern-welded Damascus blades were produced in Solingen, Germany, back to the 1500s AD. It is said that Peter Schimmelbusch, also called Peter Simmelpuss, forged the first German Damascus blade in "modern" times[1]. In Manfred Sachse's book *Damascus Steel: Myth, History, Technology, Applications*, you can read a legendary tale of how Peter Simmelpuss' father learned to forge Damascus blades. Sachse also proposed that there was likely influence from Toledo, Spain, on Solingen, as Toledo had a long history of pattern-welded Damascus.

Several innovations increased the amount of production in Germany for pattern-welded Damascus, one being the hydraulic hammer which was used by the 1400s. In addition, a patterning technique was developed in Germany which involved pressing, or "swaging" of the steel and then grinding flat. You can see this technique in pictures found in this chapter. This process differs from the method of cutting into the steel and then forging flat.

Beroaldo Bianchini published about the swaging process in German in 1829: "The drawing-out of the twice wrought and welded faggots into bars under the hammer, and then the forging of the blade between the two dies is actually performed in the same way as in the case of ordinary blades; indeed, the only difference lies in the fact that the second dies must be provided with various recesses or decorations which one wishes to apply to the blade. When striking with the hammer, the alternate iron and steel sheets of the blade are pressed into the manufactured recesses in the dies, as a result of which a certain number of elevations are produced which, when they are then files or ground, provide the required adornments."[2] There is a collection of dies used for pressing these patterns from Damascus smith Max Dinger who died in 1910[1] (image shown later in the chapter).

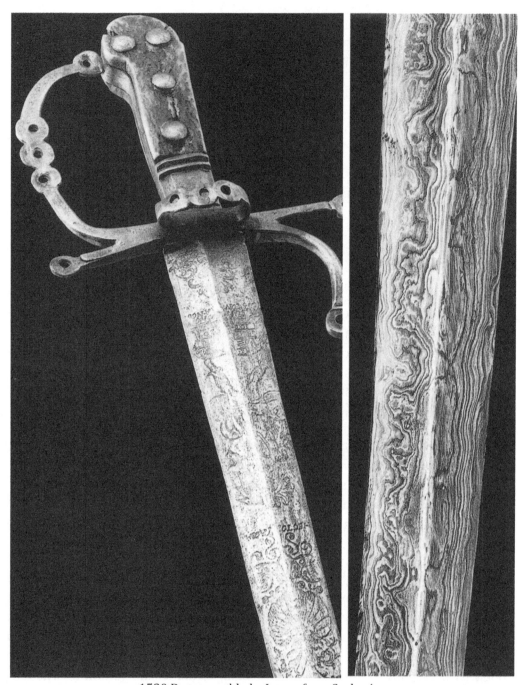

1530 Damascus blade. Image from Sachse[1].

From 1933 to 1945, there was a lot of demand for prestige weapons for the Nazi party. One of the famous smiths in this period was Paul Müller, known for his Turkish Damascus. In 1938 Müller took part in a craft exhibition in Berlin. There he met senior members of the Nazi party, who wanted Müller to establish an Imperial Forge.

Patterning dies of Max Dinger[1].

Paul Müller

In 1939 the Society for the Promotion and Support of German Cultural Heritage signed a contract with him. Müller's politics leaned toward communism and social democracy, common in Solingen. But after the flattery he received from senior members of the Nazi party (and large sums of money), he shifted his politics to match. Another unfortunate choice Müller made was that he had Karl Wester, a rival Damascus bladesmith, put in prison during the Nazi period. As explained by Sachse: "One has to acknowledge that Paul Müller was one of the great connoisseurs among Damascus steel bladesmiths. Nevertheless, he also has to be regarded as a tragic figure because of his political relationships."[1]

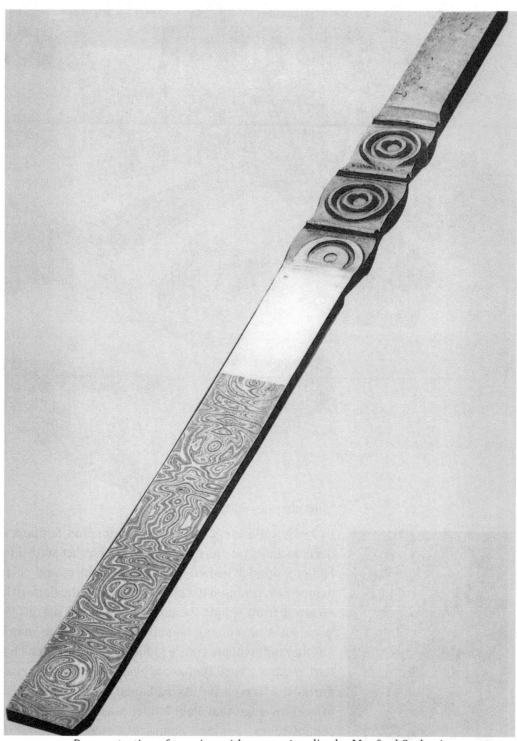

Demonstration of swaging with patterning dies by Manfred Sachse[1].

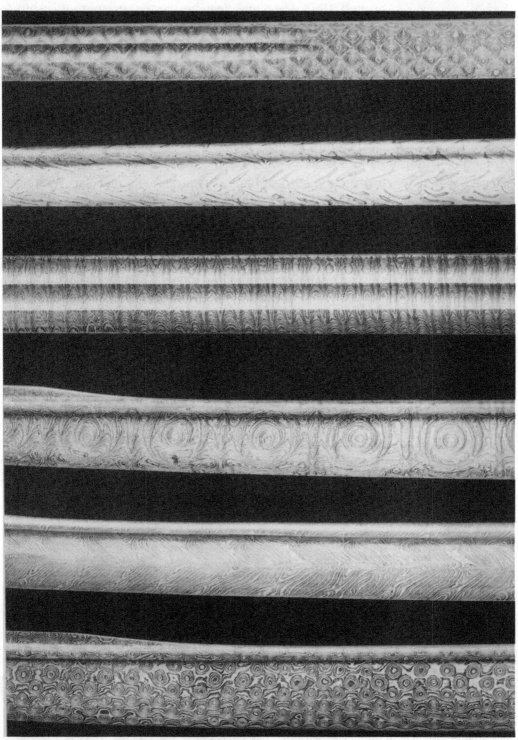

Image of German Damascus patterns[1].

Manfred Sachse, born in 1935, became interested in fencing around 1953, the same year he passed an apprenticeship for painting and decorating. He also passed his master craftsman exams as an electroplater and metal grinder in 1959. During the time spent at the Solingen Technical College, he regularly visited a company called Waffen-Schmidt which made fencing swords. There he met Paul Müller and saw his first pattern-welded Damascus blade. He developed a passion for Damascus and learned how to make it. In the early 1960s, he began receiving orders for restorations of historical weapons. He made several trips to places like Liège, Belgium, to learn about historical Damascus pieces, and met Karl Wester and Paul Dinger (son of Max Dinger). Sachse is said to have forged more than 30,000 Damascus steel blades for the Solingen manufacturing industry. He wrote several papers, presented many courses on Damascus, and in 1989 wrote an excellent book on the history of Damascus called *Damascus Steel: Myth, History, Technology, Applications* which is also available translated into English. That book covers ancient European and Japanese swords, Wootz, the Indonesian Kris, Damascus gun barrels, and other topics, and should be read by anyone with an interest in ancient Damascus.

Even though Sachse was making Damascus knives in the 1960s, he still credited Bill Moran with re-popularizing Damascus: "William F. Moran was the first smith in the USA to embark on the forging of Damascus steel; from 1970 to 1972, he made his first attempts, and offered his daggers and knives for sale. Two years later, smiths in Germany began to follow suit, and the first damask knives were offered for sale at the Dortmund Weapons Exchange in 1974. It was Gerd Kühne who was the first to have knives manufactured - with blades of damask steel - for the purpose of placing them on the market... During the period 1975-1976, Dietmar Kressler manufactured knives for Gerd Kühne using Manfred Sachse blades of damask steel."[1]

Beau Hickory

Hickory was born in 1929 and grew up in Seattle, Washington. Beau Hickory was initially a stage name though eventually, he had it legally changed in the 1960s. The stage name was for performances in Wild West Shows. Hickory said in a 2017 interview:

> My father started pretty early in show business. Dudley Smith, an uncle of his and great uncle of mine, was with the Buffalo Bill show. He rode bucking horses in the arena... Family stories

Beau and Tinnell Hickory[3]

are that he did a little bit of service on the guns that they had. My father met him in 1913 after the show closed up, and he gave my father a single-action Colt that he used in the Wild West Show. That started my father down the line of gun collecting... In any event, my father did tour the whole United States doing an act very similar to what Will Rogers had left when he got big on Broadway. He did trick roping and Western philosophy... I got started in the Wild West Show business in rodeo because my father was an excellent announcer, and he announced the show, and he'd do trick roping and aerial trick shooting. As soon as I could walk, I had a handful of trick ropes. In any event, that's how I got started... I started packing iron when I was 3½. You have no idea how hard it is to hide a revolver in a sun suit. I'd go down to the beach and shoot.

I have done show-type shooting since I was about eight. My father and I did a little bit of performance together, me throwing targets and him shooting them.

I shot on stage. We've done school shows. I think I shot in almost every school yard in the city of San Francisco. I would take a collection of antique guns for these kids to be able to actually handle, and I have stacks of letters that they would write to me after the show.[3]

Eventually, he had a show with his wife, and I was able to interview her about their past: "We were billed as Beau Hickory and Tinnell. We both did trick roping, trick shooting, shooting from the back of a running horse, you name it. Lots of excitement."[4]

Along with performing in Wild West shows, Hickory's father worked full-time as a blacksmith[4]. Tinnell told me the story:

His father was an avid collector of firearms, as Beau was growing up. His father's area of collecting was not just old western firearms, but his father was interested in hunting guns in Europe. Many of them had Damascus barrels. As a child his family had a huge collection of them, and he found every book he could read and did tons of research on how they did that. His father was a gunmaker and he was a gunmaker and they both made Damascus gun barrels. Really interesting stuff. His first blade was in 1946, his personal Damascus Bowie knife. He was the first person in the country who started with Damascus early on.[4]

Wild West shows and Damascus blacksmithing? What else did he do? Tinnell said:

If you know Beau, he was a true Renaissance man. He didn't do just one thing. If it was made out of metal, he could make it: from fine sculpture to an aluminum racecar, he even made cars that raced on the Bonneville salt flats. He could make a high-rise steel building. Ornamental iron for homes. He started in San Diego and then San Francisco in 1950. During those years he worked for a couple shops, and he finally branched off and opened his own shop. He was responsible for 3 to 10 tons of iron going through his own personal hands each month. San Francisco is covered with some of the most beautiful ornamental ironwork (that Hickory made).

During the 1970s he was building automobiles. He had his own business, Sports Racecar Lab, and he built aluminum bodied cars that were typical of the 1930s in design, but he was also a designer and pioneer in fiberglass kit cars that went on VW chassis. One of the cars was featured on the cover of Popular Mechanics. Beau was so prolific for so many years I have buildings full of boxes full of things. I can only do so many things. He had so many things going on. He did make a good living, but he never did it full time at any one thing except the iron business. For quite a few years he made a lot of money just with ironwork. In later years that was boring to him. He also had several other businesses: saddle maker, gun engraving, high fashion jewelry, design of women's clothing. He was also a fine artist; his oil paintings and sculptures were quite beautiful. If it was in the art field, he either dabbled in it or excelled in it. His favorite phrase was, "Talent is." If you're talented at one thing you can apply it to anything.[4]

Hickory also taught many classes on blacksmithing:

Beau had a school in San Francisco for many years. That was before 1984 (when they moved to Arizona), he taught classes for years, usually in the fall and winter. He would have two classes, a morning and evening class, 12 weeks per session. He had about 15 stations set up with anvils and all the tools. You could have two people per anvil if you wanted to. He put a lot of people into the iron business and knife business. Over the last five years of Beau's life, he had a lot of his old students who hunted him down and thanked him. Both for their business and their personal lives because when he taught his classes, he didn't just teach how to make things; there was a whole philosophy of how to apply these things to his life.[4]

Hickory was very active within the blacksmithing community: "Beau was one of the founding members of the California Blacksmiths Association, he drew up the corporate papers and got them their non-profit status. He was very active in putting on conventions and demonstrations."[4] In many areas, Hickory's showmanship and blacksmithing would be combined:

He was also a member of ABANA (Artist-Blacksmith's Association of North American), and in 1980 he was [the] featured demonstrator at [the] ABANA yearly convention in Santa Cruz. There was me, him, and one of my girlfriends. It wasn't just a demonstration of different techniques. We did it in full attire, he was dressed as the god of fire, Vulcan, and she and I were dressed as his golden handmaidens. (Vulcan is often portrayed with a blacksmith's hammer as he is also the god of metalwork). The legend says that when Vulcan's mother cast him off Mount Olympus, he landed below and one of his legs were injured. To have helpers he made two golden handmaidens to work with him. When it was time for the event, they had a backdrop behind Beau's anvil. People were out talking, waiting for the demonstration. We were behind the backdrop in our wardrobe getting ready. When it was time to start, Beau stepped through the curtain, a lady on each arm. All of a sudden one of the smiths looks up and sees us and gives an elbow to the next guy, and he elbowed the next guy, and like a wave everybody realized we were standing there ready to work. While Beau was working at the anvil my friend and I switched off being his striker. When I wasn't working at the anvil with him, I would go through the crowd with a bottle of wine and glasses. It caused such a sensation at the event that the two guest Japanese sword makers shut down their demonstrations and ran to watch. Everyone was quite impressed. I looked sort of like a Greek goddess, yards of golden, gauzy material. My dress was held by a single buckle at my waist. My sandals had leaves and grape clusters. I was bare from the waist up except for copper bra cups. It was so funny because that night there was a big banquet. I went wearing my costume. The older blacksmiths would lock eyes with me but not look at my chest. They would ask to look at the bra cups. You really couldn't see anything because Beau had designed it so beautifully. But they were quite fancy. The word got around so widely about the demonstration that a blacksmithing group in New York paid to see the same demonstration.[4]

Despite being an early maker of Damascus, Hickory was very proud of his patterning techniques:

> One of his Bowies had a very complex pattern in it: It has a design on one side of the blade and when you flip it over it has his signature in the blade, in the laminate. His signature was a very long process. He had to lay it all out in steel. One side of the blade has a ladder

design, and the other side has his signature. When he ground the knife the first time his signature popped out perfectly. Then he thought, "I will make it a little bit shinier," and polished it a little more and it took away that definition, but you can still see it easily. There's one knife that has a fairly long blade, 10-12 inches and it has straight laminates right down the blade and then flared out from the side like a chevron, going the other way. The original blade he made when he was young (1946), has a bold, broad tiger stripe pattern. He didn't like super fine patterns you couldn't see like a samurai sword. He liked big, large, coarse designs.

He was the best at pattern prediction. He could show you how to put the laminates together to show how it would come together. He taught classes on that. One of his favorite analogies was if you were a kid growing up buying Christmas candy, they had a little picture of a flower or a plant in the candy. That's the same principle as Damascus steel, they would make this great big roll of candy that is sometimes six inches in diameter and lay the pieces in the middle, the design, and then roll it and keep going until it rolled out to the tiny diameter of the candy, getting smaller and smaller for those candies. Same principles with Damascus.[4]

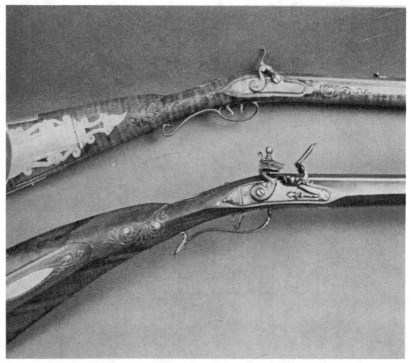

Beau Hickory Rifles with intricate ivory and silver[5].

Hickory participated in the knife community for several years before moving on to his many other interests. He advertised his Damascus knives in *Blade Magazine* in the late 1970s and early 1980s[6], and went to several knife shows. Tinnell said: "He used to go to the knife shows, quite regularly, but he could never sell anything because all the other knifemakers would come by and pick his brain about how to do stuff and he never had any time to sell knives. So, he quit going to shows. He did know Daryl Meier when Daryl was just getting started."[4] So while Hickory isn't well known anymore for knives, he was a very early maker of Damascus blades in the USA, a few decades before they were popularized by Moran. In between his Wild West shows and decorative ironwork. Oh, and the fiberglass cars and the chastity belts.

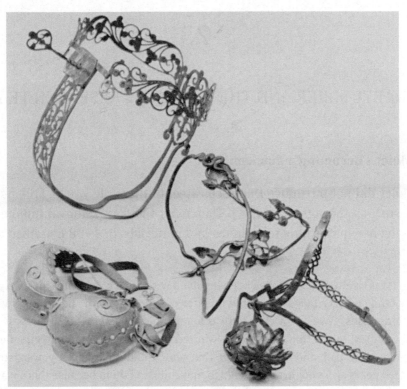

Beau Hickory chastity belts and breast cups[5].

-23-

DARYL MEIER AND THE DAMASCUS RESEARCH TEAM

Daryl Meier - Becoming a Blacksmith

Daryl Meier did not introduce Damascus steel to the knife world. Still, he pushed it further, such as through the sale of Damascus to stock removal knifemakers and introducing new patterning techniques and materials. In 1988 he talked about how he got started:

> Since I was twelve years old, I've been involved with muzzle-loading rifles, very interested in shooting, not as a target thing, but the idea of shooting them and hunting with them. In the mid-sixties, I was in a club with other people who felt the same way. It's a little different now, but at that time, if you wanted to have a good-looking gun and a powder horn, shooting bad, and all the things that you would have had if you existed a couple hundred years ago, you just about had to make most of them because there was no source for you to buy them. So everybody started making powder horns, hunting knives, and so forth. I got involved in several of those crafts, and I finally decided it would be better for me to specialize in blacksmithing and barter my blacksmith work for leather work or horn work or whatever else I might need. I started making tomahawks, throwing knives, and patch knives. Once I got started, then I did graduate, in terms of accumulation of equipment, so I could do different kinds of things: better equipment, bigger anvil, better hammers, and so on.

> I didn't have any particular person that I studied under for any length of time. I did a lot of reading - old books and how-to books. Also, when I got started, the only contemporary book that had any how-to information was the Alex Bealer book (*The Art of Blacksmithing*). That book has been influential, although I was involved in blacksmithing before I ever ran across it. It's an inspirational book, a romantic book, and it just reads well. It's uplifting to read, and it gives you a sense of - I don't know how to say it - confidence almost. He makes things seem simple enough that you feel like you could accomplish them. So I'd say that Alex Bealer was an influence.

> I don't really know (why I became a blacksmith). At the time when I got involved initially, it was a hobby. I did it some evenings but mostly weekends, and I worked in various types

of business administration jobs as a vocation. Sometime after I got involved in blacksmithing, I found that I was getting more satisfaction on a personal basis out of my handwork than I was out of my brain work and my administration responsibilities. I took another look at the concept that we're supposed to amount to something or be somebody here, make some contribution to society, or amass a bunch of wealth, or whatever the hell it is. So I chose to make a living out of blacksmithing. Over a period of years, I've gotten to the point when I was almost making a living blacksmithing... I guess that over a period of time it's just been that I'm trying to limit my activities to only doing what I enjoy and hoping that activity will generate an income. I haven't compromised very much. The only really highly successful people who make a lot of money are the ones who've achieved a very famous stature, and the number of those in relation to all the people practicing in the field appears to be less than one-tenth of 1 percent.[1]

The Damascus Steel Research Team

Meier observed a demonstration on making pattern welded Damascus from Ivan Bailey, a nationally known artist blacksmith, in 1973[2]. However, he still had to figure out much of the technique on his own:

> I just worked by trial and error and I asked people questions, not about pattern welding so much as about the techniques that are involved, like forge welding, fluxing, and fire building, and I built up information.

> It turns out that there were one or two other people who were teaching themselves, and apparently, the knowledge of how to do this stuff had been carried all along, because since then I've found people whose fathers had taught them. But I never could identify them.

> I got real interested in it and read everything I could find, which was very little. I was strongly influenced by Cyril Stanley Smith (author of *A History of Metallography*) and also by Jerzy Piaskowski from Poland. He's a metallurgist in Cracow, Poland. I ran across an article he had written and I corresponded with him. His hobby is ancient technology. He's done a lot of research on ancient artifacts and their metallurgical examination. He dissects them, does other metallographic examinations of them, finds out what their composition is, and then makes suppositions about how they were put together. I corresponded with him, and he's a very nice gentleman. I'd write a letter with three questions and receive a letter with three answers, but the answers and the questions didn't quite match 100 percent.

> And then sometime subsequent to that, I was in the SIU (Southern Illinois University) program here at the Graduate School. I was making pattern-welded steel, and then Jim Wallace and Bob Griffith came as graduate students, and at that time, neither one of them had been interested in it; that is, they hadn't done anything on their own. So I remember they saw me making it, and they stopped and looked at what I was doing and said, "Hey, you're going to have to show us how to do that." I remember that I showed them everything I knew in one afternoon in about two hours, although it had taken me nearly two years to learn.

So in other words, I was at a very elementary level. Then the three of us became just like little kids. Every time one of us would find something out, we couldn't wait to go show the other two. Between the three of us, we learned quite a bit in a very short period of time. We picked up a lot of knowledge, and then after that, we were approached by Dona Meilach to write something about it for her book (*Decorative and Sculptural Ironwork*), which we did as a committee. We each wrote sections of it, and then the other two would review the sections that

Damascus Research Team. Left to right: Jim Wallace, Daryl Meier, and Robert Griffith[3].

were done and argue around until we were all pretty well satisfied. So that was a significant thing. That was the only work on "how to make pattern-welded steel" that had ever been published in modern times. Since then we've all gone our own ways, but I stayed in it; that is to say, it's still the major portion of my activities.[1]

Damascus Steel Chapter in *Decorative and Sculptural Ironwork*

The chapter the Damascus Steel Research Team wrote is surprisingly complete for the early date in which it was published (1977). The chapter began with a brief history of pattern-welded Damascus which came mainly from Dr. Cyril Stanley Smith. The history included exciting details like the French word Liberté forged into the pattern of Damascus gun barrels in the 19th century. They also wrote about the Malaysian Kris, Japanese swords, and the Merovingian Franks and Vikings in Europe. This included images of the different types of construction the Merovingians used to have a combination of hard and soft steel for durability while maintaining cutting performance.

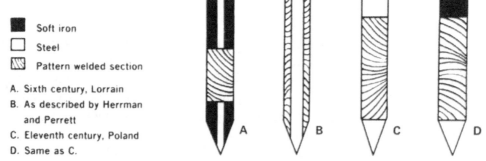

Methods of Assembling Merovingian Blades. Image from Ref. 3.

The chapter also included information on materials to use in the construction of pattern welded steel, including the first published suggestions of using A203E, a high nickel steel, and pure nickel itself, along with the already known W1, 1095, mild steel, and wrought iron. The chapter even suggested different combinations, how well they forge, weld, and etch for contrast, and potential uses for different applications. And, of course, instructions on welding, forging, cutting, and re-stacking for developing the final layer count, and final polishing and etching for revealing the pattern.

Another interesting part of the chapter is that it includes the patterning techniques that they knew at that time, which they broke down into three major types: 1) material removal (grinding, incising) followed by forging, 2) surface manipulation (punching, piercing) followed by material removal, or 3) twisting. They included diagrams for the two different methods of producing ladder pattern. The "rungs" of the ladder can be made by grinding grooves and then forging the steel flat, or by pressing/forging in groves and then grinding flat. Regarding twist, they also described the generation of a "chevron" pattern where two different twist pieces are placed together in opposite directions before being forge welded together. This is the base technique for developing the Turkish Twist pattern described in Chapter 24.

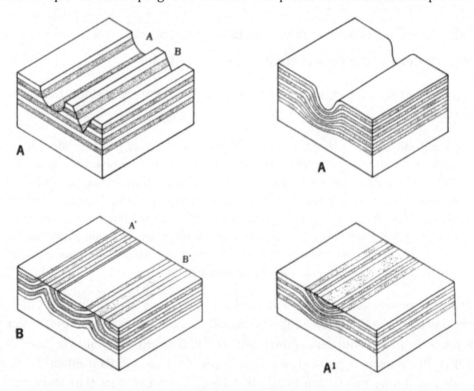

Schematic diagrams of ladder pattern construction from *Decorative and Sculptural Ironwork*[3]. The left shows cutting the steel followed by forging flat, and the right shows deforming the steel and then grinding flat.

The process of producing a "chevron" pattern by the Damascus Steel Research Team[3].

Research Beyond the Damascus Steel Research Team

Meier did not see the same mystical properties in pattern welded Damascus that Moran reported:

> I found out early on that the pattern-welded steel I was making didn't have magic properties; that is to say, when you're done making it, if you quench it and don't temper it, it's not pliable. It couldn't bend and spring back, so I figured maybe I didn't know as much about it as I'm supposed to. I got to looking into that, and the reason for it is in the carbon. Once you start out with two different kinds of steel that have a different carbon content, in forge welding or in any subsequent hot work, like more welding for other seams or forging, the conditions are right for the carbon to diffuse from the high concentration to the low, and it'll try to equalize itself.[1]

Indeed, the perception of Damascus steel's performance was evolving rapidly in the early days. In 1983[4], Meier published a poll he conducted on the performance of Damascus from 23 producers of the material: Bill Bagwell, Phil Baldwin, Gary Barnes, Sid Birt, Bill Buckner, Jim Corrado, Jack Crain, Jim Crowell, Fain Edwards, Howard Faucheaux, Jimmy Fikes, Don Fogg, Kurt Lang, Scott Lankton, Dan Maragni, Louie Mills, Bill Moran, Al Pendray, George Schepers, Jim Schmidt, Steve Schwarzer, Jim Serven, and Wayne Valachovic. In some cases, more than 23 answers were given because four makers gave different answers depending on the layer count of the Damascus. Meier asked the makers to compare the performance of Damascus to a plain high carbon steel. On the subject of edge holding, 11 believed that Damascus was superior, 11 said that it was the same, and one said that plain carbon steel was better. For toughness, 22 felt that Damascus was better, four said the same, and one said that carbon steel was better. So, it seems that opinions had shifted away from pattern welded Damascus having better edge retention than plain carbon steel, but most still believed Damascus had better toughness.

By 1988 Meier's appreciation for the ancient people making Damascus only deepened:

> The Merovingian period ... (is) the earliest example of pattern-welded steel that's highly refined that I know of, but I don't think Merovingian people invented it. There's evidence of it being done earlier. The Merovingian period runs roughly from about A.D. 400 to A.D. 650... (They lived in) South Central Europe. Merovingian is a Germanic culture but it got as far west as France and also into what is now eastern Europe. The more information I uncover about these ancient people, the more respect I have for them as craftsmen. I've had the same experience that most of us have. We go through the contemporary American education system, which gives us the idea that anybody that lived before 1925 was basically uncivilized, but I find some of the work done in the fifth and sixth centuries amazing. People didn't have a written language in Merovingian culture or if they did, it was known to so few it didn't count. Maybe a few of them could converse in a nonnative written language like Latin or whatever. There's some evidence of Latin characters appearing on some of their work, but anyway, they made some fabulous stuff and they were doing things without modern technology. I find it difficult for me to do today what they did, and in some cases, I haven't been able to.

> One thing that I've always felt in relation to blacksmithing is that you can't store it in a book. It has to be stored in the living mind and arms... The only way you can convey a color is with a color chart, which isn't as good as the real thing. When you've got a person standing there at the fire, and you say, "See that right there," then that's it.[1]

Thoughts on Damascus Steel

Daryl Meier has always been an excellent philosopher regarding pattern-welded Damascus. So even though none of these quotes fit in anywhere in particular, I wanted to include them anyway:

> I'm not researching this stuff because I see a pot of gold at the end. I just want to find out, and I can't answer that question of why I'm interested. I just am. It's like, well, I like trees, I like to look at trees. I like to be around trees. I can't explain why. I grew up around trees and I just like them. I just got hooked on this, and I'm going to stay with it until I pass out of the picture.[1]

> It's not as dynamic to work on steel at room temperature as it is to work on steel that's 2,500 degrees. I call my work the application of brain power and force, unlike knife finishing, which is finesse, not force.[5]

Once I learn the cause and effect of a pattern, I use the molds as a basis for modification. I can make any pattern in steel you can furnish me a drawing for, but I'm most inspired by movement of abstract into representational.[5]

Most of the reputation that I have in my particular field is among other people who have a special interest, and they're involved in doing it themselves rather than in being consumers. Also, they don't have the kind of money it takes to buy my work. So it's a strange thing. Almost everyone throughout the United States who is involved in pattern-welded steel for any length of time either knows me or has heard of me. But they're not the buyers or the collectors. Also, I'm not a salesman, I'm a blacksmith. I think like a blacksmith, and when I lay the hammer down, I lose interest. When I see I've accomplished what I wanted to as far as hammering is concerned, I'm done.[1]

[When you start making Damascus you are quickly infected by the] Damascus Virus, and there is no known cure.[6]

The complexity of the pattern in pattern welded steel is limited only by the imagination and skill of the maker.[7]

There is no right or wrong involved in either the patterns or the finish of welded Damascus. What a pattern should look like and how it is finished is entirely the choice of the individual smith, just as the painter may choose to do an abstract in oils or a watercolor landscape in realism. The pattern on a welded Damascus knife is the creative expression of an individual human spirit.[8]

Daryl Meier's Legacy

In the coming chapters, you will see several more areas of Meier's influence on the evolution of pattern-welded Damascus, but his influence has not consistently been recognized. Ric Furrer in 1999 said, "It's because of Daryl's continued teaching that we have the range of Damascus that we see today. His teaching has permeated the modern knife-making industry and for whatever reason he is not acknowledged for having the effects that he's had, to me he will always be one of the greatest. I don't think people understand the true genius of his work... He is the father of modern Damascus."[9] Mosaic Damascus is another topic covered in later chapters, and Steve Schwarzer had this to say on Meier's influence on it: "Daryl seems to be the only person directly connected to all the people who were first working in the modern mosaic medium."[10] And, "Is Daryl Meier the Father of Modern Mosaic Damascus? Without a doubt, he gets my vote. He was the only person I could call on for many years to help solve problems in this fantastic reborn art medium that began in the mists of time."[10] Devin Thomas said about Meier, "Daryl Meier has had a bigger influence on my work than any other smith. On everything from forge welding, patterning, materials, metallurgy, perspective, price, and love of the craft."[11]

-24-

COMPOSITE DAMASCUS AND WORDS IN STEEL

Daryl Meier described Turkish Damascus in 1983:

> Another type of pattern beginning to appear in contemporary welded Damascus knives is called Turkish Damascus by German cutlers. A more appropriate name would be European Damascus since the technique was used in Europe at an earlier time than in Turkey. In any case, the pattern is created by forging a billet into a small square rod, twisting the rod, and then welding two or more such rods together on edge. European swords from 500 AD had up to 6 twisted rods welded together to form the central part of the blade. The pattern can be varied by the number of layers in each rod, how tight the rod is twisted, alternating right and left hand twisted rods, by the amount of grinding the finished piece. In the 18th and 19th century, this technique was used in the gun barrel making business.[1]

Daryl Meier made his Turkish Twist Merovingian sword in 1975 with 203E, W1, and wrought iron, shown on this page.

Jerry Rados said, "As best as I've been able to track it down, Turkish Damascus was given its name in Germany in the 1600s-1700s to distinguish it from other patterns. It had to do with Turkish tapestry and manipulating different patterns."[2]

Indeed, this technique has been found in blades of very ancient date. Understandably, these blades are badly corroded, such as the closeup of a Viking sword on the following page. However, the pattern is still visible through the corrosion, despite the blade being from the 8th-9th century AD.

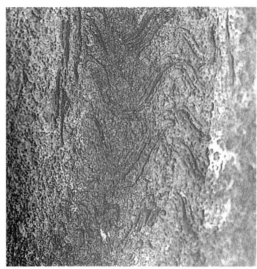

Turkish pattern in ancient Viking sword. Image from Ref. 3.

Sometimes other advanced techniques were used in ancient blades, such as using twisted and untwisted portions of forge-welded bars to create an interesting pattern. For example, the images below show a recreation of the Sutton Hoo sword, which dates to the 6th to 7th centuries AD. Scott Lankton recreated it in 1989 for the British Museum, which houses the original sword.

Sutton Hoo sword recreation from Scott Lankton. Images from Ref. 4.

Tighter multi-bar Turkish patterns are also common with blades and gun barrels (which we will discuss next). Below is a picture of the process of making a billet of Turkish twist Damascus from Jerry Rados, famous in recent decades for his beautiful Turkish Damascus. First, the billet is stacked up with alternating layers of the two steels, then forge welded, forged down, and twisted clockwise or counterclockwise. Then the alternating twist-direction bars are forged together to make the pattern. Finally, Rados ground the bar to different depths to show how the pattern changes with grinding.

Rados, who has made Damascus since 1984, was also inspired by Daryl Meier to make the steel he is best known for: "Basically, I picked the Turkish up from Daryl, and I have worked on it a bunch of years, trying to refine it, and I think I have it down pretty well now."[5]. Pretty well, indeed!

Jerry Rados Turkish Damascus. Image from Ref. 6.

Damascus Gun Barrels

Damascus barrels were often made with multiple twist bars with very tight Turkish twist patterns, which were striking. Manfred Sachse demonstrated how these barrels were made, which is shown below. First, steel is twisted, then forged into a multi-bar Turkish pattern, and then wrapped around a cylinder.

Demonstration by Manfred Sachse[3].

Barrels of this level of quality were produced in Persia dating back as far as the 16th-17th centuries AD[7]. Knowledge of Damascus gun barrels made their way to Europe around 1683[3]. Jean Francois Clouet (1751-1801) of France was a pioneer of pattern-welded Damascus gun barrels. He was a chemist, a member of the French Academy of Sciences, and a technical director of a steelworks in Daigny[3]. The city of Liège, France (now a part of Belgium) became a major center for Damascus gun barrel production by the early 19th century. Other European cities became known for Damascus production, including Birmingham, England; St. Etienne, France; Suhl, Germany; and Brescia, Italy. These European smiths also created new patterning techniques, such as putting words into the Damascus steel pattern in Liège. For example, you can see a gun barrel below that says "Zènobe Gramme" in the steel (the name of a Belgian electrical engineer).

"Zènobe Gramme" Damascus[3].

Daryl Meier learned of these gun barrels with words in the steel from Cyril Stanley Smith's *A History of Metallography*[8]. Meier also bought a book called *Damascus Barrels* by Jean Puraye[9] when he visited Liège in 1978[10]. After this trip, Meier decided to try to recreate this patterning technique. He machined the letters of his last name, "MEIER" from solid blocks of steel and then placed them into the billet[11]. The result was a 1978 dagger with "Meier" found along the spine.

Daryl Meier Damascus dagger with "MEIER" in the pattern.

To better understand the construction of this type of Damascus, I have another similar attempt by Manfred Sachse from 1984[3]. First, he created his name in the top and bottom of a billet, along with straight layers of two different steels in the center. The shape has been distorted somewhat because keeping billets perfectly square while forging them is difficult.

However, this view is from the "end" of the bar, so the letters are not visible in the "flat" of the bar. But if the bar is further forged down, twisted, and then ground flat, the letters are now twisted, but can be viewed on the ground portion of the bar, mirrored on either side. Sachse then forged multiple twist bars together, and the name Sachse can be seen where the different bars meet.

European gunsmiths also included other designs and patterns in the ends of their bars, such as by creating a "checkerboard" in the end of the billet and then twisting those into a Turkish pattern. This created a pattern called "Bernhard," which looks very interesting.

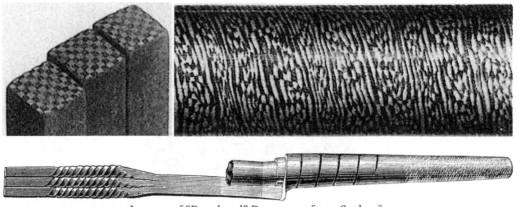

Images of "Bernhard" Damascus from Sachse[3].

In 1985-1986, bladesmith George Werth devised a method for creating different patterns in the end bar by using square wire from a friend who owned a wire factory in Chicago[11]. In March 1986, he visited Daryl Meier, who showed him his dagger with his name in the blade. Werth realized that he could do the name without machining from full bars of steel but instead with his square wire method, which he did in September-October of 1986. Werth made a similar blade for photographer Jim Weyer[12] shown on the following page. The end of the bar shows what the billet looked like before twisting, and the twisted portion was ground to two different depths. Notice that the shape of the pattern changed with further grinding. Werth also used a checkerboard pattern similar to the old gun barrels.

Rob Hudson and Composite Damascus

"Composite Damascus" was coined by C. Robbin "Rob" Hudson. Composite Damascus refers to welding several bars of Damascus to each other to create mixed patterns in a single piece. It is similar to the multi-bar construction of Turkish but is a more encompassing term for many combinations. There are examples of what could be called Composite Damascus historically, such as the European gun barrel shown with a strand of straight-layer Damascus mixed in with twisted bars.

Hudson is a bladesmith who began making knives in 1970. Hudson had a bachelor's degree in art and had been living in Baltimore as a high school art teacher, focusing on painting and sculpture[13-15]. Martha Hudson, his mother, also greatly influenced him, who was also an artist. Hudson was fascinated with knives at a young age: "I always liked knives as a little kid. Ever since I can remember. I've been fooling around with knives – using them, breaking them, fixing them, altering them, throwing, carrying and losing them and then making them."[14] "When I was 15 I saw a Moran carving knife, and something told me that I was looking at a *real* knife. When I turned 16 and got my driver's license, one of the first things that I did was to drive to Bill Moran's shop, which isn't that far from Baltimore, and I gave him a drawing of a large Bowie knife I wanted him to make for me. Six months later, I had a *real* knife."[15].

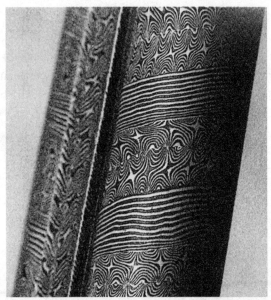

"Composite Damascus" gun barrel from Ref. 3.

Hudson first began making knives as a hobby in the evenings. In 1974 he and his wife moved from Baltimore to Maryland's Eastern Shore, which led to Hudson switching careers from art teacher to bladesmith. "I just got married and was totally broke; I had to do something,"[14] said Hudson. Hudson didn't use many power tools when producing his knives, and he appreciated historical methods for making knives. He said he was "continuously going backwards" in how he made his knives. "There are very few who are crazy enough to do it this way. It's a weird occupation. It's a neat way to build. I'm glad I'm doing this. I don't want to do anything else."[14].

This lack of large power tools helped lead to Hudson's many "Composite" Damascus pieces. Without a power hammer or press, making a large bar of Damascus was a lot of labor, which he needed for a Damascus dagger order[16]:

> I made my first composite blade for exactly the same reasons the early Vikings did. I had an order for a large Damascus dagger and didn't have a power hammer. It was the hot bicentennial summer of 1976 and the idea of trying to build a huge coal fire and pound out this massive billet by hand in my tiny sweatbox of a shop just wasn't very appealing! So I ended up making three separate billets outside, each on a different day and on the fourth day welded them together on edge before shaping the blade. It seemed so easy by comparison. I made Damascus knives by hand for another nine years before I came to my senses and got a 25-pound Little Giant power hammer. The experience left me with all kinds of tendonitis problems but forced me to experiment with the composite style construction.[17]

Hudson introduced his idea of composite Damascus blades at the 1983 New York Knife Show.[17] On the following page is a composite Damascus dagger Hudson produced sometime between 1976 and 1983, as it is missing the "M.S." stamp he began adding when he became an ABS Master Smith in 1983[18]. The knife has a bar of straight W2 in the center with two separate edge pieces of W2/203E Damascus. The center bar of W2 also has another frequent feature of Hudson Damascus knives – a "resist etch" pattern. Hudson would draw figures using a permanent marker which would resist etching, leaving a bright, unetched portion of the blade.

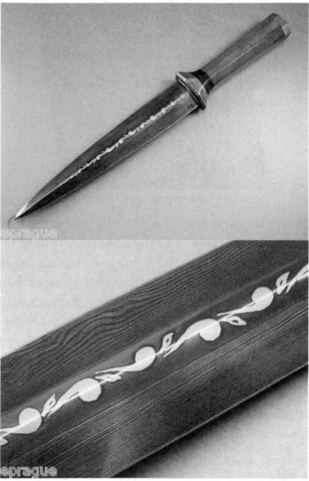

Images of early Composite Damascus blade by Rob Hudson. Images from ePrague.

The following page has a picture of a Viking dagger with his more evolved style of the 1990s, including Hudson's signature "flame edge" pattern, also created with a resist-etch. Hudson's style was very influential. Rick Dunkerley said, "His large composites had a major impact on many Damascus makers in the 1990s. Shane Taylor and I were very inspired by his work. He was a big influence on us at that time."[19] "I submitted five knives for my American Bladesmith Society Master Smith test in 1997, and four of those knives had composite-bar blades. I made the decision to submit composite-bar blades because I felt the forging technique demonstrated to the judges that I was a competent Damascus steel maker. It must have helped because I passed the test."[20] The beautiful style of Hudson's knives and Damascus reflect well on his art background; however, Hudson did not view his knives as only art pieces. He said, "I like all of my knives, even the ones destined for under-glass collections, to be totally usable and abusable, virtually indestructible. It is fine for a knife to be fancy until the fanciness detracts from the function. Then, it is an art object, not a real knife!"[18].

Knifemaker ~ Robbin Hudson, MS   Photo by Coop

Hudson 1990s dagger with "flame edge." Photo by Coop.

-25-

DON FOGG AND HOT DEFORMED PATTERNING

Don Fogg is an influential knifemaker known for introducing several Damascus patterning techniques. He has recently shared his knifemaking history in a 2021 podcast interview:

Image from Ref. 2.

> I sold a small business, and it gave me enough money to sit and not have to figure it out. In that process, I had a friend who was running a blacksmith shop. I started bringing him coffee and hanging out; watching him work. It wasn't long before I asked him to take me on, with no pay. I was a gopher, and he would give me a lesson every once in a while. He had a show in the summer where he was obligated to keep the shop open in the summer. To do the craft show he had to find someone to run the shop, and that was my payback. I'm running the shop, and you get the peanut gallery coming in and out; tourist things are strange.

> I met Jimmy Fikes. I had an order even though I hadn't been blacksmithing very long. It was for a gate closer that was a scroll with a rock in the middle. Jimmy chipped in, and he showed me his ironwork. It was medieval, just incredible work. We struck up a friendship. I don't know how much longer I ended up having my own shop in the Lakes region of New Hampshire, and he came up for a visit. We would go back and forth. That was around 1977.

> He came up in the winter with a *Blade Magazine* and had found out there was a Connecticut knife show. I had never worked with high carbon steel; I had never made a knife. Along with another friend of his, Jim Schmidt, we drove down together to our first knife show. Walked around, met people I have been friends with my whole life; Mel Pardue comes to mind. We realized this was actually a market. Jimmy came back and said, "We can do this." He had worked with Hank Reinhardt down in Atlanta, making weapons for the Society for Creative Anachronism. So, he knew a lot about medieval weaponry. He learned the basics in a practical shop.

> He gave me the lessons and I started. Jim (Schmidt) had already been making knives. The halfway point for all of us was Jimmy's shop. I basically sat on the sidelines and watched them show their stuff until I got my skills up enough that I was proud to show them something.[1]

I was the junior maker and often just sat and watched them tease each other. They were very competitive, and each would do their best to outdo the other.[3]

The Philosophy of Knifemaking and Teaching

Knifemaking was a therapeutic activity, as he said in the 2021 interview:

> What attracted me initially (to knifemaking) was the ability to quiet my mind. I had all these incessant thoughts that I couldn't quiet, other than through negative behavior. That was the first step. I didn't set up to be an artist; that's never been my goal. I guess my motivation was to stay interested. To do things that interested me. I tapped into something that was interesting in every way. It provided me mental calm, it challenged me, it was learning something from scratch and there is no shortcut. You are rewarded by seeing the work in front of you; it isn't abstract. It gives you feedback. A lot of what I found missing in my life, money wouldn't do it, I wasn't a social person, but finding something that could reflect myself in the work. I didn't realize any of that at the time; it comes with hindsight.[1]

Fogg wrote about his philosophy of knifemaking in a series of articles called "The Way of Bladesmithing" on his now-defunct website, DFoggKnives.com, reiterating some of his thoughts such as those above: "When working, the mind must be totally involved in the process. If thoughts are anywhere but on the work, it shows immediately. Most tools are dangerous if you are not focused, they demand attention. Consciousness is forced to remain in the present. Every action requires thoughtful consciousness and every action shows directly in the work."[4]

DFoggKnives.com was one of the early websites about knifemaking in the 1990s and included many topics: Damascus, making a katana, sharpening, grinding and finishing, and heat treating. His website also included a discussion forum which has since morphed into BladesmithsForum.com, which continues today. Fogg said:

> I had always been interested in the possibilities of the internet. I had an account that was mostly a collection of netservers; mostly geeks. When it went graphic a lot of blacksmiths and other visual artists started sharing. I started out with a forum because I kept getting a lot of the same questions over and over. Once AOL kicked off the graphic thing, I developed a webpage. The webpage was wonderful. I made friends all over the globe.[5]

This website and discussion forum were an extension of Fogg's early ideas about teaching, as he was also part of the group that started the Ashokan seminar. Fogg said, "We felt that the only way bladesmithing was going to gain momentum was if we taught each other and demonstrated to the public. In the early days Ashokan was a premier event and attracted everyone who was interested in the craft."[3] Ashokan seminar started September 1981 at Ashokan University, initially sponsored by the American Bladesmith Society:

> We set up a hammer-in for bladesmiths at Ashokan, our first hammer-in. It drew people from all over. Steve Schwarzer, Mel Pardue, Al Pendray, Dan Maragni, Phil Baldwin, there were a bunch of people. We had a really exciting conference; it was like sipping from a firehose. Everyone was attacking these problems differently. We all grew exponentially. We were competitors in a tight market. But by teaching each other the skills, we upped our game, and fast. It's counterintuitive; you wouldn't think you would want to share

things with your competitor. But when you're starting from the ground level, that's the fastest way for everyone to grow. The shows were competitive. There was a limited amount of money, especially in high end product. You wanted the money to fall on your table. It was kind of an unspoken rule back then that there weren't any secrets. Once a lot of people started coming in, it didn't narrow down, but you had your friend network, and a lot of people weren't in the network. To get into the network you had to share.[1]

Murad Sayen, Kemal, and Artistry in Knifemaking

Fogg spoke of learning as an essential aspect of his knifemaking:

> Once I started, I didn't know anything, so I was constantly buying books, [*Stones' Glossary of Arms and Armor*] because it had pictures of weapons from all over the world. Different blade shapes and handle configurations. It was just like eye candy for me. And then there was a museum arms catalog, and I used

Murad Sayen (left) and Don Fogg (right). Ref. 6.

to subscribe to that. Everything I could read on the subject, which wasn't much. It was sort of like bootstrapping the whole process. The more you learn about the craft, the more you can appreciate how they did it. It was intellectually interesting; it took me back to every culture in the world. And I never got to the end of it. You go through the process, and all of a sudden you have to learn the science when you start heat treating. You learn more and more about what you're doing. I never tired of that.[1]

Fogg credits Murad Sayen with helping him understand design and art. Together they formed "Kemal," and Fogg's knives from 1980 to around 1990 were primarily within this partnership:

> I had a partner for about 15 years who was an incredible artist, Murad Sayen. I didn't even understand the language of line. I'm an English major. Murad, I partnered up with him, he was making hunting knives. All of a sudden, I realized this guy is an artist. He's a photographer, oil painter, and sculptor.

> I was suddenly opened up; this stuff has even more dimensions that I was even seeing. There are so many levels to the work. I learned on my own pieces; if I didn't like it, I took it off. I didn't spend a lot of time with paper and pencil because it was more in the flow of the piece. I was learning, so I didn't know what to put down on paper. I had to make it in order to see it. It was a process of discovery in myself. If it caught my eye, I would study it. Why is this catching my eye? Why is this attractive and that is not? The disadvantage (of the partnership) is I didn't learn the handles. Then when Murad and I split up I had to start all over. That was a long period of time too. My blade work was way beyond my

handle work. I had to recreate my customer base as well. Basically, it was time to go separate ways. It was sort of like we had played all our tunes and we weren't willing to go back and play them again. We ran the music out.[1]

Pattern-Welded Damascus

"I started out with Damascus. Bill Moran was the only game in town; Hastings and Bagwell were soon to follow him. Jimmy (Fikes) taught me how to weld steel. Fikes had attended an ABANA conference where several blacksmiths had shown up with Damascus. Beau Hickory and Daryl Meier, Rob Hudson may have been in there. They had all learned that stuff. Jimmy said, 'I can do that; it's just a forge weld.'"[1] By 1977, Fikes was offering a range of Damascus patterns, including "maiden hair, ladder pattern, starburst, flower, wood grain, and others."[7] Fogg was fascinated:

> There were a lot of mistakes along the way, finding materials; it was all trial and error at the beginning. It was a hook; it was different than everything else on the table. It was interesting. To me, it was absolutely fascinating. I started, and it just took over. I started exploring; we didn't have much pattern development, mostly incised patterns and wood grain. Then I started seeing what the steel could do. I could see how many variations you could get and how visually attractive it was.

> It's such a beautiful material and so responsive to the maker. You can get it to do almost anything you want, and it still maintains its organic quality. It's a Cadillac material.[8]

W's Pattern

Fogg developed the well-known "W's" Damascus technique (demonstrated in images on the following page) in the early-to-mid 1980s:

> To me, it was one of those mistakes that was pretty. I was going pretty fast, and I got the handle orientation wrong somewhere in the process. Instead of working on the flat, I worked it on the edge. After a billet has been forged out a ways, it can be hard to tell. I drew out the bar of steel, ground it and etched it, and drove immediately to Jimmy's house to figure out what the hell I had done. Because it was pretty; it was visually interesting. We evolved it, and I said I will do that intentionally. I knew it was a success; I made up a small dagger out of it and took it to the New York Show. It was frantic when it got on the table. There was a guy waiting at the door saying, "I have to have that."[1]

Fogg named the pattern "W's" because of the angular points the layers make where they meet from the previous "C" shapes in the deformed bars. However, the W's pattern was more than just a new pattern. It also changed how Fogg, and subsequently others, view how to develop patterns with pattern-welded Damascus:

> I started running with it. If I do this, I'm not limited to inscribed ladder patterns or what you get from a straight laminate. Now I realized any change in aspect while that billet is being worked will affect the product, and you can manipulate it. I played with the W's pattern for quite a while. What it did was it changed my way of looking at what could be done. Damn, this is endless. I had a run where I made variations. Never made the same billet twice for years. A lot of those didn't turn out.[1]

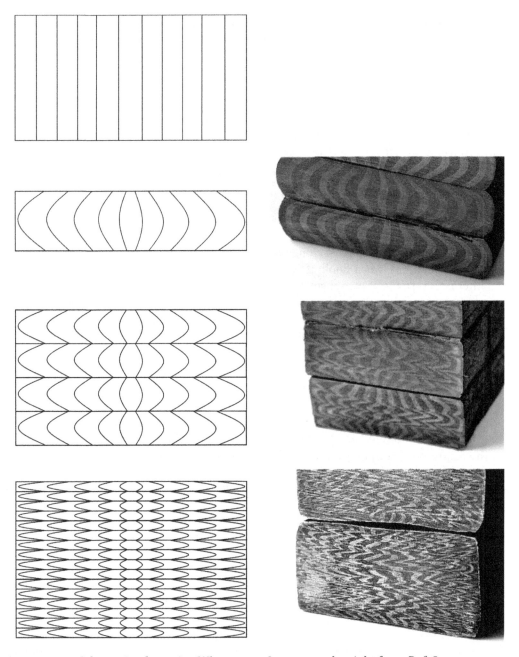

Schematic of creating W's pattern. Images on the right from Ref. 9.

Fogg sees the development of Damascus patterns as an evolutionary process:

> Damascus is so expressive. It will draw you right into a dramatic pattern and, because it's organic, the steel responds, and every blow is registered in that steel. If you go with that and see it as an asset, you might get a wood grain or tree-bark look, or even some other version of it. Pattern development can drive you crazy. You can create something and see a section that gives you an idea, and then something else that will give you another idea.[8]

Kemal (Fogg/Sayen) knife with 3-bar composite Damascus including accordion W's pattern in the center bar. Weyer photo.

Accordion

The W's pattern is created on the end of the bar, so Fogg created a method to bring the pattern onto the flat of the bar: "I needed a way to get the end grain pattern to the surface. I came up with a method where I hot cut the bar and unfolded it, resulting in wide bands interspersed with pattern. Hot cutting led to a lot of failures. The bar would split, which was heartbreaking."[5] "Instead, I started cutting the steel with a bandsaw and forging that flat. This led to a lot fewer failures."[5]

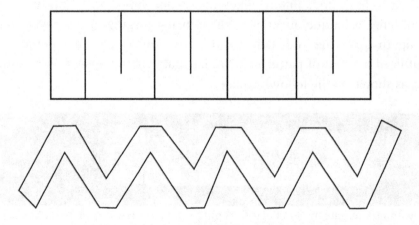

Schematic of accordion process.

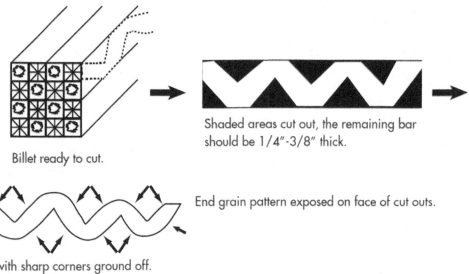

Billet ready to cut.

Shaded areas cut out, the remaining bar should be 1/4"-3/8" thick.

End grain pattern exposed on face of cut outs.

Bar with sharp corners ground off.
Ready to flatten.

Schematic of Accordion Method by Rick Dunkerley[10].

Rick Eaton Damascus folder with accordion mosaic Damascus[10]. Note the regions of distorted pattern mixed with the regions where the "end pattern" is visible.

The accordion method became common with revealing "end" patterns and mosaic patterns. The earliest published use of the term "accordion" I found was from Al Dippold in a 1995 *Blade Magazine*[11].

Jelly Rolls and Ribbons

Don Fogg also worked on various shapes during forging: "The next evolution we came up with was Jelly Roll. Jimmy Fikes and I developed it."[5] This was more of a blacksmithing technique, forming a shape using forging. A low-layer count bar is rolled up to form the roll. Other shapes can be forged, of course. This is an underutilized method of patterning. Similar patterns were seen in rare Indian gun barrels, as shown on the following page.

Jelly Roll Damascus by Don Fogg, first published in Knives '91 (1990). Weyer photo.

Image of Indian jelly roll gun barrel circa 1700[12].

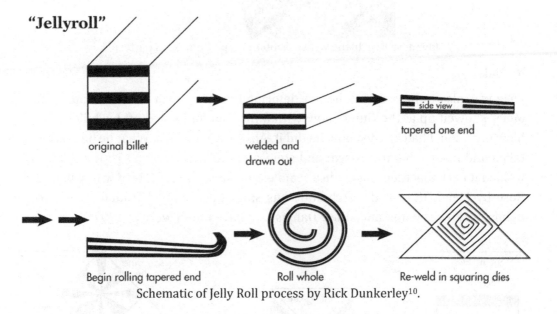

Schematic of Jelly Roll process by Rick Dunkerley[10].

Fogg also saw others doing forged shapes for patterning:

> There was a miner down near me named Larry Sandlin. Larry was a very clever craftsman; he made fiddles and guitars before he started doing ironwork. He started making knives, and then he learned that Jimmy Fikes and I lived right next door, and we showed him how to make Damascus. He didn't have a power hammer, so he came over to the shop to draw out a billet. He drew it out until it was 15 feet long, 3/8" wide, and 1/16" thick. I let him go, I didn't know where he was headed, but I had faith in him. He ribboned it up like Christmas candy. And then he welded it up; it was beautiful, it was gorgeous.[13]

The Javanese Kris also inspired this technique. Daryl Meier proposed a similar technique in a 1983 *Blade Magazine* article for the method by which a Kris blade had been constructed [14]. A picture of a Kris with this type of pattern is shown on the following page.

Javanese Kris blade with a "ribbon"-type Damascus pattern[14].

Radials

Fogg was also impressed by the hot deformed patterns that others created: "Daryl Meier showed up at the Guild Show with a tile that he lopped off his billet that just blew my mind. I had no idea how he did it. Absolutely no idea. What he had done was taken and made a blunt crosscut and squeezed the layers down to a point and then welded it back together and made a seamless mosaic."[1] This pattern was called Calico Rose by Meier, though the technique has since been named "radial." The earliest published knives with Calico Rose Damascus I have found were in 1997[15].

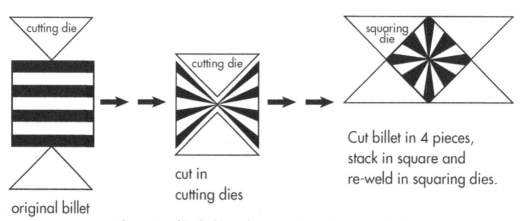

Schematic of Radial Development by Rick Dunkerley[10].

John Lewis Jensen knife with Meier's Calico Rose Damascus[16].

Feather

Don Fogg also developed a popular technique called the feather pattern:

> I got this book – it was a museum catalog out of the University of Hawaii, and it was documenting a collection which is now in Florida of Kris's. The Javanese and Malaysian Kris. Those guys did a lot with creative and innovative patterning. One of the things that they did, which took me a long time to figure out, similar to where Daryl pinched the billet in two; these guys would make up a billet and hot cut it with a dull chisel. Instead of cutting the lines, they would pull the lines down. That's where the feather pattern came from. There's a picture on my website that was done years and years ago. You don't want a sharp chisel because you'll cut through the edges; you want something pretty dull. So, it takes a lot of force. But the Javanese, they were doing that, making swirls in it, just playing with it.[13]

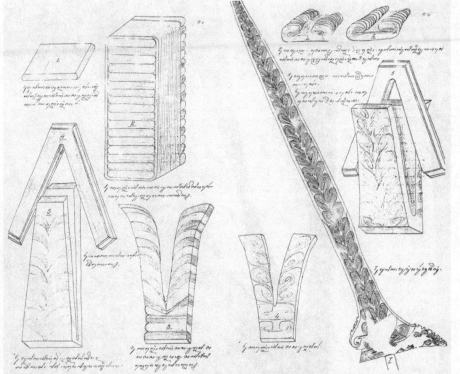

Old Javanese manuscript showing the hot cutting technique to affect the pattern[17].

Larry Sandlin dagger with feather pattern Damascus from 1996[15,18].

Don Fogg/Jim Kelso "Life and Death" Dagger made in 1997 with Feather Pattern on one side (life) and rough textured wrought iron on the other (death).

Forging on the Bias

Another technique for changing the look of Damascus with distortion is "forging on the bias" or forging on the corner. This is similar to the W's technique, though instead of rotating 90° and forging on the "side," the steel is rotated onto the corner (bias), forged, then re-squared. Rick Dunkerley told me that he first learned about this technique from Hank Knickmeyer and that the term came from J.D. Smith. Forging on the bias puts the original layers at an angle and distorts them.

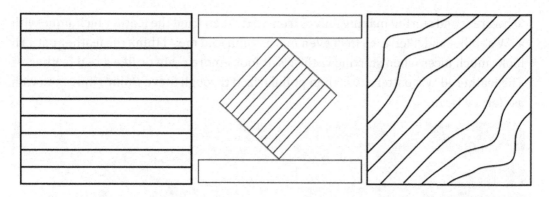

This technique can be used to generate many patterns. For example, four bars can be put together to form a pattern, called a "4-way." Four of these bias-forged pieces will form a star in the center. Then multiple of those can be put together for multiple stars, as shown in the following image.

Combining Methods and the W's "Explosion"

Of course, many Damascus methods result from combining various patterning methods, such as bias forging in combination with a 4-way above. Combining W's pattern with many other techniques has become a standard way of creating complex patterns. For example, Kay Embretsen's "Explosion" pattern was introduced in the mid-1990s (discussed in Chapter 29) which is a simple modification of twisted W's. It was from excellent knives with beautiful Damascus utilizing a few methods, starting with W's pattern, that led to the high popularity of W's-based complex patterns ever since. Another popular combination is to "feather pattern" a W's billet, which creates a very interesting pattern. Steve Dunn learned about feather pattern from Steve

Schwarzer and started making knives from it. He then tried the feather technique with a "W's" billet and liked the effect even more. Dunn told me: "I think the feather pattern looks much nicer when starting with W's. It looks more realistic like a real feather."[19] The feathered W's pattern was then popularized through Steve Dunn's knives as well as Harvey Dean.

Harvey Dean knife with "Feathered W's" Damascus. Image from Knife Treasures.

Steve Dunn also developed a pattern called "Thorns and Thistles," first shown in a knife at the 1998 Guild Show. This shows a combination of hot deformation for manipulating the pattern to demonstrate the evolution of patterning techniques. Dunn started with a seven-layer billet of alternating layers of 1075 and 203E (15N20 was not yet widely available). Then, he forged it down to a small ¾" square piece, 52 inches long. Dunn told me the story of how he developed the pattern:

> I made a bottom die, called a saddle die. I started running those ¾ x 52 inch bars through that v-die, with the layers vertical. I thought the layers would all go down to the point (like radial), but they didn't. Instead, the outer layers curve back on itself. I put the triangles together and rewelded them and then combined them with a 4-way at a time and you start getting flower petals. I continued that process, 4-waying each time until the pattern reached a size I liked. At one point I took a break and etched the end of the bar, it was in the summer, and I was looking across the field and I saw a thistle, and I thought, dang, the pattern looks like thorns and thistles.[19]

Steve Dunn 1998 "Thorns and Thistles" folder[20].

Process of making Steve Dunn's "Thorns and Thistles" Damascus.

-26-

AL PENDRAY, JOHN VERHOEVEN, AND WOOTZ

History of Recreating Wootz

Though both are referred to as Damascus, Crucible Damascus (often called Wootz) is an entirely different material than pattern-welded Damascus, as Wootz gets its pattern from the structure of the steel itself, not by welding two different materials together.

If there is anything consistent in the history of recreating Wootz, it is that many have claimed to have recreated it. Early investigations began in 1790 when Helenus Scott delivered some Wootz ingots, or cakes, to the British Royal Society for analysis. The Wootz steel was reportedly very high in quality and the best type of steel being made in India. James Stodart, a cutler, analyzed Wootz steel by forging one of the ingots into a penknife, and reported that it was better than steel produced in Britain. Those results were reported in 1795 by George Pearson[1], along with his analysis, which confirmed that Wootz was primarily iron, and that it was steel rather than wrought iron or some other form of iron.

In 1805 David Mushet presented his findings on the Wootz steel ingots[2], and he concluded that the carbon content of Wootz was higher than steel made in Britain. This was correct; Wootz ingots historically had very high carbon, between about 1.2 and 2.0%, and typically around 1.5%. Stodart and Faraday performed their experiments with alloying of steel based on their analysis of Indian Wootz[3,4], as was previously described in Chapter 1. They also reported that they had unintentionally produced Wootz when they attempted to alloy steel with titanium: "With titanium we failed, owing to the imperfection of crucibles. In one instance, in which the fused button gave a fine damask surface, we were disposed to attribute the appearance to the presence of titanium; but in this we were mistaken; the fact was, we had unintentionally made Wootz. The button, by analysis, gave a little silex (silica) and alumine (aluminum oxide), but not an atom of titanium." So, their erroneous conclusion that Wootz contained silicon and aluminum came from an analysis of steel they had made themselves rather than from an original ingot.

The work of Stodart and Faraday was published in France, and Jean Robert Bréant conducted many alloying experiments of his own with steel in an attempt to recreate Wootz. He eventually came to the same conclusion as Mushet, that Wootz was steel with a very high carbon content[5]. He also found that the pattern of Wootz comes from the carbides present in the steel, and he reported that the steel must be cooled slowly to allow the carbides to form. Reportedly he successfully made a Damascus-like pattern but did not explain the method in sufficient detail to be recreated. Pavel Petrovich Anosov, a metallurgist in Russia, next attempted to make Wootz using the reported research of Stodart, Faraday, and Bréant[6]. He continued with the hypothesis that the structure of Wootz Damascus came from slowly cooling of the ingot to form a coarse carbide structure and studied different clays and graphite to use in making the steel. These findings were also extended by Tchernoff in the 1860s and Nicolai Belaiew in the early 1900s[7]. Belaiew reported that the Wootz Damascus of Anosov could successfully cut a silk scarf in midair while European steel reportedly could not be sufficiently sharp to complete the task. Käthe Harnecker of J.A. Henckels in Solingen, Germany made Wootz Damascus blades in 1924[8] basing her methods on Belaiew. Harnecker put low carbon Swedish steel in a box at 1000-1100°C for 12-14 days with wood allowing carbon to slowly diffuse into the solid steel (it was never melted). The steel was then very slowly cooled and then forged at a low temperature and annealed. This resulted in a characteristic pattern on the blades she produced. Henckels continued to make blades with the steel until the 1950s, though the steel's appearance didn't quite match the best historical blades. In 1924, Zschokke published micrographs of ancient Wootz Damascus blades[9].

Henckels Wootz knife. Photo from Sachse.

However, there was a loss of interest in Wootz Damascus in much of the USA and Europe from the 1920s onward until Cyril Stanley Smith published *A History of Metallography* in 1960, which discussed, among other topics, Wootz Damascus, its properties, and microstructure. This revitalized interest in this ancient technique of steelmaking. In 1980[10] Oleg Sherby and Jeff Wadsworth published a historical review of Wootz Damascus and their recent experiments on recreating the steel. They reported that to recreate Wootz Damascus, they started with very high carbon steel, cast and very slowly cooled to form large grains with cementite (iron carbide) formed on the grain boundaries. The ingot would then be forged at a low temperature of 800-900°C where the cementite would spheroidize and become segregated, leading to carbide "stringers." However, while this resulted in a "macro" visible pattern, on a microstructure level it did not appear the same as the ancient Wootz Damascus. The elongated stringers developed by Sherby and Wadsworth came from rolling the steel with a rolling mill, a tool that would not have been available to the ancient smiths. Dr. Cyril Smith also argued against the method that Wadsworth and Sherby had proposed[11]. Therefore, more research was necessary to find the process by which the steel was made anciently.

Al Pendray

Alfred H. "Al" Pendray was a blacksmith and farrier in Florida[12]. He learned from his father, who was also a farrier and blacksmith, and his grandfather learned blacksmithing in Cornwall, England. When pattern-welded Damascus was becoming popular in the early 1970s, Al's father made his own Damascus blade, and told Al, "There, that's what they're talking about. It's something that any competent blacksmith should be able to do routinely."[13] Pendray was introduced to Wootz Damascus by Robert C. "Bob" Job:

The first time I really got interested was I was at a hammer-in in New York, at Ashokan. There was a gentleman that had some blades there that I looked at. The biggest thing was the mystery of it. It was so fantastic; it could do all these wonderful things. The modern metallurgists and steel people have tried to duplicate it, and they were unsuccessful. I more or less thought, well, it would be fairly simple to dig a hole in the ground and melt stuff, I can do that. Well, it turned out to be a lot more difficult.[12]

At the 1982 Kansas City Guild Show, Steve Schwarzer successfully cut a silk scarf in midair with one of Pendray's Wootz blades. Pendray said, "We've cut the silken scarf floating in the air in Kansas City. I've tried that with carbon steel and pattern-welded steel, and I don't know of anybody who's been able to cut one with any other type of steel."[13] At the 1982 Ashokan conference Bob Job, Al Pendray, and Steve Schwarzer demonstrated how to make Wootz Damascus, and bent the blade into a "U" shape to show its ductility[14]. And at the 1983 Ashokan conference, Al Pendray cut rope with one of his Wootz knives 120 times and had to let other people continue cutting because he had worn the skin off his right thumb rubbing against the cutting board[15].

The method for making the Wootz blades at that time came from Bob Job. His method for making Wootz Damascus was to put steel, cast iron (very high in carbon), and flux in a crucible and heat it to 1525°C with a coke-fired forge. Next, the steel was slowly cooled, then forged out to elongate the cementite (iron carbide)[16]. Reportedly there was a patent pending using Job's process, but I can find no record of it, so perhaps it was rejected. By 1987, Pendray was using a modified process, as reportedly he was starting with high-purity iron in a crucible which was carburized with charcoal rather than getting carbon from cast iron[13]. Pendray said in a later interview:

> I didn't have any way of analyzing material and that, and I was just guessing what they put in it. In the old writings, they all said they used these certain kind of leaves, and some of them said wood chips. They were talking about thrusting the blade through the thigh of a slave, and quenching them in blood, and the urine of a goat fed curly ferns is what made the curly pattern on it. You know how that stuff gets spread around you never know, but you read it, and, well, something is going on. Somebody did something right, because there's a sword blade, I've looked at them. Well, I worked on my own probably four or five years. Now I wasn't just solid with it because I was making a living feeding the family, shoeing the horses. I was at ABANA conference in Birmingham, and Wally Yater, who had been doing some work, trying to do it. Wally said to me, "I made two melts last year." I said, "Well I made over 200, and I virtually didn't get anything. I got some pieces, but they were not the right structure." He referred me to Dr. Verhoeven.[12]

John Verhoeven

Dr. John Verhoeven, a metallurgy professor at Iowa State, was also interested in Wootz Damascus:

> Having read Smith's book (*A History of Metallography*), I had been aware of the fascinating mystery of how Damascus steel swords were made and was mildly interested. But my interest was piqued when one of my students gave me a copy of a paper published by W. (Wally) M. Yater in the blacksmith magazine, *Anvil's Ring*, 1983[17]. Wally had done a thorough literature review of the information gathered mainly by English observers in the early 1800s of how the Wootz steel cakes had been made in India. I thought that the secret might lie in a better understanding of how the small steel ingots solidified. I had been doing research on the solidification of metals for over a decade and came up with a hypothesis on how the surface patterns might form in these steel blades. I had solidification equipment in my lab, and a graduate student, Larry Jones, also interested in this problem, so we did some experiments to check out my hypothesis.[18]

Verhoeven and Jones did a similar experiment to Wadsworth and Sherby with slow cooling and hot rolling. They found that they got a similar microstructure, but also reported that "their best surface patterns were poor simulations of the patterns on Oriental Damascus blades."[19].

Verhoeven and Pendray Join Forces

Verhoeven told the story of how he and Pendray started working together:

> When I started working on this, I was doing all this research on this Wootz steel and the blades. We had a rolling mill, and I didn't have anyone that could forge like Al. And so, the only way could take the ingot and knock it down and make sheets out of it was to heat it up and do hot rolling, so I did that. It didn't work. The pattern wasn't the same. So, then I

needed somebody to forge it. I said, "I got to get a forging guy." Al calls me on the phone. If you've ever talked to Al on the phone, well, we probably talked for an hour and a half. I told Al, if you can reproduce this, I will figure out why.[12]

Al Pendray reported that Verhoeven came to regret that statement, "(Verhoeven) said, 'I wish I never told you that. Five years later, I still can't explain what's going on.'"[12]

Pendray and Verhoeven created Wootz patterns in the early 1990s by forging ingots at relatively low temperatures. Pendray said in an interview:

> Early on I did it by eye (checking the temperature), it was very bright and hard to do it, had to put on dark glasses. But then when Dr. Verhoeven started working with me we had to have a positive deal, so he knew exactly what temperature worked. At that point, we developed a technique where the furnace is raised off of the ground, and we bring a thermocouple up through the pedestal block. And the long protective tube with a platinum-rhodium thermocouple and that touches the bottom of the crucible and that way we know exactly what's the temperature outside of the crucible.[20]

Using the low-temperature forging technique, they got the right microstructure and macro-pattern, which led to a 1992 patent[21], but the process was still inconsistent. They still didn't understand what was happening to create the aligned sheets of carbides which would form during the forging: "The authors do not fully understand the mechanism by which the Cm forms into the cluster sheets in the forging operation, but they are continuing research to better understand the process."[22] Verhoeven continued the story:

> Well, we started working on that in '89, and we didn't figure it out until '96. I'm a little bit embarrassed about how long it took me. And I thought for sure it was due to impurities. We analyzed the steel, and we found that they had manganese, sulfur, phosphorous, and a few other impurities, but we didn't analyze for impurities that were present below about 0.1% because I didn't think they'd be important, so we worked on it and worked on it and what we actually found the secret is an element that's present at 0.005 to 0.01%. And that element is vanadium metal, and it is present in some ores and not in other ores. When it is present in that ore that you are using to make the Wootz cake from which you make the blade, you're going to have a very good chance of getting a pattern on the forged blade, and if it's not present, you're not going to get it.[20]

This effect of vanadium was also why the technique for Damascus patterned Wootz was lost, because the ores with a small content of vanadium were no longer being used. Verhoeven said: "The last good blades were around 1850, 1840-1850. And after that, the technique was lost. You can make a crucible steel that will have pretty much the same composition with no pattern."[20] And: "Certain parts of India they had ore bodies that had vanadium in it. So, when the smiths used that, they got success. They were shipping these Wootz ingots to other areas to have them made into blades. So, in these other areas, these bladesmiths aren't telling anybody how they do this. So, all of a sudden, they start getting Wootz ingots that don't have vanadium in it because the Wootz is coming from another supplier. So, after two generations the magic that the fathers taught the sons, lost."[12] Pendray told the story similarly: "We finally ran into the fact that a carbide former was necessary. We went through a lot of them and finally figured out, looking at the old blades, that just about every one of them had a small percentage of vanadium."[12] Other "carbide forming" elements can work for the same effect, but their approach was to recreate the ancient methods.

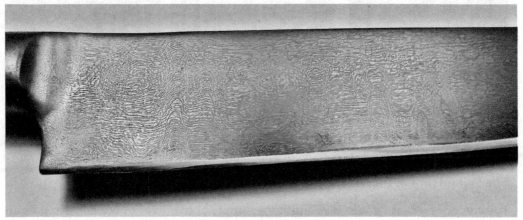

Wootz knife from Pendray[18].

The Mechanism

Unlike the previous processes from researchers like Wadsworth and Sherby, Pendray and Verhoeven's method didn't rely on slow cooling. In fact, the pattern could be eliminated by overheating the steel, dissolving all of the carbides, and then restored by doing a series of low-temperature thermal cycles. The carbides would again align in sheets that would create a pattern. Al Pendray said:

> I developed a forging technique where I could erase the patterns and then put them back in. And [Verhoeven] said, "You won't be able to do that." Because their experience was once you overheat it and take the pattern out it was done. So, I made the same experiment with two different other heats and sent it to him, and he said, "I don't understand it. What you're doing doesn't make any sense metallurgically to us."[12]

Verhoeven said, "We discovered that with the vanadium in there, you have to thermal cycle it. You can't just cool it down once; you have to keep going up and down below the transformation temperatures. After the first cycle, you don't see any pattern. After the sixth cycle, all the carbides are lined up."[12] During solidification of steel, there is segregation of different elements, which creates regions rich in that element and areas lean in that element.[18] During forging, those segregated regions become elongated parallel to each other. In the alloy-rich regions, the carbides are not simple iron carbide, but instead have some amount of other elements in them, such as vanadium. If the steel is heated too hot, all of the carbides are dissolved, and then they re-precipitate (form within the steel) in a random array with no pattern. But if the steel is heated to an intermediate temperature, the more stable carbides are not dissolved, which are those high in vanadium. Upon cooling down, those more stable carbides grow a little bit, and the less stable carbides shrink. So, after several cycles heating up to the intermediate temperature and cooling down again, the carbides are preferentially located in bands in the rich regions. Read more in Dr. Verhoeven's book *Damascus Steel Swords: Solving the Mystery of How to Make Them*[18].

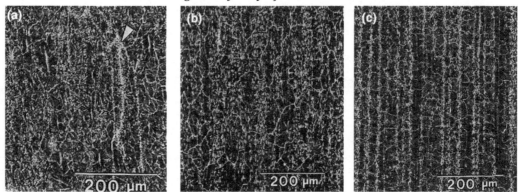

Changing microstructure with increasing thermal cycles. Image from Verhoeven[18].

The Friendship of Blacksmiths and Metallurgists

The story of recreating Wootz is another example of teamwork between artisans and metallurgists. Even though it was Verhoeven who was a Professor with a Ph.D., he still had great respect for Pendray: "He (Pendray) could have gone to college and gotten a Ph.D. just like that. I never worked with anybody that could learn faster. And he is extremely patient. I think Al is an amazing man. It's been a pleasure for me to know him."[12] Another Professor and metallurgist who encouraged them was Dr. Cyril Stanley Smith. Verhoeven said:

> When I began working on the mystery of how Damascus steel swords were made in the 1980s, I had a few letters of correspondence with Dr. Smith shortly before he died in 1992... He offered me encouragement and advice in my work... Al often spoke in a sort of reverent tone of "Dr. Cyril Stanley" and how Dr. Smith had taken him aside at the meeting, encouraged him in his work, saying that it was his opinion that the mystery of how to make these blades would probably be solved by a working smith.[18]

Daryl Meier also talked to me about the influence of Dr. Smith. Meier invited him to a blacksmithing conference, and Dr. Smith was very friendly to all and answered many questions. When driving Dr. Smith back to his lodge for the night, he told Meier, "I rather like blacksmiths"[23].

Dr. Verhoeven went on to write *Metallurgy of Steel for Bladesmiths & Others who Heat Treat and Forge Steel,* which he published as a free download in March 2005. This was a complete book-length introduction to steel metallurgy targeted at knifemakers. Later the rights to the book were purchased by ASM International and published as *Steel Metallurgy for the Non-Metallurgist* in 2007. This again demonstrated Dr. Verhoeven's interest in the knifemaking community and his willingness to educate all who wanted to learn about science, metallurgy, and engineering.

-27-

THE BEGINNINGS OF MOSAIC DAMASCUS

What is Mosaic Damascus?

The term "mosaic," in reference to pattern-welded Damascus, goes back quite far. The term was used by Frenchman Jean Francois Clouet in an article published after his death in 1803[1]. He wrote about the manufacture of pattern-welded Damascus and three methods of patterning: 1) Cutting into the pattern-welded steel and then forging flat, 2) twisting, where he also discussed use of designs viewable in the end of the bar that become visible on the flat once twisted and cut, and 3) mosaic, where designs are created in the end of the bar and those are cut off and embedded in the blade. Perhaps more importantly for our understanding of pattern-welded Damascus since the 1970s, the word "mosaic" in reference to complex patterns was also used by Cyril Stanley Smith when discussing Clouet in his *A History of Metallography*, published in 1960, which Daryl Meier owned. Smith did not really explain this term, so it was left to the bladesmiths to define it.

Steve Schwarzer described the history of the term:

Steve Schwarzer

> The term 'mosaic Damascus' was coined by Meier and yours truly. It describes the process of inlaying small bits of stone, ceramic or glass in a mortar mix to generate a pattern or image. Mosaic also was applied to the special glass made in Italy. The techniques used to generate my early mosaic images are very similar to the methods used in the Italian glass industry to generate images or geometric patterns. Early in the introduction of mosaic steelmaking - or should I say reintroduction - Daryl (Meier) and I were the only knifemakers I knew of using these techniques. We spent hours discussing which terminology should be used to describe the construction methods employed in making mosaic patterns. The development of terminology was necessary to enable us to describe the research we were doing on this very complex material.[2]

The earliest published use of the term mosaic by either of those gentlemen was in 1983 in *Blade Magazine* by Daryl Meier:

> A section of sword blade dredged up from a river in Denmark shows another method of developing pattern in a welded Damascus structure. A billet of irregular shapes and sizes of material was welded and drawn down into a rod. The rod was then cut like a loaf of bread into many thin slices. The slices were then laid, like floor tile, on a flat piece of steel and welded. Another set of slices was then placed on the other side of flat steel and also welded. The flat steel became the central part of the sword blade and the slices formed a mosaic pattern on both sides! Tedious work indeed and done about a thousand years ago during the dark ages - the time of ignorance![3]

The Damascus sword he was referring to[4] has a striking "tiled" pattern, shown in the image to the right.

Steve Schwarzer started making knives in the early 1970s. He credits Al Pendray and Jim Schmidt for helping him with early knifemaking. As he said in 1986: "If I could hammer as good as Al and finish as good as Jim, I'd be the best in the world."[5] But Schwarzer is best known for his contributions to mosaic Damascus, as he explained:

> The art of mosaic Damascus is in fact a very old tradition, one dating back well over 2,000 years. I personally began my quest to discover and rediscover these ancient techniques in about 1979. I had been pattern-welding since 1976, and began to read everything I could find about the mosaic Damascus-making method. Although the tradition had almost died out in the rest of the world, there remained pockets of information in the Viking and Indonesian cultures, as well as in the northern India, where there were a few working bladesmiths using these techniques. My research eventually took me to the mosaic gun barrel methods used in Belgium. My friend Glen Gilmore sent me a box of slides from the

factory at Liège. I noticed the similarity of technique in the glass industries in Italy and ancient Egypt, and began to apply these techniques to steel, substituting steel alloys for the other art's colored glass rods. Once the pattern was created, I began to search for a method to expose the image created and apply it to the side of the blade. Daryl Meier was also researching these patterning techniques. His first solution was to place the images in a bar and twist it, then grind away the surface until the image appeared. I tried the method, but was not satisfied with the mixed results. Many say the key to mosaic is to express the image in a tile-like fashion. I tried several methods, and most resulted in spoiled steel. I went back to my blacksmithing books and researched joining methods. I chose the scarf weld, figuring that, after 6,000 years of constant use, it should work. I sawed off some of the first end-grain sections like I'd slice a loaf of bread, and butt-welded the slices together, again with mixed results - but after adding the scarf, it went like magic.[6]

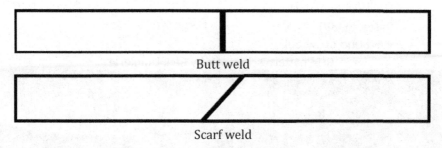

Butt weld

Scarf weld

"It was several years before my skill level was developed to the point I could use the information to develop the techniques in my own blades. After a few years of trial and error and a lot of help and encouragement from Meier and Don Fogg - and anyone else I could glean information - I began to experience a small amount of success."[7] The first published knife I found with "tiled" mosaic Damascus was from 1988 using Schwarzer's "spider web" pattern in the bolsters[8]. This pattern is created by generating a grid of steel layers. Schwarzer also wanted to include this pattern in his blades: "But then came the next problem: If the block was not made from all high-carbon steel or had nickel added, it would not cut properly. I added a center core capable of holding a cutting edge."[6] This type of construction is also called "San Mai" which means "three layers."

Image of Steve Schwarzer mosaic Damascus bolster from 1988[8].

Grid

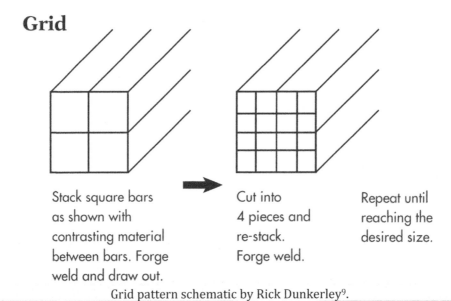

Stack square bars as shown with contrasting material between bars. Forge weld and draw out.

Cut into 4 pieces and re-stack. Forge weld.

Repeat until reaching the desired size.

Grid pattern schematic by Rick Dunkerley[9].

Schwarzer spider-web pattern knife with San-Mai core published in 1993. Weyer photo[10].

Daryl Meier and The American Spirit Bowie

Illinois Governor Jim Thompson read about Daryl Meier in a local newspaper and contacted him to commission a piece for President George H.W. Bush. Meier chose a patriotic theme for the blade: "A mid-19th century Bowie knife design was chosen for this commission because it is the 'American' knife. I tried to depict a scene in the blade of the knife that would remind President Bush of his view during his inaugural parade. There are thirteen waving flags and a banner saying 'U.S.A.' on both sides of the blade. The cross guard is partially composed of a meteorite from Texas, and the handle scales are moose antler from Alaska."[11]

Meier used a new technique for the stars and letters; he hired out wire EDM to cut those shapes out of blocks. So, each star had five points, and there were 50 stars in each flag. The flags alternated orientation while the letters "USA" read correctly from left to right no matter which side of the blade you look at. It took Meier 18 months and about 800 hours to complete the blade to present to President Bush in December 1990[12]: "I sweated blood to make that blade right. I could've made something that looked suggestive of what I made, but then it all has to do with the philosophy of how you do a thing. That's what's prevalent in our society. We're too cost conscious and few of us are willing to pay the full price for a good product. You've either got to say, 'I'm going to make something I'm proud of' or you've got to get out of the business."

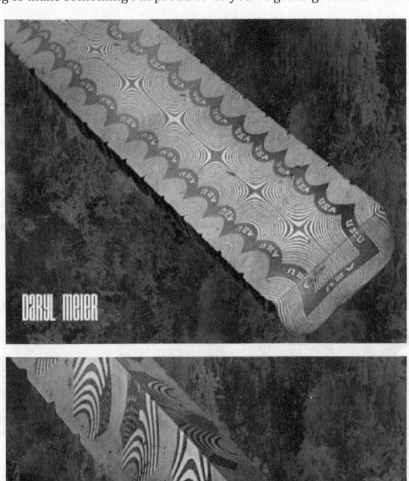

Spirit Bowie Images from Refs. 10-13.

Definition of Mosaic Revisited

Surprisingly, Meier did not consider this fantastic piece of steel to be mosaic:

> The term "mosaic," before it was associated with pattern-welded steel, involved taking little pieces of material and stacking them up or arranging them in patterns for tiles, or whatever. When you were done arranging the pieces, the pattern was accomplished. That was it. One method making a mosaic-Damascus blade is to arrange steels in a bar until a pattern is achieved and to cut slices off the end of the bar, and that becomes a knife blade through forging... From my point of view, it is legitimate to call that a mosaic pattern. What I did 20 years ago was to take a bar, and rather than slice the end off it, I twisted it. That caused the detail of the design to coil. Then, by cutting it lengthwise, I got a repeating pattern. My work went beyond the concept of mosaic. What I did was an interpretation of Turkish Damascus.[14]

Schwarzer countered: "This is a point he and I have spent many hours discussing. Daryl says it is not mosaic, I say it is. We are still debating this issue."[7] The working definition of "mosaic" has become relatively broad. Rick Dunkerley polled several Damascus makers about the definition of mosaic[9]: "At the BLADE Show in 1999, I questioned some of the best Damascus makers on hand as to their opinion on this topic and each one had a different definition for mosaic Damascus. Since there seems to be no clear definition we will refer to all end-grain patterns as mosaics." And indeed, the term mosaic is generally used among pattern-welded Damascus makers as any pattern that comes from the end of the billet, or "end grain." So, this would encompass some patterns already discussed including W's, radials, bias forged steel, Meier's "name" dagger, and the Bowie knife presented to President Bush.

Steve Schwarzer's Mosaic Continues

Schwarzer used the wire EDM technique introduced by Meier as well; his first knife using EDM had anvils throughout the blade. "My first attempt was using small anvil shapes, surrounding them with stacked material. It worked marvelously. The wire-cut parts meshed perfectly and forge-welded without much distortion."[6] This knife won "Best Damascus" at the 1992 Blade Show[15].

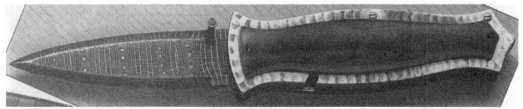

A later anvils knife from *Knives '99*.

However, Schwarzer wanted to do something more spectacular with this technique, leading to one of his best-known pieces called "Hunter's Dream," first shown in 1993. Schwarzer described the impetus behind this knife:

> The true beauty of real bladesmiths is they cannot resist a challenge. Once a technique is planted in a fertile mind, it takes on a life of its own. For instance, Daryl admonished me very early in my mosaic work to make images that represent something which actually exists, not just a distorted mass that's suggestive of an image. "Make something that looks real," he said. Those words of wisdom stuck in my head and provided me the direction I needed to produce my "Hunter's Dream" knives. They comprised a series of pieces, each with the mosaic Damascus image in the blade of a hunter shooting a covey of quail over a dog, and were a direct result of Meier's inspiration.[16]

However, wire EDM and an idea were insufficient for making a "real" image. It is difficult to forge steel without distortion; it requires slow and careful forging from all four sides. Schwarzer said he had to "knead the bar under the air hammer, basically. You have to 'feel' the material under the machine. You can teach technique but not the feel. It's something I've developed over the years from fooling with this stuff."[17] Other Damascus makers were also impressed. George Werth commented: "Even the barrel of the gun is straight."[17] The individual characters had to be forged down to the right scale and then placed together into one scene, all while avoiding distortion. And the steel did not want to cooperate: the billet cracked twice during heat treating, and he had to cut the scene out and insert it into another bar of Damascus. Schwarzer said, "That was extremely time consuming because I had to cut the scene out of the Damascus bar without cutting the scene itself."[17]

Proliferation of Mosaic Damascus

Schwarzer saw this as a start to many other possibilities, as he said in 1994: "We're going to see elaborate scenes portrayed in Damascus in the very near future and these will be the result of hammering, not etching."[18] The number of people making mosaic Damascus was very small at this point but Schwarzer believed in teaching the concepts to others: "I decided to share my knowledge with a few good bladesmiths so they could help educate the public about this 'new' art form. It was the best move I could have made. I had a lot of people wondering why I was giving away information that had taken years and lots of dollars to develop. All I can say is that by teaching you can expand your life in the knifemaking business by the number of students and friends with whom you share such information. The feedback I have received as a result will take four lifetimes to explore, feedback that never stops."[7]

Other early makers of mosaic Damascus included George Werth, Pierre Reverdy, James Batson, Hank Knickmeyer, Hugh Bartrug, and Bill Fiorini.

Hunter's Dream Knife[10]. Weyer photo.

Hugh Bartrug - Ashley Forge

Bartrug began blacksmithing in the early 1950s with a job in a steel mill blacksmith shop in southwest Pennsylvania[19]. There he learned how to forge weld and about different steels, and in 1980 he started making Damascus: "The first guy who showed me a Damascus blade was knifemaker Ed Small. He really psyched me up on making Damascus. I thought you had to be a magician to make Damascus, but I went back to the mill and started forging it."[20] Bartrug met knifemaker Greg Gottschalk, also living in Pennsylvania, at a gun show. Gottschalk told me[21]: "I taught him (Bartrug) how to make a knife, and he taught me how to make Damascus. We went to a bunch of different shows together and he would pick my brain I would pick his. He was a super guy." Bartrug lost his job when the steel plant closed in 1985 and began making knives full-time. He named his business Ashley Forge after one of his grandchildren[20].

"The Einstein Knife" by Hugh Bartrug, scrimshaw by Sandra Brady. Weyer photo.

One Damascus technique Bartrug introduced by the mid-80s was using nickel sheet in his Damascus in combination with bluing the steel[19]. Bartrug said, "I like shock appeal. I hot-blue all my blades and with the nickel in there, they really stand out."[20] His reputation grew, and at his first knife show, the Super Blade Show of 1986 in Knoxville, Tennessee, he received the award for best Damascus knife. He almost didn't make it to that show after a grinder grabbed a blade he was working on and put it through his foot. His wife Joyce said, "We weren't sure he'd be able to make the trip. When we went to that show, we prayed, 'Dear God, let us put Elizabeth, Pennsylvania, on the map today.'"[20] Hugh credited his wife Joyce with helping him succeed: "Joyce is my buddy. She's worked with me since day one. She helps me in the shop. You can't do it all by yourself. She'll assist me at the furnace while I (work on the steel), and she handles the phones. I can't get any work done when I'm on the phone."[20]

Bartrug enjoyed his knifemaking, as he said in 1987: "I like forging the steel. Every morning it's like getting up at Christmas, it is so satisfying to see many of the knives develop as I work on layering the steel."[19] In 1993, he said, "I don't take custom orders. I do what I like to do. Life's too short to do what other people want you to do all the time. I enjoy doing what I do. When you enjoy doing what you do, you can put your whole soul into it."[20] Bartrug was an early bladesmith involved in making mosaic Damascus, and in a 1993 interview, he laughed while saying, "To me it's the most interesting Damascus, though you make less money at it."[20]

His "The Einstein Knife" was sold at the 1992 Blade Show featuring scrimshaw by Sandra Brady. The formula letters are bordered with pure nickel and the knife features his signature multi-color bluing. He said, "I had to keep cutting pieces off the steel and doing a bunch of filing to keep it the right shape"[22]. The bolster was forged down from a 3-inch square piece. The knife took 2-1/2 months to complete. Bartrug followed a simple philosophy in his knifemaking[20]: "One thing I learned years ago was from the late Don Hastings. He said knives should have simplicity, power and elegance. No matter what you do in life, if you put those three things into it, it will come out right."

Ray Rybar

Another knifemaker who started under Bartrug is Ray Rybar, who received his Journeyman Smith in November 1998 and Master Smith in 2001. I interviewed Rybar in 2023:

"I was shoeing stake horses at the Meadows race track [near Pittsburgh, Pennsylvania]. A lot of those guys that were driving the trucks wanted a knife. Naturally, they would come to the blacksmith shops, and anyone that came to the shops looking for knives, those other blacksmiths, about ten of them on the track, would send them over to me. I lived across from Hugh Bartrug...The last real accomplishment that Bartrug did was the "Michael's Sword" [featuring a cast of the Archangel Michael and a quote from *Paradise Lost* etched along the edge]. Right after that, Hugh started on a mosaic with "In God We Trust" [never finished], and it took a while to get it legible. He was of the opinion, as was one of his counterparts Steve Schwarzer, that a plain old picture sometimes won't cut it. A guy might say, "I want to do a horse and rider." When you're done with it, it looks like an alligator with a turtle on its back. You don't have that option when you go to words, numbers, and letters. I chose to go to Bible verses, and you can either read it or you can't. It can't be a lizard that turns into a fishing worm. If you can't read it, it isn't worth 15 dead flies. The smaller and tighter it is, the more desirable it is for some collectors....[Bartrug's] "In God We Trust" was partially the inspiration for Bible verses, and Hugh was a Bible banger as well. Most people don't know what's in that work. It totally became an evangelizing effort to the point that unless I am specifically required by an individual to do a non-Biblical knife, I don't even think about them anymore. I'm doing the Ten Commandments in Paleo Hebrew. I might not live long enough. It might end up like Hugh's "In God We Trust."[23]

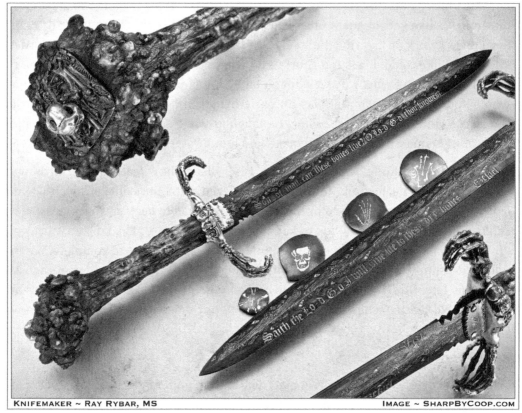

KNIFEMAKER ~ RAY RYBAR, MS   IMAGE ~ SHARPBYCOOP.COM

Ray Rybar "Bible Damascus" knife. Image by Coop.

Pierre Reverdy

Pierre Reverdy became fascinated with blacksmithing at a young age: "I fell in love with steel when I was 14 years old. In the small village where we lived there was a big fair every spring, and one year a blacksmith was showing his work in architectural blacksmithing. I watched this man for two hours and then spoke with him. Then I came home and told my father that I wanted to be a blacksmith. He said, 'Why not?' After that I came to the blacksmith's workshop and worked with him."[24] In 1985, Reverdy traveled to England for an international conference of blacksmiths, and he met Daryl Meier, who had presented his dagger with his name in it. Later that year, Reverdy spent a week learning from Daryl Meier in Sid Birt's shop. Reverdy was

Image of Pierre Reverdy knife and the original EDM-cut scrolls[25].

inspired by the French and Belgian Damascus makers who had such intricate patterns in the 18th and 19th centuries, and he made his own steel with the word "Liberté" in it in 1990.

Reverdy advanced his talents from there:

> I have developed the Damascus technique in very fine detail. I call my new way of working "poetique Damascus," and to reach this level of fine detail I bought an EDM machine. I spent some months learning how to run the machine and also how to make all the programs and drawings on the computer. The most important thing is that every step of

production is done in my workshop. Poetique Damascus gives the knife the opportunity to speak about a certain theme. Making knives takes a very big part of my life. I like the steel, sculpture and art. I have reached into a lot of fields like painting, drawing and science, and I don't really need to go outside knifemaking. A lot of people have a life and a job and in the evening they come home and do what they really like. What is tremendous for me now is that this is my life and what I want to do. There is no difference.[24]

Art Professor Damascus

Hank Knickmeyer

Knickmeyer was an art professor at a small college in St. Louis. He began making Damascus in 1989 and learned from Daryl Meier, Steve Schwarzer, Don Fogg, and others. He said:

> Teaching artists are expected to work for themselves as well as teach, and that research and work for myself translates into what I've been doing with Damascus for the past five years. I think some of the most important work I've done has been with Damascus. I'm just fascinated by this stuff. I like the control over the image that I can incorporate in it and the fact that it records everything you do to it. Somebody that knows Damascus can look at it and read right back to the beginning. In a sense you're recording the whole process... A lot of what I do in making a knife happens while I'm making it. I just follow the lead of whatever is going on in the forge.[26]

Knickmeyer liked to incorporate different shapes into his Damascus and fill around it with "filler bar" of mild steel. He also created other effects with distortion, such as forging on the bias (discussed in Chapter 25) and tiling different patterns together.

Hank Knickmeyer knife from ~1993. Image from Arizona Custom Knives.

Methods of Assembling a Mosaic Blade

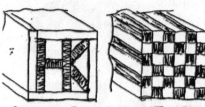 

A MOSAIC BAR IS CONSTRUCTED USING BARS OF VARYING ALLOY CONTENT
CHROME AND NICKEL ARE EFFECTIVE IN PRODUCING CONTRAST
THE IMAGE IS READ FROM THE END OF THE BAR

BARS MAY BE COMBINED INTO A MORE COMPLEX BAR

THE IMAGE ON THE END OF THE BAR MUST BE MADE TO READ ON THE SIDE OF THE BLADE. IT CAN BE DONE BY:

WELDING SEVERAL BARS TOGETHER SIDE BY SIDE — THEN SAWING OFF SLICES

OR

TWISTING THE BAR TIGHTLY, THEN SAWING SLICES LENGTHWISE — READABLE (BUT DISTORTED) IMAGES WILL BE FOUND ON CUTS PASSING NEAR CENTER

THE TWISTED BAR CAN ALSO BE FORGED FLAT THE GROUND DEEPLY FOR A RELATED EFFECT

OR

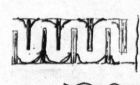

CUTTING THE BAR PARTIALLY THRU FROM OPPOSITE SIDES THEN STRETCHING IT OPEN. THE IMAGES ON THE CUT FACES WILL APPEAR ON THE SURFACE AFTER FLATTENING.

IMAGES

STRETCH & FLATTEN

THE STEELS USED IN THE BARS PRODUCED ABOVE MIGHT NOT MAKE A GOOD CUTTING EDGE. A CUTTING EDGE OF HIGH CARBON STEEL (OR DAMASCUS) MAY BE ADDED BY:

 HIGH CARBON

FORGING THE BAR TO SHAPE THEN FORGE WELDING ON A HIGH CARBON EDGE BAR

OR

SANDWICHING A HIGH CARBON BAR BETWEEN TWO MOSAIC BARS SAN-MAI FASHION

HANK KNICKMEYER U.S.A. 636-285-3210

Knickmeyer schematic of his process drawn in 1993[27].

279

Bill Fiorini

Bill Fiorini was another art professor who made Damascus. He explained his history in 1994:

> I was schooled as a goldsmith, and in 1976 I was introduced to the blacksmithing art at the ABANA Conference. From then on I started working with iron, and it was a natural transition from jewelry work into pattern-welded steel. I went to school and received a Master of Fine Arts degree in jewelry making, and I've been teaching and had my own shop since then. But now I'm just about strictly a knifemaker, even though occasionally I'll still do a ring or something. Knives were something that came along later. I've known Daryl Meier for several years and was introduced to the material by his work and the rest of the Carbondale team. I went through all the transitions of a little art and decorative work, like fireplace tools and all that, so that sort of work led me to knives. I'm kind of new at the knife world, and I have never made a carbon steel blade. All my work has been in Damascus... With the situation I'm in I cram everything into a three-day schedule at the university. That allows three or four days a week in the shop.[28]

He was very interested in new Damascus patterns: "The most involved thing I'm doing right now is mosaic with nickel and tool steel combined. Looking at the patterns and developing new ones are what I'm primarily interested in. I've gone back to the old 18th century gun barrels, and some of the patterns we're using on knives now are the same as the old gun barrels."[28]

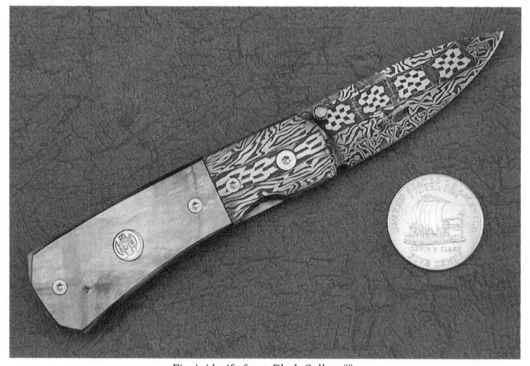

Fiorini knife from BladeGallery[29].

J.D. Smith

J.D. Smith

J.D. Smith taught bladesmithing at the Massachusetts College of Art and Design. I interviewed him in 2022:

> My first interest in Damascus was at age 10 in 1959. I was a kid growing up in New York City; I lived less than a mile of Metropolitan Museum of Art. I used to wander around there all the time on my own. I went to the Arms and Armor section, I saw a Turkish shamshir on display and it was really beautiful, and it had a really light green jade handle. The entire piece was encrusted with emeralds and diamonds. I remember, "What are all these wavy lines on there?" I put it out of my mind; being a ten-year-old, your attention goes to other things. In the early 1970s, I was at a friend's house who had an Indonesian Kris. I said, "Oh, those wavy lines again. What is going on here?" I gave it some attention but moved on. In 1978-1979 I was walking down the street and popped into a 7-Eleven. In a *Fortune* magazine, there was an article on Bill Bagwell making a Damascus fighting knife. I said, "Oh, there's those lines again." I had a name to give it: Damascus steel. I was in music school at the time, in college studying jazz composition and trombone. In college, I had taken a course in musical instrument repair, so I had learned some nonferrous metalworking.

> Then I graduated, I was dating a Russian girl who was a silversmith. And I had mentioned to her that I could use a little more work. "We need someone in the shop right now," she told me. I already had earned a Bachelor of Fine Arts. "Why don't you put together a resumé with your experience?" I got a position there and she trained me. We learned that

a blacksmith had made some specialized silversmithing tools, and we should go collect them. I went down to the smithy, and it was one of those moments that changed my life. "This is what I have to do now." He took me on after many months of badgering, as an apprentice under a special government program. I studied with him for the next 3-1/2 years, on weekends. He wouldn't allow me to do bladesmithing right now, he wanted me to learn basic methods first. I learned traditional blacksmithing. Then I said, "We have to make some Damascus." Using wrought iron and old files I made some Damascus, with him striking [with a hammer]. I got to read some books on Damascus and knifemaking from Sid Latham. He had interviewed Bill Moran, and I saw pictures of Damascus steel. I said, "I have to do this."

I got married to the woman who introduced me to it all. When she was pregnant with my daughter, I went to Ashokan in 1984. I met like-minded individuals like Tim Zowada, Jimmy Fikes, and a bunch of other people. Very interested in performance bladesmithing and making Damascus steel. I fell in with these guys, Jimmy Fikes was maybe two hours away, and so I would drive out to his house and get some mentoring from him. Eventually I studied with Jot Khalsa, he taught me everything I know about basic knifemaking. Grinds, slotted guards, tapered tangs, pins, drilling holes, all kinds of construction. I got a really solid basic knifemaking education from him. I went on to start my own knifemaking business in 1987.

I was working out of our silversmithing shop for a while. Then I rented a space in south Boston. Every sale I made, I would funnel it back into the shop. That would allow me to do more of what I would do. Within a couple years, I had a functioning bladesmithing shop. I went into the Guild, and became a member of the ABS. By 1998 I had become a Master Smith. That took five years. I developed a really close relationship with Daryl Meier, [Devin Thomas], and a whole bunch of people. Rob Hudson was a big influence on me; huge with his multi-billet construction and patterning. [He was t]he first person I saw do the "architectonic" Damascus where you build complex structures from basic modules. Making a blade composed of six or eight bars, all just different densities of twist. 2-layer twisted, 4-layer, 15, 100-layer, twisted left and right and stacked according to their constituents in an interesting arrangement. He's one of the bigger influences on me, personally, in terms of how I wanted my blades to look.[30]

Through producing these composite blades, J.D. Smith developed the "fish mouth" technique, where a triangle of steel is cut out of the blade before forge-welding the steel to a point.

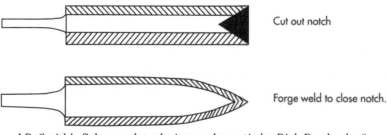

Cut out notch

Forge weld to close notch.

J.D. Smith's fish mouth technique schematic by Rick Dunkerley[9].

1994 J.D. Smith Damascus dagger with multi-bar construction. Weyer photo.

-28-

STEELS USED IN PATTERN-WELDED DAMASCUS

Early Damascus Materials

As noted in previous chapters, early Damascus materials ranged from Moran's O1 and mild steel to the Damascus Research Team's recommendations of W1, 1095, A203E, nickel, mild steel, and wrought iron. It is impossible to give a complete list of every steel used in pattern-welded Damascus as just about everything has been used. A203E is a low carbon steel with 3.5% nickel, which was popular as a "bright" layer in Damascus until 15N20 became more common. The high nickel content means the steel resists etching to provide contrast in pattern-welded Damascus. Though A203E steel was first recommended by the Damascus Research team, nickel-alloyed steels were used historically. In Indonesia, they used meteorite as one of the materials in their pattern-welded steel, which contains high levels of nickel, typically 5-10%[1]. In the early 1980s, Phil Baldwin began using a 9% nickel steel, common for cryogenic applications.

Steel	C (%)	Mn (%)	Si (%)	Ni (%)
A203E	0.1	0.65	0.35	3.5

Cable Damascus

Cable, or wire rope, has been used in knives since at least the late 1800s[2]. Bill Bagwell was the first to use cable for an intentional Damascus pattern in 1974[3]. He called the dramatic pattern "Satan's Lace." Cable Damascus was later popularized by Wayne Goddard, who first wrote about it in 1985[4]. Wire rope is all made from the same steel, not mixed wires of different materials. However, the individual wire surfaces are decarburized (loss of carbon) which gives a bright layer when etched[5]. Wayne Goddard and Ed Fowler contacted metallurgists from several companies to analyze the material: Robin Parkinson of Union Wire Rope, Gary Maddock from Dayton Progress, and Pat Wall from Pacific Machinery and Tool Steel. Pat Wall took the micrograph of forge welded cable on the following page.

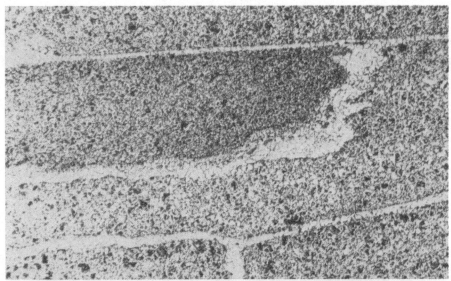

Image of decarburized surfaces in wire Damascus[5].

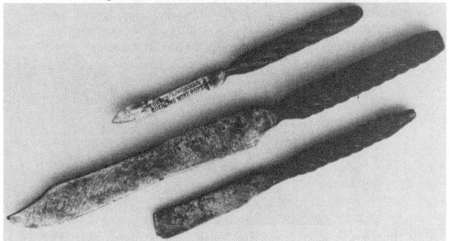

Image of early cable knives from ~1900[5].

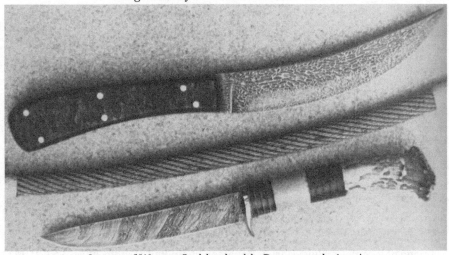

Image of Wayne Goddard cable Damascus knives[4].

Chain Damascus

Other high carbon recycled materials, such as various chain materials, came shortly after. Wayne Goddard began making knives from chainsaw chain. Vince Evans made a knife using a motorcycle chain around 1989[3].

Image of Goddard Chain Saw chain knife from *Knives '91* (~1990).

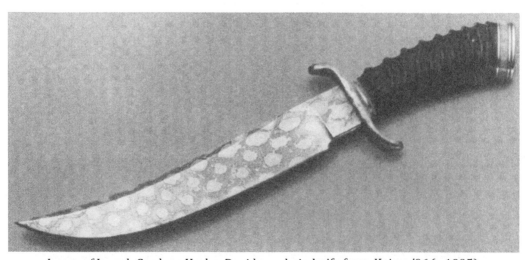

Image of Joseph Cordova Harley Davidson chain knife from *Knives '96* (~1995).

Double High Carbon Damascus

A disadvantage of using low carbon metals like A203E, mild steel, wrought iron, etc., with a high carbon steel is that carbon rapidly equalizes to make a combined medium carbon steel. When using pure nickel as the "bright" layer, little or no carbon diffusion occurs, but there is soft nickel that will become part of the cutting edge. In Jim Hrisoulas's *Complete Bladesmith*, published in 1987[6], he wrote about what he called "Super Damascus": "There is a unique way to make a pattern-welded blade that will outcut all other Damascus steel blades: pattern-weld O-1 and 1050. Welding these two materials will allow for a very sharp edge as all of the laminations involved will harden and temper." He also recommended the odd combination of Vasco Wear and 1050/1060 which he called "The ultimate Damascus steel blade." He elaborated on the reasons for "Super Damascus": "There are two good reasons to make the Super Damascus. First, the blades will out-cut all other Damascus steel, and, second, carbon migration will be all but eliminated due to the fact that you are welding a high-carbon steel to a tool steel."

In 1990-1991, Tim Zowada began using O2 and L6 as a "double high carbon" Damascus mix[7]: "I had learned about austenitizing in salt and marquenching. I just wanted a mix that would harden both constituents well in the salt as well as have a nice contrast."[7] At that time, it was common for knifemakers to refer to bandsaw steel as "L6," but Zowada was buying the steel, not bandsaws[7]: "I was getting 1/2 x 2 inch L6 decarb-free bars directly from Carpenter. I also got my O2 as 1 x 2 inch decarb-free bars from Carpenter...Eventually, Carpenter quit making O2, so I went to O1." The high manganese content of O2 and O1 gives it a darker etch while the nickel content of L6 leaves it bright after etching. L6 has ~0.75% carbon, and O2 has ~0.9%, so there was very little carbon migration between the two steels. Darrel Ralph began using O2/L6 for his Damascus[8] after connecting with Tim Zowada[9], and Ralph advertised this steel combination as "super high carbon Damascus"[10].

For over two decades, the combination of 1084 and 15N20 has been the most popular for making Damascus. 1084 (or 1080, 1075, etc.) is a simple carbon steel with higher manganese than 1095 for a darker etch in pattern-welded Damascus. The 15N20 is a 0.75% carbon steel with 2% nickel and low manganese for resisting etching. So, the two steels work similarly and have almost no carbon migration but have contrast after etching. However, 1084 and 15N20 were not always commonly used in Damascus. Devin Thomas said, "Guys were using all kinds of stuff. 5160, 52100, mild steel, A203E, and they would come to me with their forge welding issues. Some of the stuff was just difficult to weld and then wouldn't etch well after making it. I told them to switch to 1084 and 15N20 to have steels that were simple and easy to work but would have good contrast in the end product."[11] Rick Dunkerley says that Devin suggested that mix to him around 1995[12]. Dunkerley and the "Montana Mafia" then popularized that combination[13], as Dunkerley noted in 1998: "It's an all-high carbon mix. Many industrial bandsaws are made of it. What a lot of people call L6 is really 15N20. It manipulates and welds real well without delaminating."[13]

Stainless Damascus

Friedrich Schneider and Richard Hehn began developing stainless Damascus in 1983. They used a martensitic stainless steel, 1.4111 with 1.1% carbon, 15% chromium, and 0.5% molybdenum, in combination with an austenitic stainless steel, 1.4571 (316 stainless with Ti), which has nearly zero carbon, 18% Cr, 2.25% Mo, 12% Ni, and 0.5% Ti[14]. The nonhardenable austenitic stainless steel resists etching better than the 1.4111, which gives the final steel contrast.

An investigation into early (~1980s) factory-made Japanese Damascus blades revealed the core steel was roughly 440C in composition while the Damascus side steels were 1) 0.3% C and 18.4% Cr and 2) 0.4% C and 15% Cr[14]. Later, Takefu used many other core steels such as VG-10 and SG2.

Terry LaBorde, starting in 1986, used either CPM 440V or ATS-34 in combination with 316L in his pattern-welded Damascus[15]. A.J. "Bud" Hubbard in his stainless Damascus, introduced in 1990, used 440A in combination with austenitic stainless[11].

In the mid-1980s, Jim Hrisoulas[16] and Cleston Sinyard independently created a method using cast iron chips mixed with flux to weld stainless into a steel billet. The cast iron melted at a lower temperature than the steel and so would drop between the layers, bonding between the steel layers. Once the carbon diffused from the cast iron, the steel's melting temperature would drop and solidify. Sinyard presented his method in 1989 for welding D2 or 440C along with wrought iron[17].

Devin Thomas used stainless Uddeholm AEB-L[18] after it was recommended to him by Wayne Goddard, in combination with austenitic stainless 304. He also used 440C, ATS-34, and AEB-H as the "dark" layers in combination with 304, and a "double high carbon" mix of AEB-L and ATS-34. He claimed that the combination of the fine microstructure of AEB-L and the coarse microstructure of ATS-34 provided enhanced performance[19]. In the mid-2000s he was making other combinations, including CPM-3V/154CM, D2/154CM, and Vasco Wear/154CM[20].

Mike Norris used a combination of Uddeholm AEB-H, 302, and D2 in his early Damascus[18]. He began to replace the D2 with CPM-440V (S60V) in some of the steel for greater performance by 1997[21] and was making San-Mai construction Damascus with a high-performance 440V core by 1998[22]. As he said in a 1998 article: "I've done some cutting tests on hemp rope. ATS-34 will take 36-38 cuts. I got 115 cuts with CPM 440-V. I got more cuts with it than any steel I've ever tested."[22] And when used as the core steel he added: "It takes the most aggressive edge I've ever seen on any Damascus"[22].

See Chapter 30 for more on the history of stainless Damascus.

Powder Damascus

Steve Schwarzer started using a new form of steel for Damascus:

> I started looking for ways to get around the expensive EDM elements in my projects. I am constantly searching for better and more efficient ways to create the controlled images I desire. Sometime in the mid-80s, I was at a hammer-in at Jim Batson's house where I met a young guy name Gary Runyon. I was demonstrating pattern-welding, and he pulled me aside and asked if I could help him with a problem. He was trying to get nickel powder to stick to cable by mixing it with borax and then forge-welding. I asked him why he didn't put it in a piece of pipe, seal it up and weld it. That, as far as I know, is the first Damascus forged in a powder-metal canister. What followed was a discussion about powder metal and where to acquire it. As it turns out, Gary's job involves powder-metal technology.[23]

In 1992 Schwarzer found his use for powder:

> I asked my buddy Joe "Hy" Hytovick if he could wire-cut my actual signature in a block. Names in block letters had been produced in the gun industry via mosaic Damascus for hundreds of years. I wanted my actual signature. Hy cut it for me, and I tried to stuff tiny .010-inch nickel sheet into the cut, but it did not want to go. We were looking at it and he said, "Too bad you can't pour that nickel sheet in there." The light went off, and I said, "We can!" I got ahold of Gary and acquired some nickel powder. Access to this fine-grit powder opened a whole new world of pattern development. I bought an EDM unit and began making all manner of images.[23]

In 2000 Schwarzer said:

> I built a mosaic folder a few years ago called the "Blue BB Knife," the handle of which was quite striking and was written up in BLADE. The handle material came from the fertile mind of Joe Hytovick and myself. Joe wired a precision metal box of exact dimension with his wire EDM. When the box was filled with BB shot and vibrated, the BBs would stand in columns. It was a simple matter to fill the box with nickel powder and weld on the end caps. The real trick is the forging of the box of material with no distortion.[24]

Image of Schwarzer knife from 1994 with handle Damascus of BB's and nickel powder[25].

289

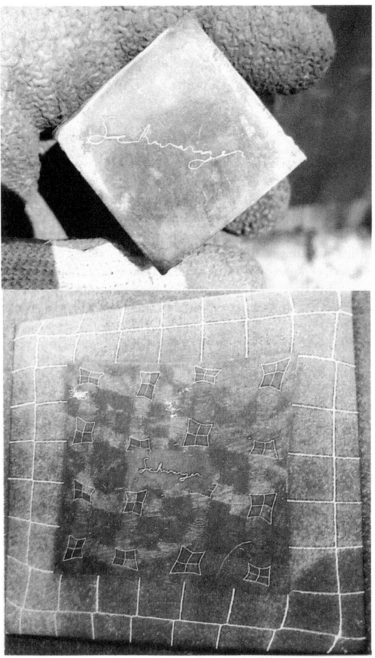

Steve Schwarzer's signature in steel, made by filling with powder.

Stainless Powder Damascus

Damasteel was first sold in Sweden in 1995[26]. The "first generation" of Damasteel used Elmax and 304L steels (see Chapter 30), but this seems to have been relatively short-lived. By 1996 they were offering the "double high carbon" mix they have been known for ever since[27]. The steels used are PMC27, a powder metallurgy version of

Sandvik's 12C27, and RWL34, a powder metallurgy version of ATS-34/154CM. The "RWL" stands for Robert Waldorf Loveless to recognize his role in popularizing the steel for knives.

Understanding the Mechanisms of Different Steels in Etching and Performance

In early Damascus of the 1970s and 1980s, it was thought that carbon differences between the two steels led to differences in etching and properties. Those Damascus smiths would talk about the hard and soft layers giving the Damascus various improved properties such as the "Damascus cutting effect" with soft layers that would wear faster leading to a serrated edge. And soft layers provide higher toughness. It was frequently claimed that the high and low carbon content of the layers would lead to the difference in etching. However, it was increasingly realized that other alloying elements would lead to differences in etching, such as the use of A203E with its high nickel content.

In 1998, Professor John Verhoeven and bladesmith Howard Clark published a study on various pattern-welded Damascus combinations where they looked at the carbon content, hardness, and etching behavior[28]. Combinations studied included 5160-1018, 1086-1018, 5160-L6, 52100-L6, and L6-wrought iron. They used simulations to look at the time it would take for carbon to diffuse from a high carbon steel to a low carbon steel and found that, depending on the forge welding temperature and layer thickness, carbon could equalize between the layers in less than a second. This is due to the very small size of carbon atoms; they are "interstitial," meaning carbon atoms sit in between iron atoms, and thus can diffuse rapidly. However, larger atoms such as manganese, chromium nickel, etc., are "substitutional" atoms, replacing individual iron atoms, and thus diffuse much more slowly. Substitutional atoms are unlikely to fully equalize between two different steels in the time scales typically used by bladesmiths. Therefore, Verhoeven and Clark concluded that the substitutional elements lead to the differences in etching behavior rather than carbon content.

To confirm the simulations of carbon equalization, they also performed microhardness measurements. They found equal hardness for the two layers, even when using a combination of L6 (high carbon) and wrought iron (very low carbon). That combination also had the best contrast, where the wrought iron etched very dark compared with L6. The next best was 5160/1018, followed by similar etching with 1086/1018 and 5160/L6, and the poorest contrast was with a combination of 52100 and L6. Contradicting the prior carbon-difference hypothesis, with the 1086/1018 combination it was the 1018 steel that etched darker because of its higher manganese content (0.75% Mn vs 0.4%). The high chromium in the 52100 and the high nickel in the L6 led to that combination having poor contrast because both had elements that led to resistance to etching and darkening.

However, Verhoeven and Clark gave a couple of caveats to this study. One is that hard and soft layers can be created in pattern-welded Damascus if there is a difference in "hardenability" of the two layers. Hardenability is a measure of how fast the steel needs to be quenched to reach full hardness, and this property mostly comes from alloying elements such as manganese, nickel, and chromium; the elements that do not fully equalize during forge welding. Therefore, if a steel with low hardenability is used in combination with one of high hardenability, and some intermediate quenching rate is used, it can be possible to have low and high hardness in the different layers. Tai Goo used air hardening steel A6 to achieve this effect, see his quote in Chapter 12.

Differences in alloying elements can also affect carbon diffusion, however. In some instances, there have even been experiments where "uphill" carbon diffusion (carbon diffusing from low carbon to high carbon) was found by utilizing steels with a significant silicon difference by Darken, published in 1949[29]. Silicon and chromium are two elements with a substantial effect on this property. Silicon causes carbon to diffuse preferentially away into the low-silicon steel, and chromium has the opposite effect.

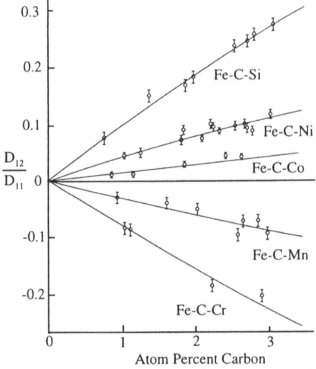

Experimental ratio of diffusion coefficients[30,31]. Positive means the carbon diffuses away, and negative means the carbon diffuses toward the steel with a higher alloy concentration.

-29-

GROWTH OF MOSAIC DAMASCUS

Rick Dunkerley

While in the Air Force in San Antonio, Texas, Rick Dunkerley became interested in knifemaking: "One of my buddies was Lonnie Hodges, and he got me interested in making knives."[1] Those knives were made by stock removal. He left the service in 1985 and worked as a ranch hand before moving to Lincoln, Montana in 1990 to take a job in logging. In 1991 he read an article by Ed Fowler, and Dunkerley began to forge blades. "The entire process involved in producing a blade by forging appealed to me. And during my early days as a smith, both Fowler and Wayne Goddard ... were extremely helpful."[1] Dunkerley also worked as a hunting guide which greatly influenced the design of his knives. As he said in 1998: "My hunting and utility blades have been influenced strongly by my 10 years as a guide. When knives are used on a daily basis, they can constantly be tested and refined. All of my field designs have been tested extensively not only by myself but also by other professional guides."[1]

However, Dunkerley's knifemaking began to turn away from "using" knives toward Damascus steel pieces. In 1997 Dunkerley won "Best Damascus" at the Oregon knife show and "Best Damascus Design" at the Blade Show, where he also became a Master Smith that year. In 1998 Dunkerley said: "Today, developing different patterns has become almost an obsession for me, and the variations for mosaic and composite bar blades are limitless. Making patterns that have not been done before is a tremendous motivation to me."[1]

Barry Gallagher

Barry Gallagher is a knifemaker living in Lewistown, Montana:

> I started making my own knives in the spring of 1993. During the first few months I just put handles on production blades, but I got bored with that really fast. Then, in 1994, I met Rick Dunkerley, and he got me started forging. I was mostly forging 52100 steel for high-performance hunting and utility knives, doing multiple quench and freeze treating. Then, I started making Damascus in Rick's shop and have been making it ever since.

I've just gotten so obsessed with forging Damascus steel and developing new techniques. Starting with an idea, following through with it and seeing the finished product is something that I find very gratifying. I'm trying to make Damascus that nobody else has made… It's planned chaos. I want to control the chaos. I do have it in my patterns but I like to plan the chaotic look in there intentionally and develop it. It's about thinking where you're heading with each piece of steel, with a lot of welding and putting bars back together in developing a pattern to get an end result.[2]

Image of Barry Gallagher Knives[3].

Shane Taylor

Shane Taylor began making knives in 1984 in his hometown of Miles City, Montana[4]. In 1988 he saw an ad in a knife magazine for Steven Brooks from Big Timber, Montana. Taylor was able to meet him, and Brooks would help him with his knifemaking. Taylor began making Damascus all on his own. Taylor read a book about Japanese swords being made of many layers. He peeled flux off old welding rods and used scrap steel, like files, to make his layered steel. "Being isolated in Montana, I was making Damascus before I realized anyone else in the country was. I picked up *Blade Magazine* and saw Damascus by other makers."[5] By 1995, Taylor was making mosaic Damascus knives, and in 1998, he went full-time as a knifemaker. He received his Master Smith title in 2000.

The Montana Mafia

In 1995, Dunkerley started a yearly hammer-in at his shop, which became an annual event. Dunkerley felt that bladesmiths in the Western USA needed to "get together and exchange ideas and techniques."[1] Steve Schwarzer and Don Fogg were two early Damascus specialists to visit. Fogg and Schwarzer found the atmosphere of those Montana hammer-ins to be infectious. Fogg told Schwarzer: "I warn you, these guys are on fire!"[6] Fogg compared the atmosphere of sharing ideas and learning to the time when Jim Schmidt, Jimmy Fikes, and Fogg were learning together.

The "Montana Mafia" in 2019. Left to right: Rick Dunkerley, Barry Gallagher, Wade Colter, Shane Taylor, and Josh Smith.

Many of top knifemakers and Damascus makers visited or presented at these hammer-ins, including Wayne Goddard, Al Dippold, Don Fogg, Steve Schwarzer, Hank Knickmeyer, Rob Hudson, Ed Schempp, Devin Thomas, and Joe Cordova. Many of the methods taught at hammer-ins from 1997-1999 were captured by Joe Olson in a series of sketches he called "Hammer Doodles." Steve Schwarzer nicknamed the group the "Montana Mafia," which initially included Rick Dunkerley, Barry Gallagher, Wade Colter, and Shane Taylor[7].

Many techniques were taught, popularized, and spread to Damascus makers, including using powder in Damascus, radials, bias forging, W's, accordion, and Jelly roll. The hammer-ins led to the spread of many ideas into the Damascus world, and there was significant growth in mosaic Damascus.

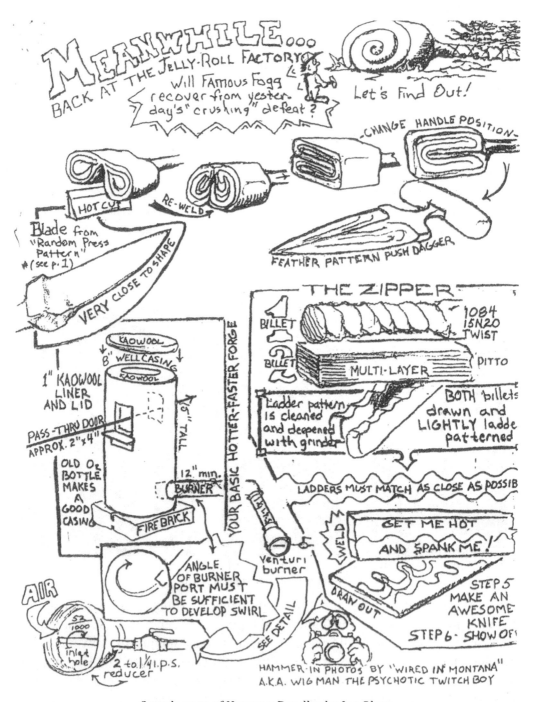

Sample page of Hammer Doodles by Joe Olson.

Damascus Folders

Another early visitor to Montana was Al Dippold, a Missouri knifemaker known for his mosaic Damascus folders. Dippold learned about Damascus with the help of Daryl Meier and Hank Knickmeyer. In about 1990, Dippold got tired of building fixed blades, as he said in a 1997 interview: "I started with friction folders and slip joints and then got into lockbacks and locking liners about two years ago. I haven't made a fixed blade since."[8] Dippold made fancy, well-finished mosaic Damascus folders with file work and other extra touches. He added file work after seeing it in other high-end Damascus folders made in the New England area such as by Jim Schmidt, a pioneer of Damascus folders. As Dippold said, "It really decorates a knife and gives it extra flair."[8] He also used high-end handle materials such as fossil ivory and pearl.

Images of Al Dippold Mosaic Folder circa late 1990s[9].

Barry Gallagher was inspired by Al Dippold and started making his own mosaic Damascus folders. Hank Knickmeyer was impressed with Gallagher's knives: "I think Barry makes a really slick-operating folder. It's slim, small and works well... If I had to pick the work of a folder maker I'd most like to own, he'd be one of them."[10] Rick Dunkerley began making mosaic Damascus folders shortly after: "A lot of the things Barry and I've learned, we've learned working together. I taught him how to make Damascus and he taught me how to make folders. Two brains on one idea come at it from several different angles, and that's beneficial for us both."[10]

Image of Jim Schmidt Damascus Folder. Weyer Photo.

Proliferation of Powder

Ed Schempp

Ed Schempp and Devin Thomas were experimenting with powder in the 1990s. Schempp said, "I had heard that Steve Schwarzer was playing with powder, and I had experience with powder hardfacing (coating of harder material over a base), and knew I could do [powder Damascus] too. I hadn't seen anything that Schwarzer had made with powder until after Devin and I had been playing with it. Devin did fish made with nickel sheet filled with powder which was the first good work I had seen with powder."[11] They couldn't find powder in small quantities, so they purchased a 750-pound drum of 1018 powder and then also sold it to other Damascus makers. "It was packaged to put in potato chips to attract oxygen. We started mixing graphite with it to carburize it and give us more variation in the color, to darken it up in the final Damascus."[11] Later, they purchased steel sandblasting grit, which was closer in composition to 1080 after a recommendation from Crucible metallurgist Ed Severson. And then Jeff Carlisle began selling steel powders in smaller quantities to interested knifemakers. Bob Kramer found a source of 4600e powder which was similar to 15N20 but without carbon. Devin presented steel powder methods at a 1999 Rick Dunkerley hammer-in. Devin developed a method of forming shapes with nickel sheet and filling them with powder, and he demonstrated this by making images of fish in the steel. Rick Dunkerley's first powder Damascus knife was also made in 1999, which he brought to the Blade Show. Dunkerley was unaware of any finished knives made using steel powder apart from Schwarzer at that time. In 2000 he made a fishing-themed knife and used Devin's fish Damascus in the handle.

Powder in Damascus became very popular. The master of the nickel foil image method, in my opinion, is Cliff Parker, who does very intricate images. Matt Diskin used powder in combination with laser-cut elements from flat steel pieces that were stacked and then filled with powder. The most famous user of this method is perhaps Shane Taylor, an example of his Dragon Damascus is on the next page. Powder is most often used now for filling in various recycled elements.

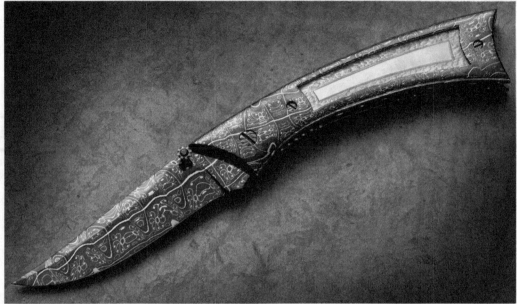

Rick Dunkerley's first powder Damascus knife in 1999. PointSeven Studios Photo.

Dunkerley knife with Devin Thomas fish Damascus in the handle. PointSeven Studios Photo.

Cliff Parker knife with Damascus made from nickel foil filled with steel powder. Image from BladeGallery[12].

Shane Taylor knife with Damascus made from laser-cut plates stacked and filled with powder. Image from BladeGallery[13].

Swedish Damascus

Kay Embretsen met an old axe smith in 1979 who taught him how to forge blades. By 1983 at 26, Embretsen went full-time as a knifemaker. Embretsen began making pattern-welded Damascus in 1984 after watching a documentary featuring ancient pattern-welded Damascus swords discovered on an island just outside of Sweden. Embretsen also traveled to a historical museum in Stockholm with ancient blades on display. Embretsen began specializing in high-end folders in 1992. He said, "If one doesn't work, you have to find out why it doesn't work. You have to make quite a few folders before you finally get them functioning properly. Many gray hairs started coming in when I first started building them."[14] One of Embretsen's signature Damascus patterns is "Explosion," which Don Fogg told me was one of the first evolutions of a W's pattern that he saw. The explosion pattern used a twisted W's bar cut down the center, and then the two pieces are flipped around and welded to themselves. Embretsen was making this pattern by 1994[15].

Image of Kay Embretsen knife with explosion pattern. Weyer photo.

Conny Persson began making pattern-welded Damascus in 1991 after he visited Kay Embretsen at a fair in Sweden[16,17]. Embretsen showed him some of his Damascus, and Persson started to make it by hand. Persson bought a power hammer in 1992 and sold Damascus knives that year. In 1994 he read an article in the Swedish Knifemakers Association newspaper about Steve Schwarzer's and Hank Knickmeyer's mosaic Damascus, and Persson started making mosaic Damascus shortly after. In 1999 Persson won the "Best Damascus Design" at the Blade Show, and he featured very intricate Damascus with Multiple pieces

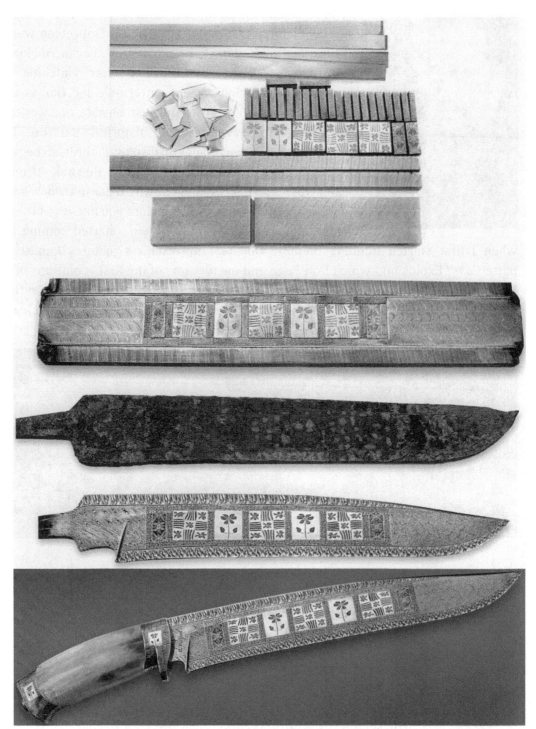

Construction of a 2004 Conny Persson Knife[18].

welded together in a single blade. Persson's work impressed many of the American Damascus smiths, including Shane Taylor and Steve Schwarzer. Devin Thomas told Schwarzer, "Persson's blades contained the cleanest patterning he had ever seen."[16]

Johan Gustafsson began making knives in 1993, went full time as a knifemaker in 1994, and started making pattern-welded Damascus in 1996[2,19]. He was influenced by knifemakers like Kay Embretsen, Roger Bergh, and Conny Persson. In addition to being known for intricate mosaic Damascus, Gustafsson became most famous for the wide range of colors he achieved in his blades. Heat bluing had been done by makers such as Hugh Bartrug, but Gustafsson's bluing was on another level. Gustafsson used a bluing solution called "nitro blue" at a boiling temperature of 280-285°F. The types of steels used, how much they were forged, and how they were polished affect the final color. Gustafsson uses Swedish tool steels, including 20C (simple high carbon), 15N20, 2550, and 17VA.

Joel Davis and W's Theory

In recent years there has been an "explosion" in W's-based patterns. Near the beginning of this trend was Joel Davis, who has made many W's patterns starting in the early 2000s. I interviewed him in 2022:

I took the last class that Bill Moran ever taught on Damascus. That was a four week class in 2000. Even at the time at 20 I knew Bill Moran's significance. He taught in a coal fire, O1, and mild steel. It was amazing. I visited Don Hanson on the way home from the Bill Moran school. Back then he was doing very interesting W's. In 2002 I got another scholarship and took Don Fogg's Damascus class. He was the guy that planted the seed. I heard Don Fogg's description of how W's actually formed. I didn't actually know; I knew about them in books, but I had no idea how that was done until I asked him. And he told me exactly how it's done. Other great makers such as Ron Newton, Chris Marks and the Montana Mafia guys had a great influence on my path... To date, I have successfully executed over 5000 individual W-theory based patterns in the span of 20 years. I've been very blessed to be able to follow a dream and do what I love to do.[20]

He also told me about his pattern-development process: "I am constantly chopping the end of the billet off to monitor what is going on, so I know how to subsequently manipulate it to get it what I want to look like. I have about 30 five-gallon buckets full of nuggets. People ask me to make more of my same stuff, and I don't want to make it again. I could live 10,000 lifetimes and not make all the stuff I've thought up."[20]

Images of 2002 Gustafsson knife[2].

Below you can see the process of making Joel's "Fire Within" pattern which starts by making W's but with all dark layers in the top half. In the final pattern this creates a dark border around the "fireball." After the "C's" are forged and restacked several times, one-quarter of the fireball is formed. At that point, the bar is bias forged to change the orientation of the flames and then stacked in a 4-way to make the final pattern. A tutorial posted in 2007 by British bladesmith Mick Maxen of a similar pattern (inspired by Davis) further popularized this type of W's based pattern[21,22].

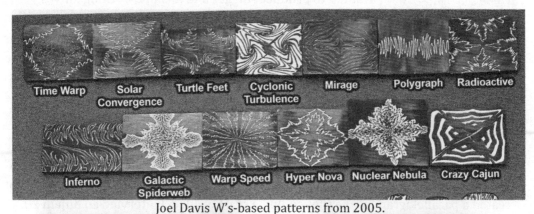

Joel Davis W's-based patterns from 2005.

Images of constructing "Fire Within" from 2005[23].

Steve Filicietti

Steve Filicietti is an Australian bladesmith who began learning forging at ten years old when he inherited his grandfather's blacksmithing equipment, and his father encouraged him to learn[24]. "I would make, use and lose knives, then make some

more."[24] By 16 years old, he was forge-welding Damascus and producing working knives. In 1998 Filicietti went full-time as a knifemaker[25,26]. At the 2001 Australian Knifemakers' Guild show, in his early 30s, he won four awards: Best Forged Knife, Best Damascus Knife, Best Utility Knife, and Best Bush Knife.

After reading many American knife magazines, Filicietti wanted to produce "bigger and flashier working knives."[24] He developed a style of large mosaic Damascus blades with tiled patterns. "I prefer to make large ornate Bowies as well as high-quality field-grade fighters and camp knives, and a range of integrals,"[25] he said in 2003. "I'm also an entirely self-taught bladesmith, which I believe has led to some of the unique techniques used in the construction of my knives and my Damascus steel."[25]

He developed a technique where the "end grain" pattern could be moved to the blade surface without the "accordion" or "loaf" technique. Instead, he cut the bar at a 45° angle and then rotated the tiles, which moved the pattern to the surface. This 45° cut created a parallelogram so the tiles could be scarf welded. I interviewed him in 2022:

> [It] was a long time ago when I started doing the flip thing (mid-to-late 90s). Well before that, I had seen pictures in magazines of the stuff that Schwarzer, Knickmeyer, Hugh Bartrug, and others were doing. I started making some cool end-grain patterns and tried forging pieces together and drawing them out. I had a couple of issues with this: 1) I had trouble getting the pattern proportion right in my finished blades (usually big Bowie-size stuff), and 2) I was concerned about the strength of the small weld area at 90 degrees (butt weld). Then one day [I] was looking at the end cut of a long flat bar and figured I could just try welding the pieces up on the flat cut at a 45 angle which would give me a longer weld in a finished blade. It worked the first time, and I never looked back. Even made those blades by hand, and possibly in a coal forge.[26]

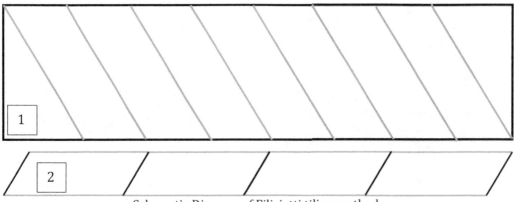

Schematic Diagram of Filicietti tiling method.

Image of Filicietti knife published in 2002[24].

Tom Ferry

Ferry forged his first billet of Damascus in 1997. In 2002 he said: "Damascus continues to be my drive in knifemaking because of the endless pattern development and artistic expression portrayed in each blade."[27] In 2002, Timascus was introduced, which Ferry described as "a titanium-based Damascus ... that Chuck Bybee came up with, and me and Bill Cottrell developed."[27]

Tom Ferry had learned about the method of cutting the bar at an angle for tiling, but he further developed it. The need to cut the bar at an angle rather than vertically meant that the pattern would be distorted. Ferry had the idea to "pre-distort" the pattern before cutting and tiling.

> I had seen the technique used by Steve Filicietti and Tim Hancock before I fully understood what I was looking at. And the tiling technique had been also used by a number of other makers during that time with only fair results as far as maintaining the integrity of the Damascus pattern. The patterns were over-exaggerated, and many of them looked like twist Damascus. In 2003 at Ed Schempp's world building hammer-in is where I did an initial test project of using my pre-form or pre-distortion technique. This was about the same time rolling mills were becoming part of the game and through a roller I could quickly estimate or calculate the distortion in large part due to a rollers one way drawing cycle. Controlling the final outcome rather than letting the steel dictate the finished look.[28]

Ferry described pre-distortion in a 2007 interview: "Pre-distortion is the act of taking a perfect image or Damascus pattern and distorting it to account for future forging operations, which will in turn correct the pre-distortion and yet again create a perfect image. This can sometimes seem like the destruction of many hours of work but, in the end, will always work out if all the calculations and details have been well thought out and planned."[29] Ferry would forge the square bar further down to a rectangular shape to "pre-distort" the pattern before the cutting and tiling technique. He figured out a simple formula to predict how much he needed to forge the bar down so that once the bar was re-forged the original "square" pattern would be seen. Damascus maker Gary House dubbed this technique the "Ferry Flip."[28]

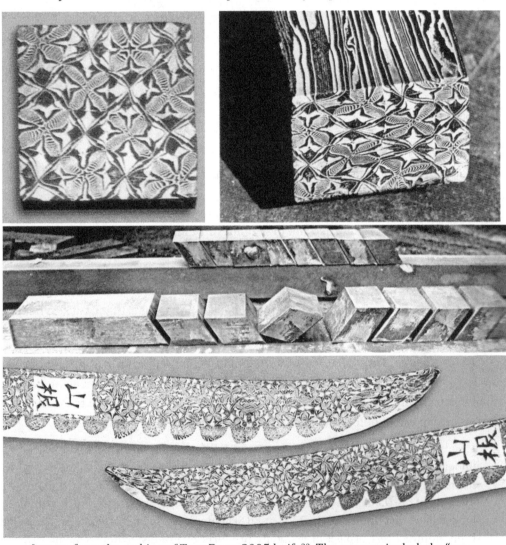

Images from the making of Tom Ferry 2005 knife[23]. The process included a "pre-distortion" step (top right) before tiling to maintain the original proportions after forging.

-30-

DAMASCUS FOR SALE

The first person to sell Damascus to other knifemakers was Daryl Meier in 1974[1], and he was advertising his Damascus in *Blade Magazine* in 1977[2]. Somewhat surprisingly, the idea of buying Damascus did not sit well with many at the beginning, as Meier reported in 1984: "Ten years ago, when I offered pattern welded steel as material to knifemakers at the Kansas City show, the idea was about as viable as lead balloons at the circus. The consensus at that time was that custom makers should do the whole thing themselves."[1] Tommy Lee and Bud Nealy were two of Meier's earliest customers. Another reason for the backlash was that pattern-welded Damascus was the domain of forging bladesmiths, while the Damascus for sale was used primarily by stock removal makers. Bill Moran himself was opposed to stock removal Damascus. He said: "The more closely the blade is forged to the finished size, the better the blade will be. It is possible to take a blade with only a hundred or so layers and grind a great deal off a blade that was forged very thick and make the layers appear finer than they really are. The only reason the grain is so pronounced is that it is greatly magnified because of the long angle of grinding across the surface of the blade."[3]

However, by 1984 this attitude seems to have evolved, at least according to Meier. He said the negative reaction to purchased Damascus was no longer found: "Fortunately, that feeling is no longer as prevalent. There are more specialists now: scrimshanders, engravers, etchers, carvers and even Damascus makers."[1]

AC Enterprises, later Charlton Ltd., and later still named Damascus USA, advertised their Damascus as "hand-forged-to-shape" rather than a flat bar that would be stock removal only. Their initial Damascus blanks were made in India from 300 layers of W1 and mild steel. Since then, they have sourced Damascus from many makers, including some in the USA. In 1986, owner Rob Charlton described his company:

We don't make Damascus, but we are the sole purveyors. It is high-grade, 100 percent hand-forged, top quality Damascus steel which we've been importing now for about four years. Charlton Ltd. Was the first to offer bar stock in volume for stock removal makers. Then we came in with rough forged-to-shape knife blanks, then swords, then finished forged-to-shape knives. Our latest item is 100 percent anvil forged-to-shape finished knife blanks.

There is nothing inherently wrong with a stock removal Damascus blade. It still retained many of the fine qualities associated with Damascus. But, going into the forge as we do at Charlton Ltd. And hammering the Damascus bar stock to shape, we pack the layers of steel more tightly. In the final analysis, the finished product has nothing to do with Damascus steel, since it is the forging that is the key. We pack those molecules tightly. At Charlton Ltd., our Damascus might be several times the price (of other Damascus), but there's darn sure 100 times more labor involved.[4]

Many top knifemakers had used Charlton's Damascus by 1986, including Buster Warenski, Paul Fox, Ray Beers, D'Alton Holder, Bob Engnath, Jim Pugh, Frank Centofante, and Bob Lum[1,4].

Another early seller of Damascus steel was Phil Baldwin of "Shining Wave Damascus Steel," made from W1 and a low carbon steel with 9% nickel. Baldwin began making knives in 1968 and was selling Damascus in 1983[5]. Other makers had asked him for his Damascus, which became a business. Sales started to fall off, and Baldwin eventually exited the pattern-welded Damascus market to focus on mokume (non-ferrous pattern-welded material) for jewelry. Baldwin told me: "The jewelry business is much larger than the knife business. Cheaper Damascus from competitors limited my market. The mokume business has been more stable."[5]

Fain Edwards began selling pattern-welded Damascus in 1984. In that year, he described the history of his company:

Until about four years ago I stayed away from making the stuff because many attributed the making of Damascus to magic, alchemy, if you will, which is of the Devil. I wanted no part of it! But my Lord Jesus Christ had other ideas. He told me in no uncertain terms that I was to make Damascus Steel and that it would be the finest that the world had ever seen. He would reveal a process – void of bat's wings and owl doo – to me, Fain Edwards. He did just that! As always, God had a man, Steve Schwarzer whom He sent to me to confirm His plan. Steve told me, "You have to make it!" "I don't have to do anything," I told Steve. Steve showed me his techniques of forge welding, which he stated he had learned from Al Pendray, who had learned from his father John Pendray, etc. etc... When I would run into a problem, God would supply the answer and solution. That brings us to the point where we are today: the world's largest producer of Damascus Steel – Ameristeel, by name. Made in America, by Americans, who still state with boldness that which is embossed on our coinage: In God we Trust.[6]

In the early 2000s, Brad Vice purchased Edwards' Damascus business and renamed it Alabama Damascus.

Tim Zowada made his first knife in 1978 and went full-time in 1983 at the age of 21. He said, "I am mostly self-taught, and I spent a lot of time reading books on knifemaking and blacksmithing. At first it was all trial and error, until at the American Bladesmith Society's Hammer-In out in Wyoming in 1983. Bill Moran and Don Hastings took me under their wings and showed me what it was really all about."[7] By 1985 Zowada was also making some Damascus steel blanks for sale to other knifemakers[7], which grew to be more and more of his business into the 1990s; he was advertising Damascus billets and mokume in Blade Magazine in 1991[8]. Zowada was also a major proponent of gas forges: "After hearing Phil Baldwin's lecture (about gas forges) at the Ashokan Seminar in about 1985, I built one shortly thereafter. I was selling shells by 1988 or 1989. I stole most of my initial ideas from [Phil Baldwin]. At the time, Bill Moran was saying welding in a gas forge was impossible."[9] Zowada also taught people how to make their own gas forges, including an article in the 1990 *Blade Magazine*[10], and gave away plans for free.

Stainless Damascus

Friedrich Schneider and Richard Hehn began developing stainless Damascus in 1983[11]. Richard Hehn wrote about this development in his 2001 book *Messer Profi-Tipps für Benutzer und Sammler* (translated from German):

> Oxide formation between the layers can lead to delamination, but this only becomes visible when the completed blade is etched in acid...It was experiences of this kind that led to stainless Damascus. The basic idea was to combine high-quality stainless Damascus steel to manufacture blades. Actually, that's impossible from the start, because stainless steels cannot be forge welded due to the formation of chromium oxide... The solution was actually quite simple and was called a vacuum. The vacuum not only prevents scale and slag, but also the formation of chromium oxide on stainless steels. A billet of stainless steel would therefore have to be forged by a heavy drop hammer under a vacuum... Sounds easy, yes, but it took a while before the first piece of stainless Damascus was on the table.

> The first problems were the technical facilities. You need a large forge ... and drop hammer. The next problem was the raw material. The initial material used was five millimeters and had to be rolled and milled to remove scale, which took machines and a lot of time. Without a partner with the corresponding facilities, it didn't go any further. When this was finally found, we could begin. At first, a maximum of 20 layers could be welded in the forge. But stainless steels are so-called "air hardening." This meant the billet had to first be annealed, milled, sawed, and re-stacked to be forged again. What a hassle, even without considering the possible reduction in quality due to uncontrolled carbon diffusion. The only way to solve this problem was to use the thinnest sheet metal strips possible to achieve the desired number of layers. Procuring thin, non-hardenable Chromium-Nickel steel in rolls was no problem (when a certain minimum order quantity was ordered), but high hardenability blade steels in a thickness of 0.3 mm were not on the

market. The only solution was Swedish AEB-L. This grade, from a performance point of view, is not exactly ideal, but it was the only way to continue the project. But even then, there were still a lot of problems, because the 50 kg rolls that were delivered were covered in oil and graphite from the mill...so at first, we had to cut hundreds of sheet metal strips to length and degrease them. The first experiment was performed in atmosphere. The billet was forged under the hammer and the sight was pathetic. The individual layers came out in all directions and formed a steel "rose" with hundreds of petals. So, it wouldn't work without a vacuum. The next test was done without air and produced flawlessly welded layers.

The first blades showed a very beautiful pattern with wonderful contrast, but the cutting left much to be desired. In the next test, the proportion of the non-hardenable chromium-nickel steel was reduced to one-third. As even this measure was not enough, the non-hardenable austenitic stainless was replaced by 1.4034, which has a low hardenability. The hardened 1.4034 had the same resistance to etching as AEB-L. This meant that both steels were uniformly etched, and the contrast was poor. In addition, the forging of this steel was laborious and material intensive. It was possible to achieve a perfect weld, but this only covered about two-thirds of the billet surface. At the edges, where the material could flow without resistance, there were inevitably defective spots. In addition, the different flow behavior of the steels used could be seen at the edges, so it was quite possible that the edge areas consisted of up to 80% of one steel grade – usually the grade with the lower carbon content. The entire edge areas were thus unusable and had to be cut away. In addition, the stainless steels turned out to be very difficult to deform at forging temperature during patterning. This meant that about two-thirds of the material was lost again when the blanks were milled. With thin blade blanks, the through-forming was sufficient, but with blades over five millimeters thick, the outer layers showed the desired pattern, but it became weaker towards the core...The best way to achieve a good-looking blade, in this case, was to use a twist pattern... Twist patterns work best with hollow-ground blades as they are preferred in fighting knives, but even with this method at least two-thirds of the material must be ground off.

The stainless Damascus project concluded with success. It was possible to produce stainless Damascus steel with properties that far surpassed those of 440C in terms of edge retention and elasticity. From an economic standpoint, the project was a failure. Almost industrial levels of production would have been necessary to reduce costs to a profitable level. Today, stainless Damascus is therefore purely a product for enthusiasts.[11]

Richard Hehn knife in stainless Damascus.

Terry LaBorde

Terry LaBorde made his first stainless Damascus in 1986[12]. "A unique aspect of my Damascus is the fusion in a pressure inert atmosphere, not in an open forge, resulting in unusual contrasts." He began advertising his stainless Damascus in *Blade Magazine* in 1991[13]: "NEW ALL STAINLESS DAMASCUS, exciting blend of ATS 34 and austenitic. Corrosion resistance. Brilliant contrast, new texture. Prepatterned billets, blades, finished knives. Also, CPM-T 440V all stainless Damascus. Extreme wear resistance, completely stain-free from the Vanadium carbide tool steel of the stainless steels. Also sheet stock." In 1993 he also advertised his patterning[14]: "100% MOSAIC STAR IN STAINLESS DAMASCUS! In either ATS-34 Damascus or Super Damascus CPMT440V. Mosaic pattern twists into stars thru the blade."

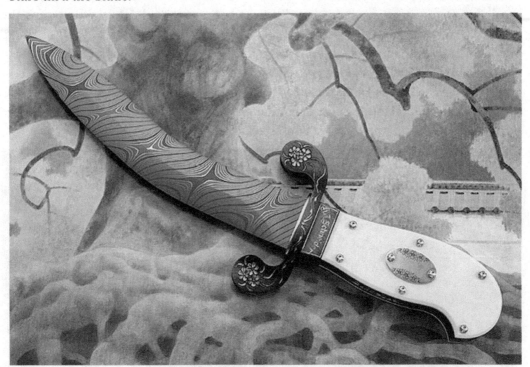

Herman Schneider knife with Terry LaBorde stainless Damascus from 1993[15].

Bud Hubbard

A.J. "Bud" Hubbard patented a method of making Damascus, which he applied for in 1988 and was granted in 1989[16]. The patent describes the use of thin strip stock that is held tightly together by steel sheet to prevent air from entering between the layers; the ends of the steel sheet were left open. Flux is therefore avoided by the layers being tightly held, and only one weld is performed because the layers are very thin to begin with. The patent also describes layers of stainless steel on the top and bottom of the stack "coated with a refractory substance" to prevent welding of the layers to the outer steel sheet box. Hubbard advertised his steel in *Blade Magazine* in 1989 as "Precision Engineered Damascus." He described it in the ad: "The steel in these knives has all the qualities you expect from Damascus. Toughness, ease of sharpening, long lasting edge and beauty. I make it by a new, patent-pending process that gives me an unusual degree of control over the properties and appearance of the blade." In December 1990, he advertised stainless Damascus as well: "No fancy talk. I have found that the P.E.D. procedure for carbon steel works equally well with stainless razor steel. The knife illustrated here, or any other knife, can now be had in completely rust-free STAINLESS STEEL DAMASCUS. It is a little more expensive to make, but for those who want carefree Damascus, here it is."

In an article on stainless Damascus in 1997[17], the process described for Hubbard's stainless Damascus was somewhat different because it said, "He tightly wraps the billet twice with .002-inch stainless foil." Perhaps the addition of this foil wrap allowed the welding of his stainless Damascus (foil prevents oxidation). He also talked about the performance of his Damascus: "As to edge holding and ease of sharpening, a stainless-Damascus blade is at least as good as one of stainless alone. P.E.D. stainless is as stainless as its components, which are all stainless. I've left blades outdoors in sun and rain for days without change."[17]

Tim Herman Knife with Bud Hubbard Damascus. David Darom photo.

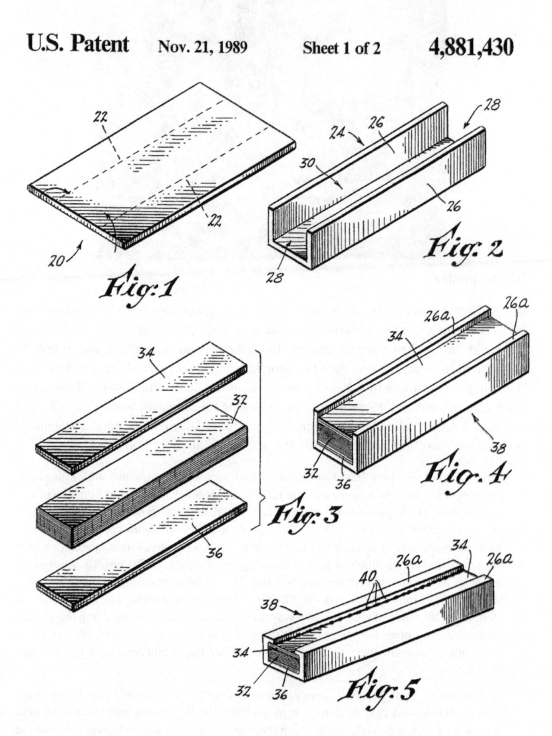

Schematic from Hubbard's Patent[16].

Devin Thomas

Devin Thomas was interested in knives from a young age. He grew up working in his father's welding shop and learned the basics of knifemaking from his uncle. At 16 years old, he worked for knifemaker Bob Lofgreen for a summer which greatly accelerated the learning process of making knives. At 18 years old in 1983, he won "Most Impressive Knife by a New Knifemaker" at the California Custom Show. At 19 years old, he went on a Mormon mission. Mormon missionaries work 6 days a week and have a day off called "preparation day" or "P-Day." Each missionary is assigned a "companion" that they proselyte with. Devin lived in Japan for two years.

> People knew I was into knives and swords, and they would show me stuff. Knives and pictures of knives. There were makers making ceremonial swords and kitchen cutlery. There were certain spots where you could watch sword polishers on this ramp, and it was in a public place, and you could see the guys rubbing swords with rocks sitting down. There were more kitchen cutlers than anything. There were guys forging yardwork tools like spade shovels, handheld ones. Sickles with curved blades, small ones. They would cut rice with them when harvesting rice by hand. There was a lot of interest in that kind of stuff, and I got to see them make it. Kitchen cutlery, blacksmithing tools, razors, wood planes, and all kinds of stuff. Cutlery was everywhere. On P-Days, my companion and I would drive our bikes 5 or 10 miles out to a guy who was making knives. We would see him work with tamahagane. He would wrap it in rice paper and cover it in clay to forge weld it.

> Back at home, I got a *Knives* Annual where Goddard talked about cable Damascus. I thought that looked easy to make. Or at least doable. Bill Moran and those guys kept emphasizing how difficult it was to make Damascus. Jim Hrisoulas and Daryl Meier would say forge welding was easy, and that gave me the confidence to try it. When I heard about cable, I thought that's something I can actually try. Then I moved on to make 1095 and nickel Damascus from a recommendation from Jim Hrisoulas. I started with a simple coal

forge that I built, but I couldn't find any coal, so I had to use charcoal. I learned about gas forges from Tim Zowada and found that to work a lot better. I made a power hammer after drawing up plans in my head. I had never actually seen a power hammer in person at that point. I worked for my dad and then Nevada Brass and then Olson Glass. Knives and steel were on weekends and after work. When I worked for Olson Glass, they had me on a swing shift so I could get up in the morning and put in several hours on knives and Damascus. That was pretty convenient. And it got me out of all the boring church meetings.

There was a guy who had an idea to sell Damascus spurs. He sold a lot of cowboy stuff like hats and things. So, I made up some Damascus spurs for the Antique Arms show and went with him. He thought I looked like Dan Blocker from the show Bonanza playing Hoss Cartwright. So, he put a 10-gallon hat on me, wrote "Hoss" on a nametag, and put it on me for the show. It stuck, and I continued to wear the hat at knife shows.

First, I learned how to weld with flux. I kept calling all these guys about welding flaws, and they would tell me different things. Daryl Meier said he would weld all around the edges of the steel. I figured I could do the same thing by welding up sheet metal around the stack instead. Plus, I had seen the tamahagane wrapped in rice paper and coated in clay, and it seemed to work by a similar principle. I would wrap the steel in sheet metal; it wrapped around all the edges; the top and bottom plates were exposed, and I welded around the top and bottom corners. It held the sheets tight. Shortly after, I moved to just welding up a box. I would put in a tiny drop of kerosene. Meier liked to use WD-40 or kerosene or other things. He called it a "hydrocarbon" added to the box.[18]

Devin made his first Damascus in 1986, his first stainless Damascus in 1989, and began selling stainless Damascus in 1991. Wayne Goddard mentioned Devin's stainless Damascus was available for sale in a 1993 *Blade Magazine* article[19]. Devin Thomas said:

AEB-L was the first steel I tried on the recommendation of Wayne Goddard. I had to call Uddeholm Strip in Ohio to order the steel. The billets were pretty small because I was focused on steel for folding knives. As the billets got larger, I started attaching a vacuum to the billet to make sure it welded. Terry LaBorde, Bud Hubbard, and Schneider were selling stainless Damascus when I started. American knifemaker Herman Schneider would talk about the German Friedrich Schneider, or otherwise I wouldn't have known about the Schneider stainless Damascus. Terry LaBorde didn't sell much steel. Someone would order a custom piece from him, and he would make it; pretty expensive. Hubbard's process was slow because he was dependent on Daryl Meier. Hubbard would stack them up, Meier would forge and weld it, then Hubbard would machine the pattern in, and then Meier would forge and roll it out again (Hubbard switched to someone else for forging the steel after a couple of years). Bud Hubbard threatened to sue me for using the method from his patent for making my stainless, but I hadn't seen it. I had no idea how to even look up a patent. I was doing my own thing.[18]

Mike Norris

At 28 years old, Mike Norris started making knives from circular saw blades in 1982 after watching his dad make some kitchen knives. His early knives were very crude, but he found a copy of *Gun Digest Book of Knives* at a bookstore, read through it, and realized there was a whole world of handmade knives. So he bought some equipment and started spending all his free time on knifemaking. He got tired of doing construction work and went full-time as a knifemaker in 1993. Norris said:

Mike Norris

> As I would go to shows, I would see a lot of Damascus, but Damascus just didn't impress me for some reason. I didn't have any desire to make knives out of Damascus, much less forge any Damascus. But during that time, during that couple years that I was actually full-time a guy had commissioned me to make some knives for him, and he was providing steel, and they were all Damascus by Devin Thomas. That really intrigued me because it was stainless, and you didn't have the rust problems. So, Devin Thomas was my inspiration, he's the one that inspired me to get started in forging stainless.

> So, in '95, I was at a show in the Spring in, I think Secaucus, New Jersey, and I was talking to some other makers about stainless Damascus and how I had been thinking and pondering about it for a while. I turned to my wife and said, "You know what? I'm going to make stainless Damascus." I knew nothing about forging except what little bit I had seen in a video here or there, but very little. I think I had been to one forging demonstration, and I went with a friend, and he wasn't really interested. I had to scramble to find all the information I could. I opted to go with a hydraulic press because of where my shop was, I couldn't put a trip hammer in there. The Guild Show was in July, and by the time the Guild Show came around, I had four knives made from stainless Damascus, and I had four billets for sale at the show. I ended up winning the Beretta award for outstanding achievement at the Guild Show that year, and that kind of started it. Whenever other knifemakers would look at my steel and they'd ask whose steel was that and I said it's mine. And they'd go, "It's stainless?" and I said, "Yeah," they'd say, "Would you make it for somebody else?" and I'd say, "Well, yeah, sure." So that's how I got started, and within two years, I was so busy making stainless Damascus that I basically had to quit making knives. At that time, me and Devin, we were the only ones (selling stainless Damascus).[20]

Norris wraps his billets in stainless heat treating foil to keep out the oxygen and thus successfully forge weld.

Other companies that have since been selling stainless Damascus include Chad Nichols and Vegas Forge. Vegas Forge traces its roots back to Devin Thomas. Rob Thomas split off from Devin to form Rob Thomas Damascus. Rob then sold his business which became Vegas Forge.

Damasteel - Stainless Powder Damascus

In the early 1990s, Pelle Billgren was the CEO of Söderfors Powder, a division of the steel company Erasteel in Sweden[21]. Billgren attended a forging event in Gysinge, Sweden, and became interested in the idea of supplying powder metallurgy steels to blacksmiths. Someone recommended that he visit knifemaker Kay Embretsen. In Embretsen's shop, Billgren saw pattern-welded Damascus for the first time, and he thought that this material was a good fit for the Erasteel powder metallurgy steel, and Embretsen and Billgren began working together in 1992 to make the new product which would be branded as Damasteel. The powder steels were "poured" into about 150 layers and HIPed under high temperature and pressure to form the solid billet of stainless Damascus. And then patterned using various standard techniques such as twisting and swaging.

At the beginning of 1994, Damasteel applied for a patent with Embretsen and Billgren as the inventors of the process[22]. Early experiments recorded in the patent used Elmax steel for the "dark" layers and austenitic stainless 304L for the bright layers. First, they tried Elmax powder mixed with 304L "chips" that were between 1 and 5 mm. The powder and chips were mixed in a steel capsule, welded closed, the air was evacuated, then given a hot isostatic pressure (HIP) process where the steel was held under pressure of 1000 bar and 1150°C (2100°F) to consolidate the steel into a solid. This solid was forged into a knife blade and, after etching, they got a random pattern from the mixed steels. The second method they tried used Elmax powder mixed with 25 layers of 2 mm 304L sheet with the Elmax powder layers being 3 mm thick, followed by the same HIP process to make a 50-layer pattern-welded Damascus. They then devised a method whereby alternating layers of Elmax powder and 304L powder were laid down in a canister prior to the HIP process. This allowed the use of two different steel powders together in a single billet. Damasteel won the "Innovation of the Year" prize at the 1994 EPMA conference in Paris (European Powder Metallurgy Association)[23]. Damasteel was being sold in Sweden in 1995[24]. Patterns available early on included twist, rose, Odin's eye, and ladder[25].

Damasteel Rose Pattern.

Production Knife Damascus

The earliest production knives made with pattern-welded Damascus were made by Böker using Manfred Sachse Damascus. This knife was released in 1980 in a limited run. Ever since they have done an annual limited edition Damascus knife. The Böker website says:

Boker Annual Damascus 1980 - Our first serious collector's knife for the German market. So to say a first test. So we wanted to put in there everything that makes the knife besides the damask blade desirable. This also included a low edition of 300 pieces. The selection of the model was easy. For reasons of cost, the knife was only allowed to have one blade and it should be lockable. I had learned this in the USA and above all I saw in it the future for Böker. The multi-piece pocket knife was already a domain of the two Swiss manufacturers Victorinox and Wenger. To attack their market position in this field was tantamount to suicide for me. So we decided to sail past the two Swiss with our model policy to the left and right... I can still remember exactly: the departments responsible for blade production could hardly wait to punch, harden, grind, polish and "extract" the first blades from damask steel. This is the name given to the acid treatment of the finished blade to make the damask structure of the 300 layers visible. One was really eager to learn new things and could hardly wait for the first delivery of about 30 Damascus plates from Manfred Sachse. The other employees in the stitching, assembling and finishing departments didn't want to be left behind but also wanted to be tickled with challenges. So the answer was ivory shells from the Odenwald, then still legal, neatly riveted to the brass plate. Our Damascus maker Müller was doubly challenged. Firstly, he had to remove the finished blade with nitric acid and secondly, using an ancient etching process, he applied a traditional engraving with a forging scene from 1690 to the ivory. The learning curves in the various stages of working on this knife, about 120 working steps without Damascus forging, only gradually increased. These 300 knives were all made twice... [T]oday we laser the blades from the delivered blanks. At that time, the punching cuts had to be used for normal running production. Since the damask was delivered much harder than the normal steel, we had to learn to heat the damask blanks glowing to make them more supple and thus not to batter our cuts.[26]

Böker Annual Damascus Knives from 1980 and 1982 with Sachse Damascus. Note the billet of steel ready to be forged in the handle of the 1982 knife.[26]

The earliest American-made Damascus production knives came from steel made by Fain Edwards of Ameristeel. In 1984 Lee Benchmade Knives introduced a line of knives made with Ameristeel. In the same year, Schrade announced a line of knives with Edwards' steel. Several production knife companies were not particularly happy with pattern-welded Damascus. Buck Knives introduced a Damascus 110 in 1989, but Buck vice president Charlie Gregory said:

> We didn't do very well with it. It was just too expensive and people who were willing to pay that price had a problem with the quality of the Damascus. As far as a mass producer,

I don't think anybody can make Damascus go. That's an element that is probably going to stay with the custom knifemaker. The market is too narrow for Buck.[27]

Case Knives also had poor experiences with Damascus. Jim Parker, a former owner of Case Knives, said in 1993:

Personally, I wouldn't make a Damascus line now. Maybe just a limited edition of 300 units or so. Fain Edwards and I started producing Damascus knives in 1985. We learned that we had to sell them at about three times the price of regular surgical steel knives. It was very difficult to be profitable at the retail level even with educated sales people.[27]

However, not everyone blamed the poor sales on the high price of the Damascus. Tim Zowada said it was the poor quality of the Damascus that was being produced for the factory knife companies. Zowada made Damascus for a Spyderco Worker model, released as a 12-year com-memorative knife (1981-1992). Zowada said, "It's not that much more expensive once a factory tools up for it, gets used to working with it

1984 Schrade with Fain Edwards Damascus[29,30].

and adapts the techniques. The problem is that in the past, when that was tried, the commercially produced Damascus was of inferior quality and it didn't go over well."[27] Daryl Meier saw the pros and cons of working with production knife companies, as he said in 1993:

I'm working with three different factories negotiating price and product right now. If I connect on a contract with any one of them it could tie up 99 percent until I get that order out. The problem with sales to factories is that it's more like a cold canvass sale – knocking on doors. They generally don't come to you. You have to go to them and convince them that they ought to be putting out a Damascus line. It's not really selling them on the material but selling the idea.[27]

Mike Haskew wrote that Damascus knives from production knife companies were making a comeback around 2001: "After a relatively short period of steady sales during the mid-to-late 1980s by such companies as Bear MGC, Böker, Damascus USA, and Buck, among others, the enthusiasm for factory Damascus knives began to wane. For a variety of reasons, including production difficulties, availability of the material, profit margins, and issues surrounding the quality of Damascus produced by some sources, some factories began to shy away from Damascus."[28] William Henry knives had won the *Blade Magazine* 2000 Investor/Collector Knife of the Year with a knife in Daryl Meier ladder pattern Damascus. Matt Conable said, "[H]e's one of the top

Damascus makers in the world, and for his ladder pattern specifically. It's a particularly beautiful choice on our gent's folders. It is an extension of what we do, making beautiful things that are viable as tools."[28] Ernest Emerson also saw the industry changing in ways that made Damascus more viable:

> The customer who was buying a $49 knife 10 years ago is now buying a $149 knife, and that allows people like us to make a Damascus-bladed version of an already expensive factory knife, and charge enough to make it worthwhile without getting into the custom knife price range. Some knives from factories are already expensive, but they are well made. So, people are already paying a higher price and the extra speed bump is not that much to get a Damascus blade. Plus, we've got guys like Devin Thomas, who we buy from, that are now making Damascus of superior quality in higher quantities.[28]

Stainless Damascus in Production Knives

The first factory production knives I can find in stainless Damascus come from a Damascus maker not mainly known for making stainless Damascus – Daryl Meier. He made a combination of 440C and 431 stainless steels for Bear MGC that were advertised in 1994[31]. I talked to Meier about his start with stainless Damascus:

> I started making stainless pattern welded steel after a time that my skill level was pretty well advanced. The early stainless that I made was done under the power hammer. In order to do that, I stacked up the material and arc-welded the edges to keep them together and to seal the contact surfaces. The thing is with stainless in forging the stuff if you get a chromium oxide formed on the stainless steel, the fluxes that we normally use won't cut it, like borax. You got to start with clean surface and keep it clean.[32]

Another company using stainless Damascus early on was Chris Reeve Knives. The first stainless Damascus used by Chris Reeve Knives was a limited run of 25 knives in 1994-1995 using Bud Hubbard stainless Damascus for the 25th anniversary of the Knifemakers' Guild[33]. Anne Reeve told me, "After that, we used some Jerry Rados Damascus (carbon steel Damascus), and probably some random other bars Chris would collect at shows, and then pretty much settled on Devin's Damascus – I'd say the latter part of the 1990s... We used his material exclusively (except for Damasteel specials) until we moved over to Chad (Nichols). Longstanding loyalty to the man (Devin) Chris considered an absolute master!"[33].

January 1998 Microtech with Devin Thomas stainless Damascus.

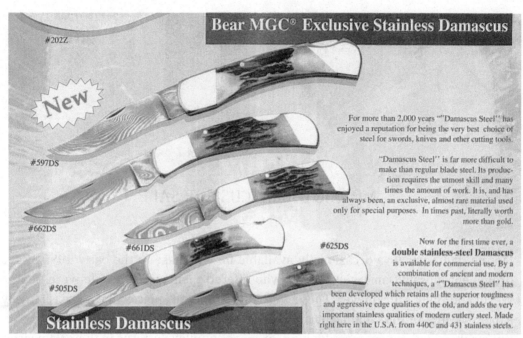

1994 advertisement for Bear MGC knives in stainless Damascus by Daryl Meier.

Devin Thomas says his earliest production customers in stainless Damascus were Chris Reeve and Microtech[18]. I found a stainless Damascus Microtech knife that was produced in January 1998[34]. In terms of larger factories, he also made stainless Damascus for Buck by 2000[35]. Mike Norris tells me his earliest sales to a production company were to Schrade in 2001[36]. Schrade announced a Joe Kious collaboration knife in Damascus in 2002[37].

1994-1995 Chris Reeve knives with Bud Hubbard stainless Damascus.

-31-

MOSAIC DAMASCUS FOR SALE

Mosaic Damascus was occasionally available for purchase in the early 1990s. Two prominent mosaic Damascus makers who offered their work for sale were George Werth and Hank Knickmeyer. These were often "one-off" pieces that were relatively expensive due to the significant number of hours for producing complex patterns. Knifemakers and buyers were looking for more complex patterns in their Damascus. Meier said in 1997: "Twenty-five years ago when I started selling Damascus, random pattern Damascus was fine. The consumer is becoming more discerning and realizes which patterns are more time consuming, risky, and involved. I'm selling more 'regular pattern' stuff, like Calico Rose, which I developed, or traditional Turkish."[1]

Turkish twist is not a mosaic pattern, but it is an example of a complex pattern available for sale at a relatively early date, such as from Jerry Rados in the 1980s. Other Damascus makers also started making mosaic Damascus for sale, including Bob Eggerling, Devin Thomas, Ed Schempp, Gary House, Chris Marks, Delbert Ealy, Doug Ponzio, and Bertie Rietveld.

Hank Knickmeyer Damascus blade with Paul Grussenmeyer carved handle. Weyer photo.

1995 Bob Garbe knife with George Werth mosaic Damascus[2].

Robert "Bob" Eggerling

Bob Eggerling began making Damascus in the early 1990s. In 1995 he started making mosaic Damascus and began selling it in 1996. He went to a knife show in 1996 in Reading, PA, and knifemaker Ken Steigerwalt was the first to buy mosaic Damascus from Eggerling. Bob talked to me in 2022:

> I didn't know anyone making mosaic Damascus at that time. I had only read about it in *Blade Magazine* and *Knives Illustrated*. I was still trying to figure out what was going on in the knife world. After reading those magazines, I wanted to get into making knives. Damascus was something that sort of snuck into my psyche reading about knives and making knives. I sold Damascus on the side, and Damascus kind of took over.[3]

Eggerling is known for his highly detailed patterns:

> I want people to know right away when you see it that that is Eggerling Damascus. I was pretty much self-taught, but I was inspired by Hank Knickmeyer. He was amazing. His work really fascinated me. Some Damascus makers, I would look at their steel and know right away how it was made. With Knickmeyer's Damascus, sometimes I never could figure it all out. Making Damascus is something you have to figure out yourself. Everybody is different and has different ideas. It's like a road, you just have to keep going on it.[3]

Eggerling makes so many patterns[4] he isn't known for any particular pattern. In fact, the pattern he is best known for is his "quilt" pattern which itself is a combination of many individual pattern "tiles" within it: "Each square of quilted steel has a different pattern in it - initials, faces, all different things. There was a place in Lancaster, Pennsylvania, that had a quilting show in the late 1990s. I used to look at books and that kind of stuff and talk to ladies that made quilts; it's a different kind of craft than working on steel."[3]

The intricate patterns and reasonable cost of Eggerling's steel made it immediately popular. In 1998 Jay Sadow of Arizona Custom Knives said, "Eggerling's stuff is hot. He's really hit the right note with his mosaic Damascus. (Knifemakers) W.D. Pease and Ken Steigerwalt introduced us to Eggerling's work about a year ago and it seems to be sweeping through the market. It just knocks your socks off."[5] In 1999, knifemaker Ted Moore said, "Eggerling has taken the world by storm with his bolster material. Everyone is going with it."[6] And knifemaker Lloyd McConnell gave a similar sentiment: "My guess is that Eggerling is probably the most exciting Damascus maker in the country. His stuff is awfully hot."[6]

A range of Eggerling patterns circa 2005[4].

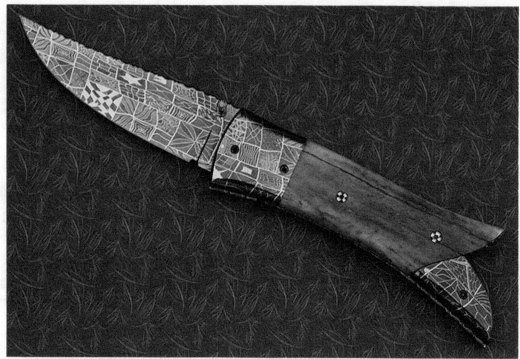

Wayne Whitaker knife with Eggerling "quilt" Damascus. Blade Gallery image[7].

Stainless Mosaic Damascus

In the mid-1990s, Devin Thomas developed mosaic patterns in his Damascus, including W's-based patterns and his own creation, Spirograph, and they showed up in print around 1999[6,8,9]. Devin said, "I was just getting bored with making the same few patterns over and over. Ladder, raindrop, and twist are fine, but you want to show a wider range in your Damascus."[10] He also offered small amounts of some picture mosaic patterns such as stars, lightning bolts, or fish by the year 2000[11]. By that year, Devin also offered spirograph and the W's patterns in stainless[12]. In 2001 Devin Thomas said, "Sharkstooth and firestorm were originated by Don Fogg, but I'm the only guy to do the patterns in stainless Damascus."[13] Firestorm was a twisted W's pattern, and sharkstooth a ladder pattern W's. But Devin was most proud of his original pattern Spirograph: "The pattern I am known for is called Spirograph, and it looks like those interlocking line patterns of a Spirograph game. I was the first to do it and developed it by changing the makeup of the billet, using some standard pattern techniques ... [and] we turned the bar stock 45 degrees once we had all our layers of steel."[14] Spirograph starts as a "grid" pattern, which is bias forged at 45 degrees, then drawn out and "ladder" patterned to raise the interlocking lines to the surface. There is also another variant of spirograph called "vines and roses" which is the same pattern but put through "raindrop" patterning dies. In 2001 Devin also had other mosaic patterns available including "reptilian" and "basketweave."[15,16]

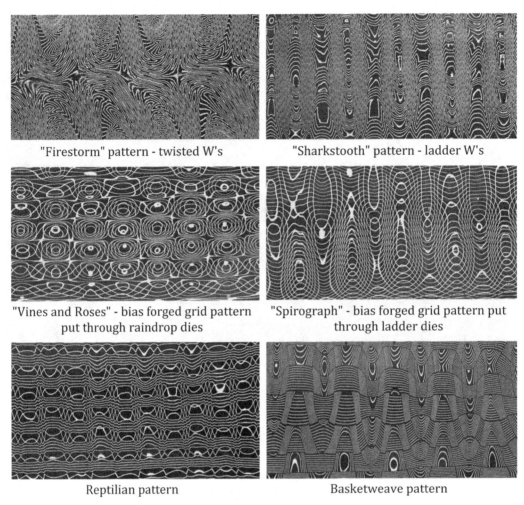

"Firestorm" pattern - twisted W's

"Sharkstooth" pattern - ladder W's

"Vines and Roses" - bias forged grid pattern put through raindrop dies

"Spirograph" - bias forged grid pattern put through ladder dies

Reptilian pattern

Basketweave pattern

Devin Thomas mosaic stainless Damascus patterns.

Mike Norris developed his own patterns available in stainless in the late 90s, and they showed up in print in 2001[17]. By 2003 he offered crazy lace, snake skin, and gator skin patterns[18]:

> When I first started, I kept thinking of different patterns, but I thought, well, everything I do is going to increase the time so much nobody will be able to afford it. So, I stuck with those basic patterns for several years before I just got bored with it. So, I started experimenting with patterns and ended up creating some really nice patterns. I will say this, I made a lot more money when I had three basic patterns; things went a lot quicker. Now some of my patterns are complex enough that there's so many steps to it, it really gets time-consuming, and I really don't charge comparable to the time. Some of it is kind of expensive, but it's really not as expensive as it should be if I got paid for my time.[19]

Ronald Best knife with Norris' Crazy Lace Damascus[20].

Combining Damascus Steels

There was also a growing trend for contrasting Damascus patterns between the bolsters and the blade. Devin Thomas reported in a 2000 interview: "For a long time we assumed that the same Damascus pattern (on both the bolster and the blade) would look better, but guys that make a lot of knives have figured out that a change can be better. Robert Eggerling started a lot of the (trend) because for a long time all he was making was bolster stock, and you had to use something different on the blade if you used his bolster stock. I've seen lots of examples of his bolsters used with my blade Damascus."[11] Daryl Meier added:

> When you use different patterns, the differences help to define the transition between the handle and the blade. The handle is one thing and the blade is something else. As the eye travels from one end of the knife to the other, then at some point you will be looking at something different. If the same pattern is in the front bolster and the blade, and the width of the bolster is about the same as the blade, then you hardly have any definition other than shadow to tell you what you're looking at when your eye is moving along. That definition is particularly difficult to see in photos.[11]

Jerry Corbit knife made in 2000 using Devin Thomas Spirograph Damascus in the blade and Eggerling Damascus in the bolster[21].

Production Knives in Mosaic Damascus

As with everything, it is difficult to pinpoint when mosaic Damascus was available in factory production knives. It also depends on how you define mosaic Damascus, and what counts as a factory knife. Microtech produced a Vector in Devin Thomas spirograph at the end of 1999[22], and ProTech was advertising knives in spirograph pattern in a November 2000 *Blade Magazine*. Spyderco announced their 25th-anniversary knife in 2002, a Delica with Ed Schempp mosaic Damascus bolsters with the Spyderco logo in them[23]. In the same year, Chris Reeve announced an Umfaan available in Gary House mosaic zebra pattern Damascus[23]. Anne Reeve told me about the zebra Damascus history:

> In 2002 or 2003, local guitar maker, John Bolin, brought Billy Gibbons of ZZ Top to the workshop. Billy liked our knives and wanted us to make something special for him and his crew. He had spent about 6 months in Southern Africa some years before and was very taken by the zebra skin we had on the wall. Conversation evolved, and Chris came up with the concept of making mosaic Damascus with a zebra pattern to it. He approached Gary House, who spent quite a while figuring out how to make it, but he finally did, and we made a handful (maybe 7-10) Umfaans for Billy. If I remember correctly, the handles

had a stylized zebra pattern graphic on them. Chris ordered a block each for Large and Small Sebenzas and Mnandis. Each block realized about 20 blades, so the number of zebra Damascus knives in the market is very limited.[24]

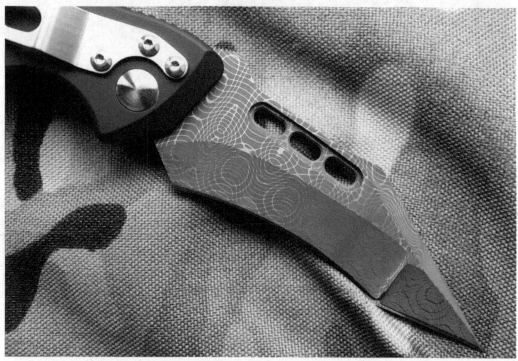

Microtech Vector with Devin Thomas Spirograph from 1999[25].

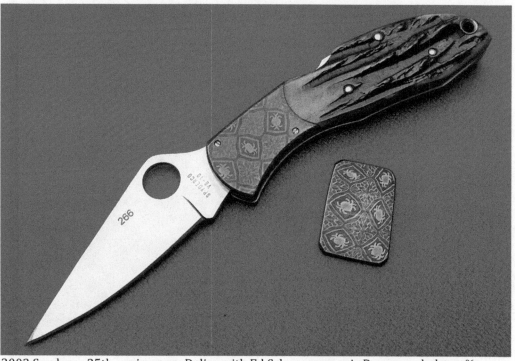

2002 Spyderco 25th anniversary Delica with Ed Schempp mosaic Damascus bolsters[26].

2002 Chris Reeve Knives Mnandi with Gary House Zebra Pattern Damascus[27].

-32-

DAMASCUS PRESENT AND FUTURE

Of course, many bladesmiths are working with pattern-welded Damascus now and I cannot write about all of them, just as I can't write about all those that contributed up until now. But I interviewed some Damascus smiths known for being innovators in recent years and asked them about what has been happening and where things are going.

Salem Straub

Salem Straub told me about his background with blacksmithing and knifemaking in 2022:

> I got into blacksmithing first, and I was living in the mountains off grid and helping my dad build a log cabin for my parents to live in. We were so far from the store, sometimes I would need to make something that we couldn't take a day round trip to get. I started to fool around, made a coal forge out of scrap car parts, and started making things from stuff lying around like leaf springs. I would just go around and chip charcoal off of trees and stuff; it was pretty rustic. I just grew up loving knives and swords, like kung-fu movies and stuff. I just got more and more hooked on it and just specialized more. To the point where I will still do some general blacksmithing, but I rarely get to.[1]

Straub is known for his kitchen knives:

> I like that chef's knives get used more than any other type of knife. Even hunters only use their knife a couple of times per year. I like to cook, and I enjoy using the chef's knives myself. I've done some professional cooking myself. It fit me as a style of knife to make. I enjoy what a big canvas they give you to express interesting pattern-welding to. I just like the northwest foodie culture here. It is fresh and cool.[1]

Straub credited several people with helping him learn knifemaking:

> I think self-taught is a misconception. I started with a lot of books. Wayne Goddard cable Damascus, 2004-2005, that was the stuff on my horizon. I would harvest old logging cable and make knives from that in my hand-cranked forge. Then I went to stacking up old

bandsaw blades and pallet strapping because that was what I could get. Just random pattern, hand-hammered, using flux. *Complete Modern Bladesmith* and *Master Bladesmith* from Hrisoulas were books I read early on. I would never put Hrisoulas forward as an authority on metallurgy, but he was great for pattern welding. The Desrosiers (Adam and Haley) taught me about multibar, and they have been very helpful.[1]

Image of Salem Straub knife[2].

Mareko Maumasi

Maumasi started his knifemaking by working for Bob Kramer, known for his kitchen knives and Damascus, among other things. Mareko told me about his early Damascus with Kramer:

> I first learned how to make Damascus while working for Bob. I started out making the standard random pattern and ladder pattern. When things were sticking really well, we moved on to other things. It would probably have been 2010 or so when I started making mosaic Damascus with a W's radial pattern. I learned about radial patterns from Bob Kramer, who got them from Ed Schempp. Bob explained to me how [radials] worked, and we rolled from there. I didn't really know of any other mosaic patterns. I had a pretty

334

good grasp of how different processes, like forging on a bias, would cause lines to move around. So, I just started playing with those processes and messing with the original stack. Exploring how the original stack affects the final pattern. What happens if you cut it into 4 or 6 pieces and then forge on a bias, and things like that. My understanding of patterns all started with Bob Kramer, who encouraged me to experiment and come up with new patterns. I was not yet poking around on the internet. Years later, everything I saw was on Facebook, and sometimes forums. This was before I was aware of Instagram. Then, the patterns I was drawing were nothing like what I was seeing on the internet. But when I worked with Bob, he encouraged me to make different patterns. He kind of let me loose to play and explore and do different things. I learned a lot, for sure. I'm very grateful for that introduction.[3]

Maumasi also says that his mentor, Bob Kramer, continues to push the envelope with exciting ideas: "I'm very impressed with what Bob is doing to create his very intricate plug welds and mixing patterns like puzzle pieces. I don't know what process he's using, whether it's EDM, water jet, or some other process. There were people using EDM in the past, but it was very different."[3]

August 2021 Kramer Damascus knife with bird and feather plug welds[4].

Maumasi found that after several years of experience, methods of patterning became easier to understand[3]: "I can look through patterns and reverse engineer any Damascus pattern, and with most patterns it takes like 20 seconds, maybe a few minutes with something complicated. Because of that, I'm constantly thinking of new ways to make patterns, new ways to weld patterns back together. Methods that are unconventional but simple. Because I want to do something that is new and different."

Growth of Damascus Through the Internet

The internet was key for Salem Straub learning Damascus[1]: "I just learned little bits from people and forums online. People like Bruce Bump would make W's or feather; Kyle Royer would share how to make twisted W's multibar, and I would just learn that online." And Straub said the internet has been helping to grow the number of Damascus makers and their techniques:

> It's interesting how trends come and go, even in the specialized area of mosaic Damascus. I think the number of mosaic Damascus makers is growing. I think it's exciting times for mosaic Damascus just because the information is more available. Mareko Maumasi with "Pattern-Welded Wednesdays" and how to build his favorite patterns on Instagram. A bunch of people just copied him. I've shared a lot of stuff on my feed. I just see a ton of that activity. Guys learning from other guys. There is a ton of youth talent. Ellis Bladeworks, he does some fantastic stuff. Jacco Van de Bruinhorst out of Canada. Another guy that is in

his early 20s doing killer stuff. There's been an explosion in South America. Rodrigo Sfreddo has had a huge influence on modern knives. He is not known as a big mosaic guy, but he can absolutely do that style of work. His whole keyhole thing, double lug guard, he just started a lot of stuff.[1]

Maumasi also talked about how the internet has changed how people learn about Damascus:

Instagram has kind of democratized the maker's ability to get information. Before, people would learn through forums or shows, but often, learning required an intermediary like a magazine. With Instagram, it allows you, through your own energy and time to develop those connections on your own. It's like 5 billion people on Instagram, and not everyone will be interested in what people are doing, but it's a huge opportunity for people that never knew you existed, or your work existed, to find you. These people are curating their own experience. They are choosing to see my work. There is a greater value in that kind of advertising. If I go to a knife show, only a fraction of 1% are even vaguely interested in what I'm doing, but people on Instagram have chosen to follow me because they wanted to see my work. Instagram has really created an opportunity to remove the middleman and allow the makers to connect to their audience.[3]

However, not everyone agreed that Damascus techniques should be shared freely, as Maumasi said:

I do get some pushback on stuff I was sharing on 'Pattern Welded Wednesdays.' That I shouldn't show the competition what to do. Or people claiming that those making cheap Damascus could copy it. Sure, I'm sharing how to make these patterns, but it takes two, three, four days to make a pattern. That doesn't make sense economically for someone making cheap Damascus. And when it comes to the idea of training my competition, I selfishly share some of my patterns or ideas I've developed because I want to see what other people do with that information when they share it back out to the world.[3]

Don Hanson also discussed with me the effects of the internet and the TV show Forged in Fire:

The internet has made the information available to everybody. You can google how to do anything as far as making knives, and it pops right up. It has exposed a lot of collectors to the knife world which is really good. And just made information right at everybody's fingertips. Forged in Fire has really blown up the knife world. It has brought more knifemakers into this world than anything. Before Forged in Fire, the Blade Show was big. Right after Forged in Fire, the Blade Show exploded. Tables sold out quicker.[5]

W's Pattern Mosaics

I asked several Damascus smiths about W's patterns in recent years because many mosaic Damascus being made uses W's as its basis in some way. Maumasi told me: "At least 75% of patterns these days have some form of W's in them."[3] Straub said, "When you get something so foundational like W's, you can combine that with other techniques like the feather cut, which goes back to the old Javanese pattern techniques."[1] Don Hanson shared his thoughts: "W's became dominant because it was interesting to look at. It's really easy to do. It's probably one of the easiest mosaics to make. You can forge it and cut it up and restack it in different ways to make just about any kind of pattern from stuff that looks like flowers to really bizarre geometric shapes and patterns. I think the main reason is it was easy, and the end product looked really good."[5]

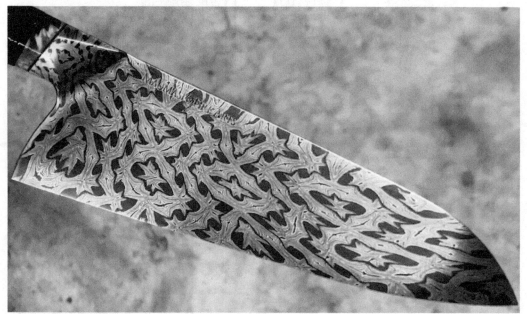

"Briar Patch" Damascus pattern from Mareko Maumasi.

The Continued Inspiration of the Past

Damascus smiths are still inspired by patterns that are 100, 500, or even 1000+ years old. Maumasi told me about some patterns he has been working on:

> I've been working on rediscovering how these ancient patterns were developed. There are some patterns that were unearthed in Scandinavian bogs in Sweden and Denmark and the Netherlands. Even some in Poland. They were mosaic Damascus swords that were built within a 500-year period, approximately 75 AD to 550 AD. They were originally unearthed in the late 1850s. People didn't understand what they were looking at. It's been so many years, and I want to know how it was done back then. I've been working on puzzling these patterns out, looking at them, and playing with ideas. I've figured out three patterns specifically. A diamond-shaped pattern, a pearl-like pattern, and a palm tree pattern. From what I understand of pattern welding, they were definitely mosaic tile-welded sword blades. How do you even do that in 100 AD?[3]

The "palm trees" that Maumasi referred to look similar to the ancient Denmark sword that Meier discussed as a type of mosaic Damascus (Chapter 27).

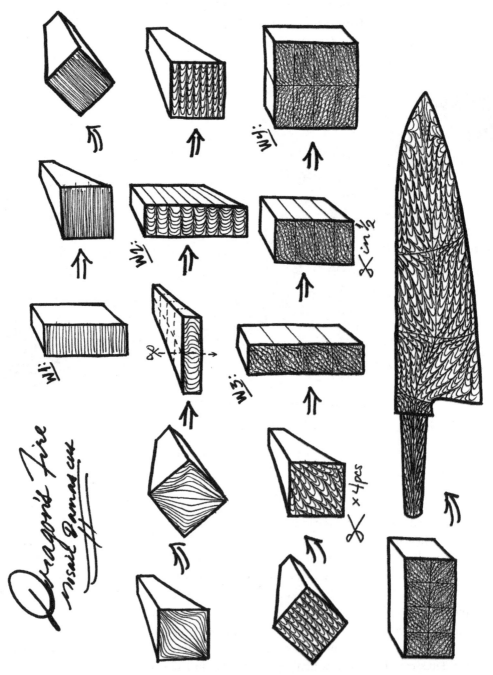

Demonstration of "Dragon's Fire" Damascus pattern from Maumasi's "Pattern Welded Wednesdays"[6].

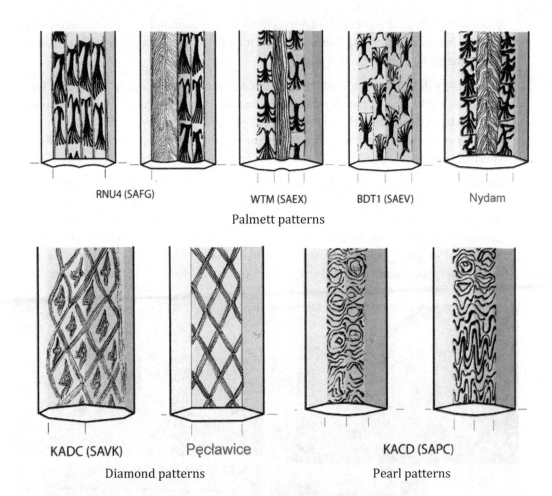

RNU4 (SAFG)   WTM (SAEX)   BDT1 (SAEV)   Nydam

Palmett patterns

KADC (SAVK)   Pęcławice   KACD (SAPC)

Diamond patterns   Pearl patterns

Illustrations of ancient bog sword steel[7-9].

Straub told me about a historical pattern he has been working on:

A guy in South America named Manuel Quiroga Güiraldes - 10 years ago, he was doing some really interesting and time-consuming pattern welding. He did the chainette pattern, or he tried it. That is the first time I've seen someone put a lot of time and effort into recreating a pattern.

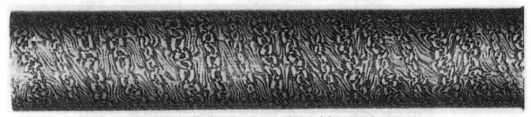

European Chainette pattern barrel from Ref. 10.

It came out pretty good, but it really highlighted what a large effect a disparity in forging equipment will make. Between the historical guys, the equipment he had at his disposal, and how it makes the steel look different. It came out a little funny, like too distorted. So, you didn't really get a good chain picture. I think he was bummed about it. He was working with big billets and working them down … with a press. It works the center more, and it pulled the picture out of his end grain.

I thought he did an awesome job, but I wanted to make it with a power hammer, which would duplicate more what the gunsmiths were making back then. The power hammer can lead to less distortion than a press if you're careful. There are no hard and fast rules; a press can distort things more than you wanted. Other times if you wanted to achieve a specific distortion like an s-curve, a press can be more controllable.

On this chain feather I just made, I used a hammer, a press, and a rolling mill; they all had their places... The chain links are formed along the meetings of the edges of two twisted bars. The end grain is visible as a repeating couple of bright, straight layers with intermediate dots of white and black to suggest link terminations. The bright, straight layers distort into a curve from the twisting. It's common that the links don't line up evenly every time along the weld seams, but the overall look is still quite effective. [1]

Production of "chainette" pattern by Manuel Quiroga Guiraldes[11].

Final "chainette" pattern by Manuel Quiroga Guiraldes[11].

Straub made his own version of the chainette pattern called jailbird with different techniques that are more typically used in Damascus today (see below). He started with a set of different layers at the center of a stack and forge welded that together. He then forged the bar on a bias to intentionally distort that center layer and create an "S" shape. The bar was then cut down the middle and flipped around to make a "C." With a four-way stack, an oval, or a single chain link, is created. Straub has used the chain links in several ways, but one striking pattern used the chain links with a feather-pattern technique to create the final look.

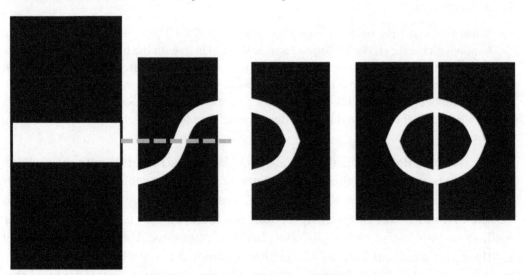

Schematic diagram of creating a "chain link."

Knife with feather "jailbird" Damascus from Straub's Instagram[12].

New Technology and Techniques

There are always new techniques being developed for pattern-welded Damascus production and patterning. 3D printing is becoming less expensive and more widely available. Designs are produced one layer at a time where the machine lays down a polymer until an entire three-dimensional plastic piece is created. Ron Hardman and Steve Schwarzer presented the use of 3D printing in the production of pattern-welded Damascus at the 2022 Blade Show West. A method they presented is the 3D printing of outlined designs that can be filled with steel powder[13].

The method was developed at Kilroy's Workshop, a metalworking school run by Hardman. He said:

> Schwarzer was here teaching a class on mosaic, canister layout, traditional methods of bending nickel sheet. We 3D printed a solid form to help with the bending process. While trying that method, I leaned over to my shop manager, Ben Bannister, and said, "If we used an organic filament we could print a thin-walled version, and then it would go to carbon in the canister." He said, "I think PLA is just cornstarch." We told Steve, and he said, "Let's give it a shot." If he had said it was a dumb idea, I probably wouldn't have done anything with it.[14]

Knifemaker Max Harder added the Blade Show West logo to a knife with the method. An exciting part of this method is the relatively low cost at which it can be done. It doesn't require any specialized equipment. Hardman said, "Any $200 printer would be able to do this. PLA is the cheapest stuff out there. $20 for a whole roll."[14] Other methods unique to this technique have also been attempted. Hardman said: "One of the things we tried was with a resin printer, castable resin. We added cornstarch to thicken it up and added nickel powder to it. When you print that and forge weld that together, you are left with nickel residue, so you have a thin white line. Instead of leaving carbon behind, the resin bakes out, and you get a bright line behind."[14] Hardman also looked at a method that has yet to be explored fully:

> I took graphite powder, and I stuck that in one of the holes instead of steel and put powder steel around it and forged that down, and it left a hollow core, you can just dump the graphite out. I could put a solid core of graphite of any shape and create almost like a channel going through. You can take that tile, forge weld that into a blade, and then cast into it. Bronze, gold, silver, etc. Something that isn't going to melt when you're heat treating, which is why you would forge weld it first. Then cast into it, and it goes all the way through. We don't have to then gouge it out or mill it out.[14]

3D printing of a canister, ready for filling with steel powder[13].

Knife by Max Harder with "Blade Show West" logo made with 3D printing and powder[13].

-33-

KNIVES AND STEEL IN SWEDEN

Iron and Steel in Sweden

Iron ore was being gathered in Sweden at least by 100 BC as "bog ore" or "lake ore."[1] This bog iron is an iron oxyhydroxide, common goethite (FeO(OH)). Nuggets of bog iron can be collected without mining, and farmers frequently gathered them. Most of the iron used in the Viking era (~700-1000 AD) came from bog iron. The bog ore could be reduced to iron with relatively simple methods, and the impurities easily removed because it does not need to be molten to remove the impurities. There was an increase in the use of "rock" ores extracted through mining in the late 1200s AD. The form of wrought iron produced and exported in Sweden was sold under the commercial name "Osmond." These were produced in spherical "balls" though also frequently cut up into smaller pieces. When the Osmonds were forged into useful shapes, this was called "bar iron." While local farmers collected bog iron, the increasing importance of ore mining led to Swedish iron being increasingly controlled by Government and industry. In the 1300s, Swedish iron was imported heavily into Germany and England because of its high quality. Gustav Vasa, king of Sweden from 1523 to 1560, prioritized producing as many products as possible domestically and importing as little as possible. At the same time, he sought to increase the number of products that could be exported. This led to significant operations in silver and copper mining, in addition to increases in iron. He wanted to increase the exports of bar iron and reduce the exporting of Osmond, because the bar iron was worth twice as much. In the 1620s, Vasa's grandson Gustav II Adolf would fully restrict the exporting of Osmond. Between the late 1600s and early 1700s, Sweden was the world's largest exporter of iron, and in the 1740s, it accounted for 35-40% of the world's output. In the Middle Ages, England began using Swedish iron to manufacture steel, and Sweden started to make steel with its own iron shortly after.

Johan Börjesson started iron production in 1640 in what is now Hagfors, Sweden[2]. In 1715 the company was purchased by Bengt Gustaf Geijer, after the company had fallen into disrepair. He named the company Uddeholm in 1720. In 1839 Uddeholm acquired the Munkfors Works, which was added to the other mills owned by the company. In 1870 Uddeholm became a public company and began using the Bessemer process in 1880.

Sandvik, a sizeable Swedish steel company, was formed by Göran Fredrik Göransson in 1862[3]. Göransson was the first in the world to use the Bessemer process on an industrial scale.

Historically important steel companies in Sweden for knife steel include Stora, Erasteel, Uddeholm, Böhler-Uddeholm, and an Austrian parent company Voestalpine. These companies are the result of many consolidations with a long history, but here are some highlights: Two steel companies started in the 1500s called Vikmanshyttan and Langshyttan[4]. Vikmanshyttan was the first company in Sweden to begin producing crucible steel in 1859[1]. Vikmanshyttan installed a new furnace in 1915 to produce high speed steel and changed its name to Wikmanshytte Bruks in 1922. Langshyttan was a steel company that also has roots back to the 1500s, and Kloster AB steel company was formed in 1871 by merging many small mills including Langshyttan. In 1927, Kloster changed its name to Fagersta Bruks. Stora purchased Wikmanshytte Bruks in 1966, which was in turn purchased by Uddeholm in 1976. In 1982 Fagersta Bruks and the former Wikmanshytte Bruks (owned by Uddeholm) merged to form Kloster Speedsteel. This company was purchased by Eramet in 1990 and changed the company name to Erasteel in 1992[4]. Uddeholm steel merged with the Austrian tool steel company Böhler to form Böhler-Uddeholm in 1991, which was in turn purchased by Voestalpine (also of Austria) in 2008.

Mora Knives and Laminated Steel

Small fixed blade knives made in the Nordic have been around since at least the Viking times and used by the Sámi people, who were semi-nomadic reindeer herding people who anciently lived in the northern parts of Norway, Sweden, Finland, and Russia. The traditional blade designs are called a "puukko" in Finland[5]. However, in Sweden, these fixed blade knives have come to be known as the "Mora" after the city in which many of the knives have been produced over the last hundred years or so[6]. The earliest person to make a large-scale knife production business in Mora was Finn-Anders Andersson, founding FA Anderssons Täljknivsfabrik in 1870. Sol-Nils Olsson was a knifemaker having difficulty making sales, so in 1877 he joined Finn-Anders Andersson. Olsson would later leave Andersson to form his own knife company. Frost Matts Mattsson formed a company making water taps, faucets, and other brass products in 1865. He died in 1914, but his company, FM Mattsson, continued on, and they began manufacturing knives in 1917.

The FM Mattsson 1945 catalog[7] tells the story of early Mora knifemakers developing laminated blades with a hard high carbon core steel surrounded by softer iron on the sides. The knifemaker's name was not mentioned, but the story probably refers to Sol-Nils Olsson or Finn-Anders Andersson. Translated from Swedish:

> The manufacture of Mora knives has ancient origins. The skilled Mora men, known for their mechanical abilities, originally produced this article for household needs and as handicrafts. Already early on, some home craftsmen realized that the knife blade obtained greater durability and sharpness if the knife edge itself was made with a thin steel blade inserted in between and joined with softer iron. It is said that the circumstance that gave a home craftsman the idea of making knife blades with inlaid steel was the following. A neighbor of his had bought a butcher's knife made in a different factory using the manufacturing methods of the time, but accidentally dropped the knife on the floor, and the blade broke off. The skilled home craftsman began to make knives with a fine steel blade sandwiched between a softer, tougher material, thereby gaining the advantages of making the knife durable and also giving the knife as an edge tool a superior sharpness. This home craftsman, who made his first knife in 1868 as a 20-year-old, later carried out this production as a craft and was thus a forerunner of the current knife manufacturers. As production and turnover increased, the closest progress was to have the material for the knife's most important part — the knife blade — produced at steelworks from inlaid special steel with suitable properties. Since it was possible to obtain a good and suitable steel for the knife blade, inlaid special steel, the development of the manufacture has progressed rapidly.[7]

Vintage FA Anderssons knives. The bottom dates to 1870-1890. Photo from Ref. 6.

Vikmanshyttan had produced laminated steel since the 1800s before supplying steel for knives. Of course, ancient Merovingian and Viking blades had laminated steel for many years before that, so inspiration was already available. Before knife steel, several Swedish steel companies supplied laminated steel for products like plows and scythes. The FM Mattsson 1945 catalog also mentions that they could purchase the steel pre-laminated by that time. In fact, it was in 1923 that the steel mill Vikmanshyttan began to produce the laminated steel WH 101[8], a "chrome vanadium" knife steel with 1% C, 0.5% Cr, and 0.1% V[9]. In 1970 the rolling mill at Vikmanshyttan was shut down, and they ceased manufacturing laminated steels. They attempted to move production to a different steel plant in Söderfors, but this was unsuccessful. Frosts continued to use laminated carbon steel, first from German steel companies and then afterward from France[8]. KJ Eriksson discontinued laminated carbon steel around 1974.

In recent years the major Mora company has been called Morakniv (Mora knives), a consolidation of several former knife companies, including KJ Eriksson and Frosts.

Razor Steel

Swedish blade steel was introduced to the world at large through disposable safety razor blades, though this invention was conceived by an American, King C. Gillette. He wrote about the history of disposable safety razor blades in 1918:

> The history of the Gillette razor is such that its reading will seem more in keeping with the tales of the Arabian Nights rather than with sober facts of the Twentieth Century, and though I have been intimately associated with its birth, growth, and development, and take much pride in the fact that I am its inventor I hardly feel mentally equipped to do the subject justice, and fear my ability to make that personal appeal which I feel would be sure to reach the public heart. It was in 1895, in my fortieth year, that I first thought of the razor, and to appreciate the causes that led to its conception, it is necessary that I should go back a little and become somewhat personal in regard to myself and my affairs.

> I was born January 5, 1855, in Fond Du Lac, Wisconsin, and from the time I was seventeen, and immediately following the Chicago Fire, in which my father lost everything, I have been the pilot of my own destiny. From the time I was twenty-one until the fall of 1904, I was a traveling man and sold goods throughout the United States and England, but traveling was not my only vocation, for I took out many inventions, some of which had merit and made money for others, but seldom for myself, for I was unfortunately situated not having much time and little money with which to promote my inventions or place them on the market. My impulse to think and invent was a natural one, as it was with my father and brothers—as will be found in looking over the records of the Patent Office, where there are a great many inventions to our credit.

> In 1891, I took the position of traveling salesman for the Baltimore Seal Company, who were manufacturing a seal for stoppering bottles. It was a small rubber disc—with a metal loop by which it was extracted, which, when compressed in a groove in the mouth of a

bottle, served as an effective closure for beers and carbonated beverages. Mr. William Painter was the inventor of this stopper, and it was at his solicitation that I took my position with the Company. At that time, they occupied a small factory on Monument Street, Baltimore, Md. Later Mr. Painter invented the present Crown Cork—the tin cap with the cork lining, now so extensively used. When the Crown was invented the corporate name of the Baltimore Company was changed to the Crown Cork & Seal Company.

From the first, I had a great liking and friendship for Mr. Painter as he apparently did for me, and when I would go to Baltimore, instead of putting up at a hotel, Mr. Painter would invite me to his home either in the city or at Pikesville where he resided in the summer. It was during one of my visits to his home that we drifted into one of our intimate talks on inventions—which always fascinated me, for Mr. Painter was a very interesting talker when interested in his subject and thoroughly conversant with all the details and possibilities of his own inventions, which though little in themselves seemed without boundary to their possibilities, when one realizes their unlimited fields of application.

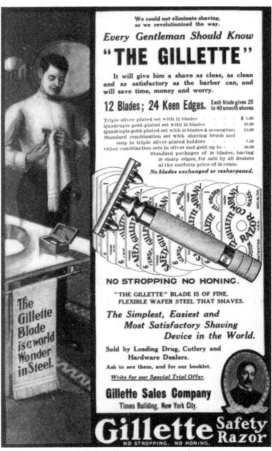

1906 Gillette razor ad.

In the course of this particular conversation, he made these remarks to me, which I have never forgotten, for after the evening was over and other days came they stuck to me like a burr. He said: "King, you are always thinking and inventing something, why don't you try to think of something like the Crown Cork which, when once used, it is thrown away, and the customer keeps coming back for more—and with every additional customer you get, you are building a permanent foundation of profit." In answer, I said: "It is easy to give that kind of advice, Mr. Painter, but how many things are there like corks, pins, and needles." He said, "King, you don't know; it is not probable that you ever will find anything that is like the Crown Cork, but it won't do any harm to think about it." That was the sum and substance of what was said, and I don't remember ever referring to the subject again to him until years after, when I showed him a model of the razor. At that time, he was ill and losing his grip on things. He said at that time, "King, it looks like a real invention with great possibilities, and I am sorry I cannot join you in its development, but my health will not permit it; but whatever you do, don't let it get away from you."

New blade resists rust!

A shake is all the drying this razor needs

NOW you won't feel guilty if you put your razor away without wiping it off after shaving—not if your razor is a New Gillette!

The last tedious step which every man hates —unscrewing the razor, taking out the blade, wiping all the parts—is banished forever by the New rust-resisting Gillette Blade, which needs no drying.

And that's only one reason for the tremendous demand for the New Gillette. It can't "pull"—thanks to the reinforced corners of the new razor cap, and the cut-out corners of the new blade. The channel guard, the new rounded-back guard teeth—these are a few of the remarkable features of the New Gillette Shave.

Ten million men can already tell you here's the biggest bargain in shaving comfort ever offered—one dollar for the New Gillette Razor in a handsome case with one New Gillette Blade. Additional new blades are priced the same as the old—one dollar for ten and fifty cents for five, in the new green packet. Drop in at your dealer's today.

The New De Luxe Razor and Blade

Both are now available at the better shops. The Gillette New De Luxe Razor has all the sweeping improvements of the popular-priced razor, plus luxury in heft, in precise machining and in delicate balance. Eight handsome styles range in price from $5.00 to $75.00.

And every De Luxe set includes ten Gillette New De Luxe Blades of Patented Kro-man steel, which combine rust-resistance and the ability to hold an exquisitely keen edge. They fit any Gillette Razor—and are available at the better shops at $2.00 for ten.

$1.00 *for ten; 50c for five. The New Gillette Blades in the new green packet.*

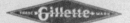

GILLETTE SAFETY RAZOR CO., BOSTON, U. S. A.

The New Gillette Shave

Gillette 1930 ad for stainless razors.

After his memorable advice about my inventing something that would be used and thrown away, I became obsessed with the idea, to an extent that made me provoked at myself, for I applied the thought to nearly every material need; but nothing came of it until the summer of 1895, when, like a child that we have looked for and longed for, it was born as naturally as though its embryonic form had matured in thought and only waited its appropriate time of birth.

I was living in Brookline at No. 2 Marion Terrace at the time, and as I said before, I was consumed with the thought of inventing something that people would use and throw away and buy again. On one particular morning when I started to shave, I found my razor dull, and it was not only dull, but it was beyond the point of successful stropping, and it needed honing, for which it must be taken to a barber or to a cutler. As I stood there with the razor in my hand, my eyes resting on it as lightly as a bird settling down on its nest—the Gillette razor was born. I saw it all in a moment, and in that same moment, many unvoiced questions were asked and answered more with the rapidity of a dream than by the slow process of reasoning. A razor is only a sharp edge, and all back of that edge is but a support for that edge. Why do they spend so much material and time in fashioning a backing which has nothing to do with shaving? Why do they forge a great piece of steel and spend so much labor in hollow grinding it when they could get the same result by putting an edge on a piece of steel that was only thick enough to hold an edge? At that time and in that moment, it seemed as though I could see the way the blade could be held in a holder; then came the idea of sharpening the two opposite edges on the thin piece of steel that was uniform in thickness throughout, thus doubling its service; and following in sequence came the clamping plates for the blade with a handle equally disposed between the two edges of the blade. All this came more in pictures than in thought, as though the razor were already a finished thing and held before my eyes. I stood there before that mirror in a trance of joy at what I saw. Fool that I was, I knew little about razors and practically nothing about steel, and could not foresee the trials and tribulations that I was to pass through before the razor was a success. But I believed in it and joyed in it.

I wrote to my wife, who was visiting in Ohio, "I have got it; our fortune is made," and I described the razor and made sketches so she would understand. I would give much if that letter was in existence today, for it was written on the inspiration of the moment and described the razor very much as you see it today, for it has never changed in form or principle involved—only in refinements. The day of its inception I went to Wilkinson's, a hardware store on Washington Street, Boston, and purchased pieces of brass, some steel ribbon used for clock springs, a small hand vise, some files, and with these materials made the first razor. I made endless sketches which have since then been used in our Patent suits—and were the basis of establishing the time and scope of my invention. These sketches are still a part of the company's records. Then came the hour of trial, for I could not interest any one in a razor, the blades of which were to be used once and thrown away, for I then thought that the razor blades could be made for very little, as I learned that steel ribbon could be had for 16 cents a pound and a pound would make five hundred blades, for my blades were slightly narrower and shorter than the blade finally introduced. I did not know then that the steel to be used must be of a particular quality and that it would cost many times what I supposed per pound, and that it was to cost the future company over a quarter of a million of dollars in laboratory tests before this question alone would be decided.[10]

Gillette found the steel he needed in Sweden, from the Munkfors Works of Uddeholm. In the late 1880s, cold rolling was introduced to Uddeholm by two engineers. One of these engineers, Gustaf Jansson, had learned the technique of cold rolling in Worcester, Massachusetts in 1876-1880[11]. Uddeholm could cold roll steel to such thin dimensions that the blades could be sharpened from the stamped steel. And their tolerances had to be incredibly tight for the production of razors. This was a simple high carbon razor steel.

Stainless Razor Steel

In 1925 Uddeholm submitted a patent for a stainless steel for "edge tools characterized by high cutting efficiency, great rust-resistance and great resistance to attacks of acids."[12] This steel was initially named "UHB Stainless 7" though it was later renamed AEB and even later AEB-H. The composition was 0.95% C, 13.5% Cr, and 1% Mn[13], though the patent had wider ranges of 0.7-1.1% C, 10-16% Cr, and 0.75-2.0% Mn. It seems that Uddeholm attempted to avoid the Brearley patent by having the carbon content above his maximum of 0.7% and claiming that high manganese would lead to superior corrosion resistance. The patent said[12], "It has been safely proved that by addition of fair quantities of the metal manganese the rust-resisting properties of the chromium steel were even increased by higher carbon content." I have never seen this claim repeated about the effect of manganese, but either the experiments performed were poor, or they were only concerned about having a claim of novelty for the patent, because manganese does not improve corrosion resistance.

Gillette introduced a blade using this steel in 1928 called "Kroman," referring to the combination of chromium and manganese in the steel. However, the blades were much more expensive than those made of simple carbon steel. Even King Gillette himself was opposed to making the stainless blades because of the high cost, saying, "all the advertising in the world will be impotent to convince the masses that it is not a deliberate move simply to get more money for our product."[14]. Unfortunately, the company had bigger problems than only high cost, as manufacturing the stainless blades was also difficult, with honing and stropping taking twice as long as the carbon steel blades. Also, the large carbides in the stainless steel would tear out in sharpening, leaving a jagged edge that was not as smooth or comfortable to shave with as the carbon steel blades[14].

The Gillette company would pull the Kroman blades from the market entirely in 1932, only a few months before King Gillette's death. The company even printed an apology for the inconsistent quality of the blades while introducing their new "Blue Super-Blade":

The Gillette Safety Razor Company feels called upon to make a confession and a statement that are undoubtedly unique in the annals of American business. It is with deep regret and no little embarrassment that we do this in order to tell you frankly what actually happened when we introduced a new Gillette razor and blade. Soon after this was done, we found that, although the great majority was pleased, some users complained quality was not up to standard. We learned why. Our equipment had not been equal to the task of producing millions of blades at high speed without a certain variation in quality that affected a small portion of our output. As a result, some blades that left the factory were not as good as you had a right to expect... We discovered and purchased for our exclusive use and at the cost of millions of dollars a manufacturing process that was amazingly superior to our own... Now we announce today's Gillette blades, made by the new management. The usual superlatives have no place in this sincere statement of ours, so we will let the quality of the blades speak for itself... You can try today's Gillette blades without risking a cent... The Kroman De Luxe blade has been withdrawn from production. We offer the Blue Super-Blade as its successor. This sensational blade is far superior to the Kroman and costs considerably less.[14]

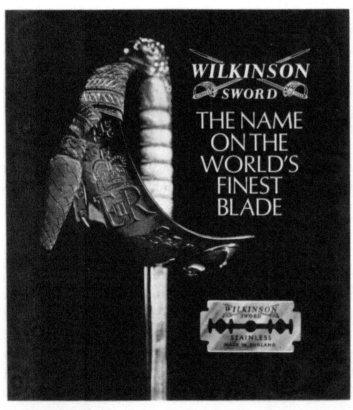

The finest razor blade in the world. By Wilkinson Sword. Stainless steel, tempered and honed with a craftsman's care. There is no smoother, sharper, longer-lasting blade. Made with a concern for quality far above the needs of commerce, a care for perfection far beyond the standards of mass production. Quite simply, the world's finest blade. By Wilkinson Sword.

Dispenser of 5 blades MADE WITH A CRAFTSMAN'S CARE - FOR SHAVING PERFECTION

Wilkinson Sword 1964 ad.

After this failure, it was some time until stainless razors became a serious contender in the marketplace. Ultimately it was not Gillette that reintroduced them but rather Wilkinson Sword, a British company, in 1956. Wilkinson Sword used the same Uddeholm AEB steel, and the blades were still rough. But in 1961, Wilkinson Sword introduced an improved version with a Teflon coating that largely took care of the smoothness issue. Wilkinson had to pay royalties to Gillette because of a patent on coated blades[15,16]. These blades were, of course, much more resistant to staining and also lasted longer thanks to the high wear resistance of stainless steel. These blades were popular enough that Gillette's business dropped 8% in 1963 and a further 11.5% in 1964[15]. Other companies quickly followed with their own stainless steel razors, including Schick in February 1963, and American Safety Razor in spring 1963, both of which also used Uddeholm AEB steel[17], giving Uddeholm 80-90% market share in stainless blade steel. Gillette would introduce its stainless razors in the fall of 1963, purchasing stainless from Sandvik steel[18]. Sandvik had entered the cold-rolled razor steel market in 1909[19], and by 1963 had 50% of the carbon razor steel market[17]. Gillette listed both AEB (Uddeholm) and 12C27 (Sandvik) as potential stainless razor steels in a patent filed in 1964[20], so it is unclear if they were purchasing Sandvik steel exclusively or if they were dual sourcing from both companies. 12C27 (0.6%) had significantly lower carbon than AEB (0.95%), along with a similar chromium content of 14%, which meant its potential hardness was somewhat lower. Still, its carbide structure was much finer, essential for achieving high sharpness for razors. The superior microstructure of this steel seems to have driven further development from Uddeholm to better compete. They patented methods for decreasing carbide size in the late 1960s, along with a composition that was modified from 12C27 for improved hardness, with increased carbon up to 0.66%[21]. This steel would be marketed as AEB-L[22], L for the lower carbon version of AEB. Sandvik would introduce its own version of this steel called 13C26 by 1975[23], and these two steels would dominate stainless razors from then on.

Steel	Year	C (%)	Cr (%)
AEB	1925	0.95	13.5
12C27	1964	0.6	14
AEB-L/13C26	1968	0.67	13
12C27M		0.52	14.5
14C28N	2009	0.6 + 0.11% N	14

Stainless Knives

Mora knives would be introduced in stainless steel by different manufacturers by the 1940s[24-26]. The steel used may have been AEB, though Uddeholm did manufacture lower carbon stainless steels similar to 420 at that time, such as "UHB Stainless 6" with 0.35% C and 13.5% C[13], or "UHB Stainless 6H" with 0.43% C and 13.5% Cr, essentially a 420HC steel[13]. By 1967 KJ Eriksson's was using AEB-L but began using Sandvik 13C26 by 1970, and in 1980 12C27 and 12C27 Mod[27]. From their earliest advertisements in the 1960s and 1970s, both AEB-L and 12C27 were being promoted as knife steels in addition to razors[22,28]. So these Sandvik and Uddeholm high carbon stainless steels became common in Mora knives.

In 1981, both Uddeholm and Sandvik were advertising their steels to American knifemakers and companies[29,30], including Sandvik advertising that their steels had a finer microstructure than other stainless steels[31]. However, the implementation of these steels remained relatively minimal, partly because minimum order sizes were large. Some custom knifemakers did take up these steels, such as Michael Walker by 1984[32], known for introducing the liner lock folding knife.

Michael Walker

Walker said of AEB-L[33]: "There are many technical reasons why I use AEB-L, such as its fine-grained structure, which takes a very fine edge. Since I do my own heat treating, I vary the hardness depending upon the intended use of the knife." The reputation of Uddeholm and Sandvik steels remained relatively low in the USA, and AEB-L was often compared to 440A, which had a reputation as being a low-end steel used only in factory knives[34]. AEB-L began to see more use through Damascus steel, such as by Devin Thomas in the 1990s because the steel was available in very thin sizes and would therefore require few forge welds. In March 2005, John Verhoeven released *Metallurgy for Bladesmiths,* where he wrote about AEB-L as being an ideal blade steel due to its very fine carbide size and high potential hardness[35]. Also, in 2005, I wrote a short summary about AEB-L for DevinThomas.com to promote its virtues[36]. In 2006 Roman Landes published *Messerklingen und Stahl* where he gave AEB-L very high scores for sharpness, ease in sharpening, toughness, and edge stability[37]. The steel continued to see little use in custom knives, partly because it was not regularly available at knife supply companies.

By 2004, Kershaw began to use Sandvik 13C26[38], and more broadly in 2007[39], due to its high potential hardnes, and can be "fine blanked" without cracking. Sandvik began attempting to get into the knife market to a greater extent at this point and began selling their 13C26 through Admiral Steel in the US in 2007[40]. Sandvik developed 14C28N for Kershaw to improve the corrosion resistance of 13C26, and those knives were being sold by 2009[41]. The 13C26 steel was used in relatively inexpensive Kershaw models where returns are likely if any rust is observed, leading to the development of 14C28N. By 2010 AEB-L was also offered in suitable sizes for knives through supply companies starting with Alpha Knife Supply[42], and its popularity grew to today, where it is relatively common in custom knives.

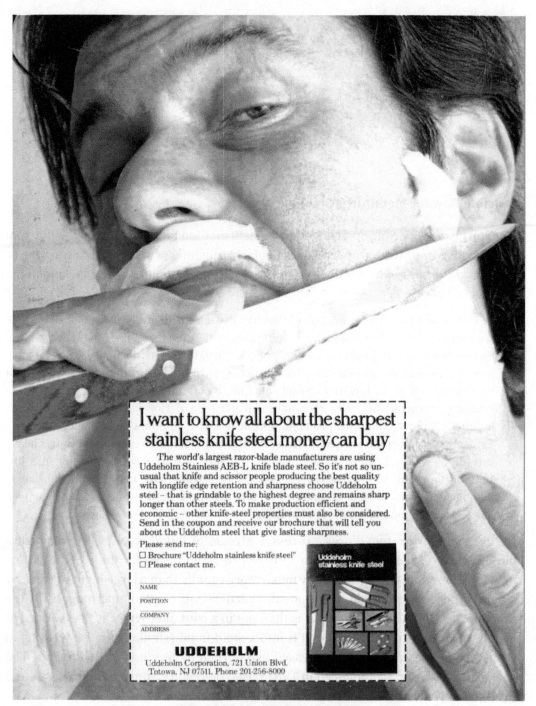

1981 Uddeholm *Blade Magazine* ad.

-34-

POWDER METALLURGY STEEL

Crucible Powder Metallurgy Steel

Metallurgists at Crucible Steel developed a new method for producing high alloy tool steels, which improved properties, particularly toughness (resistance to fracture or chipping of a knife edge). Edward Dulis and Thomas Neumeyer wrote about the process in 1970:

> In high speed tool steels, a problem that has existed for many years is that of carbide segregation in the conventionally solidified ingot that persists in relatively large-sized endproducts. The segregation of hard, brittle carbide regions in final large tool-steel sections results in undesirable characteristics such as out-of-roundness after heat treatment, regions of lowered impact toughness, non-uniform hardness, variability in grinding characteristics, etc., that have an effect on performance either directly or indirectly. To overcome this problem, production by P/M offers an excellent solution... Typical as-cast microstructure for a conventional ingot of ... high speed steel shows heavy carbide networks. In powder atomization, the very fine particle sizes result in a fast solidification rate so that micro-segregation is substantially minimized. Thus, alloy homogeneity and structural uniformity are achieved in the fine-sized pre-alloyed particles or microingots of the P/M process.[1]

During this powder metallurgy process, a stream of liquid steel passes through a nozzle which is then sprayed with nitrogen to rapidly solidify the steel as a fine powder. These fine particles solidify much more rapidly than a large ingot of steel, by several orders of magnitude. These powder particles were called "microingots" by the Crucible metallurgists. The powder is then put in a container which is pressurized at high temperature to diffusion bond it into a solid ingot (similar to forge welding in Damascus), which can the be processed like normal by forging, hot rolling, etc. The new process was first announced in 1967 based on results from a pilot process[2], and the full industrial production began in 1970[3].

Early work by Crucible on powder metallurgy was performed on M2 high speed steel with a sulfur addition, referred to as M2S[1]. This resulted in a patent from metallurgist

Gary Steven which was filed in 1967[4]. In the patent they showed that the high-temperature austenitizing treatment could be performed much more rapidly due to the fine carbides which dissolve more readily, reaching 66 Rc after quenching after holding for only 30 seconds, while the conventional steel needed 5 minutes. Also, because the steel is more uniform, size changes such as "out-of-roundness" with heat treated cylinders were reduced. This was promoted as a cost- and timesaving measure[4]: "In the case of a die where specified tolerances usually must be maintained, any nonuniformity in size after hardening treatment would also have to be in some measure corrected by additional grinding. In both the case of cutting tools and dies, this grinding operation after hardening treatment is rendered more difficult and expensive because of the increased hardness of the material." The sulfur was also added for the end user to save cost in machining, as manganese sulfides formed in the steel make machining and grinding easier. Sulfurized tool steels had already existed for many years. Still, the use of powder metallurgy means the sulfides are smaller and better distributed so that the sulfides do not negatively affect impact toughness. The fine carbide size also contributed to improved machinability and grindability. Perhaps it was thought the cost-saving benefits from shorter heat treatments, reduced size changes and distortion, and minimized grinding and machining would offset the higher costs of the powder metallurgy process. The possibility of improved performance was increasingly recognized, however. The fine carbide and grain size from the powder metallurgy process combined for improved toughness and resistance to chipping in high speed machining tools. That improved toughness led to superior tool life in "intermittent" machining, where chipping is more likely.

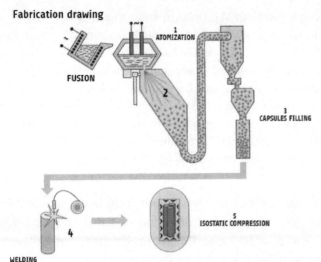

Fabrication drawing

FUSION
1 ATOMIZATION
2
3 CAPSULES FILLING
4
5 ISOSTATIC COMPRESSION
WELDING

Schematic of powder metallurgy process from Böhler.

Other steels tested with the powder metallurgy process early on included T1, T4, T5, M1, M3, M7, M35, M41, M43, and T15[1,4]. We have introduced most of these steels in previous chapters, except for M3, which is between M2 and M4 in terms of designation and composition; it has 3% vanadium as opposed to the 2% of M2 and 4% of M4. Crucible metallurgists recognized in the early studies that steels that were previously difficult to produce and often suffered from poor toughness would show the most significant improvements in properties with powder metallurgy. This particularly included steels with high vanadium like M4 and T15. Crucible metallurgists August Kasak and Edward Dulis wrote about this benefit in 1978:

> It follows that the benefits derived from the CPM process are maximized by applying it to relatively highly alloyed steels in larger product forms, although significant benefits can also be realized in smaller product forms. T15 serves well as a case in point. T15 has been recognized as the most wear- and heat-resistant grade among the standard AISI high speed steels for a long time. However, its use has been seriously limited because in conventional production this high-alloy carbide-rich steel is very segregation-prone and difficult to produce; moreover, it presents serious grindability problems to the toolmaker. These limitations are substantially reduced by CPM technology... CPM T15 is definitely superior to the corresponding conventional product in toughness and, particularly, in grindability.[5]

They even found a surprise in M4 in that its toughness went from being among the worst of the high speed steels to the best, which they found was due to vanadium carbides being the smallest with powder metallurgy production, while with conventional processing they are the largest. Kasak and Dulis said:

> As a CPM product, M4 has proven to be a pleasant surprise: its toughness, in terms of both impact and bend fracture strengths, is higher than that of any other high speed steel grade known to us... The unexpectedly high toughness of CPM M4 is attributable to an optimum combination of relatively high carbide density (number of carbides per unit volume), small size of carbides, and their uniform distribution. This combination of desirable carbide characteristics creates favorable conditions for obtaining a stable, fine-grained microstructure that leads to a high-toughness product without sacrifice of other desirable properties.[5]

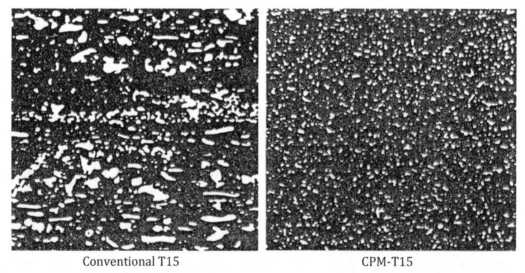

Conventional T15   CPM-T15

CPM-T15 and CPM-M4 were announced as products in 1973[6,7]. This was the start of a shift away from the original idea of cost savings through machining, heat treating, etc., toward steels with much longer tool life due to the previously unattainable combinations of wear resistance and toughness. CPM-M4 had significantly greater wear resistance than the common M2 steel, and thanks to the PM process it also had greater toughness, so chipping of the high wear resistance steel tools was no longer a concern. And while grindability was poor with conventionally produced vanadium-alloyed steels, powder metallurgy greatly improved this property. Other commercialized steels included M3 Type 2 (the higher carbon version of M3), M35, and the super hard high speed steels M41 and M42[5].

Steel	C (%)	Cr (%)	Mo (%)	W (%)	V (%)	Co (%)
CPM-M2S	0.85	4.15	5.0	6.4	1.95	
CPM-M4	1.4	4.0	5.25	5.5	4.0	
CPM-T15	1.6	4.0		12	4.9	5.0

New Powder Metallurgy Steels

It was recognized early on that powder metallurgy also offered opportunities to develop new alloys optimized for the process. Dulis and Neumeyer wrote in 1970[1]: "It is well known that particle metallurgy provides a means of making new alloys and compositions of tool steels which cannot be made by conventional melting, casting, and hot-working methods. Major advances in new tool materials will most likely come from applying the P/M process." This was increasingly recognized to lead to opportunities with high alloy contents and high vanadium carbide contents (referred to as "MC" where M means metal, such as vanadium). In 1978 Kasak and Dulis wrote:

> In the metallurgy of tool steels, it is well recognized that the performance potential of a grade depends on the alloy content and carbide content – particularly the MC-type vanadium carbide content – of the steel. As a rule, the higher the vanadium carbide content of the grade, the greater is its wear resistance. However, the conventionally produced tool steels, hot-workability in producing the steel, grindability in producing tools, and toughness in tool service are progressively reduced by increased alloy and carbide contents due to severe alloy segregation and a coarse, banded carbide structure. Therefore, as a compromise, T15 (27.8% total alloying elements, 5% vanadium) has become a practical upper limit for alloy and MC-type carbide content (in conventional ingot steels). The CPM method facilitates the production of steels with high alloy and unusually high carbide contents and thus removes some barriers that limit the alloy development effects on conventional products.[5]

Carbide Type	Hardness (Hv)	Hardness (Rc)
Iron Carbide (Fe_3C)	1000	69
Chromium Carbide ($Cr_{23}C_6$ and Cr_7C_3)	1200-1500	72-76
Molybdenum/Tungsten Carbide (Mo,W_6C)	1400	75
Vanadium Carbide (VC)	2800	87

Powder Metallurgy High Speed Steels

The first new steels developed with powder metallurgy in mind were new super high hardness high speed steels with increased overall alloy content. The first steel patented by Gary Steven was CPM Rex 71, first filed in 1969[8]. This steel had 1.3% vanadium like the popular M42 but with a different combination of Mo and W and higher Co, released by 1973[7]. The recognition that CPM allows higher vanadium contents led to CPM Rex 73 with 3.2% vanadium, patent filed in 1972[9], and the steel was released by 1974[10]. Then CPM Rex 76 trademarked in 1975[11], the same as Rex 73 but with cobalt reduced by 3%, released commercially in 1976[12]. Rex 76 became the standard grade from then on, and Rex 71 and Rex 73 were phased out.

Steel	Year	C (%)	Cr (%)	Mo (%)	W (%)	V (%)	Co (%)
Rex 71	1973	1.17	3.8	5.3	10	1.3	12
Rex 73	1974	1.55	3.8	5.3	10	3.2	12
Rex 76	1976	1.5	3.8	5.3	10	3.1	8.5

Stora/Uddeholm Powder Metallurgy Steels

Despite the many patents that Crucible filed on the powder metallurgy production of steel, Stora in Sweden also introduced its own powder metallurgy tool steels in 1970[13]. The earliest of these steels was Stora 30[13,14], later renamed ASP 30[15], a powder metallurgy version of M3 Type 2 plus 8.5% cobalt. The company also introduced ASP 23, which was M3 Type 2 without the cobalt addition[15]. A steel called ASP 60 was released by 1975[16], which had the highest vanadium yet at 6.5% along with high Mo, W, and Co. This steel had a very high combination of hot hardness and wear resistance[17].

Steel	Year	C (%)	Cr (%)	Mo (%)	W (%)	V (%)	Co (%)
ASP 23	1970	1.28	4.2	5	6.4	3.1	
ASP 30	1970	1.28	4.2	5	6.4	3.1	8.5
ASP 60	1975	2.3	4	7	6.5	6.5	10.5

Crucible was not happy with the competition and presented experiments in 1978 to show that CPM Rex 76, despite having 3% vanadium, still lasted as long as ASP 60 in machining tests[5]. They also compared with a modification of Rex 76 with 6% vanadium. They concluded that[5] "at this point, the benefits from increasing the vanadium content to 6% appear questionable." A representative of Uddeholm (which had recently purchased Stora) responded[5]: "It is very difficult to comment on the slides shown here regarding the comparison between ASP 60, Rex 76, and Rex 76 with 6% vanadium. The main reason is that the trials have been done without any coolant, and that is very unusual in commercial application in general practice. We prefer to go to the toolmaking industry and have the trials made there." Crucible would sue Stora/Uddeholm for infringing on their patented powder metallurgy process, which in 1984 was decided in favor of Crucible, and Uddeholm and Stora had to pay damages[18]. In terms of high speed steels with greater than 3% vanadium, Crucible metallurgists Andrzej Wojcieszynski and William Stasko would eventually see the benefits, as 20 years later in the late 1990s, higher vanadium versions of Rex 76 were patented, including Rex 86 (5% V) and Rex 121 (9.5% V)[19]. Rex 121 was released in 1999[20] though Rex 86 didn't come until 2005[21].

Steel	Year	C (%)	Cr (%)	Mo (%)	W (%)	V (%)	Co (%)
Rex 121	1999	3.4	4	5	10	9.5	9
Rex 86	2005	2	4	5	10	5	9

CPM-10V

Crucible also worked on a "cold work die steel" with increased vanadium and higher wear resistance. Up to that point, the highest wear resistance grades of this type were A7 and D7 developed by David Giles for Latrobe (Chapter 12). Those were limited to 4-5% vanadium because beyond that the steels were[22] "severely limited by the detrimental effects of high vanadium carbide content in conventionally ingot-cast steels on hot workability, machinability, grindability in the hardened condition, and ultimately toughness in service." Crucible metallurgists Walter Haswell and August Kasak patented CPM-10V[23], which was released in 1978[12]. Haswell had previously worked at Latrobe and worked on high wear resistance tool steels there[24], so he would have known about A7 and D7 and the design of high vanadium tool steels. CPM-10V steel used H11/A2 as its base composition with 5.3% Cr and 1.3% Mo and then they looked at different vanadium contents to find the reasonable maximum. They tested vanadium contents of 6, 10, 11, and 14% and found that the carbide size became much larger, and toughness much lower, when the vanadium content was over 10%, so they settled on "CPM-10V," the 10 referring to the percent vanadium. The powder metallurgy process cannot maintain the fine microstructure when the carbides are formed in the liquid, before passing through the nozzle and solidifying into powder. The higher the vanadium content, the higher the temperature at which the vanadium carbide forms, and for their process at that time, 10% was the functional limit. Both the toughness and wear resistance were found to be superior to the former conventionally produced A7 and D7 grades[23]. In fact, toughness was similar to the common M2 and D2 conventional steels, giving CPM-10V the potential to be a drop-in replacement for D2 in die applications. CPM-10V was the last tool steel to be given a "standard designation" by AISI; it was assigned the name A11.

Steel	Year	C (%)	Cr (%)	Mo (%)	V (%)
CPM-10V	1978	2.45	5.25	1.3	9.75

CPM-440V (S60V)

Crucible metallurgists also looked at vanadium-alloyed powder metallurgy stainless steels. Early experiments reported in 1978 looked at 6% vanadium additions (along with higher carbon) to either 440C or 154CM[12]. They found that the 6% V 154CM version had superior wear resistance but somewhat lower toughness and grindability than the modified 440C. The modified 440C version was released as CPM-440V in 1983[22]. The steel was described as[22], "The composition of this material ... is essentially that of T440C martensitic stainless steel, to which has been added approximately 5.75 wt% V with increased carbon for optimum wear resistance while maintaining toughness and corrosion resistance comparable to conventionally produced T440C stainless steel." While this steel targeted multiple industries, including plastics processing, bearings, and food processing, this steel was also thought to be well suited for knives, and samples were sent to knifemakers in 1983[25,26]. Initial use of this steel in knives was relatively limited due to its low grindability and polishability, as well as only being available in relatively thick sizes (see Chapter 40).

Steel	Year	C (%)	Cr (%)	Mo (%)	V (%)
CPM-440V (S60V)	1983	2.2	17.5	0.5	5.75

CPM-10V 1981 ad.

Uddeholm Vanadis Steels and Elmax

Uddeholm released its own powder metallurgy cold work die steels starting with "Vanadis 4," first patented in Sweden in 1986[27]. The goal with Vanadis 4 was to offer a high toughness cold work die steel[28]: "In spite of the improvements with reference to the impact strength which has been achieved through the powder-metallurgical manufacturing technique, it is desirable to offer still better tool materials in this respect and at the same time to maintain or if possible further improve other important features of the material, particularly the wear strength." They used the 8% Cr and 1.5% Mo die steels, such as Vasco Die, as their basis, and increased the vanadium content (along with carbon) up to 4%. They also showed in the patent that they were looking at versions with 10% vanadium as well, similar to CPM-10V. Vanadis 4 was released by 1990[29] and Vanadis 10 by 1994[30]. Vanadis 6 was trademarked in 1998[31] and released by 2002[32]. Uddeholm also released a stainless steel called Elmax in 1990[33]. Like CPM-440V, Elmax is a modification of 440C with added vanadium, though Elmax only has 3%.

Steel	Year	C (%)	Cr (%)	Mo (%)	V (%)
Vanadis 4	1990	1.5	8	1.5	4
Elmax	1990	1.7	18	1	3
Vanadis 10	1994	2.9	8	1.5	9.8
Vanadis 6	2002	2.1	6.8	1.5	5.4

Böhler K190 and M390

In the early 1980s, Böhler in Austria released K190[34], which was a powder metallurgy version of D7 tool steel, intended for high wear resistance cold work applications. Böhler reported that K190 had superior wear resistance and toughness compared to conventional D2 die steel. Böhler metallurgists Alfred Kulmburg, Johann Stamberger, and Hubert Lenger patented a powder metallurgy stainless steel in 1989[35]. The goal was to develop a steel for the plastics industry, where corrosion resistance and wear resistance are both important[34]: "In the transformation of plastics, the mechanical stresses to which the individual tool parts are subjected are rather low as compared to the previously discussed cold work tools. The predominant stresses are wear and corrosion. The ever-increasing use of asbestos, wood, and, more recently, glass-reinforced moulding materials results in an analogous increase of wear in both plastics processing machines and moulds... The objective, therefore, was to develop and produce a material for plastics processing machines and tools, which combines maximum wear resistance, such as [K190], and excellent corrosion resistance." Therefore, they took the base composition of K190, increased the chromium and decreased the carbon to make it stainless, and this steel was released as M390 in 1992[35]. Extra tungsten was also added for "secondary hardening" with high-temperature tempering, but the reason why this was added when it is not present in K190 was not stated. Use of M390 in knives was relatively limited for many years, the earliest example I have found of it in a knife was by German knifemaker Dietmar Kressler in 1999, which he stated was one of his favorite steels[36]. But the steel was not common at that point, even in Europe. German knifemaker Richard Hehn in his 2001 book didn't list M390 in his table of steels, even though powder metallurgy stainless steels were extensively discussed[26]. Its increase in popularity in knives is covered in Chapter 42. Böhler released a higher vanadium version of M390 called M398 in 2019.

Steel	Year	C (%)	Cr (%)	Mo (%)	W (%)	V (%)
K190 (D7)	Early 80s	2.3	12.5	1.1		4
M390	1992	1.9	20	1.0	0.6	4
M398	2019	2.7	20	1.0	0.7	7.2

CPM-15V

In the early 1990s, Crucible metallurgists William Stasko and Kenneth Pinnow looked again at the practical limits of vanadium in cold work die steels[37,38]. They found that increasing the temperature of the liquid steel in combination with appropriate ceramic refractory (to handle the higher temperature), meant that the vanadium content could be increased above 10% without large carbides. They studied versions with 15, 18, and 20% vanadium which were all successfully produced with small carbides on the laboratory scale, though ultimately, only CPM-15V was commercialized in 1994[39].

Steel	Year	C (%)	Cr (%)	Mo (%)	V (%)
CPM-15V	1994	3.5	5.25	1.3	14.5

CPM-420V (S90V) and S125V

Pinnow and Stasko, along with John Hauser, worked on high wear resistance stainless steels starting in the mid-to-late 1980s[40,41]. The first of these was called MPL-1 (also called Supracor), with 9% vanadium along with high chromium (24%), molybdenum (3%), and carbon (3.75%). However, this steel with such a high alloy content had poor forgeability and did not see much commercial use. Next, they worked on a steel which was named CPM-420V, the patent for which was filed in 1995[41]. They also found that the amount of vanadium carbide formed in MPL-1 was relatively low despite the high vanadium content, because of the high chromium. The high chromium would lead to chromium carbide formation instead. The chromium carbides were somewhat enriched with vanadium, which would be somewhat higher in hardness than "plain" chromium carbides, but much lower in hardness than vanadium carbides. Higher hardness carbides are better for improving wear resistance. The Crucible metallurgists discovered that if chromium were reduced, it would increase the proportion of vanadium carbide formed, therefore increasing the wear resistance for a given vanadium content, which also improves the toughness-wear resistance balance. They wrote in the patent[41]: "It has been discovered in this regard, that the metal to metal wear resistance of the high chromium, high vanadium, powder metallurgical stainless steels is markedly affected by their chromium content and that by lowering their chromium content and closely balancing their overall composition, a significantly improved and unique combination of metal to metal, abrasive, and corrosive wear resistance can be achieved in these materials." They found experimentally that the 420V material was an improvement over 440V in most metrics, including toughness, corrosion resistance, and wear resistance. They found that the corrosion resistance remained as good or better than CPM-440V, despite the reduced chromium, as long as the carbon and nitrogen were controlled within a specific range[41]: "to achieve good corrosion resistance at these lower chromium levels, and to obtain the hardness needed for good metal to metal and abrasive wear resistance, it is essential that the carbon and nitrogen contents of the PM articles of the invention be closely balanced with the chromium, molybdenum, and vanadium contents of the articles according to the indicated relationships." They had equations listed for the carbon and nitrogen content based on how much Cr, Mo, and V were present in the steel. They also explored versions of the steel with 12% and 14.5% vanadium, and also versions with 9% vanadium but a higher content of nitrogen. However, at least initially, only the 9% vanadium version with low nitrogen was produced commercially as CPM-420V. The name of CPM-440V was changed to CPM-S60V and CPM-420V to CPM-S90V in 2000[42]. Later the 12% vanadium version was released as CPM-S125V by 2004 (see Chapter 42). The 14.5% vanadium version was teased under the name S140V[43], but that version was never commercialized.

Steel	Year	C (%)	Cr (%)	Mo (%)	V (%)
CPM-420V (S90V)	1997	2.3	14	1	9
CPM-S125V	2001	3.25	14	2.5	12

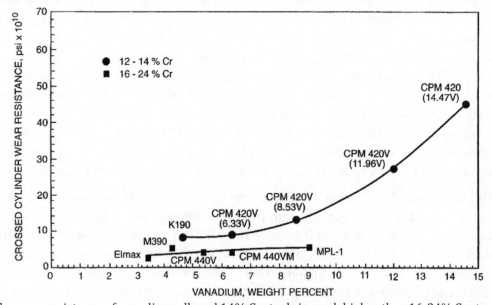

The wear resistance of vanadium-alloyed 14% Cr steels is much higher than 16-24% Cr steels.

High Toughness – CPM-3V, CPM-1V, CPM Cru-Wear

Crucible metallurgists Pinnow and Stasko began working on powder metallurgy versions of Vasco Die and Vasco Wear in the mid-1990s to develop cold work die steels with high toughness[44]. They said in the patent that Vanadis 4, CPM-10V, and CPM-15V had very good properties, but[44], "in spite of the great improvements in wear resistance or in toughness or in both of these properties offered by these PM steels, none of them offer the combination of very high toughness and good wear resistance needed in many cutting, blanking, and punching applications." They found that even though powder metallurgy keeps carbide size small, the toughness of the steel is still highly controlled by the amount of carbide: "The notable improvement in toughness obtained with the articles of the invention is based on the findings that the impact toughness of powder metallurgy cold work tool steels at a given hardness decreases as the total amount of primary carbide increases, essentially independent of carbide type, and that by controlling composition and processing so that substantially all the primary carbides present are MC-type vanadium-rich carbides, the amount of primary carbide needed to achieve a given level of wear resistance can be minimized." In other words, because vanadium carbides are harder, they contribute more to wear resistance, but since the type of carbide doesn't affect toughness, only the amount of carbide, using the very hard vanadium carbide leads to a better combination of toughness and wear resistance.

Further, they found that powder metallurgy can affect the type of carbide that forms, so even though the conventionally produced Vasco Die contained both chromium carbides and vanadium carbides, the powder metallurgy version CPM-3V contained only the desirable vanadium carbide[44]: "It has also been discovered that in comparison to conventional ingot-cast tool steels with compositions similar to those of the articles of the invention, that production of the articles by hot isostatic compaction of nitrogen atomized, pre-alloyed powder particles produces a significant change in the composition as well as in the size and distribution of the primary carbides. The former effect is a hereto unknown benefit of powder metallurgical processing for cold work tool steels, and is highly important in the articles of the invention because it maximizes the formation of primary MC-type vanadium-rich carbides, and largely eliminates the formation of softer (chromium) carbides, which in addition to MC-type carbides are present in greater amounts in ingot-cast tool steels of similar composition." Because of this change in the carbide structure, they could patent CPM-3V even though its composition was identical to Vasco Die.

They also looked at powder metallurgy versions of Vasco Wear, which they would release as CPM Cru-Wear, though that was several years later (Chapter 42). Chromium carbides would still form in these 8% Cr vanadium powder metallurgy steels if the carbon were too high, as they found with CPM Cru-Wear, which is essentially CPM-3V but with higher carbon. The high carbon of Vanadis 4 has the same effect. The toughness of CPM 3V (70 ft-lbs.), and even CPM Cru-Wear (44 ft-lbs.), was substantially higher than Vanadis 4 (27 ft-lbs.). CPM-M4 had both higher toughness (29 ft-lbs.) and wear resistance than Vanadis 4, so Crucible metallurgists did not see the need for a Vanadis 4 competitor, as CPM-M4 can also be used in cold work applications, not only high speed machining. CPM-1V, with lower vanadium and even higher toughness than CPM-3V, was introduced in 2002[45], which is a powder metallurgy version of VASCO-MA, the matrix version of M2 (Chapter 13).

Steel	Year	C (%)	Cr (%)	Mo (%)	W (%)	V (%)
CPM-3V	1997	0.8	7.5	1.3		2.75
CPM-1V	2002	0.5	4.5	2.75	2.15	1

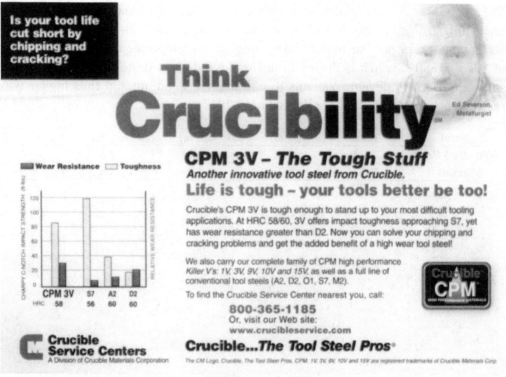

2002 CPM-3V ad.

Vanadis 4 Extra

In 2001 Uddeholm metallurgists Odd Sandberg, Lennart Jönson, and Magnus Tidesten patented a new version of Vanadis 4, "Vanadis 4 Extra" with improved properties[46]. This steel was released around 2004[47]. In the patent, they stated[46]: "Vanadis 4 has been manufactured since about 15 years and has due to its excellent features reached a leading position on the market place for high-performance cold work steels. It is now the objective of the applicant to offer a high-performance cold work steel having still better toughness than Vanadis 4 while other features are maintained or improved in comparison with Vanadis 4." They reduced the chromium content from Vanadis 4, eliminating the soft, relatively large chromium carbides, leaving only the high hardness, small vanadium carbides. This leads to a higher combination of toughness and wear resistance, as previously explained concerning CPM-3V. The researchers stated:

> As can be seen the amount and type of carbides are different in the investigated steels. Vanadis 4 Extra has a lower total content of carbides compared to Vanadis 4, and it contains only one type, the smaller and harder MC-type. A lower content of carbides and a small size both positively affect ductility. The bigger M_7C_3 (chromium-type) are relatively soft compared to MC (vanadium carbides) and have a smaller effect on wear resistance. The idea behind the development is to remove the M_7C_3 (chromium) type of carbides, but instead, in order not to lose wear resistance, the amount of MC has to be increased.[47]

They also increased molybdenum which gives higher "hardenability" (the steel can reach full hardness even when quenched slower; larger dies can be heat treated), and also giving it superior "secondary hardening" when tempering at high temperatures of 1025°F/550°C which is desirable in the die industry. Uddeholm would use this same approach to develop Vanadis 8 with higher vanadium and thus higher wear resistance[48], though it was not released until 2017[49].

Steel	Year	C (%)	Cr (%)	Mo (%)	V (%)
Vanadis 4 Extra	2004	1.4	4.8	3.5	3.7
Vanadis 8	2017	2.3	4.8	3.6	8

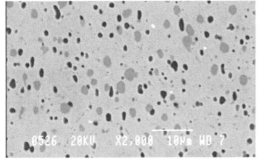 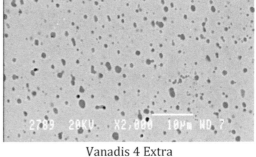

Vanadis 4   Vanadis 4 Extra

The larger light grey carbides in Vanadis 4 are chromium carbides. The carbide size is smaller and more uniform in Vanadis 4 Extra made up of vanadium carbides.

Improved Böhler Cold Work Die Steels

Böhler developed their own improved die steels such as K390 released in 2004[50,51], advertised as an improvement on CPM-10V with more elements added for "hot hardness" and hardenability. K890 was released in 2005[52,53], advertised as an improvement on CPM-3V, with more hot hardness elements. Finally, K490 was released in 2011[54,55], a competitor for Vanadis 4 Extra.

Steel	Year	C (%)	Cr (%)	Mo (%)	W (%)	V (%)	Co (%)	Nb (%)
K390	2004	2.47	4.2	3.8	1.0	9	2.0	
K890	2005	0.85	4.35	2.8	2.55	2.1	4.5	
K490	2011	1.4	6.4	1.5	3.5	3.7		0.5

-35-

KNIVES AND STEEL IN JAPAN

Tamahagane

Japan is known for its traditional production of steel used for Japanese swords called tamahagane[1]. This steel is made by smelting iron sand, which is a sand with a high percentage of iron ore, typically magnetite (Fe_3O_4). Iron ore deposits are rare in Japan, but iron sand is relatively common. The iron sand is converted to steel through a traditional furnace called a Tatara, which dates to around the 6th century AD. A clay vessel is used and fired using charcoal and a bellows. Iron sand and charcoal are layered and fired, reducing the ore (converting from iron ore to iron), and adding carbon to the iron. The steel never becomes molten, so carbon diffusion is not complete, leading to both low carbon and high carbon steel which is then separated. Swordsmiths layer the steel, forge weld, and repeatedly forge and reweld, which is said to improve the quality of the steel and to even out the carbon content. Swordsmiths also create different patterns in the forge welded material though the swords were polished, not etched, leading to relatively subtle patterns. The Japanese swords are also known for different types of laminated construction, the simplest type being called "San Mai," or three layers with the cutting edge having hard steel surrounded by softer sides. The resulting carbon content in the cutting edge steel was typically 0.5-0.7%. In heat treating, clay is applied to the sides of the blade, leaving the edge exposed, which leads to a faster quenching rate at the edge than the spine of the blade. This leads to a hard edge and soft spine, and the transition is visible in the polished steel, called a "hamon."

Yasugi Special Steels

The production of tamahagane declined as Japan turned to steel produced in the West. Swords and knives were increasingly made using crucible steel (and then electric steel), delivered from Europe. In particular, steel from Sweden became popular due to its high quality and low impurities[1]. Western-style steel production also began in Japan between the late 1800s and early 1900s.

One company that mixed Western and Japanese methods was Unpaku Steel, which built a crucible furnace in 1912. They began manufacturing steel for knives in 1913 using iron and steel blooms made with iron sand. They installed an electric arc furnace in 1915 and continued using a sinter-smelting process using iron sand developed by Fuyukichi Obana[1]. After World War I, there was a recession in Japan, and Unpaku Steel closed part of its factory in 1920. Around the same time, Unpaku changed its name to Yasugi Steelworks. In 1925, entrepreneur Yoshisuke Aikawa rebuilt part of Yasugi. He hired metallurgist Haruto Kudo in the same year who would develop methods to continue making steel from iron sand. Under Haruto Kudo, Yasugi introduced its famous cutlery steels[2] wrapped in different color papers: white paper (shirogami), yellow paper (kigami), and blue paper steels (aogami). The white and yellow series are simple carbon steels, fitting under the "W1" tool steel designation, similar to other steel manufacturers with their high carbon steels available with different carbon contents. The yellow paper steel has higher allowances for the impurities phosphorus and sulfur, however, so the white paper versions are more highly regarded and used in high-end knives. This was similar to the specifications such as from Darke with different grades based on allowable impurities, and different classes based on carbon content (see Chapter 2). The blue paper (aogami) steel came from Yonetaro Nagaoke, known as Nagahiro I, who made wood planes. He used a British steel used for the machining of cannons. He used the scrap steel from these machining operations and liked it in his wood planes, so he sent it to Haruto Kudo and Yasugi began to make a similar steel[3]. This was a high carbon steel alloyed with tungsten and chromium, like other "intra" steels such as O7, F2, and 1.2562 (see Chapter 3). Yasugi steel would be purchased by Hitachi steel in 1936, and Haruto Kudo would leave the company in 1944 to become the production manager and president of Nissan.

Steel	C (%)	Mn (%)	Si (%)	Cr (%)	W (%)	V (%)
White #1	1.3	0.25	0.15			
White/Yellow #2	1.1	0.25	0.15			
White/Yellow #3	0.85	0.25	0.15			
Blue Super	1.45	0.25	0.15	0.4	2.25	0.4
Blue #1	1.3	0.25	0.15	0.4	1.75	
Blue #2	1.1	0.25	0.15	0.35	1.25	

Stainless Steels

Hitachi would also introduce stainless steels under the name of Gingami, or "silver paper," which existed by the early 1950s[3]. The Gingami steels were based on Western stainless steels, with Gin 1 being a 440B-type steel, Gin 2 being similar to German 1.4116 stainless, and Gin 3 being similar to Uddeholm high carbon razor steel AEB. Kosuke Iwasaki was known for developing tamahagane steel and manufacturing straight razors using the steel[3]. Iwasaki got a bachelor's degree in history but was dissatisfied with the subject, so he switched to metallurgy, receiving a graduate degree. He would also study swordsmithing and Hitachi Yasugi cutlery steels. Iwasaki wrote[3] that high carbon stainless knife steels would crack when heat treating, but that a Swedish stainless steel they tried did not have this issue. This stainless steel had a cobalt addition, so in 1951 he went to Myodo Steel and requested that they make the same steel. Iwasaki did not give details of the composition apart from it having 1.1% carbon and a cobalt addition. The closest steel I could find to this description was Sandvik 18C283 with 1.05% C, 14% Cr, 0.5% Mo, and 0.5% Co, though cobalt-alloyed stainless steels trace back to Remanit 1790C in Germany (see Chapter 10). Hitachi would make a steel with nearly identical composition called ATS-55 exclusively for Spyderco in the late 1990s. Perhaps there were other cobalt-alloyed stainless steels in Sweden similar to the Remanit products as well. Takefu introduced its own cobalt-alloyed stainless steel around 1959 called VG10 (V Gold 10)[4], likely inspired by European cobalt stainless steels. Takefu is a company that hot rolls steel which was founded in 1954 and patented a method for rolling laminated metals in 1959[5]. In 1969 they patented a method for making composite steels for knives[5]. Because Takefu hot rolls steel rather than melting and casting it, perhaps VG10 was created in partnership with Myodo and Iwasaki, but these types of business relationships are difficult to figure out. Another Japanese company to introduce its own stainless steels for cutlery was Fukami Steel Company, founded in 1960. They would introduce their "AUS" series of stainless steels in 1968[6], starting with AUS-4 and AUS-6. AUS-4 was Hitachi GIN 2 with slightly less carbon and AUS-6 had slightly more. AUS-8 with yet higher carbon would be introduced in the 1970s. Fukami Steel would merge with Aichi Steel in 1978[6]. AUS-10 with ~1% carbon was used in knives by the late 1990s[7].

Steel	Year	C (%)	Cr (%)	Mo (%)	V (%)	Co (%)
Gingami 1	1933-1951	0.85	16	0.4		
Gingami 2	1933-1951	0.5	13.5	0.4		
Gingami 3	1933-1951	1.0	13.8			
VG-10	~1959	1.0	15	1	0.25	1.5
AUS-4	1968	0.42	13.8			
AUS-6M	1968	0.6	13.8	0.2	0.2	
AUS-8	1970s	0.75	13.8	0.2	0.2	
AUS-10		1	13.8	0.2	0.2	
ATS-55	~1998	1.0	14	0.6		0.4

Japanese Custom Knives

Traditional bladesmiths and knifemakers go back many years, of course, making tools like kitchen knives. However, there was a growth in the 1970s and 1980s of custom knifemakers from the influence of the growing USA custom knife market. In 1978 a group of five Japanese knifemakers and dealers visited the Knifemakers' Guild Show in Kansas City[8]. Bob Loveless was highly revered among Japanese custom knifemakers and many of them would make Loveless designs. The Japan Knife Guild would be formed in September 1980 between 40 or so individuals[8]. Loveless was selected to be the first president and held the position for several years until he insisted that a Japanese person take the position. Knife dealer Okayasu Kazuo brought 154CM to Hitachi and asked them if they could make the same steel for knifemakers.

He was told that they already make the same steel under the name ATS-34[8]. Loveless would import ATS-34 because of difficulties with procuring 154CM, and what he did obtain was "dirty," meaning a high content of impurites. Even after the costs of importing, ATS-34 was cheaper than 154CM and available in a wider range of sizes for knifemaking. ATS-34 would become the dominant custom knife steel in the late 1980s and 1990s.

Okayasu Kazuo told of a memorable request Loveless once made of him:

> One time Bob Loveless came to my shop and said, "Hey, Kazuo, I want to meet Mr. Honda." "Honda? Honda cars Honda?" "Yes, I want to meet Honda. Please, you arrange." I arranged everything, and Mr. Honda said, "I want to ... meet Bob Loveless." ... Bob Loveless and Honda-san only had five minutes to meet. They spoke [for] 30 minutes! The guy said, "Some guy waiting" and Honda said, "No, shut up, go away, I am talking to knifemaker Bob Loveless!" The knife [Loveless] showed was integral, all one piece. Next year, the new Honda car model was called Integra.[9]

There was a boom in custom knives in Japan in the 1980s from Loveless and the many Japanese custom knifemakers who had begun making knives. The Rambo movie *First Blood* also created considerable interest in knives. Families wanted custom knives as an heirloom, similar to the practice of collecting heirloom Japanese swords[10]. However, knife attacks in the 1990s changed the perception of sporting knives in Japan as weapons, while kitchen knives are seen as tools[10]. Also, the Japanese stock market crashed, and the economy stagnated for many years afterward, so there wasn't as much free cash among Japanese consumers for sporting knives. Therefore, the number of Japanese custom knifemakers shrank in the 2000s and 2010s.

Japanese Factory Knives

Japan was known for making low-cost factory knives exported to the USA and Europe for many decades. The vast majority of these companies are located in Seki City. Japanese knife factories are small businesses with relatively few employees, and many operations are done by hand. As Japan's economy developed, it became known for its high-quality products. However, the reputation for low-quality, inexpensive knives began to change as importers like Gerber sought a higher quality knife, leading to a new factory called Gerber Sakai (now G. Sakai). In 1985 Loveless was asked what he thinks about Japanese knives:

> They scare me to death even though I've gone over there and helped them do what they've done. I started preaching quality cutlery in Seki in 1975, and they said they couldn't do it. I said horse s--t you can make Nikon cameras, fine cars, and motorcycles. You've got some of the most advanced and outstanding mechanical talent in the world. Don't tell me you can't make fine knives! A man named Sakai listened. We talked by the hour. Within three years, they started to do something. Now the Silver Knight (Gerber) knife line is consisting of outstanding factory folding knives. I gave them a little bit of spirit and a little bit of help, but that doesn't mean I made them successful. They made themselves successful by being willing to work hard.[11]

G. Sakai started as "Sakai Hamono" in 1958, though they grew rapidly through the partnership with Gerber starting in 1977[12]. G. Sakai has made knives for many companies including Al Mar, Spyderco, and SOG.

Gerber Silver Knight. Photo by Arizona Custom Knives.

Another popular manufacturer in Japan is Moki[13], which began in 1907 with Mr. Sakurai, Sr. In 1952, Sakurai Sr. retired, and his son Moki Sakurai took over the company. They built an expanded factory in 1968, and the company's name was changed from Sakurai to Moki in 1987. Moki rose to greater prominence in the market when they began producing Kershaw knives in the 1970s, at that time using AUS-6 and then AUS-8. In 1982 75% of their production was dedicated to Kershaw knives[14]. Since then, they have diversified, and Moki has also produced knives for many well-known companies such as Spyderco, Al Mar, William Henry, and Fällkniven.

Masahiro knives is a company known for its kitchen knives, formed by Kazuichi Hattori. His son Ichiro Hattori worked there for 13 years before leaving to form his own company, Hattori Hamono, in 1971 to pursue his passion for hunting and sporting knives[15-17]. Since then, Hattori has also branched out into kitchen knives. Hattori has made many knives as a fixed blade OEM (Original Equipment Manufacturer) for companies such as Fällkniven, Kershaw, Browning, SOG, Al Mar, Cold Steel, and Beretta. Hattori introduced its own line of knives under the Hattori brand in 1993. Ichiro Hattori helped develop a stainless Damascus laminated blade steel for a line of Browning knives produced in 1984-1985[17]. This steel was made with stainless and nickel silver laminated over a stainless core steel. The steel manufacturer is unknown, but it would likely have been Aichi or Takefu. Hattori is also known for experimenting with many new steels such as Cowry-X and Super Gold.

1980s Browning knife with Hattori Damascus.

Koji Hara

Perhaps the best-known knifemaker and designer in Japan is Koji Hara. Hara started as an employee of G. Sakai where he would make knives such as prototypes for Al Mar, from 1982-1988[18]. In 1988 Hara would leave G. Sakai to become a custom knifemaker, and like many other custom knifemakers in Japan, he would make Loveless designs. In 1991 he did a one-man show in Seki City, and at that point, he realized he needed to make his own designs to succeed as a custom maker[19]. His son Dew Hara (also a knifemaker) said that knifemakers producing Loveless designs had[18], "a negative effect in that many Japanese knifemakers don't have their own individual styles. Why my father can sell out of knives at shows is because he has an individual design." Koji Hara is known for his striking folder designs, inlaid pearl handles, and the "Air-Step" handle design, which he introduced in 1997[20]. The "Air-Step" looks like a terraced landscape inspired by the ancient Incan city Machu Picchu or rice paddies[18]. Hara's favorite steel is Cowry-Y[19], a powder metallurgy version of BG42 produced by Daido steel in Japan, which he was using by 1995[21]. Cowry-Y is well-suited for his mirror-polished blades as the vanadium content is relatively low. Vanadium carbides are harder than many abrasives, which makes polishing more challenging.

Koji Hara knife with "Air-Step" handle. Bladegallery photo.

Powder Metallurgy Blade Steels

With the boom in Japanese knives in the 1980s, several steel companies developed new powder metallurgy stainless steels for knives which were released in the early 1990s. These included Daido Cowry-X and Cowry-Y, Takefu Super Gold, and Hitachi ZDP-189 and ZDP-247. Cowry-X and Cowry-Y were reportedly released in 1993[22]. Cowry-Y is a powder metallurgy version of BG42, which was known due to the influence of American knifemakers and Bob Loveless (see Chapter 41). Cowry-X is an unusual composition with very high carbon and chromium, 3% C, 20% Cr, 1% Mo, and 0.3% V. This steel saw relatively limited use by custom knifemakers and factories in Japan. Hattori is known for using Cowry-X which he typically produces in a stainless Damascus laminate. His popular "KD" series of knives with Cowry-X Damascus were released in 1997[23]. The similar ZDP-189 reportedly came second[24], but both Daido and Hitachi were developing high carbon, high chromium steels in the 1980s and 1990s[25-29]. These steels are claimed to be stainless while being capable of extremely high hardness, 67+ Rc. The Hitachi patent[28] states that these steels were particularly desirable because the low vanadium meant that grinding and polishing were easier than with vanadium-alloyed stainless steels. Instead, a high percentage of chromium carbide gives the steel high wear resistance, which is softer than conventional abrasives. The steel also does not require a very high temperature for heat treating as many stainless knife steels do (1832°F/1000°C vs 2000°F/1100°C or higher), thus avoiding the use of expensive high temperature furnaces.

The patent for Takefu's Super Gold powder metallurgy knife steel was filed in 1991[30]. Super Gold also looks somewhat similar to BG42 and ATS-34, though modified with somewhat lower Mo and higher V. The chromium content of 14-15% is also similar to many of Takefu's previous V-Gold steels, but the Mo content being so much higher makes the inspiration of BG42 or ATS34 more apparent. Neither of those steels are mentioned in the patent, however, so perhaps this could be called a coincidence. There are two versions of this steel, a "Super G1" and "Super G2," though only the SG2 is now available, and I have only seen SG2 used in knives. The patent lists a carbon content of 0.9-1.5% and SG2 is near the top of that range so perhaps SG1 was lower in carbon. Presumably, higher carbon and higher hardness were deemed a higher priority than higher toughness and corrosion resistance. That is the tradeoff we see with carbon content in most steels and is also what was shown in the experimental tests of the Takefu patent. Kobe makes the same steel under the brand name R2, and reportedly Kobe also manufactures the PM steel for Takefu[24], which makes sense given that Takefu does not have powder metallurgy production facilities. The earliest use of this steel I have seen was in Fällkniven knives, branded as SGPS, or Super Gold Powder Steel, in a Takefu-produced laminate with 420 steel sides released first with the U2 folder in 2004[31]. It has also been used in the Shun Elite series of kitchen knives with 410 laminated sides released in 2006[32].

Steel	C (%)	Cr (%)	Mo (%)	V (%)
Cowry-X	3.0	20	1.0	0.3
Cowry-Y	1.2	14.5	4	1.2
ZDP-189	3.0	20	1.4	0.1
ZDP-247	2.0	17	1.0	
Super Gold 2	1.4	15	2.8	2

-36-

WAYNE GODDARD AND KNIFE TESTING

Wayne Goddard talked about his history in 2000:

An old-time blacksmith gave me a formula for using an oven-tempered lathe rasp to make a knife. In 1963 I found a lathe rasp, but then I needed a grinder. So, I built one. It wasn't much of a machine, but it worked well enough to grind out my first knife and quite a few more... I didn't know it at the time, but that knife got me started on my life's work. There were those who liked that knife well enough to have me make one like it for them. If that wouldn't have happened, I might not have made any more knives. The guard is steel, the handle slabs are Oregon Myrtlewood, and I made the rivets out of 1/4-inch bolts. I ground it out very carefully from a lathe rasp, just like the formula said. I tempered it at 375 degrees F in the kitchen oven. Careful grinding was necessary to keep the edge from being softened from overheating. (The edge-holding ability of most carbon and carbon alloy steels can be ruined by heat caused by careless grinding with wheels or belts.) The grinding wheel marks were smoothed up with the disk attachment on an electric drill. Back then, my idea of a fine finish was somewhere between 60- and 80-grit.

That first knife never did sell. It seemed that everyone wanted the "improved" workmanship of the subsequent models. By the end of the first year, I decided it would be a good one to keep. I figure it is by pure luck that I still have it. I'm glad to have that knife because it helps me prove some points with new or want-to-be makers. It shows that a knife can be made with a $5 grinder and an electric drill. It clearly shows that I had no real, natural-born talent for knifemaking. I believe that the main requirement is a strong desire to do it. Years of hard work and practice will get a maker a lot closer to success than any talent they may have at the start. The only knifemakers I was acquainted with learned most of what they knew from me. They figured out some new and different ways to do things, and I would usually try their methods if it looked like a better way of doing it. I had very little money to purchase equipment, and I wouldn't have known what to get anyway. I made knives the only way I could figure out how to get the job done. This type of beginning may not be the best way to get started, but I've never regretted it. I've always been thankful for all the things I learned by doing it the hard way. I learned to solve problems on my own, and that included how to make the most of my own equipment. My primitive beginning makes me very grateful for all the tools I have to work with today.[1]

Goddard firmly believed that anyone can learn to make knives:

> I have a friend who tried to make a knife in another maker's shop. but he wasn't getting anywhere with the blade. The teacher finally said, " ... you don't have what it takes" and threw the blade away. I don't quite understand that attitude. It might be that the knifemaker/teacher had talent to the extent that he didn't have to work at developing skill. He must have assumed that training wasn't really necessary if you had the talent. You either had it or didn't have it. And he wasn't going to take the time to be a good teacher. I've always said that anyone could learn to make knives. As I get older, I have modified that statement to read like this. Anyone with the *sincere desire* can learn to make knives. My students come from all walks of life; dentists, loggers, crane operators, game wardens, roofers, welders, school teachers; physicians; and, no kidding, a butcher and a baker but no candlestick maker. Age is no barrier; my youngest student was 9, the oldest was 81.[2]

Goddard's excitement about the knife business grew thanks to Bob Loveless and the new steel 154CM, as he said in 2001:

> The pioneers of the modern handmade knife era made it easier for those of us who followed the trails they blazed. The pioneer who had the most impact on my career was Bob Loveless. I was getting my feet wet in the knifemaking business during 1971-1973, and Bob was never too busy to answer all my questions. I appreciated his no-nonsense and practical approach to knives and what they were about. His purity of design and clean workmanship gave me a goal to strive for. Bob was always a leader in finding new and improved blade materials. A lot of the excitement and growth of the handmade knife world in those early years came from his introduction of 154-CM. Thank you Bob for continuing to be a real knifemaker after all these years.[2]

"The first 10 years I made a lot of working knives. By 1974 it was becoming more of a collectors' market."[2] Goddard joined the Knifemakers' Guild in 1972 and attended his first knife show. "Up to that time, I only thought of knives as tools. Once I got it through my head tools didn't have to be ugly, I started making headway."[3] Goddard would go full-time as a knifemaker in 1973, leaving a steady job:

> I decided to specialize in folding knives starting in 1974, and that was pretty much what I made until 1983 when I got into forging and Damascus steel. My favorite knives to make are one-of-a-kind knives of Damascus steel and large Bowie knives. I am probably best known for popularizing knives that are made of forge welded steel cable. I make many, one-of-a-kind folding knives that utilize weird antler parts for the handle. My influences include ancient weapons, tribal knives, and classic Bowie knives.[2]

Goddard thought differently than many other knifemakers, as Sid Latham said[3]: "Goddard doesn't mind being called weird but feels a handmade knife should look a bit crude." Goddard would be the first to receive a Master Smith rating from the ABS under the then-new requirements in 1988.

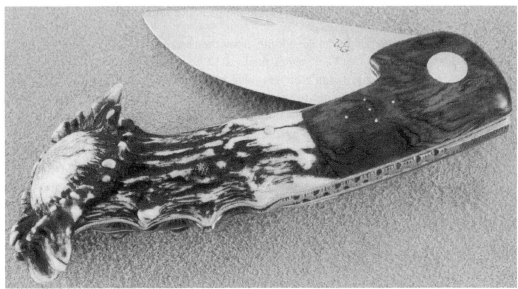

Wayne Goddard folding knife. Barry Gallagher photo.

Writing About Knife Making

Goddard's extensive experience in knifemaking and teaching led to his sharing of information through many written articles and books, first with articles such as those on cable Damascus in *Knives '86* and *Knives '87*. In *Knives '87*, he even shared the results of his rope-cutting experiments showing the effects of different steels and heat treatments. Next, he began writing *Blade Magazine* columns where he answered questions or wrote on specific topics, such as many articles on steel and heat treating. This led to his books *Wonder of Knifemaking*, gathering many of the columns he had previously written for *Blade*, and *Wayne Goddard's $50 Knife Shop*, about making knives with inexpensive tools. Goddard firmly believed that anyone could make knives at virtually any budget. Some have criticized the $50 number as unrealistically low, but Goddard expected that criticism and addressed it in his book[2]: "The figure of $50 is really arbitrary. It is the philosophy of working with the tools you have or can acquire inexpensively that's really important. If you really want to, you can create a serviceable knife shop for a lot less than you think."

In 2001, *Blade Magazine* editor Steve Shackleford wrote about what made Goddard a good teacher:

> Wayne Goddard's career making knives mirrors that of the modern custom knifemaking movement - that is, from the mid-1960s to the present. In that time, he has explored the gamut of knifemaking, from grinding blades via the stock-removal method to forging them to shape. In the interim, he has won many awards for his knives and has established

himself as one of the most respected names in handmade cutlery... Wayne is one of the rarest of blade breeds. Not only does he make a premium knife, but he also teaches others how to do it, and he does so without the slightest hint of reluctance some makers have about sharing their "knifemaking secrets." In fact, one "secret" to Wayne's success is that he has no secrets. He lays bare everything he knows about making outstanding knives and scoffs at those who guard their tricks of the trade as if they were some kind of Holy Grail. Whether in seminar form, where he turns so-called knifemaking mysteries into easily understandable, step-by-step procedures, or in the written word... Wayne stands out as one of the handmade industry's foremost instructors.[2]

Testing Knives

Goddard was an inquisitive person and often talked, and wrote about, testing of different aspects of knives. He said that "Superior knives result when better comparisons are made, and then modifying the product to make it better."[4] In 1990, he wrote about his history of testing:

> Twenty-three years ago, I started cutting cardboard as an edge-holding test. It was a boring, time-wasting process because a superior blade would cut until your arm was sore, and you were up to your knees in cut-off pieces. My first tests on rope were done in 1972 to determine the difference between 440C and 154CM. A friend and customer, Maynard Meadows, suggested using sisal rope, and successful and time-efficient tests were achieved. We started out cutting a 1-inch rope but switched to the single strands because they dulled the test blades quick enough to make the testing time efficient. Over the next eight years, I made several dozen test knives to compare various steels and to compare the same steels with different heat treatment. Maynard, an avid big game hunter, performed comparisons of the test knives on hunting trips. The field tests reinforced the results that we were getting with our rope-cutting tests. I obtained abrasive-resistance charts from two different steel companies and, by comparing their bar charts with my rope-cutting tests, I realized that I was within 5 percent of their ratings when the hardness of the steel was the same. I gained more confidence in my endurance testing as time went along.[4]

Goddard was curious about more than just steel, but also heat treating and hardness:

> One of the tests that I did with the help of Paul Bos was to determine the effect of different hardnesses on edge holding. Paul is a professional heat treater and was interested in helping with the tests. We ran test batches of D5, 154CM, and 440C steels, giving half the blades their normal working hardness. The other half were drawn back two points on the Rockwell hardness scale. The drawn-back blades cut 15-20 percent less. When comparing a blade with a hardness of 54 Rc to a blade of 60 Rc, I found the percentage loss held up. The steel that did 40 cuts at 60 Rc did 30 cuts at 58 Rc, 20 cuts at 56 Rc, 10 cuts at 54 Rc and, at a hardness of 52 Rc, would hardly cut once.[4]

Another comparison he was curious about was the effect of cryo and freeze treating; he found that the edge retention on rope scaled with the increase in hardness from the freeze treatment.

Wayne Goddard's Rope Cutting Results Reported in *Knives '87*

Steel	Rockwell Hardness	# of Cuts
154CM Freeze treated	61	44
154CM Not freeze treated	60	38
ATS 34 Freeze treated	60	38
440C Freeze treated	58	33
440C Not Freeze treated	?	25
CPM 440V	?	58
Vasco Wear	61	56
D2	60	47
D5	61	52
D5	59	40
Stellite 6K	44	58
F8	?	45
M2	64	44
52100 Handforged	60	43
5160 Handforged	60	43
O7 Handforged	61	44

For testing chopping, Goddard would cut 2x4s, but he was having difficulty making sense of what he was testing:

> To get as many comparisons as possible, I measured the depth of penetration into the wood with a single blow. However, this was not very consistent. I also performed a series of tests to determine the largest piece of wood that could be cut with a single blow. A pine or fir board 3/4-inch-by-1-1/2 inches is a good size with which to start. Knives with a dropped handle, like the kukri, always out-chopped the broom handle designs. These were useful comparisons, but I wanted to eliminate the human arm and the advantage of good handle design. I wanted to test only the efficiency of the grind. This is how the Chop-O-Matic penetration test machine came to be built... Here's how it works: Set screws hold the blade horizontally with the back against a pivoting arm. Raise the arm a given distance, allow it to fall free, and the knife edge penetrates the wood held by a clamp. Measure the depth of penetration with a vernier caliper and record it. The results are uniform, and the human element is eliminated. Initial testing was done with no adjustment for the weight of the knife, which gave the heaviest knives an advantage. One chop was done on three different materials with each knife. Total penetration on all three materials was totaled. The three materials cut were 3/4-inch mahogany, 3/4-inch cedar, and 1/2-inch Delrin. The knives were mounted to give a square hit on the test material. When there was a difference in the depth of the cut from one side to the other, the two measurements were averaged... The Chop-O-Matic machine did what I hoped it would. It accurately tested the chopping ability of blades with different types of cross sections, proving the superior chopping ability of a thin blade.[4]

Goddard also recommended the "Brass Rod Test" for testing edge strength to ensure the heat treatment and chosen hardness works for the edge geometry:

Clamp a brass rod 1/4 inch in diameter horizontally in a vise with the top third above the jaws. Lay the edge on the brass rod at the same angle used for sharpening. Apply enough pressure so that you can see the edge deflect from the pressure on the rod. The pressure works out to 35-40 lbs. Use a good light source behind the vise so that you can see the deflection. If the edge chips out with moderate pressure on the rod, the edge will chip out in use. It the edge stays bent over in the deflected area, it will bend in use and be too soft to hold an edge. The superior blade will deflect and yet spring back.[4]

Heat Treating, Forging, and Packing

Goddard made both stock removal and forged blades but did not believe that forging necessarily led to superior properties[4]: "A particular type of steel has a certain potential in cutting ability and strength. It matters not so much that it is forged or ground, but that it is given the proper thermal treatments to bring out the maximum performance." He shared a story illustrating the importance of heat treating over forging:

> It was 1983, and I was just getting into forging. I was doing a lot of experimenting with selective hardening by using a torch to soften the back of the blade. I had a lot of stock removal blades around and would use them to experiment with. I had given one of my stock removal blades made of band saw steel a soft-back draw with the torch. It was a butcher knife blade with no handle. The blade was about 9 inches long. It was what I called Mountain Man Camp Knife, the type of knife I have made ever since I got started in 1963. I had flexed that blade 90 degrees several times and straightened it out with a hammer. I had that test blade with me at one of the first Blade shows in Cincinnati. Gil Hibben was talking to me about forging and asked if I thought it made a better knife. I handed the blade to him and asked if [he] would like to break it. (He assumed it was forged.) He put the tip of the blade on the floor with his foot on it and pulled on the tang. It bent past 90 degrees but did not break. He seemed surprised, so I took the blade and straightened it out by reversing the bend. He thought forging was a good thing, and then I had to tell him that it was a stock removal blade. The strength was a combination of tough steel and proper heat treating.[1]

In the early days of forged blades, packing was one of the primary reasons why it would lead to superior blades. However, in the 1980s and 1990s, bladesmiths increasingly rejected the idea that forging would lead to the "packing of molecules." For example, Charles Bear and Ronald Koeberer wrote an article on the metallurgy side of forging called "Why Forge?" in a January 1990 *Knife World* article. They said:

> It should be said that packing is a hoax without any scientific basis. "Packing" as a concept would fit into a description of a material where there was a great deal of porosity, and compaction was necessary to eliminate that porosity. But modern steelmaking processes solve the porosity problem. Without porosity, steel cannot be compacted. The electromagnetic force that bonds the steel together also prevents the atoms and crystals from being compressed closer together... The concept that must guide us in our forging is

not packing but the concept of "grain refinement." Forging is the process used in developing a highly refined grain structure. The size of the grain is the main control variable the smith has available to work with.[5]

William W. Wood in *Knives '95* went a step further and said that grain refinement was also impossible[6]: "Refined grains could be left in the steel by reducing the temperature for the final forging sequence. This should be done for parts that will not be heat-treated, resulting in parts with a range of hardness from Rc 45 to Rc 50, much below heat-treated parts in the range of Rc60 to Rc 65. For this reason, forging can only help produce a shape in knifemaking and has little if anything to do with the final properties of the knife." Wood had a master's degree and worked as a manufacturing research engineer for the aerospace company Chance Vought, giving his views credibility. The article created quite a stir, as *Blade Magazine* editor Steve Shackleford wrote[7]: "Wood's story was one of the main topics of discussion among several makers at the recent New York Custom Knife Show and generated a call to *The Blade Magazine* from Damascus maker Daryl Meier disputing some of the story's claims. Wayne Goddard will address several of the topics covered in Wood's story in a future issue of *Blade*."

While this statement from Shackleford was in the March 1995 issue of *Blade Magazine,* Goddard had already begun a three-part series on packing starting with the December 1994 issue[8-10]. Goddard wrote:

> Some magazine articles and books claim that packing does the following: Compacts the molecules; jiggles the carbides into alignment; packs the grain; elongates the grains in the direction of maximum stress; breaks up large grains; and refines the grains... My introduction to packing came when I got serious about forging knives (1981-82). I asked several established bladesmiths to explain their forging methods. Packing was a part of the instructions I was given. There were some differences in the methods used, but the basic formula was something like this: "Hammer the edge portion of the blade with a light hammer as the edge cools down to the point that the color is barely visible in a dark room." At the same time, I asked two different metallurgists what they thought of the packing theory. Both explained that the steel recrystallizes during the hardening operation. Their opinion was that any grain refinement done in the packing process would be undone by the temperature of the hardening operation. With blade forging experience on one side and metallurgy theory on the other, it was clearly a dilemma to be worked out. Packing, as practiced by bladesmiths today on modern steels, has no basis in modern metallurgical theory.[8]

> I surveyed some well-known bladesmiths at the 1994 Blade Show & International Cutlery Fair as to whether they used the packing method and why. Some of the answers: 1) "Yes, packing is stressing the grain. Grain in the stressed condition will recrystallize smaller"; 2) "Yes, I do it because I was taught that it refines the grain, but I believe that I may get the same results with thermal treatments"; 3) "No, because I do not think it does anything"; 4) "Yes, for grain refinement"; 5) "No, for theoretical reasons"; 6) "No, because

it is no better than thermal treatments"; and 7) "Yes for 5160 steel, no on Damascus."[9]

There seems to be a mix of opinions in the preceding, and they seem to be based on theory and training. Meanwhile, I am more interested in the strength and cutting comparisons between packed and non-packed blades. In my experiments, I have learned that there is no difference between the two. The heat treatment is always the most important element in determining the quality of the finished blade. I use and teach what I call a finishing heat. It looks sort of like packing but is done for different reasons. The finishing heat consists of lightly hammering the forged blade into the final finished shape at a low heat to get the blade as straight and smooth as possible. The finishing heat is the most logical way to finish a forging. In my opinion, packing does not impart mystical properties to the blade, and I do not teach packing as such.[9]

Steel cannot be packed, compressed, or compacted. Isn't it time to retire the term "packing"?[10]

Instead, Goddard proposed that packing[10] "may have an effect on the finished blade because of the time/temperature cycle that takes place during the physical packing with the hammer." But he was not convinced that this could not also be achieved with stock removal[10]: "Will the stock removal blade that receives the same time/temperature treatments as a forged blade have the same size and type grain structure? Have the past comparisons between stock removal and forged blades taken into consideration the differences in thermal cycles?" While the term "packing" does occasionally still come up among bladesmiths, it is no longer mainstream. More and more bladesmiths have recognized that this is an outdated idea with a non-scientific basis.

Temperature in Fahrenheit

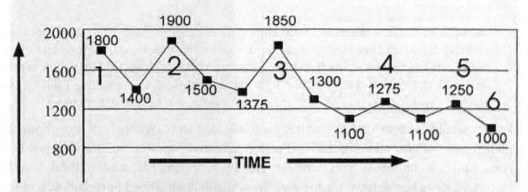

Time/Temperature Cycles For A Forged Blade

Numbers 1 through 3 in the chart mark the forging heats. The 1300° heat is the start of the packing process. This hypothetical forge/time/temperature example used three packing heats. The chart for a stock removal blade would be a straight line at room temperature. What effect does this have on the finished blade?

Goddard's schematic[10] of how forging and packing leads to many "thermal cycles" which he proposed may have a greater effect than the forging itself.

-37-

ED FOWLER AND TRIPLE QUENCHING

Ed Fowler is a rancher living in Wyoming[1]. Fowler began knifemaking with stock removal knives made from D2, which he did so for several years until he attended a hammer-in taught by Bill Moran. Fowler said in 1988[1]: "I got interested in forging blades because of what I had read. I felt that if what I was reading was true, then the stock removal blades I was making were inferior products. After three days of talking, forging, and testing blades under Moran's supervision, I was convinced that there is no finer using knife than one blessed with a properly forged blade!" At that time, he said his[1] "favorite material is 5160. I also offer Damascus blades made from steel cables."

Fowler preferred making working knives, and not art knives, from the beginning of his knifemaking:

> I sometimes feel that the true "using" knife is an endangered tool. The knives that I see sold and those that I see used lack the functional quality they need. I strongly feel that a using knife can be a work of art, but a work of art with a purpose. I will not allow art for art's sake to interfere with function. With the exception of the engraving I put on my bolsters, there is a functional basis for each and every aspect of one of my knives.[1]

Fowler would become a regular writer for *Blade Magazine*, where he wrote about his philosophies of life and knifemaking. Fowler has strong opinions about what makes a good knife, including forging, convex grinds, large guards, and natural handle materials. He is best known for his designs which include sheep horn handles, brass guards, and a blade shape he calls the "modified Michael Price grind," where the knife is thickest near the belly, or roughly the front third of the knife. He said:

> The material that I write about in *Blade* is based on time-tested methods, materials, and designs that work for me in my shop and in the environment in which I live. I don't claim that they're the only ways to make a knife, or that they'll appeal to everyone, nor are my

methods set in stone. I continue to learn more every day. I write about the lessons I've experienced over time for several reasons. First, if they can be of benefit to the world of knives, I don't want them to die with me. Second, I hope to help keep the search for the best blade man will ever know active in the minds of all who wish to join in the quest... There are many excellent knifemakers and collectors who devote their talents to the preservation of the traditional blades of the past. If this is where lies their Excalibur, I feel it both fitting and proper that they continue their edge quest ... after years of forging, I have many more questions than answers... The forged blade is based on an infinite number of variables. No single man could ever have enough time to consider all of them. Any bladesmith who claims to know them all only fools himself. Should my invitation to explore new territory challenge others to action, their contributions will be welcome. There's plenty of room for everybody... My personal Excalibur is the development of a knife that would be the most prized possession of any man of any time, past, present, or future, who lives with nature and needs a friend on whom he can count... All I can say is that the information I write is based on the trails I've followed and is presented honestly in the interest of nurturing the ultimate functional forged blade of the future.[2]

Multiple Quenching

Fowler is known for popularizing the method of the "multiple quench" or "triple quench" where the blade is heated to the austenitizing temperature and quenched three times. He would heat the edge of his knives with a torch to nonmagnetic and quench, then wait 24 hours between repeated cycles. Fowler wrote in 1998:

> When I first started making knives, it was widely believed that when hardening a blade, should the blade not harden on the first quench, it was necessary to start from scratch, and anneal the blade to its state before the failed hardening, then harden it again. To my knowledge, no one had explored the performance of blades that had been hardened more than once. Most knifemakers, including myself, considered subjecting blades to multiple hardening cycles pure folly and probably detrimental to the qualities of the blade. At first, I could not believe the superior qualities indicated by the testing of multiple quench blades. Test blade after test blade told me the same thing, "SUCCESS." As I came to believe in the process, I talked to other knifemakers, most thought I was nuts. Wayne Goddard took me seriously, and agreed to participate in experiments.

> I stumbled onto the value of multiple quenching following what I considered to be a series of errors. The following evening I called Wayne to discuss the results of my initial experiment ... the two of us have corresponded about knives and knifemaking at great lengths. Anytime I stumble upon something interesting, Wayne is the man that I contact. His vast knowledge coupled with his empirical sense of order has proven to be invaluable throughout the years... We had been planning to spend a week in my shop following the Blade Show one year. We decided that the week in my shop would be devoted to designing and conducting a series of tests that would determine the value of multiple quench blades. Wayne stated that he had read about multiple quenches in an old metallurgy book, and while at the Blade Show in Tennessee, Al Pendray produced an old handbook that recommends multiple quenches for some steels... While at the show, we discussed the possibility of designing an experiment that would compare the cutting ability of multiple

quenched blades to that of single quenched blades... The blades were made and all initial testing was done ... starting on May 23 and finishing on May 30, 1990. We selected a new bar of 5160 steel that was more than long enough to supply the steel of four blades.[2]

They found a single quench stock removal blade to make 163 cuts in single strands of 1-inch rope, a triple quench stock removal blade to make 195 cuts, a forged single quenched blade to make 269 cuts, and a forged triple quenched blade to make 509 cuts. Fowler summarized the experiment:

> The cutting order of the knives was randomly determined. The man cutting did not know which blade he was cutting with, and therefore, the probability for error should be quite low. We definitely feel that this was a fairly designed test, comparing not only the forged blade against the stock removal blade, but also the value of multiple quenches versus single quenches as it pertains to turning out a blade capable of superior performance... Through the unselfish sharing of information, we have the ability to advance the art of knifemaking to levels never possible without it... After we accidentally discovered the benefits of the multiple quench, Wayne advised me that Al Pendray was using a multiple quench on his blades. When Wayne mentioned this, I vaguely remembered reading an article to that effect about Al's knives. The trouble was, I read it and didn't integrate it to the extent that I tried it. Hopefully, the information in this article will convince more knifemakers to try multiple quenches and report their results.[2]

Wayne Goddard provided his hypothesis for the superior results in a 1993 article:

> The theory behind the triple quench is that by bringing the blade rapidly up to the hardening temperature, the grain size remains smaller than when the usual soak time is used. The soak time allows all the transformations to be made within the steel, yet the grain grows with the additional time at the soak temperature. With the rapid quench the transformation is not complete; however, the second and third quenches complete the necessary transformations. The triple-quench blades have higher strength at a given hardness yet they have more edge-holding ability than a single-quench blade of the same hardness. Ed and I do not know the reasons for this, we just know that it works.[3]

Not everyone was a fan of this heat treating method. Master Smith Dan Petersen wrote an article titled "Multiple Quench: Fact or Theory?"[4] where he criticized Fowler's methods:

> Most makers refer to the industry charts based on experimentally validated procedures to produce a given outcome for a specific type of steel. Some makers may even move ahead of the existing scientific database. When they do this, they take a light into the dark. More often than not they will fumble around in the dark and come up empty. I admire people who are willing to do that. They are the ones who are willing to take the time and effort to expand the craft. They are also the ones who are willing to risk failure. The problem is superstitious behavior. Superstitious behavior is usually defined as behavior that is empirically unrelated to the outcome achieved or desired. It usually is acquired by accident... Superstition creeps in where knowledge is absent. However, the one thing those who fumble or search in the dark must not do is mislead the less informed or naïve to believe they know the truth and see by its light. There is a fine line here that should not be crossed. Established makers owe it to those coming up in the field to give them training

and information, not superstition or folklore... For example, one article I recently read suggested that the steel would hold a better edge or be tougher if the smith waited 24 hours between heat treats on knife blades made of 52100 steel. It sort of sounded like bury a dead cat on a full-moon night at the stroke of midnight and spit three times on the wart to make it go away... I worry when I read an article from someone who describes the heat treat of a new steel in a non-empirical manner that reeks of superstitious behavior. I am particularly distressed when the wording of the article implies this is the correct way of doing things ... a theory is not fact and should not be stated as such.[4]

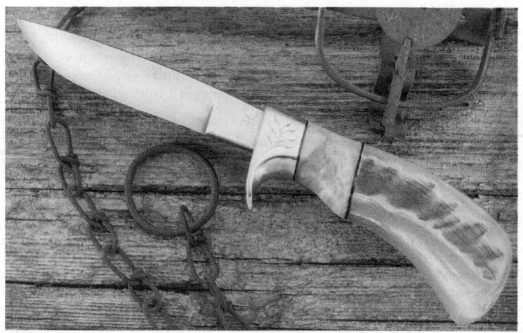

Fowler Pronghorn knife. Barry Gallagher photo.

The magazine had a corresponding response from Fowler, where he defended his recommendations. Bladesmiths that had tried the method, including Rick Dunkerley and Ed Caffrey, also defended Fowler from Petersen's criticism in the June 1994 "Readers Page."[5] Dunkerley said:

> For 11 years I've been making knives and for seven of those I've guided elk, deer, and antelope hunters. While I'm not a metallurgist, chemist, or engineer, I am a knifemaker and user and I know what works and what doesn't. In the shop and in the field, testing has shown me that Ed Fowler's heat treating methods for 52100 work! None of the blades I've forged have required an engineer to count the number of cuts of rope they can make, but I can tell you the 24-hour wait with a freeze makes a difference. That is a belief built on confidence gained from field use, not an article by a metallurgist.[5]

Blade Magazine reader Erik Kolm also wrote, "I find it incredible that one ABS Master Smith would attack another in this manner. It is apparent that Dan Petersen never consulted with Ed Fowler on the Multiple Quench method before attacking him in print... I think Petersen should spend less time being jealous of people like Fowler and Wayne Goddard and spend more time on his experimentations."

In 2005, metallurgist Dr. John Verhoeven would publish *Metallurgy of Steel for Bladesmiths & Others Who Heat Treat and Forge Steel* where he would report multiple quenching experiments with 1045, 1086, and 5150 steel and ultrafine grain sizes were obtained[6]. He did not claim that this provides large increases in edge retention or wear resistance, however, but that a fine grain size leads to improved toughness. However, Verhoeven did not wait 24 hours between each austenitize and quench because that doesn't make metallurgical sense.

52100

While Fowler's original experiments were done with 5160, he also began using 52100 in 1990[7], and by the time he and Goddard published information on multiple quenching, the main discussion was usually about 52100. Fowler became known for using and recommending 52100 steel. Fowler wrote in a 1992 article:

> Through the years I have tried many steels and heat-treating methods and have tested the results by cutting a lot of rope and breaking many blades ... I had settled upon one steel (presumably he referred to 5160) that when properly forged and heat treated cut very well. I had invested a great amount of time and effort getting the most out of that steel, so I was reluctant to change. My friend and fellow knifemaker, Wayne Goddard, had been challenging me to try some steels that he said were superior to mine. Wayne twisted my arm and bent my ear for several years. I had tried other steels and didn't get the performance that I was getting from my old favorite steel. It is pretty hard to change from something that works. It's kind of like buying a new hat; things just don't seem to fit. I started to doubt my choice of knife steel when Wayne showed me a knife that he had forged from a ball bearing (52100 steel). He had etched the blade in ferric chloride and the visible grain structure was absolutely beautiful. Not only was the blade beautiful, it was flawless and cut exceptionally well. I decided that it could be well worth my time and effort to develop a forging and heat-treating process for 52100. I didn't complete many knives for several months but the results were well worth it. Using bearing-quality 52100 steel for knives offers a significant advantage. The quality control governing the manufacture and selection of bearing-quality 52100 is rigid, producing top-quality steel. The etching of blades reveals faults that you won't normally see in a polished blade. I have etched all of my 52100 blades and have yet to find any faults in them... Forging a knife blade from a large ball or roller bearing allows you to work the steel extensively. Steel that has been forged extensively tends to respond to sophisticated heat treatment producing a significantly better blade when compared to a blade that required a minimal amount of forging.[8]

Fowler also stated, "In addition to everything else, 52100 is easy to work grind and heat treat... The high chrome content of 52100 promotes a beautiful mirror polish... 52100 also has 1.5 percent chrome to react with to make chromium carbides that can enhance cutting performance."[9] Fowler would also begin using a source of "virgin" 52100 round bars supplied by Rex Walter by 2002[9]. Fowler found that using the same steel led to an improvement in consistency.

Thermal Cycles

Fowler is generally a believer that "more is better," such as the introduction of multiple quenches or forging from larger steel stock to get more reduction. He also gradually added more "cycles" of other steps to his process. In June 1995 he mentioned that he was annealing the steel three times[10], and in July 1997 he called this process the "Multiple Bladesmith's Anneal"[11]. In both of those articles, the process included a single normalize prior to annealing. In October 1997 he stated that he would perform the normalizing treatment three times prior to rough grinding, and then normalized three times again. Fowler said[12], "Rick (Dunkerley) and I feel that this process reduces the steel's grain size dramatically, allowing future thermal treatments to push the steel to its maximum potential."

Fowler added the requirement that forging must be performed at low temperatures in July 2002[13]: "All the forge work is accomplished at temperatures not exceeding 1,625°F, which is known as low-temperature forging... When heavy scale comes off the steel during forging, you've probably exceeded 1,750°F and are very close to the temperature where grain growth can occur. Grain growth can negate the potential benefits of the forging process to the finished knife." He added, "I strongly believe that the high-performance potential of the forged blade is primarily due to and obtainable through low-temperature forging and numerous cycles. Industry, in the interest of economics, necessarily forges steel at higher forging temperatures that don't support grain refinement to the extent that can be achieved through low-temperature forging." By 2006[14], Fowler was promoting the segregated, banded microstructure which would form from the low-temperature forging and cycling, which he called "52100 Wootz." He said[14], "These past two years have been rewarding. We have achieved blades with significant banding in the transition zone just above the hardened portion of the blade. Thus, the Wootz qualities contribute to strength and toughness over the fine-grained martensite matrix in the cutting edge." He had also begun to use a torque wrench, suggested by Joe Szilaski, when bending blades 90 degrees, and he measured blades that reportedly took more force to bend in one direction than the other, which he attributed to the banded structure. Generally, banding is not considered a good thing, but it is likely difficult to avoid with such low temperature forging and thermal cycling.

Thermal cycling and multiple quenching would spread to many bladesmiths. Knifemakers "triple quenching" included names like Jerry Kennedy[15], Barry Gallagher[16], Rick Dunkerley[17], Kirk Rexroat[18], Joe Szilaski[19], Rich McDonald[20], Bill Buxton[21], Ed Schempp[22], and Bill Burke[23]. Even while "triple quenching" by bladesmiths has become somewhat less common in recent years, the "triple normalize" or "multiple thermal cycles" has remained a mainstay among bladesmiths.

-38-

DIFFERENTIAL HARDENING AND THE HAMON

Torch-Softened Spine

Moran began heat treating his knives to have a soft spine in the late 1940s (see Chapter 14). Moran would write about his process in the fourth American Bladesmith Society newsletter, published on April 11, 1980:

> The instant that [the blade] becomes non-magnetic, it has reached the exact temperature, and should at that temperature be quenched... Now, once the blade has reached this temperature, take it out of the fire by the tang and quickly quench it, point down into the oil. It is a good idea to only plunge it into the oil up to the ricasso. This will leave the tang softer and very tough, which is quite important as no one wants his blade to break off at the tang. In many commercial knives that you have all probably seen that were broken right at the tang, this was the reason why. They were hardened all the way throughout the tang, and of course, they were never as tough as they could have been if this was not done.

> Now hold the blade in the oil until the temperature drops to the point where it is still too hot to hold in the bare hand. Then take it out of the oil, wipe it off with a rag, and go immediately to the grinder and grind all the oil and scale off, until the blade is bright and shiny. Then grasp the blade with a pair of Vise-Grips and light your acetylene torch. Your purpose now is to draw the temper to the proper degree of hardness. As the blade is quenched, it is far too hard to be of any practical use. After lighting your torch, be sure to have the torch quite hot. This is very important, as you want to heat the back of the blade very quickly. This will give the temper colors time to run down toward the edge slightly. If you heat the back of the blade too slowly, the whole blade has a tendency to overdraw and become too hot all over at one time. Now, with the very hot torch flame, move it very slowly down the back of the blade, starting at the rear. Move it just fast enough that the blade is turning blue for about 1/8" or so down from the back behind the torch. As you get near the point, which, of course, is smaller, you will have to move the torch a little faster. Now take your time on this operation, because if the blade suddenly turns blue all the way down to the edge, you will have to anneal the blade and start over. You will probably have to practice this a good many times before you work out how to do it so that the blade turns the nice even color that you desire.

If your blade is to be a knife used strictly for skinning, you will probably want to temper it harder than the one we use, say, for chopping off limbs -- in other words, like a large camp knife. For the skinning knife, draw it only to about a medium-dark straw. For the knife that is going to be used as a heavy camp knife that needs to be tougher, draw it to what I would call about a purple... These colors are very accurate, although they are somewhat hard to describe, as it seems that some people see them somewhat differently than others. It is a good idea to have your quench oil near you while drawing, so that if your blade appears to be getting too hot you can quench it and stop the draw; but if it is done properly you should be able to allow the blade, after running the torch over it and getting to the purple color, to lay on the bench and cool off by itself.

Now let the blade cool to room temperature. Then take it back to the grind, and grind off bright once again, and repeat the same process. This process should be repeated, for a very fine blade, three times. This is a very old tempering technique that has been known for a long time, and it does almost the same thing as super-chilling, only it is much more simple to do in our shops. How you heat-treat your blade will give it the qualities throughout its entire life that will make the difference between a great blade and a mediocre or worthless one. Spend a great deal of time practicing and experimenting with this process, as it is the most important one to your making of the knife.

Moran's process would become common with bladesmiths because of his significant influence on the forging community and the formation of the ABS. And the importance of a "differentially" hardened blade for certain knives would also become further solidified by the 90-degree bend test required for the Journeyman and Master Smith tests.

Edge Quench

Wayne Goddard wrote about the alternative "edge quench" method in 1991:

Another method used to gain strength in the blade is edge quenching. The whole blade is heated, then quenched edge down in a pan of oil. A block of steel or aluminum is placed in the pan to regulate the depth that the edge will penetrate into the oil, effectively regulating the amount of the blade to be hardened. I learned this method when I saw Bill Moran demonstrate it at an American Bladesmith Society Hammer-in. After comparing both methods, I have come to prefer edge quenching for the majority of blades that I make. It works especially well on thin, narrow blades. Edge quenching eliminates some of the variables that cause problems when heating the back of the blade.[1]

Torch Hardening

Another method is to heat only the edge, which was popularized in part by Ed Fowler. He wrote about this process in a 1992 article:

[H]eat your blade with an oxy-acetylene torch equipped with a Victor #3 tip. Turn the oxygen and acetylene pressure down to the point that the flame is "quiet," using much less pressure than for welding with the #3 tip. To reduce carbon loss and to keep the flame as "soft" as possible, use a 2x flame (identified by an acetylene feather twice as long as the inner cone). Run just enough pressure to avoid flashback and backfire. HEAT THE BLADE SLOWLY, starting at the ricasso and working slowly and uniformly over the entire lower third of the blade. To AVOID HOT SPOTS, keep the flame moving constantly along the blade. If necessary, practice this step on a piece of scrap until you can "paint" the color into the blade with the dexterity of a master. This is probably the most difficult skill to learn. You have to work at it. Only you will know when this step is performed improperly. This is one aspect of bladesmithing where the knifemaker's ethics are of the utmost importance. It is easy to polish up and sell a blade that is of poor quality, so this is one aspect of quality that is between you and your conscience. Continuously check your blade's temperature with a magnet. As soon as the lower third of the blade is non-magnetic, QUENCH IMMEDIATELY. There is no benefit and a lot to lose by holding the blade at critical temperature for more than several seconds when hardening 5160 and 52100. NEVER HEAT THE TANG prior to quenching; it needs to be tough, not hard. HEAT SLOWLY, HEAT UNIFORMLY, QUENCH PROMPTLY. Submerge the tip of the blade into the oil first, hold for several seconds, then "rock" the rest of the cutting edge into the oil. The homemade quench table controls the depth of the quench. Hold for about seven seconds, then raise the back of the blade thus submerging the tip for a couple of seconds. Then rock back down again to submerge the rest of the cutting edge. Repeat this procedure until the color is completely out of the blade, then submerge the entire blade in the oil below the table and let the blade cool to room temperature in the oil. This slows cooling and makes for a better blade.[2]

As previously discussed (Chapter 37), Fowler repeated this procedure three times. Goddard hypothesized that Fowler's method necessitated three cycles because of the very short soak time he used with the torch[3]. This would be especially true with a steel like 52100, that due to its 1.5% chromium requires longer soak times than a simple carbon steel like 1084 or 1095. Goddard was also asked in 1999 why Fowler would quench only the edge given that only the edge was heated to begin with, and he responded that[4] "the heat often goes up farther on the blade than it appears from the color. When this happens and the whole blade is quenched, there's a loss in flexible strength from it having too much hard edge. The edge quench, properly done to the right depth, guarantees great strength in the finished blade."

Japanese Swords and Hamon

Traditional Japanese swords also contain a "temper line" referred to as the "hamon." These have been produced since around 700 AD. Legend says that the heat treatment technique was developed by swordsmith Amakuni Yasutsuna and his son Amakura after observing soldiers returning from battle with broken swords. Like edge quenched blades discussed above, the transition line is created between the hard "martensite" edge and the soft "pearlite" spine. The change in microstructure is created when the edge is rapidly quenched and the spine is slowly cooled. The transition is visible after appropriate polishing or etching. The traditional Japanese method uses clay applied to the blade, which insulates the areas where it is applied so that they cool slowly, while the edge is left exposed or with a very thin layer of clay. While even a plain, "straight" transition line is appropriately called a hamon, the term is often used in recent years to refer to different shapes and patterns that can be

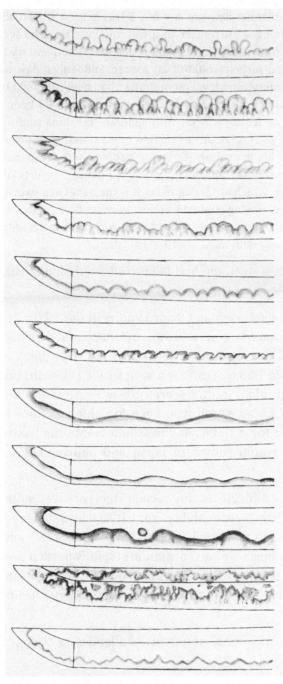

Hamon patterns from Ref. 5.

created based on how the clay is applied. These various patterns are also seen in traditional blades, and you can see schematic representations of some of these shapes in the image on this page.

Hamon in the West

Japanese-style blades and methods have fascinated knifemakers and swordmakers in the West for many decades. In 1975 Francis Boyd claimed to have made the[6] "first traditionally crafted Samurai sword blade ever produced outside of Japan." Boyd began forging by learning from his grandfather[6]: "He used to take me out to his forge - a coal burning forge, I believe - and teach me how to make cold chisels. He told me never to be satisfied with the second best in anything." In high school he read an article about swordsmiths in Japan[6]: "I knew then what I wanted to do. I guess it was a product of my grandfather's advice. I thought the old Japanese 'smiths were the greatest metalworkers who ever lived, and I decided then and there I wanted to be just like them." Boyd experimented with folding steel and making swords but did not make much progress with the traditional methods. "I guess in many ways I owe a great deal to Bernie (Bernard Levine). I met him quite accidentally, and when I learned he really knew something about edged weapons, I decided to show him my first efforts, and to let him in on my dream."[6] Levine told him about Nakajima Muneyoshi, a sword polisher and fittings maker in San Francisco, who would also teach swordsmith Michael Bell. Boyd said it was difficult to convince Nakajima that he was worthy of being taught swordmaking methods:

Meeting Mr. Nakajima was the real turning point in my life. I was scared stiff, really prepared for the worst. The worst happened. He refused to speak with me. I think he even threw me out the door. I don't know how many times I came back to see him and try to talk with him, but every time I returned, he just wouldn't say a word. Suddenly it dawned on me. He was trying to teach me humility, which was something I had never thought of. I mean, I thought I knew all the answers. As soon as I realized what a fool I'd been, I returned to him again, and this time he let me stay on. Although Mr. Nakajima confines himself to his specialties, I was able to receive an invaluable understanding of sword construction by watching him make his appraisals. So many people think a Japanese appraiser looks only at the signature. I'm certain that some of the phony "experts" do nothing else, but to a Master Appraiser, which Mr. Nakajima is, it is the detail of a sword's construction that makes the difference. Having studied so many thousands of swords for so many years, Mr. Nakajima is able to tell the provenance of a sword by noting details of shape, grain, and temper. This does away with most questions of fakery.[6]

Boyd used 1095 steel and some historical steel, and his blades contained a hamon like the traditional swords.

Some blades by USA and European knifemakers and swordsmiths in the 1970s and 1980s would be heat treated to have a hamon, almost always in Japanese style blades and swords. In 1980, there was a demonstration of traditional swordmaking techniques by Japanese swordsmiths in Dallas, Texas at a series called "Celebration of Japanese Arts."[7] Some American bladesmiths were present including Don Hastings[7]. Kuzan Oda, a Japanese-born bladesmith who lived in California for ~19 years made tanto blades with steel he forged himself and would create the hamon using traditional clay methods[8]. He would return to Japan and apprentice as a swordsmith[9].

Developing a nice hamon is notoriously difficult, as only certain steels are well suited to the technique. The steel must have low "hardenability" so that a clay-coated spine leads to sufficiently slow cooling for soft steel and a sharp transition. Steels with significant amounts of elements like manganese and chromium still fully harden even with the clay coating. But information on the exact process of producing a hamon and what steels are well suited to the process were in short supply in the 1970s, 1980s, and even 1990s and into the 2000s.

Bob Engnath, in an article for *Knives '86,* wrote about his challenges in trying to reproduce the technique:

Back about 1980, there was an article in a knife magazine about a contemporary knifemaker producing Japanese-style blades with a hardened, tempered edge and soft back. It was one of those infectious pieces that makes one say, "Hmm, all they did was put clay on the blade and scrape some off to make the edge pattern." That was the inside information from the "expert" on Japanese swords. This is the same guy who later suggested old smiths had quenched their blades in urine. In the back of my mind was the little voice saying, "If it's so easy, why aren't they all doing it?" It may have been wiser to listen to that voice, but you have already figured out I didn't. I couldn't... It didn't take long to rough out a couple of crude tantos in O1 steel, once back at the shop. Somewhere, an old samurai's favorite swordmaker was spinning like a top in his grave. About 4 ounces of powdered earthenware mixed with water made a cup of slurry, half-brushed, half-spatulaed onto the blade about 1/4-inch thick. It was thrown into the oven to dry. Half an hour later, out of the oven, the clay is full of cracks. If that isn't bad enough, it falls off in scabby chunks... So, research begins, but everything printed (in English) is sort of vague or general without any of the necessary details. There are a few contemporary Japanese swordsmiths, but they aren't giving away the inside details. It's so discouraging things stall for a while. A fellow sent me 15 pages on how-to, but it's all in very technical Japanese. One thing is very consistent. Any book with worthwhile photos or information is always expensive.

Bob Engnath

Early 1984, and there's a new clay formula to play with... It's not a simple mud clay. If you ever want to give your local chemical supply house a good laugh, just call and ask them for a couple items like straw ash or powdered grindstone ... by now Mike Bell is ready to start leaving his phone off the hook. I'd been calling him twice a week... The clay goes on smoothly, but an hour later, it's crazed like alligator leather ... when it hits the water, the whole works falls off. My teenage son says, "You haven't got the right stuff, Dad." ... Mike Bell spent 5 years in a traditional Japanese apprenticeship for sword work... I sent him a sample of the clay, and he agrees, it's not the right stuff. I didn't know Mike was acquainted with my son.

A revelation at the '84 Kansas City show: Louis Mills very generously offered his tempering clay formula and in it was something familiar, CMC Compound or "gum" as they call it in the ceramics trade... After looking at Louis' blades, it's clear that he definitely knows something about the subject... It shrinks, but patches up when cracks develop and stays on fairly well through the quench. There is an honest-to-goodness hard edge on my O1 blade. True, it's a pretty faint, shaky line that sort of wavers around, and not exactly beautiful, but the Rockwell is Rc 64 and the rest of the blade is Rc 45. Tom McClain, who shared a lot of information with me at Kansas City comes to the rescue at the Anaheim show. The O1 I'm using isn't the right stuff, he says (He must know my kid too.) What's needed is a very simple steel, no chrome, moly, or vanadium which all make steels deep hardening. Use steel with 0.25 percent or less of those items. Go with W1, W2, or a 10 series steel from 1055 to 1095... "Experiment with whatever you have available in the way of different alloys," Tom tells me. If the tempered edge stays fairly straight, no matter how crazy you do the clay design, it's a good clue that the alloy is too deep hardening. Felt so grateful I darn near kissed him on both cheeks, beard and all. With this encouragement I order about 100 pounds of W1 and try a few test blocks, starting with a batch of 10 which crack right through the hardened edge, every one of them... Next try, the W1 quenched successfully in transmission oil which worked very nicely but left a very plain, straight temper line, even with crazy clay designs. Not bad but not quite what I was after... There was some satisfaction when that first hard edge happened, but that was nothing compared to the solid glow that having half a dozen hard edges with beautiful temper lines show up after you scrape the clay off.

Now all of the proceeding may sound very scholarly and may even create the illusion the author knows what he's talking about, but I cannot begin to compare my achievements to those of the traditional smiths... Without the generous cooperation of friends like Jim Sornberger, Louis Mills, Mike Bell, Tom McClain, Sean McWilliams, Jody Sampson (sic), Bob Lum, and the mildly demented Jim Hrisoulas, who should all be awarded at least a dragon tail and ears, my modest success at this fascinating area of knifemaking would still be a bunch of cracked blades in the scrap bucket.[10]

Closeup of Bob Engnath sword with hamon.

Don Fogg, Don Hanson, and W2

In the mid-to-late 2000s, hamon blades became more and more common, and in "Western" style blades in addition to Japanese styles like a tanto or sword. In 1996, Jerry Fisk wrote about Rob Hudson's method of hardening to develop a hamon using clay[11], and Hudson himself wrote about the method several months later[12]. Both articles showed images of Rob Hudson hardening a Bowie. By 2006, there was a minor movement and growing popularity of hamon blades[13] from knifemakers like

Don Hanson and Nick Wheeler, though Hanson credited Don Fogg with generating its recent popularity. Fogg told me that in that time he had begun to be more interested in various possible patterns from the hamon than even his prior Damascus work[14].

Fogg talked about his history with the hamon in 2010:

In 1988, the Boston Fine Arts Museum hosted a demonstration and exhibition of Japanese living national treasure swordsmith, Gassan Sadaichi. Museum officials built a pavilion in the courtyard that housed his forge and workshop. There was a retinue of apprentices who worked continuously, but the real excitement came when Gassan, already an older gentleman, personally worked the steel. I was particularly interested in his forge welding a billet of steel. I had been making Damascus for nearly 10 years at the time and was able to appreciate what he was demonstrating. There was a good crowd in attendance, including local television and press. Among the privileged observers was noted metallurgist and historian, Dr. Cyril Stanley Smith, of the Massachusetts Institute of Technology. Gassan was not only revered in his own country as a living national treasure, his work and the tradition that he carried on was so significant that the museum was honored with his presence and spared no expense to make his visit memorable. My appreciation was born when I saw a sword of Gassan's on display. It was mounted in a glass case so that you could walk all the way around the piece at eye level. It was perfectly polished and lighted so you could see all the detail of the "hamon," the hardening line sought by smiths and collectors alike. In that moment, I realized the awesome power and beauty that could be achieved. I was stunned. It was as if the blade had been painted with crystals; the curve perfect, the lines elegant. In one complete expression, the blade displayed power and intent as well as harmony and beauty. It showed respect for life and the dignity of purpose.

This process can be applied to modern work with great effect and, in recent years, many top makers have experimented with and been captivated by the beauty of the hamon. There are differences in tools, materials, and techniques from the traditional approach, but still, the process of creative heat treatment has added a whole new dimension to the visual power and beauty of modern work... Bob Engnath, who was a pioneer swordmaker in the USA, created many katana blades using an air-setting, high-temperature, premixed mortar cement called #36 from Harbison Walker Refractories. The cement had the annoying problem of puffing up in the fire and popping off. Bob got around the problem by wrapping the blade with iron wire to hold the clay in place, a practical if not elegant solution. It worked well for him, though, since he hardened hundreds of swords that way.

I was first shown the technique by Jimmy Fikes, and we used another Harbison Walker product called Satanite... If the blade is properly prepared, the clay will stick through the quench and can be easily scraped off afterward... There are many options and combinations that are possible and, to be honest, they seem to change from blade to blade. The object is to get the best look that you can from the steel... The finish is fragile and those who appreciate the work have learned to respect its subtlety... The blade, held at the right angle, in the right light, jumps to life. Like snowflakes, no two are alike; the whimsy of fire and steel ensure that you will be surprised, sometimes wonderfully, by the result... For those who love steel blades, the hamon is a wonderful aesthetic addition. It is a visual record of the heat treatment, revealing some of the mystery of its making. The process is complex and demonstrates mastery at its highest level, yet it is still capricious and wonderfully organic. In its time, it was a significant advancement in edged weaponry and still has a place today, though with modern steels and heat treatments, there are other options. In this era, use of the hamon is an exciting development that adds to the visual language of the modern blade and enjoyment to those who have learned to appreciate it.[15]

In 2006 Hanson talked about the hamon:

> Hamons are like fingerprints - no two are alike and beauty is in the eye of the beholder. I personally like the more active hamon with wispy or smoky fingers dropping toward the edge, as do most collectors. A properly executed hamon will be beautiful, hard to miss. An improperly executed hamon will be unattractive and possibly drop down too close to the cutting edge. With the hamon, what you see is what you get. If it's beautiful, it was executed properly.[13]

Don Hanson

By 2012, the hamon appeared on many knives, by a range of knifemakers in various styles[16]. Hanson further popularized the hamon by becoming a source for W2 tool steel. W2, with its high carbon and low manganese, has very low hardenability and thus is suitable for developing a hamon. The small vanadium addition to W2 (as opposed to W1) further refines the grain size, somewhat reducing hardenability. I asked Hanson about what led to his selling of W2:

> W2 really changed my knifemaking. I don't know exactly why but the steel just performs so well. It might just be this old batch I came across, but it has very high hardness in combination with good toughness. Don Fogg was doing hamons before me; he was the man. There were a few doing them, but he was the prominent one. W2 wasn't widely used. I started using it around 2004 and it was getting hard to find because the manufacturers had just about quit making it. The industry wanted to move to air hardening steels. You would occasionally see somebody post something on the forums or *Blade Magazine* or something with W2, but it wasn't common. Russ Andrews was using some W2, and Nick Wheeler was using some before me. I found a big source of it and bought like 25,000 pounds. I couldn't afford to buy it all at once, but I bought about 3,000 pounds at a time. I was trying to keep the source a secret so I could buy it all. I saw the potential in the market there.

> I actually posted blades on Blade Forums and Don Fogg's forum. There was a lot of interest in the steel, and I started selling a little of it. I pretty much stopped making Damascus, or at least not near as much. I sold a little over 20,000 pounds in about 7-8 years. Putting them in 40-pound boxes, cutting them to length, and shipping through the tiny post office. It was kind of like a diversion from making knives a little bit. I had my son and my wife both helping. We were shipping an average of about 4 boxes per day. I limited it to that because the post office would complain about me bringing 8-10 boxes per day. The postmaster wasn't the problem, it was the guy that picked up the mail from her... This was a big deal for me. It got my name out there more and it connected me with W2 and hamons and it put the spotlight on me. I was getting more sales, more collectors asking for knives made with it.[17]

Don Hanson knife with hamon. Photo by Coop.

-39-

SPECIALTY KNIFE COMPANIES

There was a wave of new, small knife companies that started in the 1980s and the 1990s. These companies were spiritual successors to startups like Al Mar. And while Al Mar was dependent on Japanese manufacturing, many of the new knife companies increasingly turned to domestic USA production. Small knife companies that focused on innovation and high-end features, rather than price, could push boundaries and remain at the forefront of knives, as opposed to old, large companies focused on traditional patterns and the same 440A and simple carbon steels. These smaller knife companies were also greatly influenced by custom knives, as Sal Glesser said in 1998[1]: "Factory knives today are significantly influenced by custom makers ... the result has been an overall general quality improvement in factory knives." The knife consumer was also increasingly educated on new products, features, materials, and designs. This was the result of knife magazines, an increasing number of knifemakers and knife companies, more knife shows, more knife books, more knifemaking schools, and the Internet, which was on the rise in popularity during the 1990s. Les de Asis of Benchmade said in 1998[1]: "The hunger or demand by customers for product is much more advanced. Anybody can make a slick-looking knife out of 440-A, but if it doesn't hold an edge, people will go for another knife."

Spyderco

Spyderco founder Sal Glesser told the story of his company in 1998:

> Spyderco is a relatively new company. My wife, Gail, and I started the company in 1976, when we were living in Redding, California. That's where I first met Jess Horn (custom knifemaker). I've always been a knife hobbyist and when I heard that there was this part-time postal worker that was making these unusual knives, we just had to go check him out. And after seeing some of his knives, I decided at that time that (reproducing handmade designs) would probably be in my future somewhere. Late in 1978, we developed our Tri-Angle sharpener, and we relocated to Golden, Colorado. We'd been making sharpeners for a while when I came up with the idea of trying to produce or design a one-hand opener. I was one of those guys that learned to open a knife with one hand. By using your thumb to rotate it, it was possible to open a knife with one hand. We played

around with all the appendages you could add to a knife blade - studs and discs, and all those kinds of things to grab on to open it. That got pretty frustrating because everything we put on got in the way of cutting. So we tried to go the other way. We decided to take something away from the blade. We started off by roughing up the side of the blade so you could get a purchase using your thumb just with traction. That became a dent; a dent on both sides on a knife blade became a hole. Eventually, it became a round hole because we felt that worked best for our purposes. The clip provided easy access, something that you could get to right away, and the serrated edge was something, believe it or not, that came out of the old Ginsu knife. That was our Original Clipit.[2]

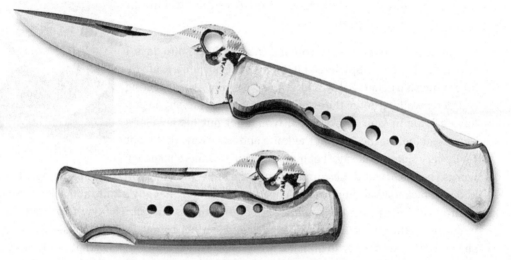

Prototype of the Original Spyderco Clipit Worker.

The original pocket clip was inspired by an unusual source, as Glesser said[3]: "I had a keychain that an artist had made some years before. It was a little frog that sat at the top of the pocket, and the keys attached to a chain and a ring were connected to the frog. I found it convenient to be able to access my keys by pulling on that frog, and I thought that would be a better solution for a pocket knife than a sheath on a belt or the knife itself in the bottom of a pocket." The first Clipit knife was later named the Worker; the first knife with a pocket clip and one-handed opening was released in 1981. And in 1982 came the Mariner, which had a serrated edge. However, Glesser said that sales were slow to build:

Sal Glesser

A lot of people thought it was an ugly knife. Retailers couldn't see any reason why they would want to put it in their store. So we actually downplayed it and sold these Clipits for years out on the show circuit before any dealers would even give it a shot. By 1984 we started to develop a little bit of a dealer base. When we decided to make the knife originally, we didn't know as much about knives as we do now. We're a pretty tough industry to break into. I had a friend in the industry; his name was Al Mar. Al didn't think much of our Clipit, at first. But I can now tell you there would never have been a Clipit if it had not been for Al's help. We introduced the first knife in 1981. It was about 1984 before the dealers started working with it and the media was not much different. The magazines didn't think it was a real knife. But in about 1985, there was an article done on river-rafting knives and that was the first time that Spyderco had actually been noted. The author talked about the cutting power of our serrations and easy access and our clip. He did mention that it was unorthodox in appearance. But it did help us get a good start.[2]

Sal Glesser would do live demonstrations of his Tri-Angle sharpener to build sales. Knifemaker Bob Terzuola told about a time he observed Sal in the 1980s:

Robert Terzuola

> Sal was right across the aisle from me and, for three solid days, I was subjected to the Spyderco pitch - how to sharpen a knife on a ceramic sharpener. Sal would grab some poor, unsuspecting character walking down the aisle and say, "Excuse me, sir, do you have a knife in your pocket?" Sal would get out his jeweler's loupe and look at the edge of the knife and say, "Boy, that's not really very sharp" and say, "Let me show you how this sharpener works," and he'd stand there and sharpen the knife. (Sal) had this pile of paper next to him and he'd pick up this slim piece of paper and he'd take that guy's knife and slice it and cut it into neat little paper dolls and circles and triangles and all. The guy's eyes would be wide and, in the meantime, other people would gather round. Sal sold a boatload of those sharpeners... I went over to him and introduced myself and said, "Sal, that really is a very impressive demonstration," and Sal said, "You know, the sharpener is really good. Otherwise, I wouldn't be selling it. But it doesn't matter how sharp the knife is. I can (cut paper) with any knife on the table... You've got to keep your eye on the bouncing ball and know what you're selling." ... His passion has always been for performance. He's always told me, whether it's in a car, or a knife, or in a person that's working for you, it's performance that you're looking for. But more than that, Sal has always ... cared for the customer. And it's a lesson that I learned over a long period of time that the customer, the person who buys (knives) and puts down his hard-earned money, expects in return integrity and expects a product that's going to work and a company that's going to back it up. And Sal has shown this in the way Spyderco works, in its success, in its performance, in its growth, and its innovations and its constant quality.[4]

Terzuola is the knifemaker typically credited with creating the modern tactical folding knife, with his ATCF (Advanced Technology Combat Folder), introduced in 1986, often considered the first. However, as with any "first," there are many evolutions on the way there; for example, Sal claims his Clipit Worker in 1981 could be considered the first tactical folder with its one-handed opening and pocket clip. Terzuola says that his ATCF was partially inspired by Al Mar's SERE (Survival-Evasion-Resistance-Escape) folder released in 1984-1985, named after the military training program. For more on the history of the tactical folder see Ref. 5. However, according to Terzuola the SERE had room for improvement:

[The Al Mar SERE] wasn't openable with one hand and it didn't have a clip on it. I learned from Michael [Walker] how to do the liner lock so the ATCF could be closed with one hand. And the thumb disc meant it could be opened one handed as well. Most folders being made were gentleman's pocket knives... [Custom makers] were making expensive one of a kind pocket jewelry, basically. They were engraved, using stone and silver, etc... They were making Maseratis and Ferraris and I was making Ford pickup trucks. They were designed for people who would be using them. Many times at shows I see collectors coming by and they would tend to favor the more expensive knives; museum quality pieces. Quite often I would see my knife in their pocket. The other ones would be part of their display collection.[6]

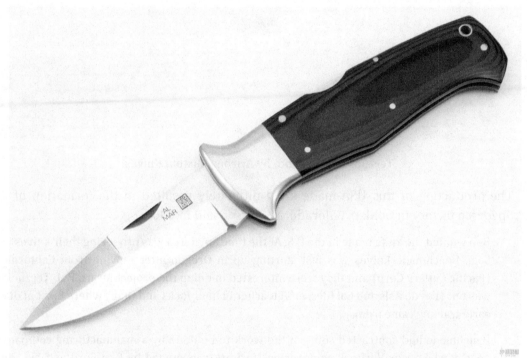

Al Mar SERE circa 1980s. Photo from Arizona Custom Knives.

Spyderco expanded beyond Glesser designs with collaborations. Sal Glesser said[3]: "The first collaboration was with Bob Terzuola. I was a member of the Knifemaker's Guild, and I used to see Bob at the Guild shows and several other shows that we worked together. He used to talk about how my knives were really ugly and how he could do a much better job designing one. I told him if he could, I would make it. Well, he did, and I made it." The Terzuola-designed Spyderco C15 was released in 1990 and had many "firsts": 1) the first Spyderco custom collaboration, 2) the first USA-made Spyderco, 3) the first production USA-made knife in ATS-34 steel, 4) the first production USA made knife with a liner lock, 5) the first production folder made with G-10, and 6) the first all-screw construction knife by Spyderco. Many of these firsts resulted from requests from Terzuola, who said[6], "Nobody could do that in the USA. At the time all of the knives were slipjoints or lockbacks and riveted together."

Terzuola ATCF. Photo by Arizona Custom Knives.

The production of this USA-made knife ultimately resulted in the formation of a Spyderco factory in Golden, Colorado, as Glesser said in 2006:

> Bob wanted the knife made in the U.S. At the time Spyderco was producing their knives in Seki. Benchmade knives was just starting up in Oregon after moving from California (Pacific Cutlery Corp) and they were interested in doing the project for us. Bob Terzuola was our (Les de Asis and Sal Glesser's) teacher in liner locks and that's where most of our early training came from.

> Benchmade had contracted some of the work to be done by a manufacturing company nearby which was Varitork engineering. Varitork was owned by Ron Ford and his son Vince Ford. They had two main machinists working with them, Craig Green and J.J. Grosmick. Ron and Vince Ford decided that they liked making knives more than the medical equipment that they were making and they felt that they could be good at it so they contacted a distributor on the east coast called Catoctin Cutlery that was run by Wayne Ramsburg. Working with Wayne they developed and built a few knives under the name of Koncept Knives and Catoctin was the exclusive distributor for Koncept Knives.

> Western Cutlery had gone out of business and they were having an auction for their equipment. Ron and Vince Ford were in Colorado looking for equipment approached me while at that auction and said that they were already doing some work for us through another company and that they would like to continue to do more work with us in the subcontracting of knife production. After a while working with them we decided to join forces and Spyderco relocated Varitork with their equipment to Golden, Colorado and formed Golden Manufacturing which was a corporation owned by Ron Ford, Vince Ford and Spyderco. Craig Green and J.J. Grosmick relocated with them and we began producing

knives in Golden about 1993. Shortly afterwards Craig Green decided to return to the Portland area for personal reasons. Ron and Vince Ford decided that they would stop making Koncept Knives because they were busy making Spyderco knives and there really wasn't enough time for both. So Koncept knives stopped production at that time.

A few years later Ron Ford decided to retire and return to the Pacific Northwest. Vince Ford and Spyderco purchased his part of the business and continued the manufacturing company. A short while after that Golden Manufacturing ran into a few problems and Spyderco purchased the remaining stock in the company, changed the name to Spyderco Manufacturing and Vince Ford came on board in Spyderco's R&D department as our chief engineer. As time went on the factory was relocated into the Spyderco building where it remains today.

I might add that both Ron and Vince were very bright individuals capable of manufacturing just about anything. Vince Ford was one of the few people I'd met that could engineer a product, go down into the shop, program all the machines and actually build it, a very unique ability in just about any industry.

At about 2000 Vince Ford decided to relocate back to the Pacific Northwest as his family missed the area and their family and wanted to return. Vince went to work with Tim Wegner at Blade-Tech making molds for Tim's sheaths and helping launch Blade-Tech's domestic knife production where he is still working today. Craig Green, another gifted machinist, is the genius behind Kershaw's manufacturing plant in Oregon and he launched the Ken Onion Speed Safe products from their Oregon facility. J.J. Grosmick is still working for Spyderco in our R&D department as a very skilled custom knifemaker. He makes all of the prototypes for our new models and helps develop the new concepts such as the locks that we work with. Mr. Ron Ford passed away several years ago but everybody else involved in the original production of Koncept knives is still in the knife industry in one area or another.[7]

Sal Glesser said that collaborations were important for Spyderco going forward:

Spyderco has gotten heavily involved in collaborations... I think this is a great thing. Actually, you end up with a win-win situation for all. The factories get better designs and they add new processes and high-tech materials. And certainly the custom knifemakers have raised the quality standards of factory knives 400-500 percent in the last 10 years. The custom makers benefit. But the real benefit comes to the end-line user. The end-line user has a bigger variety, better quality, and much more. This cooperation gives you the group powers that are just not available individually. The end-line user is a pretty important part of the equation. If you take the user out of what we're doing here, then this becomes an empty room. Everyone here is either producing a product or service for the end-line user. The end-line user oftentimes is forgotten. They're the most important part of the equation. The guy that cuts with a knife. Everybody else should be less important than that guy. That's the guy that you have to remember.[2]

The Spyderco factory opened up many possibilities for Spyderco, as Steve Jordal said in a 1999 article, then the marketing manager for Spyderco:

We've had the Golden factory up and running for about five years. It started with more of a machine-shop feel, but we're developing it into a small-to-moderate-run production facility. We like to do the tricky stuff here... We'll never get into the high-production realm of a Camillus or Buck Knives. Sal's a hobbyist at heart and prefers small-to-medium, yet high-quality production runs... Sal's crazy. He'll make up 1,200 Military models with a different steel just to try it. If it holds up, he'll have a new product, and I'll scramble to get the press releases done in a hurry. By the time I'm done, we're using yet another new steel, and I have to turn on a dime. In fact, the whole company is able to turn on a dime and roll with the changes. Hi-tech steels are our identity. Once end-line users get some experience under their belts, there's no fooling them. They know that any Spyderco knife is a real performance knife and built with the latest steel we can get our hands on.[8]

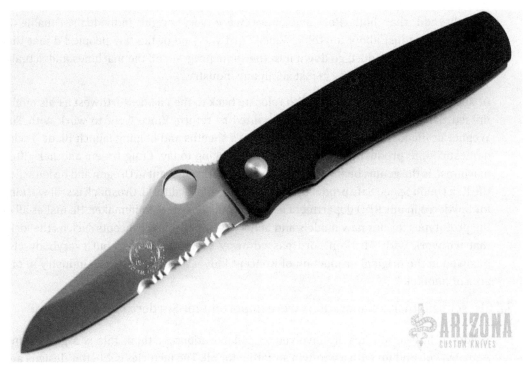

Spyderco Terzuola C15. Photo by Arizona Custom Knives.

Spyderco has continued to innovate with steel, introducing the first knives in VG-10 in 1997 with the Japanese-made fixed blade knife designed by Bill Moran[9]. VG-10 has become very popular in Japanese-made blades, especially in chefs' knives and Spyderco's Japan-produced models. It is somewhat surprising that such a popular stainless steel was first introduced in a Bill Moran knife, a bladesmith who preferred carbon steels. Spyderco would become the first knife factory to produce knives in CPM steels with CPM-440V (which will be discussed more in the next chapter). They have continued to build a reputation for using a wide range of new and high-tech steels. They have used over 80 different steel types since the company started. Spyderco would introduce a line of knives dedicated to allowing users to try new steels called the Mule Team Project in 2008, a simple fixed-blade design with no handle scales for a relatively low cost[10]. In the information about the new line of knives, Spyderco said[10]: "Knife and steel-devotees love trying out new blade steels. Spyderco leads the industry in offering, testing, and manufacturing knives using new and exotic blade steels. Taking that idea a step up, why not offer a series (team) of blades using different, exotic, and freshly unveiled steels? In doing this steel-obsessed knife knuts can try, test and use something normally not offered to the industry."

Benchmade

Les de Asis

Benchmade started as Bali-Song, Inc. in 1980, formed by Les de Asis in Los Angeles, CA, offering custom-made butterfly knives. They would change their name to Pacific Cutlery by 1983, and also won "Best American Made Knife Design" at that year's Blade Show[11]. Bali-Song and Pacific Cutlery would popularize the modern butterfly knife. In those early years, the blades were ground by custom knifemaker Jody Samson in 154CM steel, though this was changed to 440C by 1983[12,13]. The company would continue to offer custom Jody Samson-made knives until he left in 1994. With the change of name to Pacific Cutlery they also introduced a new knife:[13] "All of the Bali-Song experience in building extremely high quality knives has been utilized in the design and creation of the Model 68, the first true Bali-Song production knife. We scoured the earth for the perfect blade steel for this knife. We required outstanding edge-holding characteristics, rust resistance, and maximum durability. Swedish metallurgists furnished us a state-of-the-art surgical stainless steel which we make available to you, ready for all your cutting needs at home, at work, or in the field." It was also advertised as being 57-59 Rc. Presumably, this was made of Sandvik 12C27 steel since they would advertise that steel in a line of 1991 folding knives[14]. In 1993 they advertised some butterfly knife models in Sandvik stainless as well as others in Uddeholm stainless, presumably AEB-L[15]. Les de Asis would change the name of the company to Benchmade in 1987, and move from California to Clackamas, Oregon in 1990, and then to a larger manufacturing facility in Oregon City in 1996. While many of their production knives were imported, by 1987 they were working on reducing their imports and increasing their USA-made knives[16]. In 1989 Benchmade introduced a USA-made version of their large fixed-blade Bushmaster[17]. In 1990 they would begin offering some butterfly models in USA-made 420HC and 425M[18].

Les de Asis said in 1997 that his experience making knives himself in the early days of his company helped teach him to build his company:

> Personally, I've put more than 5,000 blades through a bandsaw. I've been a handmade knifemaker and knife builder. That makes me rather unique in the industry... All quality products require tight tolerances. We asked our vendors to meet high standards. But in

the early days, many of them looked at our product as "just a knife" and didn't meet our standards. So we started producing our own components and our own knives. Three laser-cutting machines now give us the tight tolerances we need. We started to pursue a quality product like we had been getting from Seki City, Japan... Every knife is a question of balance. How much should it "walk and talk?" How tough should the blade be? How hard does the steel need to be for the job intended? What kind of steel should we use? We found a lot of answers from the craftsmen of The Knifemakers' Guild. American ingenuity is awesome to behold. And as a result of our contacts with these handmade makers, collaborations are a large part of what we do. Quite frankly, keeping up with the creativity and innovations of the custom makers is what helps us advance. As competition heats up, only those manufacturers linked to the drive of the custom maker will succeed. The modern knife industry owes everything to custom makers. Most do not acknowledge that debt.[19]

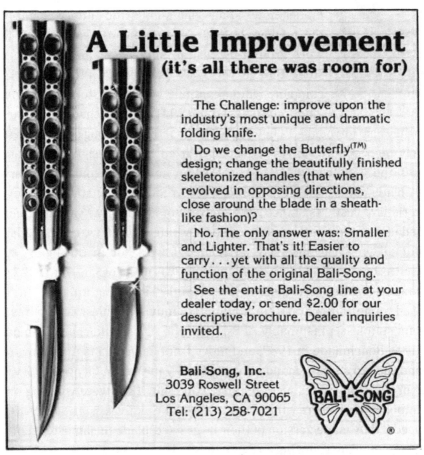

1983 ad for Bali-Song, Inc. butterfly knives.

Les de Asis focused on using state-of-the-art equipment to stay nimble, as he said in a 2000 article:

We are a CNC-controlled, precision-small-parts manufacturer operating in a flexible manufacturing environment. We can change a product at a whim, get up and going and sell it. We have complete control over the product from concept to delivery. We are virtually able to design, engineer, and manufacture a knife in a solid modeling environment. We can program our machines to duplicate and build a knife without any subsequent heavy tooling. We can make molds. Right now we are investigating outside parts that can be sourced from other manufacturing presses... We're all knife nuts here. That's the difference and that is also one of the penalties. If you're into knives, that doesn't necessarily mean you're a great businessman... We talk regularly with our customer base. We are able to really eliminate any great rift between our customer base and ourselves. We grow by being heavy on innovation and the engineering side... We really concentrate on building fun product, cool product, stuff that's innovative, exciting. We like using materials that are a little tougher to use. Each project we take on seems to be a little more difficult. If we successfully tackle increasing difficulty in tolerance, finish, and materials, it stands to reason that we are able to migrate the business forward.[20]

Microtech

Microtech was started by Anthony and Susan Marfione in 1993. Tony talked about the company's formation in 2000[21]: "My wife, Susan, and I started the company about 6-1/2 years ago. We were living in an apartment complex and had a drill press in the living room and a Burr King grinder on the back porch." They added a storage shed and then nearby buildings and created the name Microtech in 1994[22]. Tony Marfione talked about the impetus behind Microtech:

Al Mar came down to a Soldier of Fortune convention in Orlando, and he brought well over a hundred knives. All the guys I went to that show with bought knives, and I couldn't afford one. I was so upset that I couldn't buy one. The guys teased me about it and I said, "Guess what? I'll start making knives myself and you guys can take a walk." That sparked the idea.[21]

Marfione designed and built a prototype of his first knife, the UDT (Underwater Demolition Team): "That was the first knife we made and I still have it. You can see by the different extremes we've come from what a passion will do to you."[21] Marfione took the UDT to a Knifemakers' Guild Show and Les de Asis introduced him to several dealers and distributors. Microtech would grow rapidly and win Blade Manufacturing Quality awards in 1998 and 1999. Tony Marfione said:

Those awards are judged by our competition and our peers. Just knowing that our competition thinks that much of us and the people who voted appreciated what we do chokes me up. When we brought each of the awards back, we presented them to our employees at a banquet where we called each of them up and awarded them a special-run knife. The people who make up Microtech really made it happen. We also sent a letter to each of our suppliers and vendors to thank them for being a part of our success. The awards mean a great deal to us.[21]

Microtech has used many different steels, from Crucible's latest to stainless Damascus. Some of these steel choices will be pointed out in upcoming chapters.

Microtech made its name on high-quality knives, automatic folders, and aggressive lines in design. Marfione said in 2000 about the Vector model:

> To hit the masses, I feel you have to give them something cosmetically different. The blade style is freaky looking. The Vector has aluminum back spacers in different colors, which adds to the aesthetic quality... It's the little stuff, like the clean, proportionate grooves cut into the blade of the Vector. Makers are putting effort and quality into blades, and quality often equals appearance. I see radical improvements in just about every manufacturer's knife line available.[23]

Ernest Emerson

Emerson made his first knives in 1978, went full-time as a knifemaker in the 1980s, and then started Emerson Knives Inc. in 1996 to shift from custom knives to production[24]. Emerson talked about his history in a 2015 interview:

> I made my first knife in the mid-1970s when I was a student at what was then the most recognized and only, full contact fighting school in the world, the Filipino KALI Academy run by Bruce Lee's best friends and most senior Jeet Kune Do instructors, Dan Inosanto and Richard Bustillo. Part of our training curriculum included both edged weapon offensive and defensive skills. And because both Dan and Richard were Filipino, I was introduced to the Balisong (Butterfly) knife. Being a "starving student" with barely enough money to pay my monthly dues at the school I could not afford to buy one of the butterfly knives... So I asked Instructor Richard Bustillo if I could borrow his personal knife to see if I could build one. So he loaned me his and I built one with some aluminum, steel, a butane torch, and a Sears hand drill. When I brought my Balisong into school, I quickly found out that there were other "starving students" who also could not afford to buy one. So they asked me if I would build some for them. I did, just for the cost of the materials and one thing led to another, and you might say I became a knifemaker by necessity.

During the late 1980s and early 1990s I was the lead Hand-to-Hand Combat instructor for a company named Global Studies Group International (GSGI) run by a group of U.S. Navy SEALs who were former members of SEAL Team Six. By that time, my knives had become well known among the Naval Special Warfare Community, Army Special Forces (Delta), The British SAS, and various other Government agencies... This knife, the CQC-6 (introduced in 1989), became so sought after in the Special Operations community that owning an Emerson CQC-6 became a symbol of a "Made Man" in the SPEC OPS community and carried tremendous bragging rights among the operators that carried them... Benchmade Knives approached me to make a production version of the knife that I was building for all these world's elite operators, the Emerson CQC-6, the true father of the modern tactical folding knife. I told them no, but I had another design, the CQC-7, and that knife went on to become the most popular and sought-after tactical folder of all time.[24]

Emerson CQC-6. Photo by Arizona Custom Knives.

Emerson also talked about the start of his production knife company:

> I had gotten to the point in my knife making career where the demand for Ernest Emerson handmade custom knives was becoming overwhelming along with a decade-long waiting list for orders. At the same time, I had licensed several knife designs to production companies that were extremely successful. The Benchmade Emerson CQC-7 Knife was the #1 selling tactical knife in the world. I had many new designs that needed to be built and we knew that there was a worldwide demand for Emerson knives. So my wife Mary and I had a very long discussion and we decided to start a production knife company. She is the driving force behind the decision to start the company and she ran the entire business by herself for many years, which allowed me to do what I do best - make knives.

> I'm all about family and I'm all about American business. Family-owned small businesses are the life's blood that courses through the veins of this country. Small businesses are what built this country and they are the backbone of our economy... I am proud that we are doing our small part to help Americans and the American Economy. In fact, all the materials and parts of an Emerson Knife, right down to the last screw come from American companies and businesses.[24]

154CM Again

While 154CM was the first steel Bob Loveless introduced to the knife industry, the identical ATS-34 had taken over most of the market due to its better price and available sizes. In the late 1990s, Tony Marfione of Microtech contacted Crucible Steel about this lack of USA-made 154CM:

Hitachi had cornered the market in ATS-34, and small manufacturers were having a hard time getting it. We had to pay up front, and sometimes delivery might take as long as nine months. I just hate to think of American industry at the mercy of a foreign company dangling a carrot over our heads. I told Crucible that we weren't their biggest account, but that if we started using 154CM, others might follow suit. They are a very receptive company and good things are coming from it.[21]

Microtech and Emerson Knives began using the new 154CM immediately, and many others, such as Benchmade, followed shortly after. Ernest Emerson said:

All of our production blades across the board are 154CM. We used ATS-34 because it was available in sheets for mass production. Once 154CM was available in sheets, we switched to it because it's a clean steel that grinds nice, machines nice, and makes a nice knife. A deciding factor for us is that it's made in the USA. We basically do everything in-house. It's important to customers. You have to remember, knifemakers are pretty much apple pie and American flag. It's middle America, man.[25]

Microtech Kestrel in 154CM steel marked June 1999. BladeHQ photo.

Ed Severson, a Crucible Steel metallurgist, said in 1999:

We were kind of out of it and left it up to the distributors for a while. The market has changed now, and a lot more people are using full sheets of steel and laser cutting. That's the commercial guys, and looking back 10 years ago we were producing almost all bar stock in 154CM. Producing bar stock required making a pretty good bit of the stuff, and a minimum run meant quite a bit of weight. We do have material on the floor now, and if a custom maker with a shop in his garage wants to buy half a sheet, it's no problem. We can do that.[26]

As a general purpose stainless, you can't go wrong. Although CPM 420V and 440V are more wear resistant, they're also more expensive... Made in the USA doesn't hurt. We have customers who like the fact that 154CM is completely produced in the states. It's melted and rolled right here in New York state... We don't know what the overall market for ATS-34 is, and we don't care. We offer 154CM, and we haven't advertised it in any publications, yet it's doing well. The volume of knifemakers, not the quantity each is ordering, makes 154CM a small but important part of our business. Each maker might only order 10-50 pounds a year, but you have a large number of people ordering.[25]

KAI - Kershaw, Zero Tolerance, and Ken Onion

Ken Onion

Kershaw saw a significant shift in its company in 1997 when it opened a factory in Wilsonville, Oregon, and then a larger facility in 2003 in Tualatin, Oregon. The company began offering many USA-made models, pushing boundaries in many areas. In 1998 they won the Blade American Made Knife of the Year with a Ken Onion design in CPM-440V called the 1510 which included Onion's patented SpeedSafe opening mechanism. Kershaw also signed a five-year contract with Onion, resulting in many popular models[27]. Onion's SpeedSafe mechanism was significant because it offered a way to rapidly open knives without classifying them as an "automatic" knife, which is illegal in many states. Onion said in a 1999 interview:

> I had been trying to figure out a way to make a smoother assisted opening-and-closing mechanism that was legal, and a friend came over to my house with a cam for a Harley Davidson motorcycle. He wanted me to mill the cam down for him. As soon as I started thinking about what a cam does, it hit me. The SpeedSafe works off the concept of a lobe like a cam... I went to the state attorney general here in Hawaii, and he said that the SpeedSafe was legal, but he wouldn't put it in writing because it was possible that the laws could change. The SpeedSafe does not have a button. You flip the thumb stud. There are no buttons to go off in your pocket like a switchblade.[28]

Ken Onion learned how to make knives from knifemaker Stan Fujisaka: Onion said:

> I kept reading about a guy named Stan Fujisaka. I didn't want to just call him up, so I put feelers out to find somebody who knew Stan. The father of one of my wife's friends knew him, and I talked to him about Stan. He took me over to meet him, and we walked out to a picnic table. Stan showed us some of his knives and I asked where I could buy some of the machinery I needed to make knives. He knew I was serious and invited me to come the next Saturday and he would show me how. I got there at 6 a.m. the following Saturday. There were no lights on or anything, so I sat in the driveway for an hour until Stan came wandering out... He was patient and gentle with me so my ego wasn't smashed entirely. I didn't go into knifemaking to sell knives. It was just something I had to do, but Stan mailed me a form where he had already paid for a table at a local gun show. He told me I was ready and that I'd better be there with my knives. I went to the show with eight and sold seven of them.[28]

Onion first started his string of many popular factory collaborations with United Cutlery before his contract with Kershaw[28]: "I went to the SHOT Show in 1997 and I walked by the United Cutlery booth and met Gil Hibben. I shook his hand and told him I was a big fan. He wanted to see a couple of my knives and then introduced me to David Hall, the president of United Cutlery. They liked one of my folders and it became the Colt Python."

Doug Flagg of Kershaw talked about the company's new focus in 2004[29]: "The knife market is remarkably different than it was 10 years ago. Innovation continues to drive sales. New knives drive our growth and we expect that trend to continue." Ken Onion's designs gave them a string of awards, including Best Buy of the Year in 1999 with the Onion-designed Blackout, 2001 American-Made Knife of the Year with the Onion-designed Black Chive, and 2002 Overall Knife of the Year with the Onion-designed Rainbow Leek. However, Onion didn't only provide designs. I asked Doug Flagg about Onion's contributions at that time:

> We were making some of our early models and struggling as a new factory to figure it out. When we signed Ken, he not only brought SpeedSafe but he also made a huge contribution to manufacturing. Making fixtures and helping with assembly. He had experience; he knew production and he knew knives. Every time he came and visited, efficiency would go up 30%. He would fixture something to make a job easier. He was a big help with the factory.[30]

Kershaw 1510. Photo by Arizona Custom Knives.

In 2003 the Shun line of kitchen knives would be released, made in Japan, which would become some of the bestselling high-end factory kitchen knives, with VG-10 blades, or later the Shun Elite models with SG2. Onion designed the Shun Premier Chef's knife and the Shun Classic Chef's knife.

Kai USA would also start its Zero Tolerance brand of knives in 2006, made in Oregon. This brand started as a collaboration between Ken Onion and Mick Strider and Duane Dwyer of Strider Knives. Jeff Goddard of Kershaw said:

We started out a couple years ago trying to develop a line of knives in a category that we only dabbled in. We wanted to get serious about making [tactical] knives for the military and the professionals that rely on their knives as part of their everyday duty... [We wanted to] make a line of knives of the toughest materials used today in production knife manufacturing ... we followed that passion and we're trying to offer just that - the best of the best.[31]

CPM-3V, CPM-S30V, and 154CM were early steels used in "ZT" knives. Many collaboration designs with other knife designers would follow with ZT in the coming years.

-40-

RALPH TURNBULL, PHIL WILSON, AND NEW SUPER STEELS

From the 1970s through the 1990s, 154CM and then ATS-34 were the most common choices for stock removal knifemakers. Much of the rest was made up of D2 and 440C. ATS-34 had a good balance of price, ease of grinding and polishing, and recognition in the marketplace and therefore desire from buyers. ATS-34 saw a gradual increase in use in factory knives during this period, and by 1999 it was the most common choice in high-end production knives[1]. Steel companies or knifemakers introduced different alternative steels to try to take the crown as the new "super steel" or as a new ubiquitous choice. Still, none could fully take the place of ATS-34 during this period. However, some consumers were looking for new steel options, becoming frustrated with ATS-34 being the only choice for high-end knives. Knife enthusiast J. Thaddeus Hornbaker wrote to *Blade Magazine* in 1999 after an article was published that had many quotes from knife manufacturers praising ATS-34:

> I respect your magazine... But why are you inhibiting the progress of the knife industry by perpetuating lies such as the one that ATS-34 is a top steel? ATS-34 *used to be* a top steel. Today it is bottom of the rung and, with the performance I get from it (i.e., chipping and breaking as well as rusting), I wonder if it ever was a move forward in knife steels at all... The question is, why "settle" at all? There are steels that are eons ahead of ATS-34 and give strength, corrosion resistance, toughness, and supreme edge retention. Let's talk about 440V, 420V, BG-42, and tool steels such as M2 that are coated with rust inhibitors like black Teflon. For instance, 440V is shown to have 10 times the edge holding of ATS-34, and it is also more rust-resistant and tougher as well. BG-42 is not far behind 440V, showing three times the edge holding of ATS-34. Please, stop holding back progress by continuing to fool people that ATS-34 is a top steel. I do realize that you included credits to other steels in your article, but the article was filled with poor statements like, "You're not going to find a better stainless steel (than ATS-34)" and "ATS-34 is the stuff to have." Why not try to perpetuate progress in the knife industry and expose ATS-34 for the fraud that it is? Let's try to push manufacturers to move forward into the steels that are coming out, rather than stroking them and the buyers by saying that ATS-34 is the top stainless.[2]

Vasco Wear

When VASCO introduced Vasco Wear in the early 1970s (Chapter 12), it was perhaps only a matter of time before knifemakers heard about this steel with properties that exceeded D2 and A2. Ted Dowell would help popularize Vasco Wear among some knifemakers in the 1970s and 1980s, though Dowell himself learned about the steel from Bernard Sparks[3]. Sparks was one of the original founding members of the Knifemakers' Guild. In the 1960s, Sparks experimented with many steels and chose 440C, a steel that remained his favorite throughout his career[4]. After trying out Vasco Wear, Dowell added the steel to his choices for his "funny folder." Dowell wrote about the steel in his October 1977 newsletter:

> I have recently had some considerable success in testing a "new" steel called Vasco Wear. This steel has some truly remarkable qualities to offer for its use as a knife blade. For instance, it substantially betters my long-time favorite D-2 in two categories - toughness and wear resistance. This stuff is one and one-half times as tough as D-2 and has twice the wear resistance at RC 60. However, the one meaningful test of a knife steel is to make up a couple of blades and use them. To make a long story short - I have - and I'm very impressed. This is a great knife steel - not stainless (about like D-2 for stain resistance, which is not all that bad), takes a somewhat dull finish, but has all kinds of toughness, and the edge wears and wears and wears. Vasco Wear is tough to work and very difficult to finish (which is to be expected) but it seems to me to represent the first significant improvement over D-2 that I've seen. Needless to say, we have already started using it as an optional steel for some of our models. The featherweight, Full tang, and Narrow tang models are all available in Vasco Wear steel at an additional $50... Also, we're making a new Funny Folder (Type II) ... This model will be available with Vasco Wear at $195 or with D-2 or 154CM at $145.

The additional cost of Vasco Wear in a Dowell knife would increase from $50 to $75 in 1979 and then $100 in 1980. In 1979, Butch Winter published an article in *Knife World* about his experiences with a Dowell "Funny Folder" in Vasco Wear. He also shared the experience of Rita, his wife and fellow knife collector, who used a kitchen knife in the same steel. Butch wrote:

> Dowell offers the Vasco Wear for people, especially hunters, who desire a knife with superior edge holding ability. While I didn't have the occasion to dress even one deer, (I did clean a rabbit) I only cut wire, rubber hose, plastic buckets, opened tomato juice cans and scraped off gasket material and chipped ice with the Funny Folder and Vasco Wear. This is the weirdest steel I have ever used. It cuts and cuts and cuts! It looks dull and even feels dull, but it still cuts! And, it's tough. The blade is ground rather thin, and for a while, I was uneasy about applying the "pressure" to it. The episode of chipping the ice took care of that. I chipped a half-gallon milk carton that was used as a container to freeze water. I cut the cardboard away and chipped up the ice and didn't dull the point. in fact, it was only after scraping gasket material while preparing to install a thermostat in my truck that I finally dulled the knife enough that I needed to touch up the edge. And surprise! surprise! Even though Ted Dowell had warned me that Vasco Wear wouldn't be easy to sharpen, I

found it quite easy. I used a 12-inch medium Arkansas stone, and in a very short time I was able to shave hair off my arm.

All the time I was carrying the Funny Folder around and using it for various jobs on the farm, Rita was wringing out the "Kitchen Bowie." Her very first chore for the knife was cutting raw beef roast. She said her first thought was that she was cutting through butter with a hot knife. Since that first experience, all other knives in her kitchen have been "shelved" ... Since she got her knife in October, I have only had to sharpen it once (this is now the middle of April) ... Now for the one cloud on the horizon of this glowing report - Vasco Wear stains! Blood from my lone rabbit and a prime rib I used it on left small spots that would not wipe or wash off the blade. Later while cutting limes and lemons a faint blue tinge appeared. I haven't had any trouble with rust yet. If it were going to rust, by now it should have... I have a problem! I just can't be happy with a steel that stains. Edge holding, of course, is important to me, but I am willing to make what I like to call one of life's essential compromises. I'll give up several degrees of edge-holding ability for more stain resistance, but Rita is made of stronger stuff. She says she doesn't care what color the blade turns as long as it stays sharp![5]

Perhaps the experience of Butch Winter affected Dowell's opinion of Vasco Wear as well because in his November 1978 newsletter, he changed his statement that Vasco Wear was "about like D-2 for stain resistance" to "somewhat less resistant to staining than D-2."

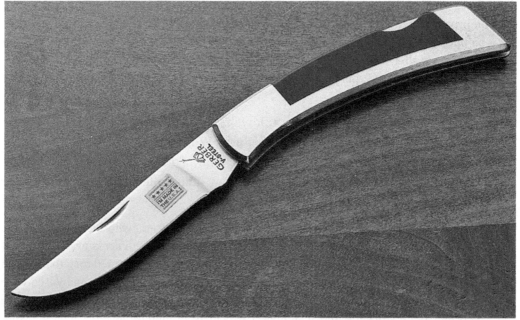

Gerber Vasco Wear knife. Photo from Nordic Knives.

Bob Loveless also tried Vasco Wear, but in 1988 he reported that he didn't like the steel, and he discussed staining using similar terms to Butch Winter:

[I]t turned out to be a disappointment for us. The first thing we discovered was that the moment this material was introduced to blood, the metal stained and discolored. That didn't make it very popular among hunters who would be cleaning game in the field. The other problem most custom knifemakers experienced was the Vasco Wear had an unpredictable way of going dull. Most steels are linear in that there is a direct proportion as to how much cutting one can do with it before it goes dull. Then you resharpen the blade and go back to cutting. With Vasco Wear, there was no linear dimension. It might be cutting perfectly well, then go dull at once.[6]

Vasco Wear would see some limited use in factory knives produced by Gerber. The FS II (Folding Sportsman II) model knives in Vasco Wear were produced from 1982 to 1988.

CPM-10V

Ralph Turnbull

CPM-10V has been used in knives since the steel was released (Chapter 34). However, its use in knives has remained relatively niche even though its combination of toughness and wear resistance is quite good. The fact that it is non-stainless, and somewhat challenging to grind and polish, has held it back. The earliest knifemaker I found who used CPM-10V was Ralph Turnbull, who used the steel in the early 1980s[7-9]. Turnbull began making knives as a hobby in 1973 and went full-time in 1980 when he was laid off from his management job at Chrysler[10]. Turnbull said: "May 2, 1980, was the day I had to make a decision. It was either hunt for another job in personnel,

or go into full-time knifemaking. I had joined the Knifemakers' Guild in 1978 with a recommender's list signed by Jimmy Lile, A.G. Russell, D'Alton Holder, and Buster Warenski. You can't do much better than that."[8] Ken Warner reported on Turnbull's 10V[7], "Ralph says you could wear out a lot of belts polishing the stuff ... anyone who is hell-bent for edge-holding will really like it, and it can be sharpened in use if it is properly shaped before heat-treat. Annealed, it grinds about like Vasco Wear... Turnbull says he won't say some dedicated and talented and very determined knifemaker won't someday get a nice finish on CPM-10V, but Ralph doubts it will happen soon."

Turnbull added in a 1985 interview[8]: "After heat-treat, the vanadium carbides are so much harder than the grinding belts it just isn't worthwhile to take the finish any farther. I could take a lot of time, wear out a bunch of belts, and mirror finish the blade, but then I'd have to charge a lot more for it." Butch Winter also used a Turnbull knife in CPM-10V and said he found it to[9] "have exceptional edge-holding. But it's difficult to sharpen and has no stain resistance. The answer to the staining problem in the future will be CPM T440V." Turnbull would later diversify his knives, making high-end folders with commercial Damascus, mother-of-pearl handles, and other fancy materials.

CPM-10V would see limited use by some knifemakers over the coming years, such as Karl Schroen in his 1985 book where he said the steel has[11] "exceptional wear resistance. The edge holding ability of this steel is amazing; however, the steel will rust quite easily. Finishing and polishing the steel is difficult and time-consuming. The advantages to me far outweigh the disadvantages." In a December 1986 *Blade Magazine,* it said that 10V was one of Bob Papp's favorite steels for hunting knives[12]. Brian Tighe was using CPM-10V by September 1994[13].

Phil Wilson was using CPM-10V by 1999. Wilson said in 1999[14]: "10V is a very specialized custom knife blade steel. It's never going to compete with the ATS34/154 CM group of workhorse steels because of the lack of corrosion resistance and finish difficulties. A very few custom makers will use it for performance reasons only and it does not appear to be economically feasible for use on a factory blade. This is a steel for an avid hunter or guide who wants a blade for field dressing, skinning, and boning, that cuts and holds an edge above everything else." Around 2000, Wilson mentioned that despite Crucible's claims that the toughness of 10V is similar to D2, he thought it was a bit better[15]: "I have been making a lot of hunting blades out of CPM 10V. The data sheet from Crucible on this steel indicates that it has about the same toughness as D-2 at the same hardness. I have found that I can heat treat 10V to about Rockwell 63 and it still has more than adequate toughness for a working knife. My experience with D-2 in the past indicates that it starts to get pretty brittle at about Rockwell 60."

Turnbull knife of unknown steel from *Knives '84.*

Phil Wilson

Phil Wilson is a knifemaker who has made his name pushing the boundaries with high wear resistance steels, high hardness heat treatments, and extremely thin edge geometry. He also wrote several articles about steels and heat treatments in the 1990s and 2000s, which we will use to get a feel for what was going on in that timeframe. I interviewed Phil in 2018 about his history:

My Grandfather was a machinist and he made knives back in the 40s and 50s with some for my uncles to take to WW2. They were made with power cut-off saw blades and phenolic, Bakelite, and Lexan-type clear plastics for handles.

I have seen knives of this era and type called "radar knives." My Grandfather could make "anything," and I wanted to be like him. He let me work in his shop with power tools as early as 8 years old and I basically made knives out of sheet metal and used walnut for handles. I worked summers as a mechanical technician during college and had access to a machine shop after hours and lunch. I made a couple of knives with 420 blades. They came out ok for looks but would not cut anything. I found out later that the blades had to be heat treated to harden. One of the machinists heat treated a couple of blanks for me and that was the start of my interest in heat treating. Later in the early 80s my son brought home a book on knife making by David Boye. I think it is still available and is a very interesting read; it has good instructions on making blades with various saw blade materials. I made a few knives in my small garage shop and caught the bug once again. Later I was trying to sharpen a commercial-made stainless-bladed knife for a fishing trip, and it was so soft that the edge just bent back and forth. I decided then that I would find good materials and make a knife that would be hard enough to hold an edge and be nice and crisp to sharpen. I found out about 440C and bought some from Jim Sornberger who had a knife supplies shop in the Bay Area back then. Later I got a copy of "Knife Making" by Bob Loveless which is an excellent reference for anyone just starting out making knives. I have done a bit of genealogy study and found out that my knife making grandfather's family goes back many generations as metal workers and blacksmiths in England. My other Grandfather's family were hunters and fishermen back to early immigrants from Scotland into Northern Ireland. The outdoors and metalworking are in the blood so I guess I can't help myself.

I am mostly self-taught from reference materials like the above books by Boye and Loveless. Bob Engnath, who had a material supply shop, had a great little guidebook on grinding and basic knife making which was also a great help. I got great advice on materials and heat treating from Ed Severson, a metallurgist and good friend who worked at both Crucible Materials and Böhler-Uddeholm. I am retired from my real job in the engineering field and have always been interested in material science so researching and testing to find the best blade materials for the application is kind of a natural tendency. One of the breakthroughs for me was when I realized I needed to do my own heat treating. At the time, I was frustrated with not being able to make knives and test them one at a time and be able to real-time measure the performance and adjust the process. I still do them one at a time even though it makes for very low production. Another key to quality control was adding a deep cryogenic treatment and I believe it adds a significant improvement to the heat treat process. I believe I have been influenced by Bob Loveless who has the most beautiful knife designs on this earth. I actually met him at the Pasadena show in the late 80s. He was sitting alone in the coffee shop at the hotel and I recognized him and decided to go by and say hello. He invited me to sit, and we had a long conversation on metallurgy and design. He encouraged me to continue with fillet knives which was very important feedback from a guy like him. I remember him saying he would make all his knives out of CPM 10V except that all his customers want stainless blades. I also admire guys like Ed Schempp, Bill Harsey, and Tom Mayo who are good friends and have the special eye for design. I also have to include Wayne Goddard who encouraged me to do my own testing and heat treating. He also managed to get me into the Oregon Show one year which started a 15-year attendance run that allowed me to meet and talk to the best in the industry. There was a lot of interest in steels and metallurgy when CPM

S90V was introduced. I held a seminar on heat treating some of these steels at the Oregon show a couple of years running. They were held on Friday morning before the show opened for set up. We had large crowds that overflowed the room and had very good interaction with many knife makers and a few knife companies. I remember the Spyderco crew there several times.

Knife testing is a matter of trying to come up with a process that can be used in the shop to predict the actual field performance. That means you have to cut something be it rope or cardboard or carpet or?? I have found over time and with the help of Wayne Goddard the best test medium is manila rope. This material comes very close to what is encountered in the field both hunting and fishing and everyday use... It is very difficult to do a test that just measures the performance of steels. All the variables come into play when cutting. Edge geometry, blade thickness, sharpening, blade finish, rope differences, ergonomics, push cut or slice. I have concluded that one can test and put blade materials in categories but without a precise measuring machine like a CATRA and exactly configured blade samples, it is not possible to cut by hand and rate one material over another with scientific precision. I cut with each knife I make and expect it to fall within parameters based on past experience and a whole lot of feel. I think when we are testing, we are really testing the performance of the overall knife that day under those test conditions and the steel grade is just part of the package. I have tried almost all the new steels as they have been introduced. Early on with Crucible, we were working together to find [a] heat treat process that lent itself to knife blades rather than larger mass and cross section normally used for tools and dies. I was writing tech articles and wanted the "scoop" on the latest grades.[16]

CPM-440V (CPM-S60V)

CPM-440V was the first powder metallurgy stainless steel released by Crucible around 1983 (Chapter 34), and samples of the steel were given to select knifemakers in the same year[17-18]. The earliest recorded knives using 440V I have found were from Karl Schroen[11] and Sean McWilliams[19], which probably makes sense because the steel was only in larger sizes and both men were forging high alloy steels. Despite knives using the steel in the early-to-mid 1980s, the growth of this steel was relatively slow. Michael Walker patented a method[20] for the "Zipper" or "Running Dovetail" with a spine of 6Al-4V titanium and an edge of steel. The earliest blades used CPM-440V, which was also the only steel referenced in the patent. This method was first reported in the December 1988 *Blade Magazine*[21-22]. The two pieces are press fit together with dovetail edges and then the joint is peened. As previously shown in Chapter 36, Goddard had tested CPM-440V in his rope cutting test in the 1980s and found it to cut very well. Goddard wrote an article in 1992[23] where he reiterated that "CPM 440V shows most promise as a superior blade material for working-type stainless knives. My tests show it to have two-to-three times more edge-holding ability than 440C, both being tested at 57-58 Rc... CPM 440V may become the most used steel of the '90s." Use of 440V by custom knifemakers did start to see an uptick starting in 1994 when Crucible began selling it in suitable sizes from their own Service Centers and through knife supply shops Sheffield Knifemakers Supply and Texas Knifemakers Supply[24].

Spyderco was the first factory knife company to begin using CPM-440V in 1996, first with the Military model[25], and 2/3 of production at the Golden, Colorado facility was in 440V in the year 2000[26]. Sal Glesser said[27], "I decided to use CPM 440V because it was the toughest, most high-tech, durable steel around." Glesser bought his first CPM steel knife from Ralph Turnbull which generated his interest in the steel[27]. In 1998 the Blade American Made Knife of the Year was a Kershaw knife designed by Ken Onion with a CPM-440V blade[28].

Phil Wilson wrote an article about 440V for the February 1997 *Knives Illustrated* magazine. Wilson started the article by asking why the steel wasn't more popular:

> CPM 440V was touted as the super custom knife steel of the 90s about five years ago, but today it is being used by only a small percentage of custom makers. A check of *Knives '96* shows about 30 makers of 1,000 listed use of this steel in their production. One might question why a steel with so much potential has such limited use? The partial answer lies in the myths that have developed over the years. I heard many of them as people stopped to chat at the recent Oregon Knife Show. "This stuff eats belts like crazy and it is impossible to get a decent finish on it." "It is very tough to heat treat and has low ductility." "The edge holding is great, but it is very difficult to sharpen." "It's good stuff, but too expensive and I can't find the sizes I work with." Is there any basis for any of this mythology? I have been grinding and heat treating CPM 440V blades for about four years and think I can shed some light on the subject. It's a wonderful steel and I believe it fully lives up to the earlier expectations. The bottom line objective is edge holding without sacrificing the ductility or corrosion resistance that we have come to expect from the 154CM/ATS-34 standard... [O]n paper we have a steel that does it all: finely dispersed, very hard, small carbides for edge holding, fine grain for toughness, high carbon content for carbide formation, and chromium for stain resistance. Does it work that way? Yes, it does, and it makes a superb blade when heat treated correctly! ... The Crucible datasheets on this steel indicate that ductility, as measured by the Charpy C notch process for CPM 440V at Rc 57 is about the same as 154CM/ATS-34 at Rc 59... I have seen this in practice with my long, flexible fillet blades. None of them have ever broken in the field. I have done some bending tests in a vise in my shop at Rc 57 and have found that CPM 440V will permanently deform a little

before it breaks... I have done a series of edge holding tests cutting 1/2-inch manila rope with 440C, ATS-34, D2, and CPM 440V using Wayne Goddard's procedures... CPM 440V blades at Rc 55 cut about twice as long as cryogenically treated ATS-34 at Rc 58/59... I use my own knives extensively and get good feedback from my customers. We have found we can fillet 20 large Halibut or about 60 Rockfish before the blade needs to be touched up.[29]

| Knifemaker ~ Michael Walker | Photo by Coop |

Michael Walker knife with "Zipper" blade. Photo by Coop.

The number of custom knifemakers and knife companies using CPM-440V would steadily increase. Wilson noted there were 30 makers listed in *Knives '96* using the steel, and by *Knives 2003* that number had approximately doubled. The primary complaint from knifemakers was difficulty in polishing. Scott Sawby said in 1997[30], "It holds an incredible edge and is a really good steel. One drawback is that it doesn't polish at all or take a good finish." In 1997 Wayne Goddard said[30], "ATS-34 has great strength in relation to its edge-holding ability. The people who have used BG-42 also seem to like it. For more money and a gain in edge holding, my choice would be the CPM metals. They are more expensive and harder to work, and whether the gain in edge holding will be worth the cost to the customer remains to be seen." Al Crenshaw said in 2000[31], "CPM 440V is terribly hard to finish, but it holds a good edge." But there were some who thought that the steel would take over, such as knifemaker Norm Beckett[31]: "Most guys prefer ATS-34, but I think they'll be switching to 440V." There were sometimes concerns about its toughness, however, as Ken Onion noted the steel "should not be abused" and that it is best at 56-57 Rc and no harder[26]. In a 2000 interview Bob Loveless said[32], "I don't like powder metallurgy results, I don't like the structure, I don't like the reduction in transverse-to-longitudinal strength... Americans are so used to new products coming along all the time with supposed improvements but it's really just advertising verbiage."

CPM-420V (CPM-S90V)

Phil Wilson would write articles about the new CPM-420V in 1996[33] and a follow-up in 1998[34] for *Knives Illustrated*. He said[33], "I first heard about it during a phone conversation with Ed Severson, a metallurgist at Crucible Materials Corporation and a Custom Knife enthusiast. He mentioned that Crucible is coming out with an upgrade for 440V and that it would be available late in 96." In 2018 he said, "I think Rade Hawkins and I made the first knife blade with CPM S90V. He had a knife supply business (now run by June and Russell Hawkins) and we went together and bought a half sheet from Crucible. I believe that RJ Martin got some material early on as well."[16]

In the preliminary article, Wilson said:

> My initial impressions are that it does very well... A fillet knife works out very well as a test blade because of the very tough environment it must live in. It must exhibit good bending strength, good ductility, corrosion resistance, and edge holding. I used the knife on a couple of fishing trips, and the corrosion resistance is very good around salt water as advertised. The edge holding so far seems to be at least equal to 440V. It may be better but some very careful cutting tests and feedback from other makers will be required before I would be ready to claim a specific percentage improvement over 440V. Grinding is about the same. Finishing on 420V is noticeably tougher than 440V. It you get a deep scratch in it you'll be a while getting it out.[33]

For the follow-up article, Phil Wilson collected anecdotes from knifemakers and also reported his own testing[34]: "Wayne Goddard ... consistently got 64 cuts on 1/2-inch manila rope with CPM-420V. Not many steels go over 45, and most are down in the 30 range according to Wayne. He also reported that it was easy to get a very aggressive, extremely sharp edge on the test blade with a Norton Silicon Carbide stone." Barry Gallagher and P.J. Tomes both reported that they liked the steel but said it was extremely difficult to grind and finish it[34]. Next, Phil Wilson made a test knife for a Kosher butcher, Rabbi Yurman. Through a week of use, killing about 1,000 chickens a day, he reported that the knife never really got dull. After that, he gave the blade to the "cut-up department where heavier work with larger animals around bones and joints are done. The report on edge holding from that department was about the same."[34] Wilson also made a fillet knife for fishermen Roger and Elena Rothschild and "Roger reports that 75 large salmon were processed before Elena felt the blade needed re-sharpening."[34] Roger also reported that restoring the edge was easy on an extra fine diamond stone. Phil also tested a fillet knife of his own and said that he filleted 22 tuna ranging from 12 to 35 lbs. and that the knives were still able to slice very thin pieces of leather and 1/2 inch manila rope easily.

In 2000 Wilson said[26], "I use 420V almost exclusively for hunting knives. Over the years I've used 440V and played around with 10V and 3V, but 420V seems to be the best overall steel for a knife blade." He also said that 420V would perform better at higher hardness than 440V and that he was heat treating his 420V to 61 Rc. "440V has more chromium and so is a tad more brittle at high hardnesses. That's why I believe 420V is the best all-around blade steel."

In 2000, Crucible would change the names of 440V and 420V to S60V and S90V, respectively, roughly matching the vanadium content of 6 and 9%[35]. This was also to help avoid confusion as the steels really do not share that much in common with 420 and 440 stainless steels. The "S" was also used to designate that the steels are stainless, while non-stainless steels like CPM-10V do not get this designation. Because of the great difficulty in grinding and finishing, S90V has never been very common in either custom or factory knives. In early tests of the steel by Spyderco, Sal Glesser said that the steel "stopped their grinders cold"[16], though in more recent years they have released S90V knives. Microtech would add S90V as an upcharge option for certain blades starting in 2000[36].

-41-

CHRIS REEVE AND CPM-S30V

Chris Reeve

Chris Reeve is a knifemaker who grew up in South Africa. He is famous for his Sebenza model knife, introducing the "frame lock," which is a modification of the liner lock where the liner is part of the titanium handle, and modernizing knife production with CNC manufacturing. An extensive interview with him was published in 2001 by *Blade Magazine*:

I was called up for military service (in South Africa), and I was a second-year apprentice for a tool-and-die maker, making a pitiful amount of money, and the only kind of good quality knife I had any knowledge of was a Puma White Hunter. That was the only exposure I had to any kind of knife other than the "cheapies." And so I went looking for a Puma White Hunter, and that was $60 back then, which was more than about two weeks salary, so I said, "Heck, I'm a tool-and-die maker, so I'm going to make one myself." I scrounged some steel, a piece of O-1. I had seen an article on Bill Moran in a very old *Gun Digest*, and there was a picture of him forging, and I thought, "Well, that's the way to make a knife." I got out a big welding torch that created a big flame, heated up this piece of steel, and beat the heck out of it! That was in 1975 and I ended up finishing up this knife, and I've still got it in my shop today.

In 1976 I saw the *Guns and Ammo Knife Annual*, which Jim Woods edited, and I looked at that magazine and I fell in love with knives. I just could not believe what people were doing. In that magazine was an article on Ted Dowell and a brief thing on Jimmy Lile. In the Lile article, there was a picture of a Jimmy Lile integral knife, and it is that knife that was very, very instrumental in me formulating the whole one-piece knife deal. The article on Ted Dowell with his integral knife also got me thinking and going on my one-piece design. After seeing that magazine, I found a piece of D-2 in the tool room, and I made my own integral. It was a different profile, but it was the same concept as Jimmy Lile's knife. That knife was in my toolbox wherever we went racing (Chris raced motorcycles in South Africa) from 1978 to 1983. When I stopped racing I then was at a loose end on weekends, and I picked up a new gun magazine in South Africa called *Magnum*. In one of their first editions, they did a big, full-color spread on custom knives, and I saw these things, and I nearly fell over. That really got me going, and I then made a little stick-tang knife with some silver inlay. I'd also seen an article on Bill Moran doing his silver inlays and thought, "I can do that!" I took that knife with me when I was doing military service up in Southwest Africa, and the wood cracked. This frustrated me and I thought, "I'm going to make a [bleep] knife that is not going to crack or do anything bad," and so that's when the one-piece idea came. That was in 1978. I didn't do much about it then, but later I was doing more military service closer to home in the city. I had a lot of time to think as I was on guard duty, and I sketched out what I wanted to do. Then, straight after I finished that military camp, I made the first one. So that's how we got started in the business. That particular first knife I sold for $60, and it was to a captain I had gotten friendly with at that camp. Just before we came to this country I got it back. I traded four of my regular knives for it, so he made a handsome profit out of that one! But it was well worth it for me to get that knife because it was very much part of the history of the one-piece concept.[1]

Reeve went full-time as a knifemaker in 1984. When asked why he moved to the USA, Reeve said, "Opportunity. South Africa was going nowhere and we were selling probably 80 percent of our production in the USA. We've always loved the USA, and came here to attend several knife shows from 1986 to 1988. Then in 1989, we emigrated."[1] Reeve's original knives were fixed blades, but he is best known now for his folding knives:

[T]he priority I've always worked under is the tool must be useful, it must work for its intended use. Now, that's a broad spectrum when you come to my fixed-blade knives, because they are general-purpose, in-the-field tools. You can't make the blade thin like a fillet knife to cut spuds up around camp, because it's not going to chop your wood also, or do other heavy tasks. It's got to be something that will work for a wide variety of tasks ... on our fixed-blade knives we find it's function, function, function.

When we came to design the Sebenza, I had looked at a LinerLock. The LinerLock was very popular then. Michael Walker popularized it, and everyone was making it. I looked at the LinerLock and I thought that I didn't want to be one of the sheep running through the same gate. Also, I had a reputation worldwide for making an indestructible fixed-blade knife. I cannot come out with a folding knife that has this willy-wally thing in the middle there that flips around. I thought, if we're going to do it, we're going to do it so that it is strong. The other thing I always say is that I'm too [bleep] lazy and too mean to make a complicated knife. I just don't like putting that amount of work into it. I said, "Let's throw these bolsters away, these liners away, let's throw everything away, and start with a clean sheet of paper and see what we come up with."

I took two pieces of titanium and made slits down the sides [for the Frame Lock]. The difficulty of making the slits properly on the Sebenza is that it made it necessary for us to have CNC equipment. To make them the way they needed to be made, you couldn't do it profitably or as accurately on manual machines... [A CNC machine] will sit and do this 24 hours a day over and over. The thing is, with a manual-milling machine, you have to physically position things, and our accuracy is not anywhere near what it is on the CNC. To machine the Sebenza profile on a manual milling machine would take three days. CNC has changed the machine tool industry completely. There is still great skill involved because the CNC machine is no smarter than the guy punching the buttons. You put garbage in and garbage comes out. But it enables me to make the Sebenza at a price that I can sell it. If I have to spend three days just milling out the handle, the thing is going to cost $2000, and no one's going to pay that. The key factor in making the Sebenza work properly is accuracy. That is the kicker with the [frame lock] mechanism. You've seen many copies on the market and all of them have problems. Anything that is simple to make has other little hidden problems that offset that simplicity, and this knife has to be precision. We are working to tenths of thousandths of an inch on the Sebenza. No other knife company is working to those tolerances - period. I know they aren't because I see their knives. We are working to tolerances that Ron Lake and Michael Walker and a lot of the other top custom makers are working to. They are achieving these tolerances by careful handling, hard work, and time. We are getting those tolerances by using extremely accurate equipment.[1]

Small Sebenza in BG42 steel. Photo by KnifeTreasures.com.

BG42

As noted in Chapter 13, BG42 has been around since the early 1960s, though it didn't begin to catch on in knives until much later. Knifemaker Lew Booth used the steel in 1974[2], which he called "a new improved 154CM," and Loveless first tried some in 1975[3]. In 1976 Loveless said they had[4], "300 pounds of another grade that's very similar (to 154CM), only with 1.2% Vanadium added, which we have just begun working with." In 1985 Loveless said:

I've been trying to get Latrobe to quote for five years on whatever quantity it insists I buy of a grade called BG42, which is just like 154CM and ATS. I bought 400 pounds of it some years ago but I could only get it one-quarter inch thick. They say they can't roll it any thinner... Of course, our standard gauge for a hunting knife is three-sixteenths inch thick. The first thing I did was throw 25 percent of it away with the trash. It has been sitting. I have only made about 12 knives out of it for friends I know are active hunters. I'm not gonna have a BG42 knife on some collector's wall.[5]

In his May 1989 newsletter, Ted Dowell added BG42 as an option: "We've added some fillet/slicing knife styles which are excellent in the kitchen or in camp. These are primarily available in our 'new' steel, BG42. This steel is a super-clean (vacuum re-melted) version of 154CM with 1-1/2% vanadium added. It holds an edge about as well as Vasco Wear and it is stainless." Later in 1999 he would call BG42 the "best of the stainlesses."[6]

ATS-34 was climbing in price in the 1990s, and Goddard recommended several alternatives in 1995, including BG42[7]. The rising price of ATS-34 may have driven some of the push for BG42 in the late 1990s. At the 1996 California Custom Knife Show, Bob Loveless announced that he was going to start replacing ATS-34 with BG42[8]. However, this change does not seem to have been fully put in place, as in 2000 Loveless said in an interview[9], "I still got a lot of it but we don't suggest it for everybody" since it was only "an 8-to-10 percent improvement in edge holding. It's a harder steel to grind. It's useless for the average customer really because, (bleep), some of our people can't sharpen these knives now."

In 1998 Phil Wilson published an article in *Knives Illustrated* about the steel:

BG42 is again unique because it has a small amount of vanadium that works to refine [the] grain structure. Some vanadium carbide is also present, which may further enhance the wear resistance. The main story on BG42, however, is not so much the chemical composition but the manufacturing methods used to produce it. The VIM-VAR designation is the key to the manufacturing method. In simple terms it means double vacuum melted... The end result is a tool steel that is extremely clean. Impurities and slag inclusions will be essentially absent in this material. This makes it very attractive for a knife blade because the strength and finish capabilities are greatly enhanced... Based on the data sheet information furnished by Latrobe, the edge holding, cutting performance, ductility, and corrosion resistance of this grade should be equal to, or slightly better than, ATS-34... I ordered a small amount ... and made a utility hunter for initial evaluation... The results so far are very encouraging. Compared to the very high wear resistance steels I have been specializing in, it is a dream to grind and finish... Preliminary cutting tests show that it is very similar to ATS-34. It is too early to say that it is going to be better, but the potential is there.[3]

BG42 also grew in sales with custom knives when it was released in sheet for knifemakers from knife supply shops, such as Tru-Grit by 1998[3] and Admiral steel in 1999[10]. Chris Reeve was selling knives in BG42 starting in 1997[11]. In a 2001 interview Reeve said:

> I was introduced to BG42 by Ted Dowell, who had been using it for years, but in round bar stock. I was instrumental in getting Latrobe Special Steels to roll it into flat bar. The reason I wanted to go with it is because it had a very nice mix of alloys. ATS-34 and 154CM are both excellent steels, and have a good stainless alloy mix. BG42 is the same as those steels except (it has) 1.2 percent vanadium. This is about the right amount of vanadium that you want. It is just enough to keep the material easy enough to work with, but gives it excellent edge retention comparable with any of the (CPM) steels.[12]

In 1998 Buck knives released a line of knives called their "Master Series" which had BG42 steel and emerald green Diamond Wood handles[3,13]. Robert Morgan, Buck's VP of sales and marketing said in a 1998 interview[14]: "Consulting custom makers helps. It starts out there in the pits. Eventually, materials like BG42 become known to dealers. We need to tell the story about the steel. To the best of my knowledge, we're the only factory using BG42. It holds an edge, it's shock-resistant, corrosion-resistant and packed with high levels of carbon, molybdenum, vanadium, and chrome." In 2000, SOG would introduce the "first BG42 production knife under $100."[15] Also in 2000, Schrade was offering BG42 on the "New Millennium" hunter designed by D'Holder and a folder designed by Ron Lake and Michael Walker[16]. Tim Faust of Schrade said[16], "We tried to find the best materials we could, and we heard great things about BG42."

Buck "Master Series" in BG42 steel circa 1998.

CPM-3V

CPM-3V, the powder metallurgy version of Vasco Die, was also released by Crucible around 1999 (see Chapter 34). Ed Severson of Crucible said in an interview published in 2000[17]: "It's ideal for a cutlery steel. It has good edge holding and high toughness, properties you usually don't hear in the same sentence for a steel. Both custom and factory makers have shown interest in 3V because of its unique properties." Phil Wilson said about the steel in 2000:

> 3V is twice as tough as 154CM and is the best choice for a chopper, something you can really abuse. A Rockwell hardness of 58 RC is best for it. That gives it the best overall balance between toughness and edge holding. Toughness is the impact force and 3V is a lot more forgiving for impact forces. You trade that out against it not having quite as good corrosion resistance and not as good at holding an edge as 420V. Everything is a compromise. The trick is to pick the right steel for the right application.[18]

KNIFEMAKER ~ JERRY HOSSOM   RETRIBUTION   IMAGE ~ SHARPBYCOOP.COM

Jerry Hossom "Retribution." Photo by Coop.

In a follow-up letter based on his prior quotes, Wilson said: "It is true that CPM 3V has less corrosion resistance than most of the popular stainless tool steels used for knife blades. I have less experience with 3V but found it to have better corrosion resistance than you would expect just looking at its chrome content. I suspect that because of its vanadium content, 3V has more free chrome to resist corrosion."[19]

Knifemaker Jerry Hossom was an early user of 3V in his knives:

> I think it's the best steel I've ever touched. Its impact resistance is phenomenal. You can bang on it and not hurt it. If you're chopping something and hit a nail, the nail will lose. Impact resistance is important in holding a fine edge because one of the components of wear is resistance to chipping, and that's one of the things impact resistance deals with. The quality I like about CPM 3V for big knives is it has a springiness in it that is really quite remarkable. You've probably read about the great swords having a spring-like quality and, by golly, [the big blades I make of CPM 3V] do have a springiness to them.[20]

Hossom did note negative aspects to 3V, however, saying that 3V "has a very unpleasant way of manifesting corrosion. When the rust develops, it develops in tiny spots ... so that, by the time you see the rust, it is a pit or hole... Still, there are far more positives than negatives to this steel."[20] Ultimately, while CPM-3V has seen consistent usage in a small number of knives since its release, it has never caught on in large scale production. Perhaps this is because 3V is not a stainless steel, and the overall market prefers stainless.

Development of CPM-S30V

Crucible Steel was becoming increasingly involved with the knife industry, attending knife shows, doing interviews for knife magazines, and getting steel available in sheets for knifemakers and knife companies. Crucible metallurgists also began dropping hints about a new steel developed specifically for knives. Most steels used in the knife industry had been developed for something else, such as 154CM for bearings. Ed Severson of Crucible said in a 2000 interview[17]: "Take CPM 420V. When Crucible did the design criteria for it, the criteria had nothing to do with cutlery." But Severson also said they had a new steel in the works with properties optimized for blade steels. He said S30V "is the first time Crucible has tailored the criteria for a steel specifically for the cutlery industry."[17] There were also hints about what properties they might shoot for, as both Bruce Divita and Ed Severson of Crucible said in 2000-2001 that the new steel would be a "stainless version of CPM 3V."[20,21]

Crucible metallurgists involved Chris Reeve in the development of CPM-S30V early on, as Reeve said in 2002:

> When Dick Barber and Ed Severson of Crucible Steel spoke to me about using their blade materials, my return was that the company didn't make an alloy I wanted to use. They made alloys [S60V and S90V] with too much vanadium and I couldn't work with them. The steels are wear resistant, but they are too difficult for me to machine and grind. The upshot of it was that I wasn't looking for wear resistance, but corrosion resistance and toughness in a blade steel. Dick and Ed looked at each other, debating whether to tell me that they were developing such a blade steel - S30V.[22]

Metallurgist Dick Barber, who headed the development of S30V, told the story similarly:

> While at Crucible I was involved with directing our alloy development as chairman of the technical review committee. During that time I attended the Oregon knife collectors meeting in Eugene where I met Chris Reeve who at that time was using BG 42. I asked Chris why he was not using our materials and he told me that if we made something better he would buy it. This conversation was the beginning of a project that I shepherded through our alloy development process to make a better knife steel. During that time, Ed Severson and myself conducted interviews with many other knife makers but Chris was the only one who had access to the early alloys which were made. In fact material for the first heat of S30V we made in the research lab was given to Chris to work with. From that material he made 4 knives, one was given to me, one to Ed Severson, Chris kept one and Scott Cook, who ground the blades, kept the fourth. Those knives were tested and compared to other materials and several suggestions were made for improvement in the alloy. From that point, a second set of materials were melted and tested in the lab... As a reward for his interest and feedback on the product, Chris was the first maker who had access to commercial quantities of the alloy... I can say as one who was there that were it not for Chris and his persistence in pushing me, there would not have been a CPM S30V or the alloys that followed.[23]

Barber also talked about the other people they interviewed:

> In addition to working closely with Chris, I also consulted makers like Sal Glesser, Ernie Emerson, Tony Marfione, Phil Wilson, Bill Harsey, Mike Jones, Steve Ingrim, Tom Mayo, Jerry Hossom and Paul Bos. I also met with hundreds of users at various knife shows on both coasts. The general consensus was as follows. The current top dog in the stainless knife world was BG42 and they wanted something that was more corrosion resistant, tougher, and better edge retention than BG42. We used this as our benchmark as well as 154CM and 440C.[24]

As to the balancing of those properties, Barber told of the design goals:

> One comment made during the data gathering phase of S30V's development really hit home with me. While talking with Ernie Emersion he said one thing: "The knife can't break because if it does the wrong guy might die." I took this to heart and the manufacturing methods used make the S30V as tough as you can make a stainless steel which is capable of being used in the low HRC 60s and hold an edge. We were concerned with not only longitudinal toughness but transverse as well. This is why we used the CPM process to get as close to homogeneous properties as possible. 25 years of steelmaking experience has told me that processing variables in steelmaking will have a big effect on end-use properties. With respect to corrosion resistance, we looked at alloy balance and in particular some tricks that have been used in other stainless steel families to improve corrosion resistance and in addition to the common Cr and Mo alloying we also added

Nitrogen which not only improves pitting resistance but also heat treat response. This enabled us to greatly increase the breakdown potential in anodic polarization testing versus our benchmark alloys (improved corrosion resistance). With respect to edge retention, replacement of Mo and Cr carbides with V carbides allows considerable abrasion resistance improvement which will show up in a controlled test such as the CATRA testing done by people like Spyderco and Benchmade. Another consideration given was fabricability for the knife maker. Since many exotic alloys require extreme heat treatment cycles we had to be sure to design S30V so the average maker could use it. To this point, I worked with Paul Bos to see what his capabilities were and we designed the alloy so Paul could successfully heat treat blades with existing equipment. One complaint which was made for some of the higher alloy materials like S90V was the difficulty in grinding and this was corrected by lower carbide volumes so the average maker could make a knife and still sell for a profit. All in all many things were considered in alloy design and according to the objective testing which we and other people have done we were successful. It is important to remember that it is possible to create conditions where any manufactured item can fail. These can be through improper design, fabrication, or use. When you need a hammer you should use a hammer and when you need a knife you should use a knife.[24]

Release of CPM-S30V

CPM-S30V was released with a lot of fanfare, and Phil Wilson wrote three articles[25-27] in one year about the steel for *Blade Magazine*:

> There's been a revolutionary growth in custom knifemakers over the past decade. The demand for a new specialty steel exclusively for knife blades has been a long time in the making. The metallurgists from Crucible are knife enthusiasts and have attended some major knife shows over the past three years. Both Ed Severson and Richard Barber have given technical seminars at the Oregon Knife Show and Blade Show to bring some knowledge to the knife public. They've seen a need and it looks like the new steel is going to fill it.

> It's finally here! The knife industry has been rife with rumors for almost three years about a "super steel" being developed specifically for knife blades by Crucible Materials Corp. The original idea was to take the basic alloy composition of CPM 3V and add enough chromium to make it stainless. CPM 3V has a reputation for being the steel to use if toughness is the main criterion. The trick was to add enough chrome to improve the corrosion resistance but not compromise the superb toughness to a significant degree.[25]

Dick Barber, for his part, told me that developing a stainless 3V was not the original design goal, and the composition instead started with 154CM, dropping the molybdenum from 4% to 2%, and adding in 2% vanadium to go with it. After Chris Reeve asked for higher wear resistance for more differentiation from BG42, the vanadium content was increased to 4%. However, as previously noted, some people in the Crucible organization promoted S30V as a "stainless 3V," which was an exciting concept for many knifemakers and enthusiasts. Ultimately, S30V has significantly higher wear resistance than 3V but substantially lower toughness.

April 2002 article from Phil Wilson.

Wilson would publish the results of his testing of S30V three-four months later:

I had a standing order of the first S30V and received some stock this past Nov. 5 (2001). It was like Christmas in November for this knifemaker! The day I received it I ground a 9-inch fillet blade from it, heat treated it, and had a completed knife to work with the day after... In addition to the fillet knife, I made a simple slab-handle semi-skinner with the new S30V and two other similar knives from CPM S90V and 3V... At Rc 61, S30V does as well or slightly better than 154CM or ATS-34 at the same hardness (for edge holding). It has the same aggressive cutting nature as 3V... The corrosion resistance of S30V is excellent. I purposely left the fillet knife uncleaned and covered with a combination of fish slime, salt water, and blood in the back of my truck camper shell overnight in a damp coastal environment. There was no evidence of corrosion or pitting the next day. The blade cleaned up like new. The same thing would be true of 154CM and ATS-34 in my experience. By comparison, D-2, 3V, and 10V - all of which have some percentage of chromium - would have shown evidence of corrosion.[26]

Impact toughness is the real question... I didn't make a chopping-type knife on which to experiment, but I did chop a Douglas fir 2x4 in half with the fillet knife. It was difficult to put much chopping energy into the task because the knife is so lightweight, yet there was absolutely no damage on the thin blade. Did the S30V make the A-2 toughness target I set for it? I'll have to wait for the final data from Crucible to see how close it comes but I think that in time it will be evident that this steel is the toughest true stainless around... Compared to CPM S90V or CPM 10V, S30V is a pleasure to work... As in all CPM steels with a high percentage of vanadium carbide, the final finishing operations take more effort than 154CM or D-2, for example. A hand-rubbed satin finish or a 240-grit belt finish is the best choice for all these steel grades. Mirror polishing S30V would take all day, though S30V takes about half the effort to mirror polish that S90V does... In time, S30V may hold the title of best all-purpose "workhorse" stainless steel.[27]

Steel	Year	C (%)	Cr (%)	Mo (%)	V (%)	N (%)
S30V	2002	1.45	14	2	4	0.2

Phil Wilson also said about the new steel[25]: "First, it should be noted that CPM S30V is meant to be a steel that most custom and production makers would want to use for a large portion of their output. It's not intended to replace the super alloys such as CPM S60V, CPM S90V, and CPM 10V. However, it could be an alternative to, or eventually even replace, such steels as ATS-34, 154CM, BG42, and D-2." Indeed, S30V did become a standard high-end blade steel for many knife manufacturers. In addition to Chris Reeve Knives, by mid-2002 Microtech and Spyderco had both announced knives manufactured with S30V, and Sal Glesser of Spyderco believed that the new steel would outperform his previous top choice of S60V[28]. Marfione of Microtech said[22]: "Our standard steel for folding knives now is 154CM, and our premium steel is S30V. By the end of the year, S30V will be our standard steel, and S90V will be our premium steel." Benchmade also announced knives in S30V in 2003[29].

By mid-2003, S30V was already the top-selling knife steel from Crucible[30]. In a September 2003 article, Joe Kertzman wrote[31]: "All new specialty knives under development at Lone Wolf Knives, the company's gent's knives, and the Harsey Tactical Folder have S30V. All Chris Reeve factory folding knives and The Green Beret Knife feature blades of S30V. Spyderco has decided to use S30V for all of its U.S.-made knives with the exception of the Maddox model... The 2003 S.H.O.T. Show confirmed that S30V is being embraced by many of the industry's major players." Other knives listed included[31] the Benchmade Switchback; Buck/Mayo TNT; Camillus Dominator; Kershaw Bump; and Strider MK1. Sal Glesser of Spyderco said of the steel:

> There is sufficient chrome and nitrogen in its makeup to provide good corrosion resistance for such a high-carbon product. The cutting tests we perform are done on our CATRA machine. We've tested many, many blade steels and have gathered considerable data for comparison. CPM S30V tested better than CPM S60V, BG42, any of the ATS's and VG-10 in sharpness and edge retention. In serrated-blade form, the graph was nearly flat, which means CPM S30V tested higher than any other serrated edge we've previously tested.[31]

Glesser did add that S30V would be expensive, however:

> The cost of using super steels in a knife generally adds a measurable increase. This is due, in part, to the cost of the steel. CPM S30V can run three-to-four times as much as other commonly used steels, partly due to processing costs... We go through wheels and belts at a faster rate, and reamers don't last as long. The metal removal process becomes slower. Using a steel like CPM S30V can add anywhere from $10-$40, or more, to the retail price, depending on the size of the knife, and labor involved in building the blade.[31]

Not every knife company was on board for the new steel, however. For example, Matt Conable of William Henry Knives did not immediately sell knives in S30V:

> S30V is more difficult to grind and does not, according to everyone I've talked to, polish as well as some of the other high-grade steels. William Henry is arguably known for the best production blade finish in the industry. I will not mess around with blade steel that does not have exceptional finishing qualities... The other thing is that every couple of years, the next, great blade steel comes around. Since I've been paying attention, the industry has jumped from 440C to AST-34 to BG42 to CPM 440V (S60V), and now to S30V. Give it two years, and it will be something else.[31]

-42-

COMPETITION FOR CRUCIBLE

20CV

Latrobe's BG42 had been building in popularity, though it had lost most of that momentum after the release of the very popular S30V. To offer a more compelling alternative to S30V, Latrobe introduced Duratech 20CV. This was a copy of Böhler M390, which at that time was highly uncommon in knives, especially in the USA, but someone at Latrobe must have seen the potential for this steel in the market. Microtech was making limited knives in 20CV by the end of 2002[1]. The earliest announcement of a knife I found in the steel was in December 2003, a Microtech UMS available in either S30V or 20CV[2]. Microtech must have been producing some of the very earliest products in the steel as the Latrobe website did not even list 20CV until 2005[3,4]. Latrobe would partner with SOG (a USA-based specialty knife company) on 20CV, and they began a marketing push for the new steel in 2005 along with the release of SOG knives in 20CV. The ad from Latrobe compared 20CV against "14Cr/4V/2Mo" (S30V) and showed that 20CV had 35% better edge retention and that SOG was introducing two new knives in the steel.

I contacted SOG in 2020 and asked about what led to their use of the steel and heard back from Beren McKay, at that time the Sr. Director of Product Development:

> I was just starting at SOG back then, so I reached out to Spencer Frazer (our now-retired founder) to pick his brain on what he remembers. What he recalls is that [the] story started when he first heard of BG42 from the custom knife maker world. He made the Recondo and an Auto Clip out of this steel and it was amazing from a performance standpoint. From a manufacturing standpoint though it was a nightmare, (I spoke with several of the people who worked on those knives and they confirmed that working with that material was awful, but the performance was incredible). Spencer still wanted to produce [the] best and highest quality knives he could at the time, so he reached out to a friend of his at Latrobe. Latrobe suggested that instead of BG42 why doesn't he try Duratech (20CV). Spencer bought a minimum quantity of it and tested it out with the TeamLeader knife. The performance of the steel was quite impressive, and it appeared to be a perfect more manufacturable substitute for BG42. Due to the recent experience with

BG42, the factories that he usually worked with didn't have much desire to work with another "new" steel, so Spencer found a new USA-based factory to work with it. I reached out to them as well and they confirmed that working with 20CV was not bad at all. On the downside, the market acceptance just wasn't there. At that same time, we experimented with S30V and it also met with lackluster consumer response. Sadly, this dampened Spencer's belief in using what he called "exotic" blade steels and it wasn't until recently that I was able to bring back using better steels. My personal feeling is that like a lot of things, SOG was just a bit ahead of the times from a consumer acceptance point of view.[5]

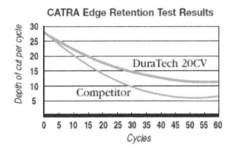
2005 ad for Latrobe's Duratech 20CV.

20CV would build in popularity, with Timberline also announcing knives in the steel in 2005[6], and Rick Hinderer introduced his popular XM-18 model in 20CV in 2007[7]. Hinderer has an interesting history, where he started with stock removal knives, then moved to forged Damascus pieces while learning from Hugh Bartrug, before moving back to sporting knives and tactical knives. Next, he designed knives for Gerber and

Benchmade before beginning his own CNC manufacturing with the XM-18 for larger batches at a more reasonable cost than his former custom knives[8-10].

CPM-154

Crucible's S30V was rapidly gaining popularity with high-end knife manufacturers, but its popularity with custom knifemakers was much more muted. In custom knives, hand-sanded satin finishes were, and are, common, and the typical abrasives in sandpaper are softer than the vanadium carbides in S30V. This makes hand sanding much more difficult and time-consuming. Because of this difficulty, many custom knifemakers stayed with ATS-34 or other steels that are easier to finish. To offer a higher-end powder metallurgy product for these knifemakers, a powder metallurgy version of 154CM was introduced at the end of 2005. Scott Devanna of Crucible said:

> It's a premium, quality stainless steel because it's made using the CPM process. It is more user-friendly than CPM S30V and will take an excellent finish. The only real complaint we've had about S30V is the difficulty in giving finished blades a high polish. Your standard stainless steel grades, like 440C, 154CM, and ATS-34, don't have vanadium. When you add 4 percent vanadium, like we did with CPM S30V, you get carbides in there that give you much-improved wear resistance, but the trade-off is that the steel is more difficult to finish... Tests from our research facility in Pittsburgh show that CPM 154 is twice as tough as 154CM. It surprised everybody. We knew it would improve toughness, but we didn't know it would be that much... You can use a CPM 154 knife blade at a Rockwell hardness of a point or a point and a half higher than ATS-34 or 154CM and still have the same toughness. And you have more wear resistance with CPM 154 because it doesn't have the tendency to chip or break, even at a higher Rockwell hardness... Cost-wise, CPM 154 is a good alternative to CPM S30V for production knife companies. But quite frankly, it isn't as good an overall blade steel as CPM S30V, and I think knife companies know that. If one of them wants to offer something different, CPM 154 is one more grade to consider, and it makes their finishing operations easier.[11]

Bill Ruple, known for his slipjoint folders, said:

> With ATS-34 or 154CM, you can put a nice hand-rubbed finish on a knife blade, and if you hold it in just the right light, you can still see squiggly little lines in the finish (from the banded microstructure). You wouldn't see the lines under normal light, but you'd pick them up under a halogen light. CPM 154 takes a lot nicer finish. The finish is important. I sell a lot of knives to collectors who expect finishes to be perfect, and I've gotten to the point in my knifemaking career of putting a high finish on all my knives, even those I know will be scratched up.[11]

Knifemaker Darrel Ralph reported his experience[11]: "In testing this blade steel, I found it to be very tough, to take an extremely sharp edge and hold it, and it re-sharpens well." Tom Mayo said[11], "The whole industry is looking for a balance between edge holding and toughness. I admire the fact that Crucible is looking to progress in blade-making technology. I still value CPM S30V blade steel, but it's difficult to work. S30V gets all these scratches in it that don't want to come out. You have to rub them and rub them, and belt them and belt them. That's part of the tenacity of the material. It takes me twice as long to put a nice finish on a 154CM knife blade as it does a CPM 154 blade, and CPM S30V takes longer yet."

Some production knife companies would also use CPM 154, such as Kershaw with a Spec Bump folder[11]. However, CPM 154 has become chiefly known as a custom knifemaker steel because of its good finishing characteristics.

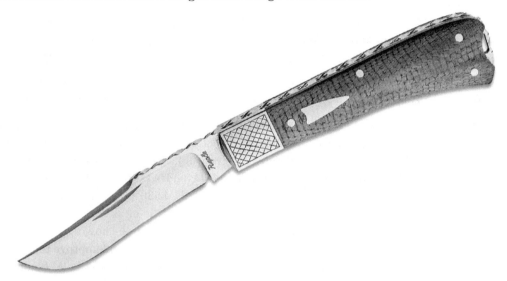

Bill Ruple knife in CPM-154. Photo by KnifeCenter.

CPM-M4

Crucible Steel has sold CPM-M4 since the early 1970s. However, its use in knives had been relatively limited. The earliest mention of it being used in knives I found was knifemaker Timothy Wright, who reportedly used it by 1983 according to *Knives '84*[12]. I contacted Tim and he told me about it:

> I was probably the first knifemaker to use [CPM] M4, first around 1980. I had read about CPM steels way back. I contacted the Crucible warehouse in Chicago... The closest stock they had was 5/32" and 1-1/4" wide. I took it and milled it down on my milling machine, to 3/16" or something less. A big pain in the ass way to make a knife. I wanted to try the stuff. Finer carbides, everything was very appealing to me... I remember that it ground pretty decently, it wasn't a nasty grind like some steels were. Then I started polishing it... I put on a brand new Vitex belt and all it did was skate over the surface... I rubbed that thing out by hand and it took something like 8 hours... Then I sharpened it and I noticed that even with the powder process it wasn't a lot of fun to sharpen. It kind of skated over the surface of certain stones. I got it sharp, but I was disappointed.[13]

CPM-M4 began to see new popularity in the mid-2000s, especially in competition chopping knives. This steel has an excellent combination of properties, though being non-stainless has somewhat limited its use in knives. Crucible metallurgist Scott Devanna recommended CPM-M4 to several knifemakers for the Blade Sports cutting competitions. The steel was then promoted by Texas knifemakers Gayle Bradley, Jerry Halfrich, Tom Overeynder, and Warren Osborne. Gayle Bradley won the cutting competition at the 2006 Guild Show with a CPM-M4 knife. This spread the popularity of the steel for its excellent properties, and soon many competition knives and folding knives were made with CPM-M4. Ed Tarney of Crucible said[14], "With most steels, you get either toughness or wear resistance. M4 balances those properties very well." Scott Devanna said[14]: "With some companies that already use M2 or D2, M4 is a significant upgrade." The four Texas knifemakers all complained about grinding and finishing the steel, with Overeynder saying, "It's a booger to work with!"[14] Still, they also agreed the properties were excellent, with Bradley saying[14], "I've never used a steel that I can get as sharp as M4... It's the king in cutting contests. It's some tough stuff, it really is." Halfrich added, "The edge holding is considerably greater than other steels I've used."[14]

CPM-M4 would be used in factory knives by Benchmade, which announced models in the steel for 2009[15], and Spyderco would release a Mule at the end of 2008[16] and announce a Gayle Bradley design with CPM-M4 for 2010[17].

Benchmade 940 (Osborne design) in CPM-M4. BladeHQ photo.

CPM-D2

Crucible would also introduce a CPM version of D2 in 2007[18], though it has remained a relatively niche product. Kershaw used it in a few knife models, often in a composite construction, with a less expensive spine steel brazed to a different steel for the edge. Jeff Goddard of Kershaw talked about the composite construction in 2009[19]: "The best advantage for consumers is they can have the best cutting steel where it is needed, on the edge, but not have to pay for the whole blade to be made out of the material. For example, Leek retail is $69. With the whole blade in CPM D2, it would run $140-ish. Here we can make the blade spine out of Sandvik 13C26, keeping the price down, and the edge will be the CPM D2."

Kershaw Dividend with CPM-D2 edge and N690 spine. Kershaw photo.

CPM-S35VN

Chris Reeve was dissatisfied with the machinability of S30V and had asked Crucible to come up with a version that was more manufacturable[20]: "Crucible had begun working on S30V and S35VN as early as 2000. When we settled on the final version of S30V, I challenged Crucible to come up with an improved version of S30V within two years waiting in the wings. Crucible agreed." It was not until 2008 that Crucible made up samples of the new version for testing. Reeve "found it to be tougher, polish and machine better, and have excellent edge retention."[20] Crucible's Maria Sawford designed the modified composition by reducing the vanadium by 1%, and partially replacing it with 0.5% niobium. The niobium helped to refine the microstructure and improve toughness. The nitrogen addition in S30V was removed from S35VN (the "N" in S35VN refers to niobium, not nitrogen). The improved machinability and grindability led many knife companies to switch from S30V to the new steel.

Steel	Year	C (%)	Cr (%)	Mo (%)	V (%)	Nb (%)
S35VN	2009	1.4	14	2	3	0.5

CruForgeV

Crucible did not want to leave out forging bladesmiths, but also wanted to make an improved steel for them. Scott Devanna told me that in 2006 a group of forging bladesmiths met with Crucible Steel representatives at the Blade Show. Unfortunately, good sources for low alloy high carbon steels were difficult to come by, and many makers instead relied on bearings or leaf springs. James Batson, who helped with the development of the steel, told me in 2021[21]: "Dan Farr spearheaded this effort for Crucible steel to make a steel for bladesmiths. Dan and his brother took their forge and anvil to Crucible steel in Pittsburgh, Pennsylvania (the Research center), and forged some blades. A Crucible Steel vice president and the head and assistant metallurgist were assigned to the project." Maria Sawford was one of the metallurgists assigned to work on the project. Dan Farr wrote about the impetus for the project in 2009:

> I took some knives I made of 1084 carbon steel on a hunting trip to Namibia. The fine dirt and sand in the animal hides took the edge off my blades so quickly, I swore I would stop using 1084 altogether. I have had a lot of great service from knives I've made of 1084, but in that particularly abrasive environment, I was terribly disappointed. That day was the catalyst for my search for a more abrasion-resistant, forgeable blade steel. Yes, wear resistance, that's what I needed!
>
> If you visit my shop, you will notice I use a lot of the knives I make. You also will notice I always have a folding knife clipped in my pocket. A friend of mine made the folder and gave it a blade of a highly wear-resistant steel. I can't match the wear resistance in the knives I make because wear resistance is overwhelmingly determined by the chemistry of the steel, and high-alloy tool steels just don't forge well by hand. I work hard at getting as much wear resistance out of simple steel as I can by making a blade absolutely as hard

as I can without sacrificing too much toughness. But I cannot come close to the wear resistance of blade steels that contain vanadium carbides.[22]

A popular steel with some knifemakers was 1086, which bladesmith Howard Clark had popularized, so this steel was selected as the basis for development. A 0.75% vanadium addition (along with the necessary increase in carbon) was made to the steel to improve its wear resistance and edge retention. Test samples of the steel were sent to Dan Farr, Kevin Cashen, and James Batson in 2007. Batson said[21], "I was sent a sample of 1086V. This steel was difficult to harden (the steel would be soft after quenching). I did some quenchant and hardness tests... Dan [also] thought the metal was too hard to harden. He asked Crucible Steel to make the steel easier to harden." Crucible increased the manganese content and made a chromium addition so that the steel would be better for oil hardening. Batson tried out two new versions with different combinations of manganese and chromium:

> Crucible steel sent me samples of two different steels that they recommended. I forged two blades so I could conduct rope cutting tests... The test results showed the blade made of 1.0% carbon, 0.76% vanadium, 0.70% manganese, and 0.53% chrome provide a superior cutter over all the knives. When tempered at 425 degrees F this blade cut 1.7 times the knife made of 52100 steel... I recommended that the new Crucible steel be made of this formula.[21]

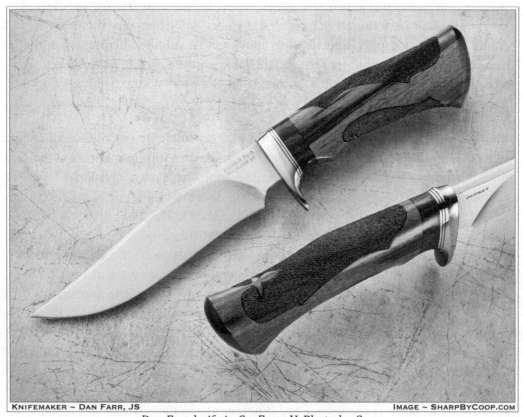

KNIFEMAKER ~ DAN FARR, JS   IMAGE ~ SHARPBYCOOP.COM

Dan Farr knife in CruForgeV. Photo by Coop.

Steel	Year	C (%)	Mn (%)	Cr (%)	V (%)
1086V-1		1.0	0.3	0.25	0.75
1086V-2A		1.0	0.7	0.5	0.75
1086V-2B		1.0	0.85	0.15	0.75
CruForgeV	2009	1.05	0.75	0.5	0.75

Crucible Bankruptcy

Crucible's CruForgeV was released almost simultaneously with another troubling announcement: Crucible filed for Chapter 11 bankruptcy[23]. The economic downturn in 2008-2009 hit them particularly hard. Crucible was purchased and restarted as Crucible Industries, LLC. The company sold off several divisions, including the Research center and pilot mill to Allegheny Technology Inc (ATI)[24], and the Service Centers to SB International which became SB Specialty Metals[25]. This would leave Crucible without distribution and warehouse capabilities, so they developed a partnership with Latrobe Steel Service Centers for the distribution of plate and bar steel in North America[26], and Zapp for distribution everywhere but North America and Japan[27]. The sale of Crucible sheet steels would go to Niagara Specialty Metals[28], the company that had been hot rolling sheet steel for knives for many years but done as contract work. The new agreement gave Niagara the right to purchase Crucible slabs, hot roll them, and sell the sheet steel directly. Scott Devanna in 2010 said[29]: "Niagara for years has taken slab steel, rolled it and converted it to sheet. Plus, Crucible doesn't have the service centers anymore and needs distributors. The knife grades and finished knife steel sheet product will be supplied by Niagara and other licensed Crucible distributors ... the goal is to keep the Crucible knife steel program going forward as well as possible. It does seem like we're doing that."

Böhler-Uddeholm

Ultimately there was minimal interruption in the availability of Crucible knife steel. However, there was an increased marketing campaign from competitors at the same time, including Böhler-Uddeholm USA, the tool steel sales division of the two companies in the USA (separate from the strip division selling AEB-L). The Kershaw Speedform was the American Made Knife of the Year at the 2009 Blade Show, which had an Uddeholm Elmax cutting edge with CPM-D2 spine[30]. Böhler N690, a less expensive conventional steel, was being used in Europe in the 2000s, similar to Remanit 1790C (see Chapter 10). By 2009 several companies were using it, including Böker USA, Fox, TOPS, and Ontario Knife Company[31]. After the success of 20CV, Benchmade and Kershaw announced knives in Böhler M390 in 2010[32,33]. By the end of 2010, Böhler-Uddeholm USA was advertising Elmax, M390, N690, and N680 to USA knifemakers and knife companies[34,35]. John Steedman of Böhler-Uddeholm USA said[34], "Now in the U.S. is a full line of plate thicknesses of Uddeholm Elmax, a super-tough, corrosion-resistant powder steel that reaches 60-plus Rc on the Rockwell hardness scale. With 18 percent chromium and 3 percent vanadium, Uddeholm Elmax offers good workability and straightforward heat treatment at a competitive price. It's available in 1/8-inch sheets and thicker, and you can buy bars for individual knives from your local knife supplies dealer." They also advertised their "Third Generation" powder metallurgy technology, saying it has "Maximum cleanliness and a minimum powder size... Our process and powder properties minimize the risk for edge chipping. For knife manufacturers, Uddeholm PM steel means excellent grinding and polishability properties, trouble-free production, and sharper edges."[36]

In September 2011, Böhler-Uddeholm would add D2, Vanadis 4 Extra, K294 (a copy of CPM-10V), K390 (a unique 9% vanadium non-stainless), and Vanax steel (see the next section) to the list of grades being advertised for knives[37]. They also published results of experiments they had performed of toughness and CATRA edge retention testing of their steels, with results showing superior edge retention for M390 and Elmax when compared with S30V, and N690 outperforming 154CM and 440C. In addition, their toughness experiments showed Elmax to have superior toughness to S30V, S35VN, and CPM 154.

Böhler-Uddeholm also worked with Phil Wilson to test the steels and promote them[38]: "When Böhler Uddeholm decided to introduce Elmax and M390, K390 and K294 I was fortunate to receive samples to try out. I was in a position to be able to take the time and had the heat treat equipment required for precision and repeatability. I reported back on results and heat treats that had worked out. I also had done enough testing in the shop and field to be able to know when I had a process that worked." Phil Wilson made the samples for their CATRA testing. Böhler-Uddeholm published his comments on the steels to their website as well.

Vancron and Vanax

Kobe metallurgists in Japan developed a technology in the 1970s where high nitrogen contents could be introduced to powder metallurgy steels by nitriding the powder[39]. This overcame the limitation of adding nitrogen using conventional technology, where the solubility of nitrogen is very low in liquid steel. When adding the nitrogen to the solid powder instead, the nitrogen content can be much higher, and the small size of the nitrogen atom means that the nitrogen becomes evenly distributed when the powder is "HIPed" into a solid ingot. Uddeholm would develop their own high nitrogen tool steel Vancron 40 and file a patent for the steel in 1999[40]. They would also use this technology to develop stainless steels, resulting in a patent filed in 2005. Replacing carbon with nitrogen in stainless steels results in better corrosion resistance, as nitrogen is less prone to forming chromium nitrides as carbon is to form chromium carbides, leaving more "free" chromium for corrosion resistance. Also, chromium nitrides are less detrimental to corrosion resistance than chromium carbides. The steels released from this patent were Vanax 35 and Vanax 75. The two steels were similar, but Vanax 75 had higher vanadium for more wear resistance. The first knife released in this steel was the Kershaw Tilt, announced in 2010, and it

Use it like you hate it
UDDEHOLM VANAX®

Knives made from Uddeholm Vanax steel barely need maintenance. It's a bold statement, we know – but the outstanding corrosion resistance and excellent edge retention of the steel makes for a blade that can handle the roughest conditions.

By exchanging carbon for nitrogen in our third generation powder process, we have given Uddeholm Vanax a different steel structure with superior properties. You can basically use your knife like you hate it. It won't mind.

www.uddeholm.com/knife

October 2011 Blade Magazine ad for Uddeholm Vanax featuring Kershaw Tilt.

received the Blade American Made Knife of the Year award in that year[41]. A Zero Tolerance knife with a Rick Hinderer design was released in Vanax 35 in 2011[42]. Uddeholm would make a new Vanax steel (sometimes called Vanax 37), which has lower chromium; it is essentially Elmax with the carbon partially replaced with nitrogen. This steel had a patent filed in 2014[43]. The prior Vanax 35 and 75 were dropped, and now only "Vanax Superclean" is available. Similarly, a modified version of Vancron 40 would be released as "Vancron Superclean" in 2020.

Steel	Year	C (%)	Cr (%)	Mo (%)	W (%)	V (%)	N (%)
Vancron 40	2004	1.1	4.5	3.2	3.7	8.5	1.8
Vanax 35	2005	0.2	20	2.5		2.8	1.9
Vanax 75	2005	0.2	21.2	1.3		9	4.2
Vanax	2014	0.36	18.2	1.1		3.5	1.55
Vancron	2020	1.3	4.5	1.8		10	1.8

LC200N, Cronidur 30, N360

Another "nitrogen steel" used in knives is Zapp LC200N used by Spyderco, first announced for the Tusk model at the end of 2013[44]. This steel actually dates back much further, originally released as a bearing steel called Cronidur 30, developed in the early 1990s by Hans Berns and Werner Trojahn[45]. This steel has also been sold as Böhler N360 and Z-Finit. They add more nitrogen into the steel by using a technology where the steel is melted and cast under high pressure. They took the German stainless 1.4116 and partially replaced carbon with nitrogen, increasing the corrosion resistance of the steel while maintaining good wear resistance and excellent toughness.

Steel	Year	C (%)	N (%)	Cr (%)	Mo (%)	V (%)
1.4116		0.5		15	0.65	0.15
LC200N	1993	0.3	0.4	15	1	

H1, X15TN, and H2

Another steel often erroneously called a "nitrogen steel" is H1, made by Myodo in Japan, which in fact does not have an intentional nitrogen addition. This is an unusual steel because it is also very low in carbon. It is a type of "austenitic stainless," the same kind of nonmagnetic stainless steel used in table cutlery, pots and pans, and other applications where strength requirements are not high. However, the austenite transforms to martensite when the steel is cold rolled, and carefully controlling the composition of the steel dictates how easily this transformation occurs. H1 was designed to be cold rolled to a relatively high hardness (55-60 Rc), thus becoming usable in knives, while also having excellent corrosion resistance. This steel was first used by Benchmade in their 100SH2O model, announced in 2002[46]. Benchmade reported[46]: "The steel used in the New!! Model 100SH20 is so corrosive resistant that we have not been able to make it rust even in our salt chamber!!" The steel has also seen use in Spyderco's salt series knives since 2004[47]. In 2005, Benchmade would switch the steel in the 100SH2O model to X15TN instead[48], a steel which had been announced by French company Aubert & Duval in 2003[49]. It has a small 0.2% nitrogen addition which can be made using conventional steelmaking technology without powder metallurgy or high pressure. X15TN is a more traditional martensitic stainless steel unlike the austenitic H1. In May 2022, Spyderco would announce that they were no longer able to obtain H1, so they worked with another Japanese steel company to develop a new steel with similar characteristics called H2[50].

Steel	Year	C (%)	Cr (%)	Ni (%)	Si (%)	Mo (%)	V (%)	N (%)
H1		~0.1	15	7	3.75	1		~0.05
X15TN	2003	0.4	15.5			2	0.3	0.2
H2	2022	~0.1	14	8.25	2.5	2.2		~0.05

Carpenter CTS Steels

Carpenter Steel would begin pushing their available knife steels in 2009[20,51], especially their powder metallurgy XHP. This steel comes from the earlier conventionally cast 440-XH steel developed by Paul Novotny, Thomas McCaffrey, and Raymond Hemphill with a patent filed in 1993[52]. These Carpenter Steel metallurgists wanted to combine the best properties of both 440C and D2:

> 440C alloy has been used in applications, such as bearings and bearing races, where both high hardness and corrosion resistance are required. Type 440C alloy has good corrosion resistance and provides the highest strength and hardness of the known martensitic stainless steels. Although Type 440C alloy is capable of providing a hardness of 60HRC in the as-tempered condition, the alloy provides a case hardness of only about 57-58HRC when it is hardened by induction heating. This limitation on the induction-hardened hardness of Type 440C alloy leaves much to be desired for applications that require a hardness of at least 60HRC. D2 alloy ... steels provide very high hardness, for example, 60-64HRC, when properly heat treated. However, because of their lower chromium compared to stainless steels such as Type 440C, [D2 is] less than desirable for applications that require good corrosion resistance.[52]

So, they increased the chromium content of D2 up to 16% to give it enhanced corrosion resistance and advertised their new 440-XH as a steel that combines the best properties of both 440C and D2. At the 2009 Blade Show, Carpenter announced CTS-XHP as a powder metallurgy 440-XH ideally suited for knives.

Steel	Year	C (%)	Cr (%)	Mo (%)	V (%)
CTS-XHP	2009	1.6	16	0.8	0.45

Carpenter was also pushing their powder metallurgy 440C, called CTS-40CP, and less expensive conventionally-produced steels 40C (440C) and BD1. BD1 was announced in 2009[53], a copy of Hitachi GIN-1 based on Spyderco's previous experience with that steel in their Japanese knives (Chapter 35). Carpenter partnered with Spyderco to work on their CTS line of knife steels[54]. They would also give samples of CTS-40CP and CTS-XHP to custom knifemakers like Bob Dozier, Gayle Bradley, Warren Osbourne, and Tom Overeynder, who all had positive things to say about the steels, especially how well they polished[54]. Carpenter also partnered with SB Specialty Metals who were given an exclusive agreement to distribute the CTS steels[55].

Carpenter also had powder metallurgy steels CTS-204P (a copy of M390), CTS-20CP (a copy of S90V), and CTS-B75P, a powder metallurgy version of BG42. Carpenter would purchase Latrobe in 2011[56], though Carpenter would continue to sell CTS-204P and not Latrobe's 20CV. Crucible had made the powder for Latrobe's Duratech 20CV, and Crucible would continue to make the steel under CPM-20CV branding. Unfortunately, Carpenter has developed a reputation for slow delivery and limited commitment to knife steels. For example, Cold Steel announced in 2018 that they would be moving from XHP to S35VN because of difficulty with sourcing the XHP[57]. Both Zero Tolerance and Spyderco have released knives in Carpenter Maxamet, a powder metallurgy super hard high speed steel with 6% vanadium for extreme wear resistance.

Steel	Year	C (%)	Cr (%)	V (%)	W (%)	Co (%)
Maxamet	2000	2.15	4.75	6	13	10

CPM S125V and S110V

CPM-S125V, initially teased in 2000, was finally released in 2004[58], though it wasn't available to knifemakers for a few years after that[59]. CPM S125V has the highest wear resistance of any available stainless knife steel. RJ Martin says he has made more knives from this steel than anybody. Even Phil Wilson, who has worked with many high-vanadium steels, found S125V too challenging to work with. RJ Martin has a degree in Materials Science and worked as an engineer in airframe design before becoming a full-time knifemaker in 2001. From that background, he has been interested in new steels, heat treatments, and other engineering aspects of knives. I interviewed RJ Martin about using S125V in 2022:

> I was delighted in the fact that no one wanted to play with it. It was a bear to work with. I was scrapping one out of four blades out of it... I just said, ok, it's an extra 100 bucks. A nightmare to grind, and extra belts. I certainly didn't make money using it at first. It was worth it because there's a certain group of collectors that just have to have whatever they consider the ultimate... Other knifemakers didn't want to be bothered with it because it was too difficult. To me, it was very attractive because if you wanted S110V or S125V you had to come to me.[59]

One of the last powder metallurgy stainless steels developed by Crucible prior to the bankruptcy was CPM-S110V, patented by Alojz Kajinic along with Andrzej Wojcieszynski and Maria Sawford[60]. This was a modification of S90V by adding niobium and molybdenum for higher corrosion resistance. The initial steel had manufacturing issues and was modified and re-released in July 2010[61]. Phil Wilson and RJ Martin were the first knifemakers to purchase S110V in 2009 and Wilson finished the first knives in it[62,63]. Wilson reported that he got similar performance from S110V as S125V while finding it much easier to work. Kershaw would release a Shallot model in S110V in 2009[64] and then a Zero Tolerance was announced using CPM-S110V in 2012[65].

RJ Martin Q36SS in CPM-S125V. KnifeCenter photo.

Steel	Year	C (%)	Cr (%)	V (%)	Mo (%)	Nb (%)	Co (%)
S110V (initial)	2006	2.8	14	9	3.5	3.5	2
S110V (final)	2010	2.8	15.25	9	2.25	3	2.5

PD#1, CD#1, Z-Wear, Z-Tuff, and CPM Cru-Wear

Carpenter Steel would introduce a powder metallurgy version of Vasco Wear called PD#1 (Punching Die #1) in 2001[66]. Carpenter would also release CD#1 in 2004[67], a powder metallurgy version of Atlas Steels Beaver (see Chapter 12). Zapp Tools USA was a USA branch of the German materials company formed after Crucible's bankruptcy. Gary Maddock would bring PD#1 to Zapp under the name Z-Wear in 2010[68]. Maddock and Evelin Ratte would slightly modify CD#1 with lower Cr and higher Mo to make Z-Tuff in 2012[69]. Crucible was making the powder for Z-Wear and would release their own version of powder metallurgy Vasco Wear called CPM Cru-Wear around 2013[70]. Spyderco released a Mule in 2011 using the conventional version Cru-Wear[71]. Spyderco remembered the history of this steel, saying it is[71] "very similar to Vasco Wear, a steel used by Gerber Legendary Blades in many of their past production knives." They would then release a Military model in CPM Cru-Wear at the end of 2013[72] and since then the steel has gained a reputation for having an excellent balance of properties including high toughness and good edge retention. Strider released a knife in PD#1 in 2013 as well[73]. Alpha Knife Supply would begin selling Z-Wear in 2014 which would increase its popularity with custom knifemakers[74].

Steel	Year	C (%)	Cr (%)	V (%)	Mo (%)	Ni (%)
CD#1	2001	0.7	8.25	1	1.4	1.5
Z-Tuff	2012	0.7	7.5	1	2	1.5

Post-Bankruptcy Crucible Steels

When Crucible had its Research center, it had dedicated research and development metallurgists, materials modeling software like Thermo-Calc, and lab-scale powder metallurgy equipment so that test heats could be produced of new compositions. Post-bankruptcy, none of these were available for helping with the development process. In 2011, Crucible introduced CPM-4V[75], a minor modification of Uddeholm Vanadis 4 Extra.

Frank Cox of Niagara Specialty Metals had an idea for a new Crucible powder metallurgy stainless steel with increased chromium for enhanced corrosion resistance. Bob Skibitski, the mill metallurgist at Crucible, came up with the final composition by modifying S35VN with increased chromium and nitrogen. The new steel CPM S45VN was announced in 2019[76].

Sal Glesser of Spyderco would request an exclusive CPM steel with a cobalt addition to it like VG10, one of their favorite steels manufactured in Japan. This steel also had a composition finalized by Bob Skibitski. This steel was announced in January 2020 by Spyderco under the name CPM-SPY27[77].

In March 2021, CPM MagnaCut would be announced[78], which I designed. This steel would further reduce chromium content from prior CPM stainless steels like S90V, S30V, and S35VN, along with balancing other elements like carbon and nitrogen. These composition changes eliminated chromium carbides for improved properties. When chromium carbides are present in steel, the regions surrounding the carbide are lower in chromium, because when the carbide forms it takes chromium from the local region. Thus the chromium carbides act as sites for corrosion to initiate, and when the chromium carbides are eliminated the steel has better corrosion resistance. Toughness was also improved through a design having no chromium carbides, as chromium carbides are larger and therefore more detrimental to toughness. Using hard vanadium and niobium carbides instead means that the same level of wear resistance can be achieved with less carbide, again providing better toughness. This is similar to how Vanadis 4 was redeveloped to become Vanadis 4 Extra (see Chapter 34). This gives the steel properties similar to non-stainless powder metallurgy steels while also having improved corrosion resistance when compared to those prior 14% chromium steels.

Steel	Year	C (%)	Cr (%)	V (%)	Mo (%)	Nb (%)	N (%)	Co (%)
4V	2011	1.35	5	3.85	2.95			
S45VN	2019	1.48	16	3	2	0.5	0.15	
SPY27	2020	1.25	14	2	2	1	0.1	1.5
MagnaCut	2021	1.15	10.7	4	2	2	0.2	

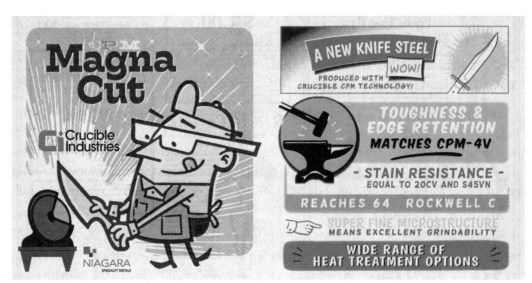

"Vintage" ad for MagnaCut by Ralph Thomas.

-43-

KNIVES AND STEEL IN CHINA AND TAIWAN

Original Equipment Manufacturing (OEM) in China and Taiwan

In the 1970s and 1980s, Seki City, Japan was the place of choice to have knives made for import into the USA. However, the exchange rate became less favorable when the value of the yen spiked in 1987 and again in 1995. This led to many knife companies moving their manufacturing of less expensive imports from Japan to China. By 1997, Les de Asis of Benchmade said[1]: "Now only about 40 percent of the cottage knife factories are left in Seki City. And if not for the commitment of Sal Glesser of Spyderco, that percentage would be even lower." The exchange rate is more favorable with China, and the cost of labor remains low. More and more American knife companies added lines made in China including Buck, Kershaw, Spyderco, Benchmade, Ka-Bar, Gerber, W.R. Case, and SOG. In the 2000s a couple of old establishment American knife companies would go bankrupt, including Imperial Schrade and Camillus. Once other companies purchased the rights to the name and trademarks, they would have most of the new knives made in China as well.

Columbia River Knife & Tool (CRKT) was formed in 1994 by two former Kershaw employees, Paul Gillespi and Rod Bremer. This company was built on knives constructed in China and Taiwan, such as the Taiwan-produced Apache released in 1998, complete with an ATS-34 blade and anodized aluminum handle[2]. Doug Flagg of CRKT said in 2007[3]: "The cost of labor is the main advantage. Anything that requires intensive labor to make will cost less overseas." However, there are some difficulties, as Flagg said[3]: "The time to market can be longer due to many overseas manufacturing issues... Communication is the biggest disadvantage. What [normally can be accomplished in hours in America] takes days overseas, and you never really know if they completely understand."

1998 ad for CRKT Apache.

For many years manufacturing in Taiwan was more advanced than in mainland China. Vance Collver of Benchmade said in 2007[3]: "There is a big difference currently between most Taiwan manufacturers and the Chinese factories, although the gap is closing rapidly. The Taiwanese have fine blanking, precision stamping, and conventional stamping capabilities, as well as a mature knife manufacturing workforce." Jeff Goddard of Kershaw said in 2007 of Chinese factories[3]: "Some of the machinery is very basic, from hollow-grinding and flat-grinding machines to forging steel... [Chinese and Taiwanese factories] do not have laser-cutting machines or the level of machinery that our USA factory does."

China-produced knives have remained controversial for a number of reasons: 1) Many USA and European customers prefer domestically produced knives, 2) Anti-Chinese and anti-Communist sentiment, 3) Several companies have copied designs from more popular knife companies and designers, 4) Concerns about workplace conditions in China, and 5) Criticisms about the quality of knives or heat treating in China.

Steel in Taiwan and China

By 2007 there was also a divide in the quality of steels that could be obtained in China vs Taiwan, as Jeff Culver stated[3]: "The Taiwanese can source and produce [knives] out of AUS-8, 440C, and N690 with no problem. There are even a few vendors that will source and produce knives out of S30V... [With Chinese factories t]he highest quality of steels they can work with are 440A, 9Cr14CoMov, 8Cr13MoV, 8Cr14MoV, 420 and other 'mystery' stainless steels." These letter-number designations come from the naming convention used in Europe, and the majority of the steels produced in China are also copies of European alloys. For example, 7Cr17 refers to 440A, 3Cr13 is 420, and 9Cr18MoV is European 440B, also called 1.4112. The most common China-produced knife steels come from Ahonest Changjiang, founded in 1974, which is "the largest high carbon stainless steel producer in China."[4] There have occasionally been issues with the steels reportedly being used. When Spyderco introduced their Taiwan-produced Byrd line of knives in 2005, the factory told them they were using 440C, but instead it was actually 8Cr13MoV[5]. 8Cr13MoV (sometimes called 8Cr14MoV) is a copy of Aichi AUS-8. Another steel used in some knives is 10Cr15CoMoV which is a copy of Takefu's VG-10.

Chinese Knife Manufacturers Market Their Knives

The quality of knife manufacturing in China continued to improve, with many factories adding CNC machining and other advanced tooling, which is used to create more expensive, but also higher-end and higher quality knives. Some of these OEMs began to see opportunities to make their own brands to sell directly and cut out the middleman. Companies that started marketing their brands include Kizer (2012)[6], Reate (2014)[7], Rike (2015)[8], WE knives (2016)[9], and Artisan Cutlery (2018)[10]. Some of these have also introduced their own budget brands, including WE Civivi[11], and Artisan Cutlery CJRB knives[12]. Many of these companies have become known for having comparable quality to the best USA-made knives. The cost of some of these knives has also increased as the quality of manufacturing has grown. This is part of why the budget lines were introduced with less expensive manufacturing. Many of the high-end knives use imported steel from the USA, Europe, and Japan such as Böhler M390, Crucible CPM-S35VN, and Takefu VG-10. Part of the reason why there has been growth in the number of new brands is also from the increasing numbers of knife enthusiasts in China and Taiwan. Many knifemakers have reported a growing number of sales to such customers[13,14], and companies like Rike (Richard Wu) and Reate (David Deng) were started by Chinese knife enthusiasts.

Spyderco Byrd Flight knife. KnifeCenter photo.

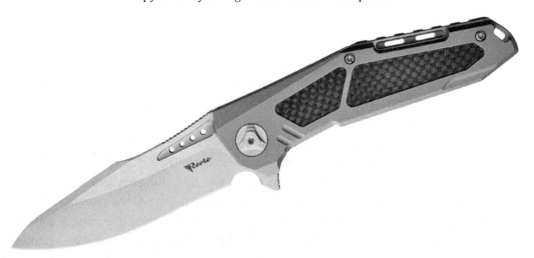

Reate K-3 model knife. KnifeCenter photo.

D2 Steel

D2 steel has maintained its popularity since the beginning of USA custom knives. Since D2 is a conventionally produced ingot steel, it costs less than powder metallurgy steels. So the excellent reputation of D2 means it serves as a lower cost "upgrade" over steels like 8Cr13MoV and 420. D2 has been used in many China-produced knives over the past several years from companies like Böker, CRKT, Bestech, Steel Will, Artisan Cutlery, WE Knife/Civivi, and Kershaw. The positive reputation of D2 continues due to a resurgence in the USA in the late 1990s and early 2000s. This was led in part by knifemaker Bob Dozier, who talked about his steel history in a 2002 interview:

> When I attended my first Guild Show in 1971, all the knives on my table were made from files. It was about that time I started using some 1095 and O-1 steels. In retrospect, the knives seem crude, but the hunters in Louisiana liked them because they would cut and hold an edge well. At the time, I had a hard time acquiring knifemaking tools and made many of [the tools] myself. I became acquainted with A.G. Russell at that show in 1971. He'd heard about me from Bucker Gascon and Jim Mustin, both knifemakers from southern Louisiana.[15]

Dozier would move on to working with 440C and 154CM, preferring to offer stain resistance over simple carbon steels like 1095. Dozier began experimenting with D2 around 1992 and was advertising knives in it in 1993[16]. Dozier said in 2002:

> I had read and heard about D-2. Very few knifemakers were using it at the time. It's hard to work and doesn't take a pretty mirror finish, but I was amazed at the razor edge I could put on a properly heat-treated D-2 blade. I was even more amazed at the edge holding it exhibited. Consequently, I use D-2 almost exclusively now. I'm no metallurgist, but I have done a lot of research and experimenting with metals. I do my own heat treating and keep close tolerances in hardness by Rockwell testing my blades at every stage of heat treating. The quality of steel and the heat treat are of paramount importance in making a knife. A poorly designed knife with a good steel, properly heat treated, will be functional. The best design, however, is useless if the steel is of poor quality or the heat treat is inadequate. Although I admire great art knives, I design knives that are extremely functional and straightforward. I make a no-nonsense "cutting tool" built to use and last.[15]

Dozier was not impressed by new steels that were released in the 1990s and early 2000s[17]: "[440V] won't cut. I don't think the carbides are the whole story. It's the size of the carbides. There's nothing better than D-2."

Some production knife companies would also begin to use D2, such as Ka-Bar in 1999[18], Queen Cutlery in 2002[19], and Swamp Rat in 2004[20]. Eric Isaacson of Swamp Rat said:

> If D-2 were introduced today, it would most certainly be considered as one of the "super steels." The fact that it has been around for so many years should not act as a deterrent to the customer, but rather as a motivator. D-2 is definitely a hi-tech steel. Its advantages include exceptional edge holding and nearly the same level of corrosion resistance as ATS-

34. Although it's a fairly expensive steel in comparison to many other cutlery grades, its reputation for edge holding commands a premium price. It's a difficult steel to grind, but D-2 is a good choice for certain applications.[20]

Wayne Goddard said in 2004[20]: "I make a fair amount of hunting knives in stainless, and I consider D-2 to be a stainless. It doesn't have as good a reputation as it should. I think there's a lot of it that isn't 'up to snuff' because it's not heat treated properly. The young guys look at that and don't like it, but I'd rate it right up there with ATS-34 and 154CM for edge holding."

KA-BAR 1999 ad for USA-made D2 knives.

Artisan Cutlery AR-RPM9 and Exclusive Chinese Steel

In 2020, Artisan Cutlery announced a new exclusive powder metallurgy stainless steel they named AR-RPM9[21]. This steel was marketed as a more budget-friendly powder metallurgy steel than better-known grades such as S35VN and M390. This steel is a powder metallurgy version of European 1.4112, also called 440B or 9Cr18MoV. The "9" in AR-RPM9 refers to the number of elements in the steel, but in fact, this is based on a misunderstanding by the company, as several of the elements included in the list of 9 are "residual" elements not intentionally added. This steel likely performs similarly to other low vanadium (<0.5%) powder metallurgy stainless steels such as Carpenter's 40CP and Crucible's CPM-154. Though with somewhat lower carbon (0.9 vs 1.05%), AR-RPM9 has lower wear resistance. The location of manufacture for this steel has not been disclosed. I have seen many assume that the steel is made in China, but the origin of the steel has not been announced. The common knife steel company Ahonest does not have PM facilities[4]. Heye Special Steel has some patents on powder metallurgy steel, but they also have a partnership with Erasteel in Sweden[22]. So it could be that Artisan Cutlery is importing PM steel from another country, either as rolled sheets, or ingots that are rolled in China. It will be interesting to see if more China-exclusive steels are developed for knives.

CJRB Ruffian in AR-RPM9 steel.

-44-

THE FAMILY TREE OF KNIFE STEEL

With so much history covered in this book, it is easy to lose track of the connections between various developments. To look at things from a high level, this chapter will make those connections more explicit. No steel is developed in a vacuum, and the many discoveries about steel design will go into any new product. So here I have presented a somewhat simplified timeline to show the evolution of knife steel to some of the popular grades that are used today. I have four different evolutionary "trees" which show from the top down how we got from the early steels to today. These four trees are:

1) Low alloy steels, which includes steels like O1, Blue Super, W2, chrome-vanadium cutlery steel, and ApexUltra,

2) Chromium-alloyed die steels like A2, D2, Vasco Wear/Cru-Wear, Vasco Die/CPM-3V, and Vanadis 4 Extra,

3) High Speed Steels like M2, M4, Maxamet, and Rex 121, and

4) Stainless steels like 440C, 154CM, M390, Elmax, S30V, and MagnaCut.

These evolutionary trees look a bit like an overly complicated evidence board with yarn tied to way too many thumbtacks, but hopefully, the corresponding explanation (and the rest of this book) will help.

Low Alloy Tool Steels

The three starting points for this tree are 1) Adamantine, the early 0.5% chromium-alloyed carbon steel, 2) Mushet, the tungsten-alloyed steel which led to high speed steel, and 3) W2, the early vanadium-alloyed steel. Starting with the Adamantine branch, the next evolution was 52100, a bearing steel that has a further increase in chromium to 1.5% for higher hardenability so that larger balls could be hardened all the way through. The chromium also helps to give the steel better wear resistance since the chromium-enriched iron carbides are higher in hardness. Chromium in this amount also helps keep the carbide size small for good fracture and fatigue resistance. Chromium also helps prevent "plate martensite" which is detrimental to toughness. The chromium-vanadium cutlery steel used for many years by companies like W.R. Case is essentially W2 plus the chromium addition from Adamantine.

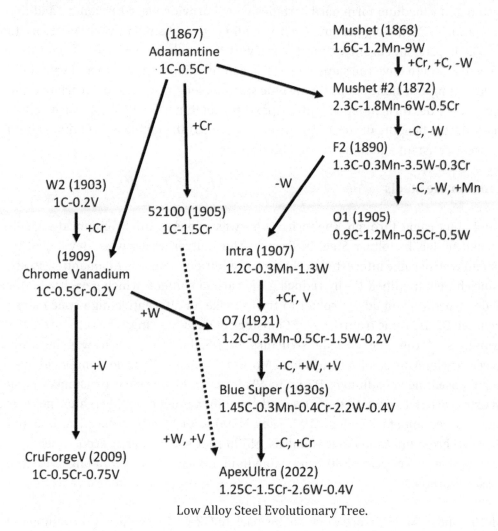

Low Alloy Steel Evolutionary Tree.

That small chromium addition increases the hardenability so that oil hardening is possible. The more recent CruForgeV is similar to Chrome Vanadium but with increased vanadium for enhanced wear resistance. In the tungsten-alloyed line, we have F2, an early modification of Mushet steel with reduced tungsten and manganese/chromium. That medium amount of tungsten gives it high wear resistance for a water/oil hardening steel and the high carbon allows it to reach very high hardness. O1 is also a modification of Mushet with reduced alloying elements across the board to make it "oil hardening" rather than air hardening. Intra was a modification of F2 with further reduced tungsten giving it more intermediate wear resistance and toughness. O7 is a combination of the intermediate tungsten and carbon in Intra with the 0.5% Cr and 0.2% vanadium of chrome-vanadium steel, which was also used as a die steel and for cutlery. The Blue/Aogami-series of steels are O7-type steels, with Blue Super being somewhat higher in carbon and tungsten, plus an addition of 0.4% vanadium, enough to give it enhanced wear resistance. Both tungsten and vanadium form hard carbides which provide wear resistance, and this combination of 2.2% tungsten and 0.4% vanadium gives it similar wear resistance to the 3.5% tungsten in F2. Since tungsten is a very heavy element, it doesn't require as much vanadium to have the same effect. The combination of tungsten and vanadium instead of tungsten alone keeps the carbide size smaller since vanadium carbides and tungsten carbides form independently. ApexUltra combines the W-V approach of Blue Super with the benefits of the 1.5% chromium of 52100, making a steel that is both more wear resistant and tougher than Blue Super.

Chromium Die Steels

D3 tool steel came through Kuehnrich who was exposed to chromium die steels produced by Jacob Holtzer Steel in France. The Holtzer steels were developed by Brustlein who became interested in chromium additions from the Adamantine steels. Kuehnrich then modified D3 by reducing the carbon content, which gave it a better carbide structure, and added cobalt to try to make an alternative high speed steel, designated D5. D2 came from Gregory Comstock of Firth-Sterling, who recognized the possibilities of the fine carbide structure of D5, but the cobalt was basically unnecessary for a die steel, and he added Mo and V instead. D7 came from David Giles who increased the vanadium of D2 up to 4% for even higher wear resistance, along with extra carbon to go with the increased vanadium. Böhler would create a powder metallurgy version of D7 called K190. Then M390 is K190 with reduced carbon and increased chromium to make it stainless. M398 (not shown), is a recent steel from Böhler which is simply M390 with vanadium increased to 7% along with the necessary carbon.

CPM-10V and CPM-15V trace their origins back to C.Y.W. Choice, a 4% chromium die steel initially used as a hot work steel. Modifications of C.Y.W. Choice with 0.5% Mo were produced for large dies. A2 had even higher Mo so that it could fully harden to 60+ Rc and be used as a cold work die steel. Another way to look at A2 is as an evolution from H13 with higher carbon, a hot work steel with 5% Cr plus 1.3% Mo. This combination of Cr + Mo became a common base for many die steels. CPM-10V and CPM-15V steels are A2 with very large vanadium additions and the necessary carbon. Somewhat surprisingly, Crucible would describe 10V as a modification of H13, which is an unusual comparison since CPM-10V is designed to be heat treated to high hardness like A2. The Mo content of CPM-10V is 1.3% like H13 rather than 1.1% like A2, which partially explains the different comparisons.

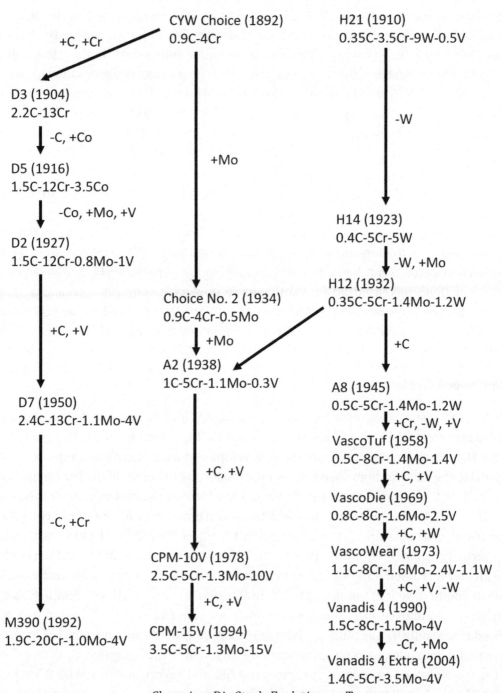

Chromium Die Steels Evolutionary Tree.

A somewhat complicated evolution comes with the Vanadis and Vasco die steels, starting out as hot work die steels and becoming cold work die steels. This started with H21, basically T1 high speed steel with the tungsten and carbon cut in half because much less "hot hardness" is necessary for a hot work die steel. H14 further reduced the tungsten content, recognizing the fact that H21 still had unnecessarily high hot hardness for the application, lowering cost and improving toughness. H12 steel partially replaced tungsten with molybdenum, making the steel better for air hardening. H11 (not shown), came out at a similar time which is the same thing without the tungsten addition. H11 and H12 were increasingly alloyed with vanadium, and H13 is H11 with 1% vanadium. A8 tool steel was a higher carbon version of H12 for hot work applications requiring very high strength. Vasco Tuf (initially called Jet Forge) was a modification of A8 and H13 with 8% chromium instead of 5%, along with 1.4% vanadium for wear resistance. Johnstin of Vasco further increased the carbon and vanadium from this base 8% Cr-1.5% Mo composition for Vasco Die. The patent referred to it as a hot work steel, but commercially it was only marketed as a cold work steel. CPM-3V, released in 1997, is a powder metallurgy version of Vasco Die. Vasco Wear was Vasco Die with increased carbon and a small tungsten addition. Z-Wear, PD#1, and CPM Cru-Wear are all PM versions of Vasco Wear. Vanadis 4 was a modification of these 8%Cr-1.5%Mo steels with vanadium increased up to 4% thanks to powder metallurgy technology. The recognition that the many chromium carbides in Vanadis 4 reduced its toughness led to Vanadis 4 Extra, which dropped the chromium down to 5% and increased Mo to 3.5%.

High Speed Steels

High speed steels were invented after Taylor and White discovered the capabilities of Mushet steel when given heat treatments from high temperatures. The grade T1 with 18% tungsten resulted from rapid development from this base composition to optimize the steel for high speed machining. In Germany, Reinhold Becker found that a cobalt addition would further increase hot hardness at the cost of some toughness leading to T4. Greater vanadium additions were made to high speed steels for improved wear resistance, the earliest being T7, where they found the tungsten could be partially replaced with vanadium. With T15, they discovered that vanadium could be further increased if the carbon was raised along with it. The steel also included a cobalt addition, so it is connected to T4. In the 1930s, metallurgists developed many molybdenum-alloyed high speed steels. Molybdenum can replace tungsten in high speed steels; only half as much molybdenum is required for the same effect. The first successful molybdenum high speed steel was M1, where the "tungsten equivalent" of T1 was maintained by reducing tungsten to 1.5% and making up for it with 8.5% Mo. High Mo steels still had issues with decarburization, so new steels were developed with low enough Mo to avoid this issue, resulting in M2. M2 steel has 6% Mo and 5% W, plus 2% vanadium. This replaced T1 as the most common high speed steel due to its superior properties and the desire to use molybdenum rather than tungsten. M4 was M2 plus higher vanadium and the corresponding carbon. M50 is M1 with the tungsten removed and the molybdenum cut in half, which gave the steel a finer

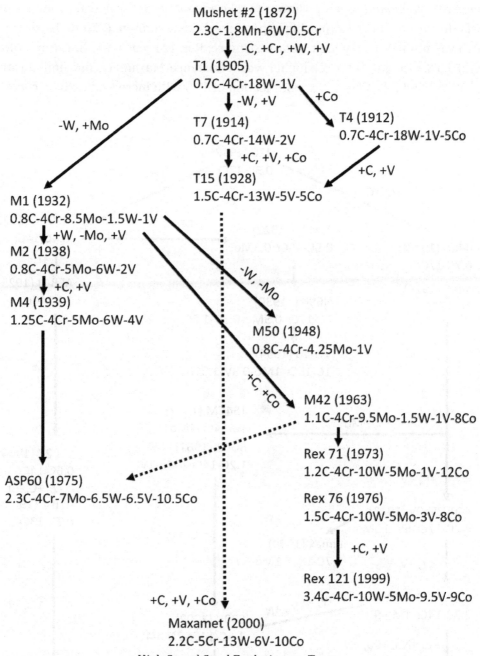

Mushet #2 (1872)
2.3C-1.8Mn-6W-0.5Cr
↓ -C, +Cr, +W, +V
T1 (1905)
0.7C-4Cr-18W-1V
↓ -W, +V   +Co
T7 (1914)   T4 (1912)
0.7C-4Cr-14W-2V   0.7C-4Cr-18W-1V-5Co
↓ +C, +V, +Co
-W, +Mo
T15 (1928)
1.5C-4Cr-13W-5V-5Co
+C, +V

M1 (1932)
0.8C-4Cr-8.5Mo-1.5W-1V
↓ +W, -Mo, +V
M2 (1938)
0.8C-4Cr-5Mo-6W-2V
↓ +C, +V
M4 (1939)
1.25C-4Cr-5Mo-6W-4V

-W, -Mo

M50 (1948)
0.8C-4Cr-4.25Mo-1V

+C, +Co

M42 (1963)
1.1C-4Cr-9.5Mo-1.5W-1V-8Co
↓
Rex 71 (1973)
1.2C-4Cr-10W-5Mo-1V-12Co

ASP60 (1975)
2.3C-4Cr-7Mo-6.5W-6.5V-10.5Co

Rex 76 (1976)
1.5C-4Cr-10W-5Mo-3V-8Co
↓ +C, +V
Rex 121 (1999)
3.4C-4Cr-10W-5Mo-9.5V-9Co

+C, +V, +Co

Maxamet (2000)
2.2C-5Cr-13W-6V-10Co

High Speed Steel Evolutionary Tree.

microstructure, required for bearing applications. The "super hard" 70 Rc high speed steels were developed in the early 1960s. All of these were modifications of prior high speed steels with increased carbon and high cobalt. The most common was M42, which was M1 plus higher carbon and 8% cobalt. With the invention of powder metallurgy, super high speed steels with even higher alloy contents were developed in Rex 71/76 and APS60; I have arrows drawn to previous steels though both of these were relatively unique compositions to take advantage of the new technology. ASP60 was the earliest of the super high speed steels that combined 70 Rc hardness with high vanadium for very high wear resistance. Rex 121 was a modification of Rex 76 with 9.5% vanadium for even higher wear resistance. Maxamet is in a similar category as ASP60 and Rex 121; it is a modification of T15 with increased carbon, cobalt, and vanadium.

Stainless Steel

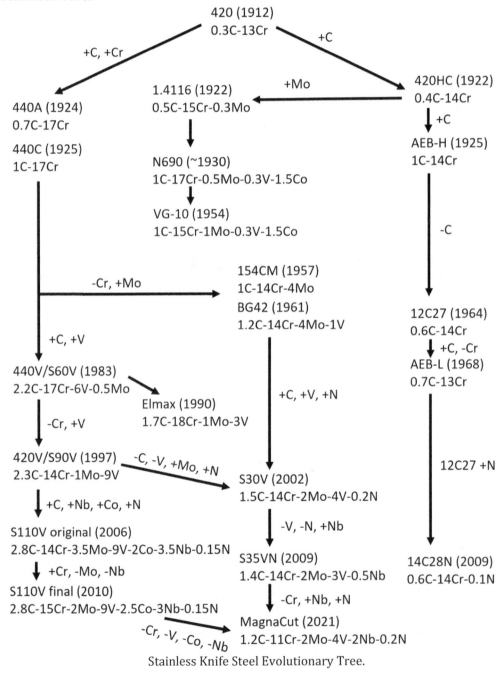

Stainless Knife Steel Evolutionary Tree.

At the top of the stainless steel tree we have 420, though technically this steel started as a modification of prior chromium-alloyed steels. When it was discovered that chromium steels with the correct range of carbon content would have high stain resistance, many knife steels followed. The simplest branch of our tree starts with 420HC, the first of these was Krupp V3M, which had increased carbon for higher hardness. Uddeholm's AEB-H had even higher carbon, both to get around the patent for 420 and also to give it very high hardness for razors. Unfortunately, the very high carbon meant that AEB-H had a coarse microstructure, making it difficult to have high sharpness and a smooth shave. 12C27 had significantly reduced carbon, giving it a much finer carbide structure and was better optimized for razors. AEB-L and 13C26 came from a small increase in carbon and decrease in chromium from 12C27 to maximize the hardness for razors while maintaining the fine carbide structure. 14C28N was a modification of 12C27 and 13C26 with nitrogen and rebalanced C-Cr to maintain similar hardness and carbide structure but with enhanced corrosion resistance.

A small detour in the center of our tree is the German 1.4116, which started as Remanit 1540, a modification of V3M/420HC with a small molybdenum addition for enhanced corrosion resistance. The original Remanit 1540 had only 0.4% carbon which made it even more similar to the original V3M. The same metallurgist that developed Remanit 1540, Wilhelm Oertel, developed Remanit 1790C, which was a 440-series steel with 1.5% Co. The cobalt addition was made for unusual reasons discussed in Chapter 10. I have 1790C listed as N690 just because it is the more commonly recognized steel at this point. Remanit 1790C also had a Mo addition which was becoming more common in 440-series steels around 1930. VG-10 is a modification of Remanit 1790C steel with chromium reduced to 15%.

The 440-series of stainless steels with 440A, 440B, and 440C resulted from a desire to have higher hardness than the original 420 stainless. So, the carbon was increased, and the chromium was raised along with it to maintain similar corrosion resistance. BG41, which I have listed as the more commonly known 154CM, was 440C but with chromium reduced to 14% and molybdenum increased to 4% for hot hardness in bearings. This also took cues from the then-new M50 bearing steel which has 4% Mo. BG42 was an early progression from BG41 with a vanadium addition, the same amount as M50 for increased wear resistance. CPM 440V/S60V and Elmax were modifications of 440C with a significant vanadium addition for wear resistance. These steels are produced with powder metallurgy to maintain toughness with that high vanadium. CPM 420V/S90V was developed when Crucible metallurgists discovered that the chromium of 440V/S60V could be reduced while maintaining the same corrosion resistance, and that reduction in chromium would lead to a higher proportion of vanadium carbides. The high chromium in 440V and Elmax means they primarily have chromium carbides, which are "enriched" in vanadium. These carbides are larger than vanadium carbides, reducing toughness. The high chromium also prevents the harder vanadium carbides from forming, reducing wear resistance. By avoiding the formation of chromium carbides, such as through increased vanadium, the corrosion resistance of CPM-S90V was maintained at the same level as CPM-S60V.

S30V took cues from both 154CM/BG42 and S90V in its development with its 14% chromium plus vanadium design. The vanadium was reduced from S90V down to 4% for better toughness and machinability/grindability. Molybdenum was set to 2%, which was also explored in the S90V patent though ultimately not used. Another alloy addition experimented with in the S90V patent was replacing part of the carbon with nitrogen, improving corrosion resistance. While this wasn't ultimately used in S90V, it showed up in S30V. S110V was a modification of S90V with increased Mo and N, and a niobium addition, all intended to improve corrosion resistance. Niobium improves corrosion resistance because it is such a strong carbide former that it mostly leaves chromium alone, unlike vanadium which enriches chromium carbides. However, there are limits to how much niobium can be added before manufacturing problems occur. Niobium also refines the overall carbide structure. The original S110V was challenging to manufacture, leading to a small reduction in Nb and Mo and an increase in Cr. S35VN was a modification of S30V with the vanadium partially replaced by Nb for a refined microstructure, slightly reduced wear resistance, and improved machinability/grindability for knife manufacturers. The high nitrogen of S30V was also dropped, presumably because the niobium helped compensate to maintain corrosion resistance.

MagnaCut further reduced chromium down to levels where chromium carbide no longer forms. This was a logical extension of the previous reduction in chromium for S90V, S30V, and S110V. The properties were improved just like with that series of steels when compared to the earlier high chromium steels like 440V/S60V, Elmax, and M390. This is a similar modification to what happened with Vanadis 4 Extra, where its chromium was reduced from the prior Vanadis 4. The lack of chromium carbides also gave a boost to corrosion resistance for the same amount of chromium "in solution" (the chromium not tied up in carbides) vs prior steels like S35VN. MagnaCut also used the niobium and nitrogen additions previously established from S110V, S30V, and S35VN for improved corrosion resistance and carbide size (see Chapter 42).

The Future of Knife Steel

As always, it is difficult to predict the future. After developing ApexUltra and MagnaCut, I would like to develop more steels. I have been working on NioMax since before either of those steels, a conventionally produced ingot stainless steel. We have not yet seen whether this steel will be successful once it is produced on a large scale. However, its properties in lab-scale production have been excellent. I would also like to make modifications to MagnaCut for different property combinations. The approach of microstructure containing only vanadium and niobium carbide, free from chromium carbide, gives it improved properties vs. prior steels. But this is an alloy design "approach," not only a single steel, so different balances of hardness, corrosion resistance, wear resistance, and toughness can be created within that paradigm. What other tool steel companies may develop I cannot predict.

In years past, most steels advertised as unique or exclusive in the knife industry have been pre-existing steel grades that are re-branded. However, in recent years, there has been a trend for companies offering knives in truly "exclusive" knife steels. Two examples are Spyderco's CPM-SPY27 made by Crucible and Artisan Cutlery's AR-RPM9. It usually does not make financial sense for a steel company to develop a unique product for a specific small customer. So I am curious if this trend will continue or if the cost-prohibitive nature of these developments will ensure this remains a rare event.

Spyderco Shaman in CPM-15V with BBB logo on the back side.

Another unique steel-related idea which was recently released is a series of knives in CPM-15V by Spyderco with a special heat treatment by knifemaker Shawn Houston[1]. Rather than a typical "custom collaboration" where the knifemaker designed the knives, these are pre-existing models with his heat treatment. These even have his "Triple B Handmade" logo on the knives to advertise this heat treatment. I have not seen this type of collaboration in the past. A small number of custom collaborations in the past have advertised that they worked with the custom knifemaker on the heat treatment, such as the Schrade-Loveless 154CM knives, but again that was for the knives designed by the maker.

While Rockwell hardnes of knives has always varied, there has been a recent trend for higher hardness, especially in production folding knives. Consumers have been requesting high hardness for enhanced edge retention at the cost of reduced toughness. As a result, more production knives are available in 62+ Rc than at any time in prior history. Some knife companies have pushed back, either because of the difficulty of grinding and machining hard steels, or because they want to offer knives with more balanced properties. However, the trend for more high hardness knives will likely continue.

There are more steels being used by custom knifemakers and production knife companies than at any prior time. More "niche" steels with extreme wear resistance or high cost are being used as well. Individual knife companies typically have a long list of steels that are used across their product portfolios. I think this is beneficial in terms of consumer choice, where it is easier than ever to purchase knives in the steel of your choice. While some companies will sometimes reduce the number of steel options for cost or logistics reasons, I don't see the number of available steels in the marketplace significantly decreasing. The number of knife steels available will continue to grow.

REFERENCES

1 The Invention of Tool Steel

[1] Pearson, George. "XVII. Experiments and observations to investigate the nature of a kind of steel, manufactured at Bombay, and there called Wootz: with remarks on the properties and composition of the different states of iron." *Philosophical Transactions of the Royal Society of London* 85, 1795: 322-346.

[2] Smith, Cyril Stanley. "The discovery of carbon in steel." *Technology and culture* 5, no. 2, 1964: 149-175.

[3] Mackenzie, Roderick, and John Whiteman. "Why pay more? An archeometallurgical investigation of 19th-century Swedish wrought iron and Sheffield blister steel." *Historical metallurgy* 40, no. 2, 2006: 138-149.

[4] Walker, Keith. "Stealing the secret of Sheffield's crucible steel." *Steel Times International* 34, no. 5, 2010: 48.

[5] Craddock, Paul, Phil Andrews, and Michela Spataro. "'Not even if we had offered him £50': Early crucible steel production and the history of the Huntsman process." *Historical Metallurgy* 50, no. 1, 2016: 28-42.

[6] Mushet, David. "VIII. Experiments on wootz." *Philosophical Transactions of the Royal Society of London* 95, 1805: 163-175.

[7] Percy, John. *On Iron and steel.* 1864.

[8] Faraday, Michael. "An analysis of wootz, or Indian steel." *Quarterly Journal of Science, Literature and Art* 7, 1819: 288-90.

[9] Stodart, James, and Michael Faraday. "V. Experiments on the alloys of steel, made with a view to its improvement." *The Philosophical Magazine* 56, no. 267, 1820: 26-35.

[10] Hadfield, Robert. "Metallurgical Researches of Michael Faraday." *Nature* 129, no. 3245, 1932: 45-47.

[11] Bessemer, H. "Improvement in the manufacture of iron and steel." US Patent No. 16,082. Patented November 11, 1856.

[12] Anstis, Ralph. *Man of Iron-Man of Steel: The Lives of David and Robert Mushet.* Albion House, 1997.

[13] Mushet, Robert Forester. *The Bessemer-Mushet Process, Or Manufacture of Cheap Steel.* JJ Banks, 1883.

[14] Jeans, William T. *The creators of the age of steel.* New York, Scribner, 1884.

[15] Science Museum Group, Object No. 1972-330.

[16] Osborn, Fred Marmaduke. *The story of the Mushets.* T. Nelson, 1952.

[17] Science Museum Group, Object No. 1889-164/3511/2.

[18] Townsend, A. S. "Alloy Tool Steels and the Development of High-Speed Steel." *Trans. Am. Soc. Steel Treat* 21, 1933: 769-795.

[19] Bauer, Julius. "Improvement in the Manufacture of Steel." US Patent No. 49,495. Patented August 22, 1865.

[20] Kanigel, Robert. *The One Best Way: Frederick Winslow Taylor and the Enigma of Efficiency.* MIT Press Books. 2005.

[21] Taylor, Frederick Winslow. *On the art of cutting metals.* Vol. 23. American society of mechanical engineers, 1906.

[22] *Iron Age*, vol. 26, 1891: 577.

[23] Gill, James Presley, *et al. Tool steels.* 1944.

[24] Mathews, John A. "Modern High Speed Steel." *American Society for Testing Materials* 19, 1919: 141.

[25] Simcoe, Charles R. "The Toolmakers." *The History of Metals in America.* 2018.

[26] Carpenter, H.C.H. "Tungsten in High Speed Steel," *Journal of the Iron and Steel Institute,* 1905.

[27] Carpenter, H.C.H. "The Heat Treatment of High Speed Steel," *Journal of the Iron and Steel Institute,* 1906.

[28] Edwards, C. A. "Function of Chromium and Tungsten in High-Speed Tool Steel." *Journal of the Iron and Steel Institute.* 1908: 104.

[29] Edwards, C.A, and H. Kikkawa. "The Effect of Chromium and Tungsten upon the Hardening and Tempering of High Speed Tool Steel," *Journal of the Iron and Steel Institute.* 1915: 6.

[30] Bain, E.C. and Z. Jeffries. "The Cause of Red Hardness of High-Speed Steel, *Iron Age* 112, 1923: 805.

[31] Grossman, M.A. and E.C. Bain. *High Speed Steel,* John Wiley & Sons, 1931.

[32] Cohen, M. and P.K. Koh, "The Tempering of High Speed Steel," *Transactions of American Society for Metals* 27, 1939: 1015.

[33] Mathews, John A. "Tool Steel Progress in the Twentieth Century," *Iron Age* 126, 1930: 1672.

2 Carbon Steel, Alloys, and Specifications

[1] Krauss, George. "Martensitic transformation, structure and properties in hardenable steels." *Metallurgical Society AIME,* 1978: 229-248.

[2] Seebohm, Henry. "The Manufacture and Uses of Cast Steel." *Engineering Magazine*, June 1881.

[3] Thallner, O. *Tool Steel.* 1902.

[4] Darke, J.M. "Buying Carbon Tool Steel on Specification." General Electric Review, April 1909.

[5] "Words 'Silver Steel, Sheffield,'" E.C. Atkins and Co., No. 2733 - Application filed 2nd July, 1875.

[6] "The Story of Silver Steel." *The Lumber Trade Journal*, June 15, 1911.

[7] Lucie, James R. *Scagel Handmade*. 2010.

[8] Lief, Alfred. *Camillus: The Story of An American Small Business*. 1944.

[9] "The Steel in Our Knives." *Camillus Digest*, May 1952.

[10] "Razors." *Tariff Information Surveys Series C*, 1921.

[11] Woldman, Norman Emme, and Albert Julius Dornblatt. *Engineering Alloys; Properties, Uses*. 1936.

[12] *YSS High-Class Cutlery Steels*, 2006.

[13] "Tool Steel Specifications for the United States Navy." *The Iron Age*. August 12, 1909.

[14] "Specifications for Steel." *Transactions of The Society of Automobile Engineers*, 1911.

[15] "Specifications for Steel." *Transactions of The Society of Automobile Engineer*, 1912.

[16] "Fifth Report of the Iron and Steel Division." *S.A.E. Bulletin*, June 1914.

[17] Brown, Charles M. "Standardizing Specifications for Tool Steel." *Transactions of American Society for Steel Treating*, August 1921.

[18] "Iron Steel and Metal Tables." *Raw Material*, September 1920.

[19] "Comparative Brands of American Tool Steels." *Transactions of American Society for Steel Treating*, Dec. 1921.

[20] Jansson, Anders. "Standard Classification for High Speed Cutting Tools. *The Tool Engineer*, February 1945.

[21] "How to Work Steel ... Tool and die steel." *American Machinist*, November 1954.

[22] Cobb, H. M. *History of Stainless Steel*. 2010.

[23] "Stainless Steel Revisions." *The Welding Engineer*, March 1943: 138.

[24] Tweedale, Geoffrey, and Tweedale Geoffrey Tweedale. *Sheffield Steel and America: A Century of Commercial and Technological Interdependence 1830-1930*. Cambridge University Press, 1987.

[25] Hadfield, Robert. "Manufacture of Steel." US Patent no. 303,150. Patented August 5, 1884.

[26] Hadfield, Robert. "Process of Making Steel." US Patent no. 342,869. Patented June 1, 1886.

[27] Hadfield, Robert Abbott. *Manganese-steel: I. Manganese in its application to metallurgy: II. Some newly-discovered properties of iron and manganese*. Institution, 1888.

[28] Hadfield, Robert Abbott. *On alloys of iron and silicon*. Ballantyne, Hanson & Company, 1889.

[29] Hadfield, Robert Abbott. "Alloys of Iron and Chromium, Including a Report by F. Osmond." *J. Iron Steel Inst*. 42, 1892: 49.

[30] Hadfield, Robert Abbott. *Alloys of Iron & Tungsten*. 1899.

[31] Hadfield, Robert Abbott. "ALLOYS OF IRON AND NICKEL. (INCLUDING APPENDICES AND PLATES AT BACK OF VOLUME)." In *Minutes of the Proceedings of the Institution of Civil Engineers*, vol. 138, no. 1899: 1-125.

3 New Alloy Additions in Steels

[1] Hengerer F., "The History of SKF3," *Ball Bearing Journal* 231, no. 1: 2- 11.

[2] Houdremont, E. and H. Kallen. "On Ball-Bearing Steel." *Transactions of the American Society for Steel Treating*. November 1926.

[3] Kjerrman, B. "Heat Treatment of Two Ball Bearing Steels." *Transactions of the American Society for Steel Treating*, 12. 1927: 759-777.

[4] Robinson, T.L. "Comparative Tests on Ball Bearing Steels," *Transactions of the American Society for Steel Treating*, 11, 1927: 607-618.

[5] Woldman, Norman Emme, and Albert Julius Dornblatt. *Engineering Alloys; Properties, Uses*. 1936.

[6] Lucie, James R. *Scagel Handmade*. 2010.

[7] Fowler, Ed. *Knife Talk: The Art & Science of Knifemaking*. 1998.

[8] Warner, Ken. *Knives, '84*. DBI Books, 1983.

[9] Edmondson, J.R. "The Gospel According to D.E. Henry." *Blade Magazine*. April 1993.

[10] The general history of these grades comes from: Gill, James Presley, *et al. Tool steels* 1944.

[11] Crucible's "Champion Double Special" with 7% tungsten was advertised in 1899: *The Iron Age*. December 28, 1899: 23.

[12] The composition of Champion Double Special isn't available until 1936, see [5].

[13] The earliest reference I have found to a Double Special steel was Carpenter in 1891. *Poor's Manual of Railroads*, Vol. 24. 1891: 1342.

[14] Later switched to using Double Special to refer to "Reading Double Special", a different steel altogether (L6).

[15] Because of the short date that Carpenter used the Double Special name, I cannot find a composition. And they later referred to their F2 grade as "K-W" by 1904, which either means that the earlier Double Special is a different grade or that it was renamed to avoid confusion with RDS: "K-W" trademark Serial No. 134,577, Carpenter Steel

Company, Filed 1920.

[16] Halcomb Double Special is an example of the F2 type with the name "Double Special" advertised by 1906: "Trade Publications: Steel – Halcomb Steel Company." *The Iron Age*. May 3, 1906. Composition can be seen in [5].

[17] Mathews in 1930 said that Ketos was released 25 years earlier. Mathews, John A. "Tool Steel Progress in the Twentieth Century," *Iron Age* 126, 1930: 1672.

[18] The earliest mention of Ketos (1906) in a publication is found in [16].

[19] "Reuben Miller Now in Charge." *The Pittsburg Press*. June 5, 1902: 11.

[20] "C.H. Halcomb Out of Crucible Steel Co." *Boston Evening Transcript*. June 5, 1902: 6.

[21] Mathews, John A. "Alloy Steels for Motor-Car Construction." *Journal of the Franklin Institute* 167, No. 5. 1909.

[22] "Halcomb Steel Plant." *The Iron Trade Review*. February 9, 1911: 293.

[23] "Crucible is Buying Big Plant." *The Pittsburgh Post*. February 11, 1911: 1.

[24] "Obituary: John Alexander Mathews." *Science*. February 22, 1935: 190.

[25] "Stentor." Trademark Serial No. 72,181,009, Carpenter Steel Company, Filed 1963.

[26] The first catalog was released in 1940: https://www.randallknives.com/randall-history/

[27] The first catalog lists the steel as "imported Swedish tool steel."

[28] The 1985 catalog calls the steel "imported O1 Swedish tool steel."

[29] *Hardware Age*. July 19, 1956: 311.

[30] *The Railway Age*. December 27, 1907: 15.

[31] *Machinery*. August 1921: 205.

[32] Sanderson, Michael. "The professor as industrial consultant: Oliver Arnold and the British steel industry, 1900-14." *The Economic History Review* 31, no. 4,1978: 585-600.

[33] Weiner, Aug. F. "Letters to the Editor: Ideal Steel." *The Engineer*. July 1, 1904: 19.

[34] Camp, James McIntyre, and Charles Blaine Francis. *The Making, Shaping and Treating of Steel*. 1920: 596.

[35] Sankey, H. Riall, and J. Kent Smith. "Heat Treatment Experiments with Chrome-Vanadium Steel." *Proceedings of the Institution of Mechanical Engineers*. December 1904.

[36] Ford, Henry, and Samuel Crowther. *My Life and Work*. 1922.

[37] Mathews, John Alexander. "Alloy Steels." *The Mineral Industry, Its Statistics, Technology and Trade* 11. 1903.

[38] *The Motor World*. May 16, 1907: 329.

[39] *Vanadium Steels: Their Classification and Heat Treatment*. 1912.

[40] "Introducing Vanadium Steel Dies for Heavy Forging." *Industrial World*. November 15, 1909.

[41] Coyle, John A. "Making High Grade Steel – VI." *Iron Trade Review*. February 25, 1926.

[42] Mathews, John Alexander. "Manufacture of Tool Steel." US Patent No. 779,171. Patented January 3, 1905.

[43] Mathews, John A. "Modern High Speed Steel." *American Society for Testing Materials* 19, 1919: 141.

[44] Mathews, John A. "Manufacture of Modern High Speed Steel." *The Iron Age*. July 3, 1919.

[45] *The Iron Age*. December 29, 1910: 30.

[46] Giles, James S. *First 100 Years: A Pictorial and Historical Review of W.R. Case & Sons Cutlery*, Smoky Mountain Knife Works, 1989.

4 Carbon Steel USA Factory Knives

[1] Levine, Bernard R. *Knifemakers Of Old San Francisco*. 1998.

[2] Howe, Henry Marion. *The Metallurgy of Steel*. Vol. 1. 1891.

[3] "Genuine Chrome Steel." *Financial and Mining Record*. January 10, 1891: 23.

[4] Dick, Steven. "The Story Behind America's First Hunting Knife." *Field & Stream* (website). October 28, 2019. https://www.fieldandstream.com/story-behind-americas-first-hunting-knife/

[5] Levine, Bernard R. "When Webster Marble Invented the American Hunting Knife." *Knives '90*. 1989.

[6] Schreier Jr., Konrad F. "Marble's Ideal." *Knives '94*. 1993.

[7] Schreier Jr., Konrad F. "The Woodcraft: Marble's Best Seller." *Knives '95*. 1994.

[8] Gagner, Larry. "Webster L. Marble: Gladstone's Sporting Goods Genius." *Knife World*. April 1996.

[9] Marble, Webster L. "Front Sight for Fire-Arms." US Patent No. 451,499. Patented May 5, 1891.

[10] Webster, Marble L. and Frank H. Van Cleve. "Safety-Guard for Axes." US Patent 604,624. Patented May 24, 1898.

[11] Marble, Webster L. and Frank H. Van Cleve. "Match-Box." US Patent No. 650,9444. Patented June 5, 1900.

[12] Levine, Bernard. "Webster L. Marble and Marble's Knives and Axes." *Levine's Guide to Knives & Their Values*. 2001.

[13] Marble Webster L. "Blade for a Hunting Knife." US Design Patent No. 48,624. Patented February 22, 1916.

[14] Hornblad, Jim. "The Marble Woodcraft Knife." *Knife World*. October 1996.

[15] Kertzman, Joe. "Bob Loveless's Knifemaking Heroes." *Blade Magazine*. February 2004.

[16] Lief, Alfred. *Camillus: The Story of an American Small Business*. 1944.

[17] "Trade Items." *The Iron Age*. December 25, 1902: 50.

[18] Wallace, William D. "Yesterday and Today: Progress in Producing Cutlery." *Camillus Digest*. April 1953.

[19] "Navy Knife #5541 L 46." Specification September 1, 1942.

[20] Grant, Marlow. "Camillus Cutlery Company." *Knife World.* March 1980.

[21] 1946 Camillus Catalog.

[22] "The Steel in our Knives." *Camillus Digest.* May 1952.

[23] 1972 Camillus Catalog.

[24] Landis, Henry S. and Reuben S. Landis. "Improvement in Pocket-Knives." US Patent No. 221,467. Patented November 11, 1879.

[25] Levine, Bernard. "Makers of Picture Handle Pocketknives." *Levine's Guide to Knives & Their Values.* 2001.

[26] "Miniature photos and designs used for decorating transparent handled pocket knives, razors, etc." Copyright no. 7477. Copyrighted April 18, 1901.

[27] *The Saturday Evening Post.* December 7, 1901: 25.

[28] "Canton Cutlery Company." *American Vanadium Facts* vol. 1, no. 5. July 1911.

[29] *Canton, a City of Homes, Diversified Industries and Real Opportunities.* 1921: 37.

[30] "Alloy Steel in Pocket Knife Blades." *American Exporter.* September 1921: 107.

[31] *The Rotarian.* December 1915: 576.

[32] *The Iron Age* vol. 119, no. 20. May, 19, 1927: 1505.

[33] Burton, Walter E. "Hobbyist Creates Modern Swords of Damascus." *Popular Science.* June 1939: 66.

[34] Giles, James S. *CASE, The First 100 Years.* 1989.

[35] Tingley, Jim. "The History of the Case Family in Cutlery." *Knife World.* February 1982.

[36] Winter, Butch. "Who Makes Your Cutlery Steels?" *Blade Magazine.* February 1984.

[37] Schrade, George. "Pocket-Knife." US Patent No. 470,605. Patented March 8, 1892.

[38] Schrade, George. "Pocket-Knife." US Patent No. 812,601. Patented February 13, 1906.

[39] Latham, Sid. "Schrade Cutlery Corp." *American Blade Magazine.* September-October 1974.

[40] Miller, Burton T. "A short history of Schrade." *Knife World.* August 1978.

[41] "Cost of Production of Cutlery in Different Countries Compared." *The Jewelers' Circular.* February 1, 1899: 22.

[42] *Supplement to the Directory to the Iron and Steel Works of the United States: Containing a Complete List of the Consolidated Iron and Steel Companies in the United States.* 1910.

[43] "English Cutlery Steel Makers Launch U.S. Campaigns." *The American Cutler.* November 1918: 4.

[44] "Wardlow Returns to England." *The American Cutler.* December 1920: 33.

[45] "Tariff...Hearings...on H.R. 7456." *Sixty Seventh Congress First Session.* August 30, 1921.

[46] Bement, Lewis D. *The Cutlery Story.* 1950.

[47] "United States Tariff Commission: Pocket Cutlery." 1939: 1, 10.

[48] *The Iron Age* vol. 158, no. 25. December 19, 1946: 54.

[49] *The Iron Age* vol. 163, no. 18. May 5, 1949: 74.

[50] Goddard, Wayne. *The Wonder of Knifemaking.* 2000.

[51] *Design News* vol. 28, no. 4. 1973: 63.

[52] https://www.bladeforums.com/threads/0170-6c-what-is-it.148569/

[53] https://www.jerzeedevil.com/threads/paying-it-backward-case-vs-gec-and-the-rest.106900/

5 Projectile, Saw, and Spring Steels

[1] Riley, James. *Alloys of Nickel and Steel.* Iron and Steel Institute, 1889.

[2] Devlin, Ron. "History Book: Carpenter Steel gains global reputation." *Reading Eagle.* February 14, 2018.

[3] Sargent, Dr. "Alloy Steels." Bulletin No. 24-C, *Mechanical Branch of the Association of Licensed Automobile Manufacturers.* January 1907.

[4] "Specifications for Steel." *SAE Transactions* vol. 6. 1911.

[5] "Specifications for Steel." *SAE Transactions* vol. 5. 1910: 169-170.

[6] "Fifth Report of the Iron and Steel Division." *SAE bulletin,* vol. 5. 1913: 167.

[7] *The Boiler Maker.* September 1909: 20.

[8] "Reading Double Special." Trademark Serial No. 134,575. Filed July 6, 1920.

[9] Palmer, F. R. *Tool Steel Simplified.* 1937.

[10] Wills, Child Harold. "Alloy Steel." US Patent No. 1,278,082. Patented September 3, 1918.

[11] *Forging and Heat Treating.* July 1922: 1.

[12] French, H.J. "Comparison of the Alloying Elements Chromium, Nickel, Molybdenum and Vanadium in Structural Steels." *Transactions of American Society for Steel Treating* vol. 11, no 6. June 1927.

[13] https://coloradoencyclopedia.org/article/climax-molybdenum-mine

[14] "Iron, Steel and Nonferrous Alloys Listed by Tradenames." *Machine Design* vol. 9 no. 10. October 1937.

[15] "Iron, Steel and Nonferrous Metals Listed by Tradenames." *Machine Design* vol. 18 no. 10. October 1946.

[16] *Metals Handbook.* American Society of Metals, 1939.

[17] "Chart of Comparable Tool Steels." *The Iron Age* vol. 155 no. 5. February 1945.

[18] Seabright, Lawrence H. *The Selection and Hardening of Tool Steels.* McGraw-Hill, 1950.

[19] Gill, James Presley, *et al. Tool steels.* American Society for Metals, 1944.

[20] *Simonds: Saw Knives Files Steel.* 1916: 34.

[21] *Simonds: Saw Knives Files Steel.* 1916: 171.

[22] Howe, Henry Mario. *The Metallurgy of Steel.* 1891: 47.

[23] Woldman, Norman Emme, and Albert Julius Dornblatt. *Engineering Alloys; Properties, Uses.* 1936.

[24] *The Timberman* vol. 21, no. 8. June 1920: 127.

[25] "The First American Saw Steel." *The Disston Crucible* vol. 1, no. 2. March 15, 1912.

[26] Tweedale, Geoffrey. *Sheffield Steel and America: A Century of Commercial and Technological Interdependence 1830-1930.* 1987.

[27] "A Word on Various Kinds of Disston Steel." *The Disston Crucible* vol. 18, no. 4. September-October 1929.

[28] "Improvement in Saw Steels." *Steel* vol. 87, no. 19. November 6, 1930: 54.

[29] Allen, Henry B. "Improvements in Steels for Wood-Cutting Saws and Knives." *American Society of Mechanical Engineering Transactions* vol. 53. 1931.

[30] Yeo, R.G.B. and O.O. Miller. "A History of Nickel Steels from Meteorites to Maraging." *The Sorby Centennial Symposium on the History of Metallurgy.* October 22-23, 1963.

[31] Browne, David H. *Nickel-Steel; A Synopsis of Experiment and Opinion.* 1899.

[32] Sergeson, Robert. "Behavior of Some Irons and Steels Under Impact at Low Temperatures." *Transactions of American Society for Steel Treating* vol. 19. February 1932.

[33] Brophy, Gerald Robert and Arthur John Miller. "Steels and Structural Embodiments Thereof for Use at Low Temperatures." US Patent no. 2,451,469. Patented October 19, 1948.

[34] Brophy, G.R. and A.J. Miller. "The Metallography and Heat Treatment of 8 to 10% Nickel Steel." *Transactions of American Society for Metals* vol. 41. 1949.

[35] Seens, W.B., Jensen, W.L. and O.O. Miller. "Notch Toughness of Four Alloy Steels at Low Temperatures." *American Society for Testing and Materials* vol. 151. June 18-22, 1951.

[36] Allen, Henry B. "Alloy Steel." US Patent no. 1,681,797. Patented August 21, 1928.

[37] *Chicago Journal of Commerce* vol. 61 no. 1. July 7, 1892: 13.

[38] *Svensk industry-och exportkalender.* 1912: 288.

[39] Jatczak, C.F. "Bearing Steels of the 52100 Type with Reduced Chromium." *Critical Mterials Use by the Steel Indsutry.* 1982.

[40] "User Report No. 20 on Experience with NE (National Emergency) Alloy Steels." *Steel* vol. 113. Sept. 20, 1943.

[41] Chandler, Harry. *Heat Treater's Guide.* 1995.

[42] "Nickel Alloy Tool Steels: A Practical Guide to the Use of Nickel-Containing Alloys." Nickel Institute. 1968.

[43] Woldman, Norman Emme, and Robert C. Gibbons. *Engineering alloys.* 1973.

[44] Woldman, Norman Emme, and Roger J. Metzler. *Engineering alloys.* 1945.

[45] *Steel* vol. 132, no. 14. April 6, 1953: 153.

[46] *Disston Crucible* Vol. 19, no. 5. November-December, 1930: 101.

[47] Loveless Knives Catalog, 1972: 6.

[48] Hollis, Durwood. *Knifemaking with Bob Loveless.* Krause Publications, 2010.

[49] Buerlein, Robert A. "Vietnam Remembered: A Jungle Fighting Knife." *Blade Magazine,* Oct. 1982.

[50] Holtzer, L.J.D. Great Britain Patent No. 189,213,137. Application July 18, 1892.

[51] "Eine neue Art von Federstahl." *Polytechnisches Notizblatt* vol. 48, no. 16. 1893: 127.

[52] Payson, Peter. "Low Alloy High Tensile strength, High Impact Strength Steel." US Patent No. 2,447,089. Patented August 17, 1948.

[53] Payson, Peter and A.E. Nehrenberg. "New Steel Features High Strength and High Toughness." *The Iron Age* vol. 162, no. 17. October 21, 1948.

[54] Allten, A.G. and P. Payson. "The Effect of Silicon on the Tempering of Martensite." *Transaction of American Society for Metals* vol. 45. 1953.

[55] Horn, R. M., and Robert O. Ritchie. "Mechanisms of tempered martensite embrittlement in low alloy steels." *Metallurgical Transactions A* 9, no. 8, 1978: 1039-1053.

[56] Nam, W-J., and H-C. Choi. "Effects of silicon, nickel, and vanadium on impact toughness in spring steels." *Materials science and technology* 13, no. 7, 1997: 568-574.

[57] Guillet, Leon. "Metals and Alloys Employed in Automobile Construction." *The Horseless Age* vol. 18, no. 19. November 7, 1906.

[58] "Specifications for Steel." *SAE Transactions* vol. 5. 1910: 171.

[59] "Specifications for Steel." *SAE Transactions* vol. 7. 1912: 46.

[60] "Fifth Report of the Iron and Steel Division." *SAE bulletin,* vol. 5. 1914: 170.

[61] "Specifications for Steel." *Transactions of the Society of Automotive Engineers* vol. 10, pt. 2. 1915: 31.

[62] Shimer, William Robert and Fred C.T. Daniels. "Silico-Molybdenum Steel and Article Made Therefrom." US Patent no. 1,893,004. Patented January 3, 1933.

[63] *Railway Mechanical Engineer* vol. 109 no. 12. December 1935: 11.

[64] "Tool Steel Brands, Alphabetically Arranged." *The Iron Age* vol. 129, no. 24. June 16, 1932.

[65] "Solar." US Trademark Registration Number 512,179. Trademarked July 12, 1949.

[66] Loveless, R.W. "Steel and Knives: A Bladesmith's Bible." *Knife Digest*. 1975.

[67] Goddard, Wayne. *The Wonder of Knifemaking.* Krause Publications, 2000.

[68] "Fifth Report of the Iron and Steel Division." *SAE bulletin*, vol. 5. 1914: 168.

[69] "Standards Committee Meeting." *The Journal of the Society of Automotive Engineers* vol. 10, no. 2. February 1922.

[70] Multiple trade journals had these ads in 1949 such as *American Machinist.*

[71] "Obituary." *The Iron Trade Review* vol. 32, no. 37. September 14, 1899.

[72] "C.Y. Wheeler Dead." *Pittsburgh Commercial Gazette.* September 7, 1899: 3.

[73] "Furnace, Mill, and Factory." *The Engineering and Mining Journal* vol. 39, no. 1. January 3, 1885: 10.

[74] *Railway World* vol. 12, no. 27. July 3, 1886: 643.

[75] "James Todd and Frank L. Slocum v. C.Y. Wheeler and The Sterling Steel Company, Appellants." *Pennsylvania State Reports Containing Cases Decided by the Supreme Court of Pennsylvania* vol. 173. 1896.

[76] "Annual Report of Inspector in Charge Ordnance Proving Ground, Annapolis." *The Executive Documents of the House of Representatives for the First Session of the Fifty-Second Congress.* 1891-1892.

[77] *Annual Report of the Secretary of the Navy.* 1892.

[78] "Manufacturing: Iron and Steel." *The Iron Age.* July 20, 1893.

[79] Wheeler, Charles Y. "Manufacture of Chrome-Steel Alloy." US Patent No. 597,869. January 25, 1898.

[80] *Journal of the American Society of Naval Engineers* vol. 6 no. 4. November 1894.

[81] "Important Consolidation of Steel Interests." *The Age of Steel* vol. 81, no. 4. January 23, 1897: 12.

[82] "To Make Projectiles Near Washington." *The Iron and Machinery World* vol. 99, no. 1. January 6, 1906: 12.

[83] *The Iron Trade Review*. May 2, 1918: 13.

[84] Gill, J. P., and M. A. Frost. "The Chemical Composition of Tool Steels." *Trans. Am. Soc. Steel Treat* 9, 1926: 75-88.

[85] Gill, James P., *Tool steels: a series of five educational lectures on the selection, properties and uses of commercial tools steels presented to members of the ASM during the 16th National Metal Congress and Exposition, New York City, Oct. 1 to 5, 1934.*

6 Early 20th Century Knifemakers

[1] Levine, Bernard. "20th Century Knifemakers." *Levine's Guide To Knives & Their Values.* 2001.

[2] Lucie, James R. *Scagel Handmade.* 2010.

[3] Lambert, Glen. "William Wales Scagel: Master Craftsman." *Knives '83.* 1982.

[4] McEvoy, Harry. "William Wales Scagel -- First Knifemaster of the 20th Century Part I." *Knife World.* April 1980.

[5] McEvoy, Harry. "William Wales Scagel -- First Knifemaster of the 20th Century Part II." *Knife World.* May 1980.

[6] Suedmeier, Harlan. "More About Richtig Knives." *Knives '88.* 1987.

[7] Lambert, Glen. "F.J. Richtig: Believe It or Not Cutler." *Knives '84.* 1983.

[8] https://history.nebraska.gov/blog/stronger-steel

[9] Wadsworth, Jeffrey, and Donald R. Lesuer. "The knives of Frank J. Richtig as featured in Ripley's Believe It or Not!®." *Materials characterization* 45, no. 4-5, 2000: 315-326.

[10] Suedmeier, Harlan. Phone conversation, 2022.

[11] "Ryerson Tool Steel Name Contest." *The Tradesman.* November 1915.

[12] Woldman, Norman Emme, and Albert Julius Dornblatt. *Engineering Alloys; Properties, Uses.* 1936.

[13] Teague, J., R. LeMaster, J. Rinksc, A. Winkelmannd, and L. Bartlett. "Attempted Replication of Frank Richtig's Forgotten Steel Heat Treatments."

[14] https://knifesteelnerds.com/2019/07/08/frank-j-richtig/

[15] Suedmeier, Harlan. Phone Interview, 2019.

[16] Krumholz, Phillip. "The Forgotten Knifemaker." *Knife World.* January 1989.

[17] Krumholz, Phillip. "The Forgotten Knifemaker Part II." *Knife World.* February 1989.

[18] Martin, Robert E. "Novel Workshop Produces Custom-Built Knives." *Popular Science Monthly.* September 1937.

[19] Krumholz, Phillip. "The Forgotten Knifemaker Part III." *Knife World.* March 1989.

[20] Towsley, Frank, Stephen Smith, and Frank Towsley. *Hear the Hammer: The Story of Montana's Ruana Knives.* 1995.

[21] Roberts, Vincent. "Ruana Knives - Past, Present, and Future." *Knife World.* October 2003.

[22] Beall, James R. and Gordon Morseth. *The Keen Edge of Perfection: A History of the Morseth Knife.* 2007.

[23] 1950 Morseth Catalog.

7 New Hot Work and High Speed Steels

[1] Mathews, John A. "Tool Steel Progress in the Twentieth Century," *Iron Age* 126, 1930: 1672.

[2] "Customs: Drawback on steel." *Treasury Decisions under the Custom, Internal-Revenue, and Other Laws* vol. 23, no. 12. September 19, 1912.

[3] Gill, J.P. "High Speed Steel." *Modern Machine Shop* vol. 10, no. 3. August 1937.

[4] Becker, Reinhold. "High-Speed Tool-Steel." US Patent No. 1,099,531. Patented June 9, 1914.

[5] *Machinery* vol. 21, no. 6. February 1915: 18.

[6] Chandhok, V. K., J. P. Hirth, and E. J. Dulis. "Effect of cobalt on tempering tool and alloy steels." *Trans. ASM* 56, 1963: 677-693.

[7] Furness, Radclyffe and Richard H. Patch. "Tool-Steel Alloy." US Patent No. 1,206,833. Patented December 5, 1916.

[8] Kuehnrich, Paul Richard. "Steel." US Patent No. 1,27,431. Patented September 3, 1918.

[9] Giltinan, D.M. "The Metallurgy of High Speed Steel." *Transactions of American Society for Steel Treating* vol. 1, no. 12. September 1921.

[10] Norris, George L. "Alloy Steels." *Transactions of the International Engineering Congress: Metallurgy.* September 20-25, 1915.

[11] "Gyro." US Trademark Serial No. 102,929. Filed April 13, 1917.

[12] "Trade publications. *The Iron Age* vol. 95, no. 22. June 3, 1915: 1276.

[13] *The Official Handbook of the Sheffield and Rotherham Chambers of Commerce with Classified Trade Index and Trade Mark Section.* 1919: 74.

[14] *Machinery* vol. 26, no. 12. August 1920: 318.

[15] Gillett, H.W. and E.L. Mack. "Ferro-Uranium." *The Journal of Industrial and Engineering Chemistry* vol. 9, no. 4. April 1917: 342.

[16] *Machinery* vol. 23, no. 12. August 1917: 175.

[17] *Machinery* vol. 34, no. 12. August 1928: 185.

[18] "Firth Sterling Circle C steel." *Railway Mechanical Engineer* vol. 102, no. 3. March 1928: 176.

[19] "Rex 440." US Trademark Registration No. 772,669. Trademarked July 7, 1964.

[20] Oertel, Wilhelm, and Arthur Grützner. *Die Schnelldrehstähle.* Verlag Stahleisen mbH, 1931.

[21] Gill, J. P., and Robert S. Rose. "Molybdenum High-Speed Steels." *Iron Age* 148, no. 13. 1941: 33-35.

[22] Gill, James Presley. "High speed steel-its history, development, manufacture, metallography, and constitution, including an extended bibliography." 1922.

[23] Gorham, L.C. "The Development of High Speed Steels." *The Tool Engineer* vol. 7, no. 8. December 1938: 10.

[24] Ritchie, S.B. "Molybdenum in High Speed Steel," *Army Ordnance* vol. 11, no. 61. July 1930.

[25] Emmons, Joseph V. "Some Moybdenum High Speed Steels." *Transactions of American Society for Steel Treating* vol. 21. 1933.

[26] Emmons, Joseph V. "Steel Alloy and Articles Made Therefrom." US Patent No. 1,937,334. Patented November 28, 1933.

[27] Kingsbury, Arthur Howard. "Alloy Steel." US Patent No. 1,996,725. Patented April 2, 1935.

[28] Gill, James P. "Alloy Steel Tool." US Patent No. 2,105,114. Patented Jan. 11, 1938.

[29] Herzig, Alvin J. "High Speed Alloy Steel." US Patent No. 2,109,746. Patented March 1, 1938.

[30] Breeler, Walter R. "Development in Molybdenum High Speed Cutting Steels." *Transactions of American Society for Metals* vol. 27. 1939: 289.

[31] De Vries, Ralph P. "High Speed Tool." US Patent No. 2,212,227. Patented August 20, 1940.

[32] Roberts, G. A., J. C. Hamaker, and A. R. Johnson. *Tool Steels.* 1962.

[33] "Wide Variety of Proprietary Irons and Steels for Industrial use Offered by 300 Suppliers." *Steel* 100. 1937: 74.

[34] "Classification of High-Speed Steels by Manufacturer and Trade Name." *American Machinist* vol. 84. February 21, 1940: 96.

[35] "Directory of Tool and Die Steels." *American Machinist* vol. 98, no. 24. November 1954.

[36] *The Tool Engineer* vol. 12, no. 12. December 1943: 57.

[37] The British patent says that the German application dated to December 30, 1926. "Improved Manufacture of Steel." Great Britain Patent No. 282,744. Patented April 15, 1929.

[38] Oertel, Wilhelm. "Steel Manufacture." Canadian Patent No. 284,428. Patented October 30, 1928.

[39] Schallbroch, Henrich and Hans Schaumann. "Kennzeichnung der Schnittleistung von Drehmeißeln durch Standzeit-Schnittemperatur-Kurven." *Stahl und Eisen* vol. 17, no. 57. April 29, 1937.

[40] Gill, James P. "German Tool and Special Steel Industry." *Combined Intelligence Objectives Sub-Commitee* Item No. 31. June 12, 1945.

[41] Kinzel, A. B., and C. O. Burgess. "Effect of Vanadium in High-speed Steel." *Transactions of the American Institute of Mechanical Engineers* vol. 100. 1932: 257-263.

[42] Keen, William Herbert. "Alloy Steel." US Patent No. 1,621,886. Patented March 22, 1927.

[43] "New Tool Steel Features Improved Hot Hardness." *The Iron Age* vol. 158, no. 24. December 12, 1946: 63.

[44] Cary, Robert A. "Further Postwar Tool Steel Developments." *Tooling & Production* vol. 14, no. 10. Jan. 1949: 56.

[45] *The Tool Engineering* vol. 26, no. 6. June 1951: 171.

8 Kuehnrich, D2, and Stain-Free Razors

[1] Tweedale, Geoffrey. "The Razor Blade King of Sheffield: The Forgotten Career of Paul Kuehnrich." *Transactions of the Hunter Archaeological Society* vol. 16. 1991: 39-51.

[2] Kuehnrich, P.R. "The Early History of Stainless Steel." *The Foundry Trade Journal* vol. 41 no. 682. Sept 12, 1929.

[3] "Correspondence: High Speed Steel." *The Locomotive* vol. 26, no. 330. February 14, 1920: 43.

[4] Perar, U.C. "Steel for Cutters, Punches, and Dies: High Chromium Steel Used in France for Casting into Milling Cutters, Punches, Etc." *American Machinist* vol. 33, no. 1. January 6, 1910.

[5] Kuehnrich, Paul Richard. "An Improved Steel." Great Britain Patent No. 191,111,063. Patented Sept. 21, 1911.

[6] "Neor." US Trademark Serial Number 93,989. Filed March. 30, 1916.

[7] Wills, W.H. "Practical Observations of Some High Carbon High Chromium Tool Steels." *Transactions of American Society for Metals* vol. 23, no. 2. June 1935: 488.

[8] Patch, Richard H. and Radclyffe Furness. "Tool-Steel Alloy." US Patent 1,206,902. Patented December 5, 1916.

[9] Gill, J. P., and M. A. Frost. "The Chemical Composition of Tool Steels." *Trans. Am. Soc. Steel Treat* 9, 1926: 75-88.

[10] Woldman, Norman Emme, and Albert Julius Dornblatt. *Engineering Alloys; Properties, Uses.* 1936.

[11] "Giant 'Razor King' A Suicide – Never Shave in His Life." *The Brooklyn Daily Eagle.* May 22, 1932: 73.

[12] Kuehnrich, Paul Richard. "An Improved Steel." British Patent no. 106,187. Patented May 17, 1917.

[13] The Sheffield Daily Independent. June 27, 1917: 7.

[14] "Making Cobaltcrom Steel Tools." *Machinery.* August 1919: 1173.

[15] Knowlton, H.B. "Facts and Principles concerning Steel and Heat Treatment – Part IV." *Transactions of the American Society for Steel Treating* vol. 9. January 1926: 125.

[16] Comstock, Gregory J. "Alloy Steel." US Patent no. 1,695,916. Patented December 18, 1928.

[17] Loveless, R.W. "And Testing Retort." *American Blade Magazine,* Jan-Feb 1977.

[18] Warner, Ken. *Knives '84.* DBI Books, 1983.

[19] Hughes, B.R., and Jack Lewis. *Gun Digest Book of Knives.* 1973.

[20] Bates, John Davis, and James Henry Schippers. *The Custom Knife.* 1973.

[21] *Blade Magazine,* Feb. 1997: 73.

[22] *Blade Magazine,* June 1997: 16.

[23] *Blade Magazine,* July 1997: 68.

[24] *Blade Magazine,* June 1999: 11.

[25] *Blade Magazine,* Sept. 1999: 32.

[26] Kleinman, John F. "Joe Funderburg: A Personality Profile." *American Blade Magazine,* Sept-Oct 1977: 29.

[27] *American Blade Magazine,* Dec. 1978: 32.

[28] Goddard, Wayne. "D-6: the Steel of Brazil." *Blade Magazine*, September 1999: 40.

[29] Goddard, Wayne. "Good, Better, Best: Results of Basic Knife Testing." *Blade Magazine*, December 1990.

[30] Warner, Ken. "The Story of Steel and Sharp." *Knives '87.* 1986: 139.

[31] "A new Industry: Thousands of Sheffield's unemployed Absorbed." *Sheffield Daily Independent.* Mar. 2, 1927: 6.

[32] "Razor Blades Revolution." *Nottingham Evening Post.* November 14, 1924: 7.

[33] "Big Plans of New Sheffield Company." *Sheffield Daily Independent.* April 5, 1926: 5.

9 Harry Brearley and Stainless Steel

[1] Brearley, Harry. *Steel-Makers and Knotted String.* Maney Pub, 1995.

[2] Armstrong, P.A.E. "Corrosion-Resistance Alloys - Past, Present, and Future - With Suggestions as to Future Trend." *Symposium on Corrosion-Resistant, Heat-Resistant and Electrical-Resistance Alloys. American Society for Testing Materials.* 1924.

[3] Zapffe, Carl A. "Who Discovered Stainless Steel?" *The Iron Age* vol. 162, no. 16. October 14, 1948.

[4] Guillet, L. "Chromium Steels." *Le Genie Civil.* March 1904.

[5] Guillet, L. "Nickel-chrome steels." *Rev. Metal* vol. 3. 1906: 332.

[6] Borchers, W. and P. Monnartz. German patent no. 246,035. Patented January 22, 1910.

[7] Monnartz, P. "The Study of Iron-Chromium Alloys With Special Consideration of Their Resistance to Acids." *Metallurgie* vol. 8 no. 7. 1911.

[8] Maurer, E. and B. Strauss. German patent no. 304,126. October 1912.

[9] Strauss, B. "Non-Rusting Chromium-Nickel Steels." *Symposium on Corrosion-Resistant, Heat-Resistant and Electrical-Resistance Alloys. American Society for Testing Materials.* 1924.

[10] Daves, Karl. "Rostfreie Stähle." *Stahl und Eisen.* August 24, 1922.

[11] Woldman, Norman Emme, and Albert Julius Dornblatt. *Engineering Alloys; Properties, Uses.* 1936.

[12] Armstrong, Bernard and Geoffrey Tweedale. "Percy Armstrong (1883-1949): A transatlantic pioneer of alloy steels." *Historical Metallurgy* vol. 27, no. 1. 1993.
[13] Armstrong, Percy A.E. "Stable-Surface Alloy Steel." US Patent No. 1,322,511. Patented November 25, 1919.
[14] Armstrong, Percy A.E. "Heat Treated Stable Surface Alloy Steel." US Patent 1,456,088. Patented May 22, 1923.
[15] White, Albert E. and Claude L. Clark. *Stainless Steel: A Digest with Abstracts and Bibliography.* Department of Engineering Research University of Michigan Ann Arbor. November 1926: 42
[16] *Iron Trade Review.* April 8, 1920: 1085.
[17] Armstrong, P.A.E. "U.S. Enterprise Forestalls German Competition." *Michigan Manufacturer and Financial Record* vol. 27, no. 26. June 25, 1921: 27.
[18] *Ludlum Steel.* 1924: 125

10 Developing Stainless Knife Steels
[1] Parmiter, O.K. "The Long Search for Stainless Steel." *American Cutler.* December 1922.
[2] "The Passing of Stainless Steel Cutlery in England." *American Cutler.* January 1918.
[3] "The State of the Cutlery Trade." *American Cutler.* December 1917.
[4] "Stainless Steel, Eliminated by War, Will Soon Return." *American Cutler.* January 1919.
[5] "New Cutlery Makers Feature Stainless Steel Knives." *American Cutler.* March 1920.
[6] Parmiter, O.K. "Stainless Steel and Stainless Iron." *Transactions of the American Society for Steel Treating.* September 1924.
[7] "Will Stainless Steel Knives Really Cut?" *American Cutler.* June 1919.
[8] "Selling Points of Stainless Steel." *American Cutler.* June 1922.
[9] "Stainless Cutlery." *Liverpool Echo.* October 10, 1930: 13.
[10] Jackman, Edwin T. "Opportunities in Stainless Steel." *American Cutler.* December 1922.
[11] Levine, Bernard. *Levine's Guide to Knives & Their Values.* 2001: 25
[12] *The Knife and Its History: Written On the Occasion of the 100th Anniversary of Victorinox.* 1984.
[13] Levine, Bernard. "Pocketknife History." *Levine's Guide to Knives & Their Values.* 2001.
[14] *American Cutler.* July 1921: 30.
[15] Alvord, Danforth Fletcher. "Design for a Blade for Grape-Fruit and Orange Knives." US Design Patent no. 48,236. Patented November 30, 1915.
[16] *American Cutler.* October 1922: 17.
[17] *Hardware News.* November 1922: 16.
[18] Lief, Alfred. *Camillus: The Story of an American Small Business.* 1944.
[19] "Sweet are the Uses of Technology." *Camillus Digest.* October 1951.
[20] "A 'Stainless' Reputation." *American Cutler.* April 1925: 26.
[21] "Stainless Criticism." *American Cutler.* May 1925: 15.
[22] Levine, Bernard. "Stainless Cutlery Co. Pocketknives." *Knife World.* May 1994.
[23] Levine, Bernard. "American Hunting Knives." *Levine's Guide to Knives & Their Values.* 2001.
[24] *Forest and Stream.* April 1923: 220.
[25] Sheehan, Thomas C. "The Lesson of Sheffield Spirit." *Hardware Review.* June 1920: 39.
[26] "New Stainless Table Product." *American Cutler.* April 1922.
[27] *Hunter - Trader - Trapper.* June 1925: 59.
[28] https://www.kabar.com/customer/history.jsp
[29] https://web.archive.org/web/20101003062833/https://www.kabar.com/name.jsp
[30] Grimm, D.Y. *Identification Guide to Remington Sheath Knives - 1925 to 1940.* 1991.
[31] *Cutlery Handbook: Season of 1937.* Remington. January 1, 1937.
[32] Brearley, Harry. *Steel-Makers and Knotted String.* Maney Pub, 1995.
[33] *Industry Illustrated.* October 1923: 72.
[34] "Tabular Presentation of Chemical and Physical Properties of Corrosion Resisting Alloys." *Chemical and Metallurgical Engineering* vol. 31, no. 2. July 14, 1924: 82.
[35] *Ludlum Steel.* 1924: 136.
[36] Strauss, Jerome and J.W. Talley. "Stainless Steels: Their heat Treatment and Resistance to Sea-Water Corrosion." *Symposium on Corrosion-Resistant, Heat-Resistant and Electrical-Resistance Alloys. American Society for Testing Materials.* 1924.
[37] Woldman, Norman Emme, and Albert Julius Dornblatt. *Engineering Alloys; Properties, Uses.* 1936.
[38] Giles, David J. "Steel Alloy. US Patent no. 1,650,707. Patented November 29, 1927.
[39] "Hy-Glo." US Trademark serial no. 71,208,300. Trademarked June 9, 1925.
[40] "David J. Giles, Latrobe Steel Executive, Dies." *The Daily Courier.* January 8, 1958: 17.
[41] "Latrobe Steel Company is One of Boro's Big Industries." *Latrobe Bulletin.* December 19, 1952: 94.
[42] *The Iron Age* vol. 117. January 7, 1926: 98.
[43] *Chemical & Metallurgical Engineering* vol. 41, no. 10. October 1934: 553.
[44] Clark, Donald G. "Uses of Stainless Steel Grow Rapidly." *Iron Trade Review* vol. 77, no. 18. Oct. 29, 1925: 1089.
[45] "Corrosion-Resistant Steel – The Metal of a Thousand Uses." *Machinery* vol. 40, no. 2. October 1933.
[46] Campbell, William. *A List of Alloys.* 1930: 62.
[47] Krauss, David A. *American Pocketknives: A History of Schatt & Morgan and Queen Cutlery.* 2002.

[48] Voyles, J. Bruce. *Antique Knives.* 1990.
[49] https://queencutleryhistory.com/index.php/historical-documents/
[50] Winter, Butch. "Who Makes Your Cutlery Steels?" *Blade Magazine.* February 1984.
[51] Petzal, David E. "The 3rd Generation from Schrade." *Mechanix Illustrated.* January 1983: 106.
[52] Oertel, Wilhelm. German patent no. 486,285. Patented January 11, 1922.
[53] Oertel, Wilhelm. "Steel Alloy." US Patent no. 1,630,448. Patented. May 31, 1927.
[54] Oertel, W. and Karul Wurth. "Ueber den Einfluß des Molybdäns und Siliziums auf die Eigenschaften eines nichtrostenden Chromstahls." *Stahl und Eisen* vol. 47, no. 18. May 5, 1927.
[55] Maurer, Eduard. "High Chromium Irons and Steels." *Metal Progress* vol. 31, no. 5. May 1937: 535.
[56] Thum, Ernest. *The Book of Stainless Steels.* 1935.
[57] *Corrosion and Heat Resisting Steels: Crucible Steel Company of America.* 1937.

11 World War II Custom Knives

[1] Gaddis, Robert L. *Randall Made Knives - The History of the Man and the Blades.* 1993.
[2] Hunt, Robert W. "The Manufacture of Bessemer Steels." *Journal of the Franklin Institute.* May 1889.
[3] Latham, Sid. "Bo Randall Custom Knife Maker." *American Blade Magazine.* May-June 1975.
[4] Williamson, Jim. "Randall's First Half Century." *National Knife Magazine.* June 1988.
[5] Gaddis, Robert L. "The First Randall Stainless Steel Carver." *Blade Magazine.* May 1994.
[6] Coulbourn, Keith. "A Cut Above The Average." *The Tampa Tribune.* 12 Feb 1967.
[7] http://www.knifetalkforums.com/ubbthreads.php?ubb=showflat&Number=137218
[8] Ables, Tom. *The Story of Buck Knives.* 1991.
[9] Hughes, B.R. and Jack Lewis. "The Almighty Buck." *The Gun Digest Book of Knives.* 1973.
[10] Baker, L.V. "Knifemaker Profile Al Buck." *American Blade Magazine.* November-December 1975.
[11] Hollis, Durwood. "Al Buck: 1910-1991." *Blade Magazine.* October 1991.
[12] Buerlein, Robert A. *Allied Military Fighting Knives And the Men Who Made Them Famous.* 1984.
[13] Levine, Bernard. "World War II: First Custom Knifemaker School?" *Blade Magazine.* November 1998.
[14] Murphy, Dave. "The Murphy Combat Knife." *Knife World.* December 1984.
[15] Warner, Ken. "The Murphy Knife Company Again." *Knives '83.* 1982.
[16] Levine, Bernard. "The Next Great Factory Collectible Part I." *Blade Magazine.* December 1996.
[17] https://www.usmilitariaforum.com/forums/index.php?/topic/150554-john-ek-commando-knives-1941-1976/

12 New Die Steels

[1] "Lumdie." US Trademark serial no. 71,483,226. Trademarked November 20, 1945.
[2] "Latrobe at Work, A Roundup of Local Industrial Plants." *Latrobe Bulletin.* October 18, 1952: 5.
[3] Gammeter, Elmer. "Die steels and the heat-treatment of dies for use in die casting aluminum base alloys." Ph.D Thesis University of Missouri. 1932.
[4] "Hotform." US Trademark serial no. 71,622,874. Trademarked March 10, 1953.
[5] Gill, James P. "Steel Alloy." US Patent No. 1,938,221. Patented December 5, 1933.
[6] Hamaker, J.C., Jr. and G.A. Roberts. "An appraisal of...Steels for Aluminum Die Casting Dies." *Die Casting Engineer* vol. 2, no. 2. June 1958: 11.
[7] "How to Work Tool and Die Steels Part 2." *American Machinist* vol. 94, no. 25. December 11, 1950: 147.
[8] Stotz, Norman I. "Jobbing Die Steels." *The Iron Age.* December 15, 1938.
[9] Sanderson, L. "A New Die Steel." *Metallurgia.* January 1941: 80.
[10] "New Air-Hardening Tool and Die Steel." *Machinery.* March 1940: 136.
[11] Luerssen, George V. and Carl B. Post. "Steel Alloy." US Patent no. 2,355,224. Patented August 8, 1944.
[12] Goo, Tai. Phone Interview, 2023.
[13] Culp, Neil J. "Wear Resistant Steel." US Patent no. 2,923,619. Patented February 2, 1960.
[14] Roberts, G.A. J.C. Hamaker, Jr., and A.R. Johnson. *Tool Steels.* 1962.
[15] Giles, David J. "Ferrous Alloys and Abrasive-Resistant Articles Made Therefrom." US Patent No. 2,575,218. Patented November 13, 1951.
[16] Giles, David J. "Ferrous Alloys and Abrasive-Resistant Articles Made Therefrom." US Patent No. 2,575,219. Patented November 13, 1951.
[17] "Condensed Review of Some Recently Developed Materials." *Machinery* vol. 67, no. 2. October 1960: 127.
[18] Giles, David J. "Ferrous Alloys and Articles Made Therefrom." US Patent No. 2,575,216. Patented November 13, 1951.
[19] Giles, David J. "Ferrous Alloys and Corrosion and Wear-Resisting Articles Made Therefrom." US Patent No. 2,709,132. Patented May 24, 1955.
[20] Gill, James P. *Tool Steels.* 1944
[21] "K-Rustless." US Trademark no. 249,571. Trademarked November 20, 1928.
[22] Woldman, Norman E. and Roger J. Metzler. *Engineering Alloys: Names, Properties, Uses.* 1945.

[23] Nekervis, R.J., C.H. Lund, and A.M. Hall. "Survey Hot-Work Tool Steels for Aircraft and Missiles." *The Iron Age* vol. 181, no. 9. February 27, 1958.

[24] "Exploring the Thermal Barrier. *Steel* 140, no. 10. March 11, 1957: 157.

[25] *Iron Trade and Western Machinist* vol. 153. 1963: 116.

[26] "Guide to Tooling Materials." *Steel* vol. 164, no. 17. April 28, 1969.

[27] "Alphabetical Index of Tool Steel Brands." *The Iron Age* vol. 186, no. 6. August 11, 1960: 309.

[28] Hughes, David P. "Ferrous Alloys and Articles Made Therefrom." US Patent 2,949,356. Patented Aug. 16, 1960.

[29] Fletcher, Stewart G. and David P. Hughes. "Ferrous Alloys and Articles Made Therefrom." US Patent no. 3,163,525. Patented December 29, 1964.

[30] *The Iron Age* vol. 189, no. 19. May 10, 1962: 131.

[31] *The Iron Age* vol. 202, no. 23. December 5, 1968: 50.

[32] Johnstin, Harry G. "Alloy steels and articles thereof." US Patent no. 3,219,442. Patented November 23, 1965.

[33] Woldman, Norman E. and Robert C. Gibbons. *Engineering Alloys 5th Edition.* 1973.

[34] Matsuda, Yukinori. "Cold tool steel." Japanese Patent 6411945, issued January 17, 1989.

[35] Kada, Yoshihiro, Atsushi Kumagai, Atsusuke Nakao, Toshio Okuno, and Takehiro Oono. "High hardness and high toughness cold tool steel." Japanese Patent 02277745, issued November 14, 1990.

[36] Leban, Karl and Herbert Schweiger. "Cold-Worked Steel of High Compressive Strength and Articles Made Thereof." US Patent no. 5, 160,553. Patented November 3, 1992.

[37] Sandberg, Odd, and Boerje Johansson. "Cold Work." Sweden Patent 511747, issued March 27, 1998.

[38] *Pulp & Paper* vol. 24, no. 2. February 1950: 13.

[39] "Hearing Before the Subcommitte on Trade of the Committee on Ways and Means." *Ninety-Sixth Congress, First Session."* July 27, 1979.

13 Bearings and Super Hard High Speed Steels

[1] *Tooling & Production* vol. 14, no. 9. December 1948: 29.

[2] Anderson, William J. "Performance of 110-millimeter-bore M-1 Tool Steel Ball Bearings at High Speeds, Loads, and Temperatures." *National Advisory Committe for Aeronautics, Technical Note 3892.* January 1957.

[3] Johnson, Birger L. "High Temperature Wear Resisting Steels." US Patent no. 3,167,423. Patented January 26, 1965.

[4] "Three Stainless Steels retain hot hardness better than standard AISI grade 440C." *Materials in Design Engineering* vol. 53, no. 3. March 1961: 19.

[5] "Heat Treatment of a Special Stainless Alloy." *Metal Treating* vol. 12, no. 3. June-July 1961: 34.

[6] "Parts Fabrication Panel." *Report of the high-strength high-temperature materials for standard part symposium.* June 4-5, 1957: 176.

[7] "Metal Selector." *Steel* vol. 145, no. 17. October 26, 1959: S-5.

[8] Blickwede, D. J., and Morris Cohen. "The isolation of carbides from high speed steel." *JOM* 1, no. 9, 1949: 578-584.

[9] Blickwede, Donald J., Morris Cohen, and George A. Roberts. "The Effect of Vanadium and Carbon on the Constitution of High Speed Steel." *Transactions of American Society for Metals* vol. 42. 1950: 1161.

[10] Roberts, George A., and John C. Hamaker, Jr. "Alloy steels." U.S. Patent 3,117,863, issued January 14, 1964.

[11] Zackay, VICTOR F., and WILLIAM M. Justusson. "The properties of martensitic steels formed from strain-hardened austenite." *ISI Special Report* 76, 1962: 14-21.

[12] "New Materials." *American Machinist* vol. 108, no. 6. March 16, 1964: 137.

[13] http://www.imsteel.com/matrix_high-speed_steels/matrix_high-speed_steels_index.htm

[14] Wright, Timothy. Phone Interview, 2022.

[15] Fletcher, S. G. and Wendell. C. R. "The new Generation of High speed steels – Their metallurgical characteristics." *Metals Engineering Quarterly American Society for Metals.* February 1966.

[16] Harvey, Richard F., and Charles W. Schuck. "High speed steel." U.S. Patent 3,113,862, issued December 10, 1963.

[17] Nehrenberg, Alvin E., Thoni V. Philip, and Steven Gary. "Low alloy steel having high hardness at elevated temperatures." U.S. Patent 2,996,376, issued August 15, 1961.

[18] Hamaker, Jr John C., James R. Handyside, and Daniel H. Yates. "Ultra hard high speed steel." U.S. Patent 3,259,489, issued July 5, 1966.

[19] Mackay, William BF, and Robert O. Carson. "High speed tool steel." U.S. Patent 3,388,986, issued June 18, 1968.

[20] *Steel* vol. 147, no. 24. December 12, 1960: 100.

[21] "Metalworking Week." *Steel* vol. 147, no. 26. December 26, 1960: 18.

[22] *Steel* vol. 149, no. 20. November 13, 1961: 125.

[23] "New Materials." *American Machinist* vol. 106, no. 8. April 16, 1962.

[24] Handyside, J.R., J.C. Hamaker, and D. Yates. "Rc 70 High-Speed Steel – Its Development, Properties and Performance" *Metal Progress* vol. 83, no. 6. 1963: 78-81.

[25] Steven, G., A.E. Nehrenberg, and T.V. Philip. "High-Performance High-Speed Steels by Design." *Transactions of the American Society for Metals* vol. 57 no. 4. 1964: 925.

[26] Roberts, George A. and Robert A. Cary. *Tool Steels.* 1980.

14 Bill Moran and Traditional Knifemaking

[1] **Hughes, B.R. and C. Houston Price. *Master of the Forge, William F. Moran, Jr. and His Classic Blades.* 1996.**

[2] Lewis, Jack and Roger Combs. "A Feel for the Steel." *The Gun Digest Book of Knives, 3rd Edition*. 1988.
[3] McEvoy, Harry K. "A living legend in modern cutlery William F. Moran, Jr." *Knife World*. February 1979.
[4] Phillips, Jim. "William F. Moran." *American Blade Magazine*. March-April 1979.

15 D2, 440C, and A2 in Custom Knives
[1] Hughes, B.R. and Jack Lewis. "D.E. Henry: Man & Bowie." *The Gun Digest Book of Knives*. 1973.
[2] Edmondson, J.R. "The Importance of Being Henry." *Knives '84*. 1983.
[3] Edmondson, J.R. "The Gospel According to D.E. Henry." *Blade Magazine*. April 1993.
[4] "D.E. Henry." Bates, Jr., John Davis and James Henry Schippers, Jr. *The Custom Knife*. 1973.
[5] Miloné, Kathie. "D.E. Henry The Bowie Knife Maker." American *Blade Magazine*. September-October 1973.
[6] Haskew, Mike. "D.E. Henry: Life Over the Edge." *Blade Magazine*. July 2014.
[7] Shackleford, Steve. "A Pirate Looks at 70." *Blade Magazine*. January 2001.
[8] Duhalme, G. Scott. "Hibben is Back...And He's Better Than Ever!" *Knife World*. October 1980.
[9] Draper Fight Knives advertisement. *The Leatherneck*. January 1967.
[10] Corsi, Del. "Gil Hibben: 65 Years in Knifemaking." *Knife Magazine*. January 2022.
[11] Lobred, Phil. "Gil Goes Golden." *Blade Magazine*. September 2007.
[12] Lewis, Jack and Roger Combs. "The Versatile Gil Hibben." *The Gun Digest Book of Knives 2nd Edition*, 1982.
[13] Hughes, B.R. and Jack Lewis. "The Stone Age is Now." *The Gun Digest Book of Knives*. 1973.
[14] Williamson, Jim. "Stone Knives." *American Blade Magazine*. August 1979.
[15] Bates, Jr., John Davis and James Henry Schippers, Jr. "Chubby Hueske." *The Custom Knife*. 1973.
[16] Hughes, B.R. and Jack Lewis. "Tale of the Texas Tang: Chubby Hueske turns out quality knives on weekends only!" *The Gun Digest Book of Knives*. 1973.
[17] Lewis, Jack and Roger Combs. "Of Makers and Metals." *The Gun Digest Book of Knives 4th Edition*, 1992.
[18] Lucarelli, Jack and John H. Hill. *James B. Lile Arkansas Knifesmith*. 2007.
[19] Winter, Butch. "Jimmy Lile: Knifemaker." Knife World. May 1995.
[20] Bates, Jr., John Davis and James Henry Schippers, Jr. "James B. Lile." *The Custom Knife*. 1973.
[21] Lachuk, John. "Ted Dowell - Master Knifemaker." *Guns & Ammo Guidebook to Knives & Edged Weapons*. 1974.
[22] Bates, Jr., John Davis and James Henry Schippers, Jr. "T.M. Dowell." *The Custom Knife*. 1973.
[23] Baker, L.V. "Knifemaker Roundtable." *American Blade Magazine*. August 1979.
[24] Shackleford, Steve. "Ted Dowell: Integrals & Integrity." *Blade Magazine*. September 1999.
[25] Lachuck, John. "Ted Dowell - Knifemaker." *Blade Magazine*. July-August 1973.
[26] Hughes, B.R. and Jack Lewis. "Ted Dowell: Sharp Operator." *The Gun Digest Book of Knives*. 1973.
[27] https://tmdknives.com/the-tmd-story/
[28] Winter, Butch. "Dowell's Funny Folder." *Knife World*. May 1979.
[29] Winter, Butch. "Knifemaker T.M. Dowell." *Knife World*. December 1996.
[30] Hughes, B.R. "Cutlery Steel Revisited." *Knife World*. March 1978.
[31] Loveless, R.W. "And Testing Retort." *American Blade Magazine*. January-February 1977.
[32] Winter, Butch. "Cutlery Steels: Another Opinion." *Blade Magazine*. October 1984.
[33] Campbell, Robert J. "The Best Cutlery Steel." *Blade Magazine*. June 1984.
[34] Sornberger, Jim. "A Few Kind Words About 440C Steel." *Knives '87*. 1986.
[35] Haskew, Mike. "Mr. Folding Knife (Ron Lake)." *Blade Magazine*. June 2008.
[36] Bates, Jr., John Davis and James Henry Schippers, Jr. "Ron Lake." *The Custom Knife*. 1973.
[37] Lewis, Jack and B.R. Hughes. "The Folding Knives of Ron Lake." *Gun Digest Book of Folding Knives*. 1977.
[38] Lake, Ron. Phone Interview, 2019.
[39] Bates, Jr., John Davis and James Henry Schippers, Jr. "Ron Lake." *The Custom Knife II*. 1974.
[40] Warner, Ken. "Custom Knifemakers Directory." *Knives '86*. 1985.
[41] Latham, Sid. "The Most Expensive Folding Knives in the World." *Blade Magazine*. August 1982.
[42] Haskew, Mike. "The Clever Lever." *Blade Magazine*. January 2012.
[43] https://www.all4shooters.com/en/shooting/knives/interview-to-ron-lake/
[44] Shackleford, Steve. "Loveless: There'll Never Be Another." *Blade Magazine*. October 1997.

16 Bob Loveless and 154CM
[1] Williams, Al and Jim Weyer. *Living on the Edge: Logos of the Loveless Legend*. 1992.
[2] Hollis, Durwood. *Knifemaking with Bob Loveless*. 2010.
[3] "The American Blade Interview: R.W. Loveless Knifemaker." *American Blade Magazine*. January-February 1974.
[4] Spangenberger, Phil. "R.W. Loveless: Leader in the Field." *Guns & Ammo Guidebook to Knives & Edged Weapons*. 1974.
[5] Loveless, R.W. "Steel and Knives: A Bladesmith's Bible." *Knife Digest*. 1975.
[6] Loveless, R.W. "And Testing Retort." *American Blade Magazine*. January-February 1977.
[7] Warner, Ken. "Loveless Looks Back." *Knives '87*. 1986.
[8] Shackleford, Steve. "A Pirate Looks at 70." *Blade Magazine*. January 2001.

17 A.G. Russell and the Knifemakers' Guild

[1] Hughes, B.R. and Jack Lewis. "Birth, Death, Resurrection: History of the Knife." *The Gun Digest Book of Knives.* 1973.

[2] Davis, William C. *Three Roads to the Alamo: The Lives and Fortunes of David Crockett, James Bowie, and William Barret Travis.* 1998.

[3] Thorp, Raymond. *Bowie Knife.* 1948.

[4] Hughes, B.R. and C. Houston Price. *Master of the Forge: William F. Moran Jr. and His Classic Blades.* 1996.

[5] Gaddis, Robert L. *Randall Made Knives - The History of the Man and the Blades.* 1993.

[6] Wellman, Paul. *The Iron Mistress.* 1951.

[7] Warner, Ken. "The Best Knives Made." *The Gun Digest.* 1966.

[8] Latham, Sid. "A.G." *American Blade Magazine.* January-February 1974.

[9] Russell, A.G. "Foreword." *Knifemaking with Bob Loveless.* 2010.

[10] Hughes, B.R. "The Knifemakers Guild: Beginnings." *Blade Magazine.* August 1986.

[11] Lewis, Jack and Roger Combs. "The Knifemakers Guild." *The Gun Digest Book of Knives 2nd Edition.* 1982.

[12] https://tmdknives.com/the-tmd-story/

[13] Williams, Al and Jim Weyer. *Living on the Edge: Logos of the Loveless Legend.* 1992.

[14] Lile, James B. "Directions of the Guild." *American Blade Magazine.* April 1981.

[15] Shackleford, Steve. "Ted Dowell: Integrals & Integrity." *Blade Magazine.* September 1999.

[16] Lile, James B. "An Attempt to Establish Some Ethical Standards for the American Knifemakers Guild." *American Blade Magazine.* July-August 1976.

[17] Beall, James R., and Gordon Morseth Sr. *The Keen Edge of Perfection a History of the Morseth Knife.* 2007.

[18] Warner, Ken. "The Knife." *The Gun Digest.* 1971.

[19] Wootters, John. "The Age of Superblade Arrives." *Shooting Times.* December 1971.

[20] Hughes, B.R. and Jack Lewis. "Tops in the Trade." *The Gun Digest Book of Knives.* 1973.

[21] Hughes, B.R. and Jack Lewis. "Are Custom Knives Right for You?" *The Gun Digest Book of Knives.* 1973.

[22] Hughes, B.R. and Jack Lewis. "A Matter of Demand." *The Gun Digest Book of Knives.* 1973.

[23] Bates, Jr., John Davis and James Henry Schippers, Jr. *The Custom Knife.* 1973.

[24] Bonte, Frederick B. "Ferrous alloys and method of manufacture." U.S. Patent 2,087,764, issued July 20, 1937.

[25] Loveless, R.W. "And Testing Retort." *American Blade Magazine.* January-February 1977.

[26] Shackleford, Steve. *William F. Moran, Jr. Forever of Legend.* 2015.

[27] Haskew, Mike. "The Knifemaker's Life Line." *Blade Magazine.* February 2000.

[28] "The Man Who Carried Knifemakers." *Blade Magazine.* November 2001.

[29] Shackleford, Steve. "Knifemaking Equipment - And Where to Get It." *Blade Magazine.* June 1986.

[30] Shackleford, Steve. "5 Ways To Recharge The Guild." *Blade Magazine.* December 2000.

[31] Terzuola, Robert. Phone Interview, 2023.

18 A New Generation of Factory Knives

[1] Gerber, Joseph R. "Birth of an Enterprise." Found in Phil Rodenberg's *A Chronology of Gerber Legendary Blades: 1939-1986.* 2010.

[2] Levine, Bernard. "The Next Great Factory Collectible Part I." *Blade Magazine.* December 1996.

[3] Gerber, Francis. "Origins of G.L.B." Found in Phil Rodenberg's *A Chronology of Gerber Legendary Blades: 1939-1986.* 2010.

[4] Levine, Bernard. "The Next Great Factory Collectible Part II." *Blade Magazine.* January 1997.

[5] Levine, Bernard. "The Next Great Factory Collectible Part III." *Blade Magazine.* February 1997.

[6] Cassidy, William L. *The Complete Book of Knife Fighting.* 1975.

[7] Latham, Sid. "Gerber Legendary Blades." *American Blade Magazine.* November-December 1976.

[8] Latham, Sid. "NSGA Show." *American Blade Magazine.* March-April 1974.

[9] Lewis, Jack and Roger Combs. "Peter Kershaw's Philosophy is Simple." *Gun Digest Book of Knives, 3rd E.* 1988.

[10] Voyles, Bruce. "Japanese Cutlery Factories. " *American Blade Magazine.* April 1982.

[11] *American Blade Magazine.* May-June 1976: 45.

[12] *American Blade Magazine.* December 1978: 37.

[13] Latham, Sid. "1981-1982 Roundup Factory Folding Knives." *American Blade Magazine.* February 1982.

[14] https://kershaw.kaiusa.com/history

[15] Lewis, Jack and Roger Combs. "Al Mar Knives." *Gun Digest Book of Knives 2nd Edition.* 1982.

[16] Voyles, J. Bruce. "Knives in the Northwest, Part II." *Blade Magazine.* August 1985.

[17] Shackleford, Steve. "10 Years That Changed The Knife Industry." *Blade Magazine.* May 1998.

[18] Walker, Greg. "Al Mar: Remembering the Father of Specialty Cutlery." *Blade Magazine.* March 1993.

[19] Ables, Tom. *The Story of Buck Knives.* 1991.

[20] https://www.buckknives.com/blog/buck-lifetime-knives/

[21] Hughes, B.R. and Jack Lewis. "The Almighty Buck." *The Gun Digest Book of Knives.* 1973.

[22] https://www.bladeforums.com/threads/110-blade-steel.232779/

[23] https://www.buckknives.com/about-knives/heat-treating/

[24] Latham, Sid. "The Schrade-Loveless Knife." *American Blade Magazine.* March-April 1976.

[25] Hollis, Durwood. *Knifemaking with Bob Loveless.* 2010.

[26] Winter, Butch. "Loveless Left 'Em Gawking." *Knives 2002*. 2001.
[27] *American Blade Magazine*. July-August 1977: 2.

19 Furnace Heat Treating
[1] Osmund, F. "On the Influence of Low Temperatures Upon Certain Steels." *The Metallographist* vol. 2, no. 4. October 1899.
[2] "Sverker 21 datasheet." Uddeholm. 2016.
[3] "Verfahren zur Herstellung von Messerschmiedewaren aus nichtrostendem Stahl." Switzerland patent no. 217,568. Accepted October 31, 1941.
[4] *Business in New York State*. 1965: 9.
[5] Ehlers, W.A. "Hardening Steel in Gas Fired Furnaces." *Scientific American*. October 30, 1920.
[6] Hotchkiss, A.G. and H.M. Webber. "Modernizing with Protective Atmospheres." *General Electric Review* vol. 57 no. 6. November 1954.
[7] Lewis, Jack and Roger Combs. "Heat Treating." *The Gun Digest Book of Knives 2nd Edition*. 1982.
[8] "Do You Remember the Guild Show When..." *Blade Magazine*. September 2006.
[9] Hollis, Durwood. "The Boss of Heat Treatment." *Blade Magazine*. June 2007.
[10] https://www.pacificheattreating.com/about-us/
[11] Barney, Richard W. and Robert W. Loveless. *How to Make Knives*. 1977.
[12] *Blade Magazine*. August 1983: 9.
[13] "Holt-Sornberger: Supplies and Service to the Custom Knife Trade." 1985-1986 Supply Catalog.
[14] https://blademag.com/blade-show/cutlery-hall-of-fame/keeslar-and-sornberger-join-cutlery-hall-of-fame
[15] Sornberger, Jim. Phone Interview, 2022.
[16] https://petersheattreat.com/about-us/
[17] Kertzman, Joe. "Is D-2 All It's Cut Out To Be?" *Blade Magazine*. March 2004.
[18] Stallsmith, Brad. Phone Interview, 2022.
[19] *Blade Magazine*. May 2005: 61.
[20] *Blade Magazine*. August 1986: 9.
[21] https://corp.paragonweb.com/about-us/
[22] https://www.bladeforums.com/threads/old-paragon-kiln-issue.1064518/#post-12154065
[23] Smith, Robert. "How to Heat Treat Knives." *Blade Magazine*. February 1987.
[24] https://kilnsandovens.com/about-evenheat-kiln-inc/
[25] *Blade Magazine*. August 1991: 68.

20 Bill Moran's Reintroduction of Damascus
[1] Hughes, B.R. "Magic in the Steel: 50 Years of Bill Moran." *Blade Magazine*. April 1990.
[2] Material handed out at the 1973 Knifemakers Guild Show written by Bill Moran.
[3] Hughes, B.R. "William F. Moran Knifemaker Profile." *American Blade Magazine*. April 1979.
[4] Shackleford, Steve. *William F. Moran, Jr. Forever of Legend*. 2015.
[5] McEvoy, Harry K. "A living legend in modern cutlery William F. Moran, Jr." *Knife World*. February 1979.
[6] Lewis, Jack and B.R. Hughes. "Bill Moran: Damascus Midwife." *Gun Digest Book of Folding Knives*. 1977.
[7] Hughes, B.R. "Happy 25th, Damascus!" *Blade Magazine*. November 1998.
[8] Hughes, B.R. and C. Houston Price. *Master of the Forge: William F. Moran Jr. and His Classic Blades*. 1996.
[9] Figiel, Leo S. *Damascus Steel*. 1991.
[10] Moran Knives brochure 1973.
[11] Latham, Sid. *Knifecraft: A Comprehensive Step-By-Step Guide to the Art of Knifemaking*. 1978.
[12] Shackleford, Steve. "The Steel that Changed Knives Forever." *Blade Magazine*. April 2013.

21 American Bladesmith Society
[1] https://www.americanbladesmith.org/abs-history/
[2] Hughes, B.R. and C. Houston Price. *Master of the Forge: William F. Moran, Jr. and His Classic Blades*. 1996.
[3] Hughes, B.R. "Interview: Bill Bagwell." *Knife World*. February 1978.
[4] Overton, Mac. "On the Verge of Knifemaking Renown." *Blade Magazine*. August 1986.
[5] https://www.americanbladesmith.org/wp-content/uploads/2021/03/ABS-ByLaws-2-7-2013-Signed.pdf
[6] Overton, Mac. "Forged vs Stock Removal: The Makers Sound Off." *Blade Magazine*. August 1989
[7] Overton, Mac. "Stainless Or Just Higher Prices?" *Blade Magazine*. October 1989.
[8] Szilaski, Joe. "Forging Stainless: Benefits and Drawbacks." *Blade Magazine*. February 2007.
[9] Hughes, B.R. "Steel." *Knife World*. December 1977.
[10] Petersen, Dan. "The Knife Steel Debate: The Final Chapter." *Blade Magazine*. April 1985.
[11] Schroen, Karl. *The Hand Forged Knife: An Introduction to the Working of Modern Tool Steels*. 1985.
[12] McWilliams, Sean. "Book Review: The Hand Forged Knife by Karl Schroen." *Blade Magazine*. August 1986.
[13] "Sean McWilliams' Hand-Forged Stainless Steel Bowie." *Blade Magazine*. February 1986: 75.
[14] McWilliams, Sean. "Forging Stainless Steel." *Knives Illustrated*. Spring 1988: 38.

[15] McWilliams, Sean. "Microphotographing Steel." *Knives Illustrated.* Spring 1989.

22 Was Pattern-Welded Damascus Ever Lost?
[1] Sachse, Manfred. *Damascus Steel: Myth, History, Technology, Applications.* 2008.
[2] de Beroaldo Bianchini, Natale. *Abhandlung über die Feuer-und Seitengewehre.* 1829.
[3] https://www.cowboysindians.com/2017/04/trick-shooter-real-gun/
[4] Hickory, Tinnell. Phone Interview, 2022.
[5] Meilach, Dona. *Decorative and Sculptural Ironwork.* 1977
[6] "Classified Ads." *American Blade Magazine.* December 1977.

23 Daryl Meier and the Damascus Research Team
[1] Reichelt, Richard. "Daryl Meier." *Heartland Blacksmiths: Conversations at the Forge.* 1988.
[2] https://web.archive.org/web/20190416090441/http://meiersteel.com/whoiam.html
[3] Meilach, Dona Z. *Decorative and Sculptural Ironwork: Tools, Techniques, Inspiration.* 1977.
[4] Meier, Dayl D. "Damascus Steel." *Blade Magazine.* October 1983.
[5] Kertzman, Joe. "Damascus for the Masses." *Blade Magazine.* November 1997.
[6] Goddard, Wayne. "What You Should Know About High Speed Steels." *Blade Magazine.* December 1994.
[7] Meier, Dayl D. "Damascus Steel." *Blade Magazine.* August 1983.
[8] Meier, Dayl D. "Damascus Steel." *Blade Magazine.* April 1983.
[9] https://web.archive.org/web/20101201131515/http://swordforum.com/summer99/ricfurrer.html
[10] Schwarzer, Steve. "Father of Modern Mosaic Damascus." *Blade Magazine.* July 2002.
[11] Thomas, Devin. Phone Interview, 2022.

24 Composite Damascus and Words in Steel
[1] Meier, Dayl D. "Damascus Steel." *Blade Magazine.* June 1983.
[2] Kertzman, Joe. "The Ultimate Guide to Damascus Steel." *Blade Magazine.* November 2001.
[3] Sachse, Manfred. *Damascus Steel: Myth, History, Technology, Applications.* 2008.
[4] Lankton, Scott Michael. "A Replica of the Sutton Hoo Sword." *The Anvil's Ring.* Fall 1989.
[5] Haskew, Mike. "Practicing Steel Manipulation." *Knives 2008.* 2007.
[6] http://www.culverart.com/Word%20Damascus%20Gun-Barrel%20Patterns.pdf
[7] Figiel, Leo S. *On Damascus Steel.* 1991.
[8] Smith, Cyril Stanley. *A History of Metallography.* 1960.
[9] Puraye, Jean. *Damascus Barrels.* 1968.
[10] Meier, Daryl. Email, 2022.
[11] Levine, Bernard. "'Werth' His Weight in Damascus." *Blade Magazine.* April 1988.
[12] Shackleford, Steve. "Belt Buckle Blues." *Blade Magazine.* September 1993.
[13] Humphrey, Theresa. "Rock Hall resident is one sharp craftsman." *The News Journal.* Oct. 16, 1996.
[14] Johnson, Kenneth. "Master bladesmith lives in the past and likes it." *The Star-Democrat.* Jan. 29, 1989.
[15] Hughes, B.R. "Knives That Look and Feel Good." *Blade Magazine.* August 1988.
[16] https://www.worthpoint.com/worthopedia/vtg-robbin-rob-hudson-abs-ms-custom-456703037
[17] https://web.archive.org/web/20101201184513/http://swordforum.com/summer99/robhudson.html
[18] Dagenais-Renkoski, Ann. "Knifemaker Rob Hudson." *Knife World.* May 1998.
[19] Dunkerley, Rick. Email. 2022.
[20] Dunkerley, Rick. "His Forge Burns Hot for Mosaic Damascus." *Blade's Guide to Making Knives.* 2005.

25 Don Fogg and Hot Deformed Patterning
[1] Don Fogg Interview. *Josh Smith Show.* 2021.
[2] Fogg, Don. "Dedicated to the Study of Sword Making." *Blade's Guide to Making Knives.* 2005.
[3] Shackleford, Steve. "He Was The Master." *Knives 2009.* 2008.
[4] https://web.archive.org/web/20130403030617/http://dfoggknives.com/wayof.htm
[5] Fogg, Don. Phone Interview, 2022.
[6] Lewis, Jack and Roger Combs. "Kemal: The Damascus Balance." *The Gun Digest Book of Knives: 2nd Edition.* 1982.
[7] "Classified Ads." *American Blade Magazine.* Sept-Oct 1977.
[8] Haskew, Mike. "Charging Headlong Into the Edge Evolution." *Blade Magazine.* June 1998.
[9] https://www.bladeforums.com/threads/w-pattern-Damascus-wip-my-way.956157/
[10] Dunkerley, Rick. "His Forge Burns Hot for Mosaic Damascus." *Blade's Guide to Making Knives.* 2005.

[11] "The Accordion Technique." *Blade Magazine.* July 1995.

[12] Figiel, Leo S. *On Damascus Steel.* 1991.

[13] Fogg, Don. *Damascus Steel* (DVD). 2009.

[14] Meier, Dayl. "Damascus Steel." Blade Magazine. August 1983.

[15] Warner, Ken. *Knives '98.* 1997.

[16] Darom, David. *Art and Design in Custom Folding Knives.* 2006.

[17] Sachse, Manfred. *Damascus Steel.* 1994.

[18] Warner, Ken. *Knives '97.* 1996.

[19] Dunn, Steve. Phone Interview, 2022.

[20] "Hot Handmade." *Blade Magazine.* January 1999.

26 Al Pendray, John Verhoeven, and Wootz

[1] Pearson, George. "XVII. Experiments and observations to investigate the nature of a kind of steel, manufactured at Bombay, and there called Wootz: with remarks on the properties and composition of the different states of iron." *Philosophical Transactions of the Royal Society of London* 85, 1795: 322-346.

[2] Mushet, David. "VIII. Experiments on wootz." *Philosophical Transactions of the Royal Society of London* 95, 1805: 163-175.

[3] Faraday, Michael. "An analysis of wootz, or indian steel." *Quarterly Journal of Science, Literature and Art* 7, 1819: 288-90.

[4] Stodart, James, and Michael Faraday. "V. Experiments on the alloys of steel, made with a view to its improvement." *The Philosophical Magazine* 56, no. 267, 1820: 26-35.

[5] Breant, M. "Description of a process for making damasked steel." *Journal of the Franklin Institute* 1, no. 6, 1826: 334-338.

[6] Anossoff, P. "On the Bulat," *Annuaire du Journal des Mines de Russie*, 1843: 192-236.

[7] Belaiew, N.T. *On the Bulat*, Thesis, Michael Artillery Academy, St. Petersburg, 1906.

[8] Von Harnecker, K. "Contribution to the question of the Damascus steel." *Stahl und Eisen* 44, 1924: 1409-1411.

[9] Zschokke, B. "Du damassé et des lames de Damas." *Revue de métallurgie* 21, no. 11, 1924: 639-669.

[10] Wadsworth, Jeffrey, and Oleg D. Sherby. "On the Bulat-Damascus steels revisited." *Progress in Materials Science* 25, no. 1, 1980: 35-68.

[11] Smith, Cyril Stanley. "Damascus steel." *Science* 216, no. 4543, 1982: 242-244.

[12] Oppenhemier, Todd. *Alfred Pendray, Wootz Hunter - An Homage to a Remarkable Craftsman.* YouTube video, 2018.

[13] Phillips, David and Linda Phillips. "The Wizard of Wootz Al Pendray." *Blade Magazine.* August 1987.

[14] Meier, Daryl. "Damascus Steel." *Blade Magazine.* December 1982.

[15] Meier, Daryl. "Damascus Steel." *Blade Magazine.* December 1983.

[16] Meier, Daryl. "Damascus Steel." *Blade Magazine.* October 1982.

[17] Yater, Wallace M. "The Legendary Steel of Damascus. Part III Forging, Pattern Development and Heat Treatment." *The Anvil s Ring* 11, no. 4, 1983: 2-17.

[18] Verhoeven, J.D. *Damascus Steel Swords: Solving the Mystery of How to Make Them.* 2018.

[19] Figiel, Theo. *On Damascus Steel.* 1991.

[20] Loades, Mike. *The Secrets of Wootz Damascus.* YouTube video, 2017.

[21] Verhoeven, John D., and Alfred H. Pendray. "Method of making" Damascus" blades." U.S. Patent 5,185,044, issued February 9, 1993.

[22] Pendray, Al, *et al.* "How The Ancients Made Real Damascus." *Blade Magazine.* Aug. 1992.

[23] Meier, Daryl. Phone Interview, 2022.

27 The Beginnings of Mosaic Damascus

[1] Clouet, A. Antiens. "Instruction sur la fabrication des lames figurées, ou des lames dites Damas." *Journal des Mines, Ann.* 15, no. 12, 1803: 421-35.

[2] Schwarzer, Steve. "Father of Modern Mosaic Damascus." *Blade Magazine.* July 2002: 67, 70.

[3] Meier, Dayl D. "Damascus Steel." *Blade Magazine.* August 1983.

[4] Thomsen, Robert. *Et meget mærkeligt metal: En beretning fra jernets barndom.* Varde staalværk, 1975.

[5] Phillips, David. "Damascus that Really Works." *Blade Magazine.* December 1986.

[6] Schwarzer, Steve. "The History and Presentation of Mosaic Damascus." *Knives 2019.* 2018.

[7] Schwarzer, Steve. "The Mosaic Whiz Kids." *Blade Magazine.* November 2000.

[8] Weyer, Jim and Corey Gray. "Handmade Gallery." *Blade Magazine.* February 1989.

[9] Dunkerley, Rick. "His Forge Burns Hot for Mosaic Damascus." *Blade's Guide to Making Knives.* 2005.

[10] Weyer, Jim. *Knives Points of Interest Book IV.* 1993.

[11] https://web.archive.org/web/20060613025227/http://meiersteel.com/

[12] Burdette, Nathan. "President George Bush's American Spirit Knife." *Blade Magazine.* June 1991.

[13] Warner, Ken. "Meiered Steel." *Knives '92.* 1991.

[14] Kertzman, Joe. "The Ultimate Guide to Damascus Steel." *Blade Magazine.* November 2001.

[15] "Best Handmade Knives of the Blade Show." *Blade Magazine.* November 1992.

[16] Schwarzer, Steve. "Father of Modern Mosaic Damascus." *Blade Magazine*. July 2002 :71-72.
[17] Shackleford, Steve. "Schwarzer's Mosaic Damascus Scene a First." *Blade Magazine*. September 1993.
[18] Hughes, B.R. "How to Tell One Damascus From Another." *Blade Magazine*. September 1994.
[19] Trout, Ruth. "Bartrug Ashley Forge." *Knife World*. November 1987.
[20] Shackleford, Steve. "Simplicity, Elegance, and Power Hugh Bartrug." *Blade Magazine*. August 1993.
[21] Gottschack, Greg. Phone Interview, 2022.
[22] "E = MC^2: The Einstein Knife." *Blade Magazine*. November 1992.
[23] Rybar, Raymond. Phone Interview, 2023.
[24] Haskew, Mike. "The French Connection: Pierre Reverdy." *Blade Magazine*. March 1995.
[25] Darom, David. "Pierre Reverdy." *Custom Fixed Blade Knives*. 2007.
[26] Haskew, Mike. "If Not Knifemaking, He'd Rather Be Sailing: Hank Knickmeyer." *Blade Magazine*. January/February 1995.
[27] Knickmeyer, Hank. *Damascus Steel* (DVD). 2009.
[28] Haskew, Mike. "The High Tech Damascus Man Bill Fiorini." *Blade Magazine*. November 1994.
[29] https://www.bladegallery.com/shopexd.asp?id=3498
[30] Smith, J.D. Phone Interview, 2022.

28 Steels Used in Pattern-Welded Damascus
[1] https://vacaero.com/information-resources/metallography-with-george-vander-voort/153771-metallography-of-iron-nickel-meteorites.html
[2] Kelley, Gary. "Wire Damascus: An Old Steel With New Possibilities." *Blade Magazine*. April 1988.
[3] Hughes, B.R. "Damascus Patterns: the many faces of Damascus." *Blade Magazine*. April 1990.
[4] Goddard, Wayne. "State of the Art Forge-Welded Cable." *Knives '86*. 1985.
[5] Goddard, Wayne. "State of the Art Welded Wire Damascus." *Knives '87*. 1986.
[6] Hrisoulas, Jim. *The Complete Bladesmith*. 1987.
[7] Zowada, Tim. Phone Interview, 2022.
[8] Briar Custom Knives Advertisement. *Blade Magazine*. September 1994.
[9] Hughes, B.R. "You won't get 'Stuck' with a Briar Knife!" *Knife World Magazine*. November 1996.
[10] Breed, Kim. "Darrel Ralph Bird & Trout." *Blade Magazine*. July 1995.
[11] Thomas, Devin. Phone Interview, 2022.
[12] Dunkerley, Rick. "His Forge Burns Hot for Mosaic Damascus." *Blade's Guide to Making Knives*. 2005.
[13] "Cover Story." *Blade Magazine*. July 1998.
[14] Sachse, Manfred. *Damascus Steel*. 2008.
[15] "Knifemaker Showcase." *Blade Magazine*. October 2000.
[16] http://www.knifenetwork.com/forum/showthread.php?t=17124
[17] Overton, Mac. "Stainless or Just Higher Prices?" *Blade Magazine*. October 1989.
[18] "Winter, Butch. "Kingpins of Stainless Damascus." *Blade Magazine*. March 1997.
[19] Haskew, Mike. "The Hoss of Damascus." *Blade Magazine*. July 2001.
[20] Haskew, Mike. "The Knifemakers' Top Damascus Makers." *Blade Magazine*. April 2007.
[21] Kertzman, Joe. "Damascus for the Masses." *Blade Magazine*. November 1997.
[22] Kertzman, Joe. "State-Of-The-Art Stainless." *Blade Magazine*. February 1998.
[23] Schwarzer, Steve. "The History and Presentation of Mosaic Damascus." *Knives 2019*. 2018.
[24] Schwarzer, Steve. "The Mosaic Whiz Kids." *Blade Magazine*. November 2000.
[25] "Technique So New, Even Maker Can't Divulge It." *Blade Magazine*. Jan/Feb 1995.
[26] "Readers Respond." *Blade Magazine*. August 1997.
[27] Goddard, Wayne. "Blade Materials For the 21st Century." *Blade Magazine*. March 1997.
[28] Verhoeven, John D., and Howard F. Clark. "Carbon diffusion between the layers in modern pattern-welded Damascus blades." *Materials characterization* 41, no. 5, 1998: 183-191.
[29] Darken, L. Sr. "Diffusion of carbon in austenite with a discontinuity in composition." *Trans. AIME*. 180, 1949: 430-438.
[30] Kirkaldy, J. S. "Diffusion in multicomponent metallic systems." *Canadian Journal of Physics* 35, no. 4, 1957: 435-440.
[31] Brown, L. C., and J. S. Kirkaldy. "Carbon diffusion in dilute ternary austenites." *Transactions of the Metallurgical Society of AIME* 230, no. 1, 1964: 223.

29 Growth of Mosaic Damascus
[1] Hughes, B.R. "Rick Dunkerley." *Knife World*. December 1998.
[2] Haskew, Mike. "Mosaically Inclined." *Blade Magazine*. April 1998.
[3] Darom, David. *Custom Folding Knives*. 2007.
[4] Waring, Suzanne. "Shane Taylor - Bladesmith." *Knife World*. January 2013.
[5] "Knifemaker Showcase." *Blade Magazine*. February 1998.
[6] Schwarzer, Steve. "Mosaic Masters of Montana." *Blade Magazine*. July 1998.
[7] Schwarzer, Steve. "Mountain-Man Mosaic." *Blade Magazine*. March 1999.
[8] Tieves, Bruce. "Master of the Mosaic Folder." *Blade Magazine*. March 1997.
[9] https://windycityknives.com/products/al-dippold-the-scarab-Damascus-liner-lock-folder-one-of-a-kind

[10] "Custom Forecast." *Blade Magazine*. March 1998.
[11] Schempp, Ed. Phone Interview, 2022.
[12] https://www.bladegallery.com/shopexd.asp?id=2436
[13] https://www.bladegallery.com/shopexd.asp?id=1088
[14] Tieves, Bruce. "How Swede It Is!" *Blade Magazine*. January 1997.
[15] "Trends - The Folding Knife." *Knives '95*. 1994.
[16] Schwarzer, Steve. "Persson to Persson." *Blade Magazine*. January 2000.
[17] https://www.connyknives.com/
[18] Darom, David. *Custom Fixed Blades*. 2007.
[19] http://www.johanknives.eu/about.php
[20] Davis, Joel. Phone Interview, 2022.
[21] https://web.archive.org/web/20090626085242/http://www.britishblades.com/forums/showthread.php?t=4
6899
[22] https://www.bladesmithsforum.com/index.php?/topic/9581-mick-maxens-Damascus-tutorial/
[23] Darom, David and Dennis Greenbaum. *Custom Knifemaking*. 2006.
[24] Spencer, Keith. "Hammerin' Hot Steel Aussie Style." *Knives 2003*. 2002.
[25] Ewing, Dexter. "Filicietti - Australian for Knife." *Blade Magazine*. October 2003.
[26] Filicietti, Steve. Email, 2022.
[27] "Knifemaker Showcase." *Blade Magazine*. December 2002.
[28] Ferry, Tom. Email Interview, 2022.
[29] Farr, Dan. "The Golden Age of Mosaic Damascus? Part I" *Blade Magazine*. August 2007.

30 Damascus for Sale

[1] Meier, Daryl. "Damascus Steel." *Blade Magazine*. June 1984.
[2] "Classifieds." *American Blade Magazine*. May-June 1977.
[3] Material handed out at the 1973 Knifemakers Guild Show written by Bill Moran.
[4] Miller, Burton. "Forging His Way to The Top - Rob Charlton." *Blade Magazine*. August 1986.
[5] Baldwin, Phil. Phone Interview, 2022.
[6] "Fain Edwards Dream Come True AmeriSteel Damascus." *Blade Magazine*. December 1984.
[7] McEvoy, Harry K. "Blades of Tim Zowada." *Knife World*. July 1985.
[8] *Blade Magazine*. June 1991: 60.
[9] Zowada, Tim. Phone Interview, 2022.
[10] Zowada, Tim. "How to Make Your Own Gas Forge." *Blade Magazine*. December 1990.
[11] Hehn, Richard and Norbert Klups. *Messer Profi-Tipps für Benutzer und Sammler*. 2001.
[12] "Knifemaker Showcase." *Blade Magazine*. October 2000.
[13] "Classified Ads." *Blade Magazine*. December 1991.
[14] "Classified Ads." *Blade Magazine*. August 1993.
[15] https://www.bladegallery.com/shopexd.asp?id=86873
[16] Hubbard, Arthur J. "Method of making heterogeneous blade-like metallic cutter member." U.S. Patent 4,881,430,
issued November 21, 1989.
[17] Winter, Butch. "Kingpins of Stainless Damascus." *Blade Magazine*. March 1997.
[18] Thomas, Devin. Phone Interview, 2022.
[19] Goddard, Wayne. "Exotic Steels: Blink and You'll Miss the Latest One." *Blade Magazine*. November 1993.
[20] "Mike Norris Damascus." *Bladeology Podcast*, 2020.
[21] https://damasteel.se/about-us/
[22] Billgren, Per, and Kaj Embretsen. "Method relating to the manufacturing of a composite metal product." U.S.
Patent 5,815,790, issued September 29, 1998.
[23] https://damasteel.se/story-of-Damascus-steel/
[24] "Readers Respond." *Blade Magazine*. August 1997.
[25] Goddard, Wayne. "Blade Materials For The 21st Century." *Blade Magazine*. March 1997.
[26] https://www.bokerusa.com/annual-Damascus-knives
[27] Haskew, Mike. "The Factory Damascus Game." *Blade Magazine*. November 1993.
[28] Haskew, Mike. "The Return of Factory Damascus." *Blade Magazine*. March 2001.
[29] https://www.allaboutpocketknives.com/knife_forum/viewtopic.php?t=11287
[30] https://www.allaboutpocketknives.com/catalog/schrade-knives/65328-vintage-1985-custom-Damascus-
schrade-lb5-mexican-rosewood-folding-hunter-pocket-knife-knives-sheath
[31] *Blade Magazine*. May 1994: 95.
[32] Meier, Daryl. Phone Interview, 2022.
[33] Reeve, Anne. Email, 2022.
[34] https://www.microholics.org/post/old-school-98-Damascus-ludt-for-sale-price-drop-8383109
[35] https://web.archive.org/web/20010722210148/https://www.bladeforums.com/ubb/Forum54/HTML/0023
64.html
[36] Norris, Mike. Email, 2022.
[37] Shackleford, Steve. "'02 Factory Knives Are Oh Too Cool." *Blade Magazine*. June 2002.

31 Mosaic Damascus for Sale
[1] Kertzman, Joe. "Damascus for the Masses." *Blade Magazine*. November 1997.
[2] "Study of Steel and Shape and Stuff." *Knives '96*. 1995.
[3] Eggerling, Robert. Phone Interview, 2022.
[4] Darom, David. *Custom Folding Knives*. 2006.
[5] Shackleford, Steve. "Damascus Naturals Knock 'Em Dead." *Blade Magazine*. December 1998.
[6] Kertzman, Joe. "Worthy of The King of Steels." *Blade Magazine*. September 1999.
[7] https://www.bladegallery.com/shopexd.asp?id=83812
[8] "Hot Handmade." *Blade Magazine*. July 1999.
[9] "Hot Handmade." *Blade Magazine*. August 1999.
[10] Thomas, Devin. Phone Interview, 2022.
[11] Haskew, Mike. "The Bolster Boys: Fittest of Fittings." *Blade Magazine*. May 2000.
[12] "Damascus Bolster Folders." *Knives 2001*. 2000.
[13] Kertzman, Joe. "The Ultimate Guide to Damascus Steel." *Blade Magazine*. November 2001.
[14] Haskew, Mike. "Practicing Steel Manipulation." *Knives 2008*. 2007.
[15] https://web.archive.org/web/20020204135445/http://www.devinthomas.com/pages/products.html
[16] https://www.microholics.org/post/rare-microtech-ultratech-2001-sn-008-devin-thomas-reptilian-Damascus-1250-8454436
[17] Kertzman, Joe. *Knives 2002*. 2001.
[18] Haskew, Mike. "The Steel People." Blade Magazine. *June* 2003.
[19] "Mike Norris Damascus." *Bladeology Podcast*, 2020.
[20] https://knifetreasures.com/viewKnife.php?arg_id=578&arg_selected_menu=3
[21] Darom, David. "Jerry Corbit." *Custom Folding Knives*. 2006.
[22] https://www.bladeforums.com/threads/microtech-socom-camo-handle-info-needed.156053/
[23] Shackleford, Steve. "'02 Factory Knives Are Oh Too Cool." *Blade Magazine*. June 2002.
[24] Reeve, Anne. Email, 2022.
[25] https://www.microholics.org/post/vector-8401392
[26] https://spydercollector.wordpress.com/2017/07/09/spyderco-25th-anniversary-delica-and-40th-anniversary-native-5/
[27] https://www.bladegallery.com/shopexd.asp?id=1635

32 Damascus Present and Future
[1] Straub, Salem. Phone Interview, 2022.
[2] https://www.prometheanknives.com/gallery
[3] Maumasi, Mareko. Phone Interview, 2022.
[4] https://kramerknives.com/knives/gallery/
[5] Hanson, Don. Phone Interview, 2022.
[6] https://www.instagram.com/maumasifirearts/
[7] Marcin, Biborski. *Illerup Adal, Vols. 11-12: Die Schwerter und Die Schwertscheiden, 11: Textband; 12: Katalog, Tafeln und Fundlisten*. Vol. 25. Aarhus Universitetsforlag, 2006.
[8] https://www.tf.uni-kiel.de/matwis/amat/iss/kap_b/advanced/tb_3_3a.html
[9] https://www.tf.uni-kiel.de/matwis/amat/iss/kap_b/illustr/ib_3_3.html
[10] Sacshe, Manfred. *Damascus Steel*. 2008.
[11] Manuel Quiroga Güiraldes Facebook.
[12] https://www.instagram.com/salemstraub/
[13] https://www.instagram.com/kilroysworkshop/
[14] Hardman, Ron. Phone Interview, 2022.

33 Knives and Steel in Sweden
[1] Löwegren, Gunnar. *Swedish Iron and Steel; A Historical Survey*. 1948.
[2] https://www.uddeholm.com/en/about-us/history/
[3] Fagerfjäll, Ronald. *The Sandvik Journey: The First 150 Years*. 2012.
[4] https://www.erasteel.com/history/
[5] Ruusuvuori, Anssi. *The Puukko: Finnish Knives From Antiquity to Today*. 2020.
[6] https://oldmora.blogspot.com/
[7] https://morakniv.se/en/this-is-morakniv/historical-catalogs/
[8] https://morakniv.se/en/morakniv-stories-en/woodcarving-morakniv/
[9] https://gamlavikmanshyttan.se/gamlaindustrin/compoundstal/
[10] Gillette, King C. "Origin of the Gillette Razor." *The Gillette Blade*. February 1918.
[11] Kastrup, Allan. *Swedish Heritage in America*. 1975.
[12] Veges, Arved Eduard Gaston Theo. "Manufacturing edge tools and special composition of steel for same." U.S. Patent 1,644,097, issued October 4, 1927.
[13] Woldman, Norman Emme, and Albert Julius Dornblatt. *Engineering Alloys; Properties, Uses*. 1936.
[14] Adams, Russell B. *King C. Gillette, The Man and His Wonderful Shaving Device*. 1978.
[15] "The Blade Battle." *Time*. January 29, 1965: 78.

[16] Fischbein, Irwin W. "Razor blade and method of making same." U.S. Patent 3,071,856, issued January 8, 1963.
[17] "The Steelmakers' Edge." *Time.* September 20, 1963: 93.
[18] "Steel that serves the world." *International Management Digest.* January 1964: 73.
[19] https://www.home.sandvik/en/about-us/our-company/history/history-years/1900-1939/
[20] Fischbein, Irwin W. "Razor blade and method of making same." U.S. Patent 3,518,110, issued June 30, 1970.
[21] Jakenberg, Klas-Erik. "Method in the manufacture of stainless, hardenable chromium-steel strip and sheet." U.S. Patent 3,660,174, issued May 2, 1972.
[22] "Steel strip for knives and scissors." *Iron Age.* February 11, 1974: 70.
[23] Bergquist, B. *et al.*, SANDVIK 13C26 Stainless Razor Blade Steel, Bulletin 66-4E, Steel Research Centre, Sandvik, Sandviken, Sweden, Oct. 1975: 1.
[24] Bröderna Jönssons knivfabrik 1947 catalog.
[25] AB Carl Anderssons Knivfabrik 1942 catalog.
[26] KJ Erikssons Sax & knivfabrik 1950 catalog.
[27] https://oldmora.blogspot.com/2020/07/steels.html
[28] *Purchasing.* April 20, 1967: 36.
[29] *American Blade Magazine.* February 1981: 52.
[30] Lile, James B. "Directions of the Guild." *American Blade Magazine.* August 1981.
[31] Dahlman, Anders. "EKA: The Swedish Manufacturer." *American Blade Magazine.* October 1981.
[32] Latham, Sid. "Tanto: A Classic Renewed." *Blade Magazine.* February 1984.
[33] Overton, Mac. "Walker's Lockers." *Blade Magazine.* April 1987.
[34] https://web.archive.org/web/20040208155352/http://ajh-knives.com/metals.html
[35] Verhoeven, John D. "Metallurgy of steel for bladesmiths & others who heat treat and forge steel." Iowa State University. 2005.
[36] https://web.archive.org/web/20060213062223/http://www.devinthomas.com/pages/faq.html
[37] Landes, Roman. *Messerklingen und Stahl.* 2006.
[38] https://www.bladeforums.com/threads/kershaw-onion-question.299530/#post-2553240
[39] https://www.bladeforums.com/threads/420hc-stainless-steel-or-sandvik-13c26-cpm-d2.606650/#post-6300374
[40] https://www.bladeforums.com/threads/sandvik-steels-and-reps-at-the-blade-show.478482/
[41] https://www.bladeforums.com/threads/420hc-stainless-steel-or-sandvik-13c26-cpm-d2.606650/#post-6301962
[42] https://web.archive.org/web/20100416090322/http://www.alphaknifesupply.com/bladesteel.htm

34 Powder Metallurgy Steel

[1] Dulis, E.J. and T.A. Neumeyer. "Particle-metallurgy high-speed tool steel." *ISI Paper 126.* 1970: 112.
[2] Korbin, C.L. "Tool steels take the powder route,' *Iron Age.* December 7, 1967: 71.
[3] Obrzut, J.J. "PM tool steels come out swinging," *Iron Age.* January 14, 1971: 51.
[4] Steven, Gary. "Sintered Steel Particles Containing Dispersed Carbides." US Patent 3,561,934. Patented February 9, 1971.
[5] Kasak, A. and E.J. Dulis. "Powder-metallurgy tool steels." *Powder Metallurgy* no. 2. 1978: 115.
[6] *Metal Progress* vol. 104, no. 3. August 1973: 18.
[7] *Mechanical Engineering* vol. 95, no. 9. September 1973: 109.
[8] Steven, Gary. "High-Speed Steel Containing Chromium, Tungsten, Molybdenum, Vanadium and Cobalt." US Patent 3,627,514. Patented December 14, 1971.
[9] Steven, Gary. "Vanadium-Containing Tool Steel Article." US Patent 3,809,541. Patented May 7, 1974.
[10] Lenel, F.V. "New Processes for P/M Components." *Metals Engineering Quarterly* vol. 14, no. 1. February 1974.
[11] "CPM Rex 76." Trademark no. 1,043,948. Filed April 10, 1975.
[12] Bratt, R.W. "New Wear Resistant Alloys by P/M." *SME Technical Paper* EM78-270. 1978.
[13] Schwartz, N.B. "ASEA-Stora Push PM Tool Steels." *Iron Age* vol. 206, no. 5. July 30, 1970.
[14] Wastenson, Erik Goran, George Heinrich, and Arthur Gerhard Bockstiegel. "High Alloy Steel Powders and Their Consolidation into Homogeneous Tool Steel." US Patent 3,704,115. Patented November 28, 1972.
[15] "PM high-speed steel outperforms competitors." *Iron Age* vol. 214, no. 17. October 21, 1974.
[16] Hellman, P., and H. Wisell. "Effect of Structure on Toughness and Grindability of High Speed Steels." In Colloquium on High Speed Steels (Saint Etienne, France). 1975.
[17] Hellman, Per and David King. "ASP high speed steels - new steels for cold work tools." *Sheet Metal Industries* vol. 54, no. 5. 1977: 463.
[18] "Crucible wins patent suit against two Swedish firms." *Iron Age.* November 19, 1984: 18.
[19] Wojcieszynski, Andrzej, and William Stasko. "High-speed steel article." U.S. Patent 6,057,045, issued May, 2000.
[20] *Advanced Materials & Processes* vol. 159, no. 1. January 2001.
[21] https://www.crucible.com/History.aspx?c=7
[22] "Recent Developments in Crucible Particle Metallurgy Tool Steels." *Metal Powder Report.* September 1983: 476.
[23] Haswell, Walter T., and August Kasak. "Powder-metallurgy steel article with high vanadium-carbide content." U.S. Patent 4,249,945, issued February 10, 1981.
[24] Fletcher, Stewart G., and Walter T. Haswell Jr. "Abrasion resistant ferrous alloy containing chromium." U.S. Patent 3,489,551, issued January 13, 1970.

[25] McWilliams, Sean. Phone Interview, 2022.

[26] Hehn, Richard and Norbert Klups. *Messer Profi-Tipps für Benutzer und Sammler.* 2001.

[27] Roberts, William and Börje Johansson. "Verktygsstaal intended Foer cold working." Sweden Patent 457356, issued December 30, 1986.

[28] Roberts, William and Börje Johansson. "Tool Steel." US Patent 4,863,515. Patented September 5, 1989.

[29] *Advanced Materials & Processes* vol. 138, no. 4. October 1990: 72.

[30] "Plastics Equipment News." *Plastics World* vol. 52, no. 4. April 1994: 58.

[31] "Vanadis 6." US Trademark 75-579,559. Filed October 29, 1998.

[32] Tengzelius, Jan and Olle Grinder. "Powder Metallurgy in the Nordic Countries." *International Journal of Powder Metallurgy* vol. 38, no. 4. June 2002.

[33] "PM Newsgram." *Plastics World* vol. 48, no. 9. August 1990: 3.

[34] Kulmburg, A., and B. Hribernik. "The Heat Treatment of P/M Tool Steels." In *Materials Science Forum*, vol. 102. 1992: 31-42.

[35] Kulmburg, A., J. Stamberger, and H. Lenger. "Use of an Iron-Base Alloy in the Manufacture of Sintered Parts With a High Corrosion Resistance, a High Wear Resistance as Well as a High Toughness and Compression Strength, Especially for Use in the Processing of Synthetic Materials." European patent 348,380, issued November 19, 1992.

[36] https://web.archive.org/web/20070204133426/http://www.kresslerknives.com/Messermagazin.pdf

[37] Stasko, William, and Kenneth E. Pinnow. "Prealloyed high-vanadium, cold work tool steel particles and methods for producing the same." U.S. Patent 5,238,482, issued August 24, 1993.

[38] Stasko, W., K. E. Pinnow, and W. B. Eisen. "Development of ultra-high vanadium wear resistant cold work tool steels." *Advances in Powder Metallurgy and Particulate Materials–1996.* 5, 1996: 17.

[39] "CPM 15V." US Trademark 1,889,146. Filed April 26, 1993.

[40] Hauser, John J., William Stasko, and Kenneth E. Pinnow. "Wear and corrosion resistant articles made from pm alloyed irons." U.S. Patent 4,765,836, issued August 23, 1988.

[41] Pinnow, Kenneth E., William Stasko, and John Hauser. "Corrosion resistant, high vanadium, powder metallurgy tool steel articles with improved metal to metal wear resistance and a method for producing the same." U.S. Patent 5,936,169, issued August 10, 1999.

[42] "2000 sharp." *Blade Magazine.* March 2001: 69.

[43] Goddard, Wayne. *Wayne Goddard's $50 Knife Shop.* 2001.

[44] Pinnow, Kenneth E., and William Stasko. "Wear resistant, powder metallurgy cold work tool steel articles having high impact toughness and a method for producing the same." U.S. Patent 5,830,287, issued Nov. 3, 1998.

[45] https://web.archive.org/web/20020419023209/http://www.crucibleservice.com/eselector/prodbyapp/tool die/t&dapptitle.html

[46] Sandberg, Odd, Lennart Jonson, and Magnus Tidesten. "Cold Work Steel." Swedish Patent No. 519,278. Patented February 11, 2003.

[47] Tidesten, Magnus, Odd Sandberg, and Lennart Jönson. "PM Tool Materials: An Optimised PM Produced Cold Work Tool Steel." In *European Congress and Exhibition on Powder Metallurgy. European PM Conference Proceedings*, vol. 3: 1. The European Powder Metallurgy Association, 2004.

[48] Hillskob, Thomas *et al.* "Cold Work Tool Steel." US Patent 10,472,704. Patented November 12, 2019.

[49] Tidesten, Magnus, Anna Medvedeva, Fredrik Carlsson, and Annica Engström-Svensson. "A New Cold Work PM-Grade Combining High Wear Resistance with High Ductility." *BHM Berg-und Hüttenmännische Monatshefte* 162, no. 3, 2017: 117-121.

[50] Wildhack, Helmut *et al.* "Cold Work Steel with High Wear Resistance." Austrian Patent 411,534. Patented February 25, 2004.

[51] Schemmel, I., W. Liebfahrt, A. Bärnthaler, and S. Marsoner. "Böhler K390 Microclean–A New Powder Metallurgy Cold Work Tool Steel for Highly Demanding Applications." In *Proc. of the PM2004 Powder Metallurgy World Congress & Exhibition.* 2004: 773-778.

[52] Schemmel, Ingrid *et al.* "Cold work steel article." US Patent 7,682,417. Patented March 23, 2010.

[53] Makovec, Heinz, and Ingrid Schemmel. "PM Tool Steels: K890 MICROCLEAN-A New, Powder Metallurgy Cold Work Tool Steel Combining Wear Resistance and Highest Ductility." In *European Congress and Exhibition on Powder Metallurgy. European PM Conference Proceedings.* 2005: 197.

[54] Jesner, Gerhard and Devrim Caliskanoglu. "Cold-forming steel article." US Patent 8,298,313. Patented October, 30, 2012.

[55] https://web.archive.org/web/20110907144834/http://www.bohler-edelstahl.com/english/166_ENG_HTML.php

35 Knives and Steel in Japan

[1] https://web.archive.org/web/20150729221657/http://www.hitachi-metals.co.jp/e/tatara/nnp01.htm

[2] https://ja.wikipedia.org/wiki/工藤治人

[3] Iwasaki, Kosuke. *How to Look at Cutlery* (Japanese). 2012.

[4] https://web.archive.org/web/20190702222249/http://www.e-tokko.com/v_gold_10.php?lang=en

[5] https://www.e-tokko.com/profile.php?lang=en

[6] http://www.atm-fukaumi.co.jp/en/company_en

[7] Kertzman, Joe. "Steel Superstars: State-Of-The-Art Stainless." *Blade Magazine.* February 1998.

[8] "History of the Japanese Knifemakers Guild – Part 1." https://www.youtube.com/watch?v=AAIOIg-Pj6M
[9] "History of the Japanese Knifemakers Guild – Part 2." https://www.youtube.com/watch?v=FDYXYAmIr7Y
[10] "History of the Japanese Knifemakers Guild – Part 3." https://www.youtube.com/watch?v=PHnuJpwzPkI
[11] "A Knife Talk With Bob Loveless." *Blade Magazine.* August 1985.
[12] https://www.countryknives.com/brand-history/g-sakai/
[13] https://www.countryknives.com/brand-history/moki/
[14] Voyles, Bruce. "A Tour of Japanese Cutlery Factories." *American Blade Magazine.* April 1982.
[15] https://japanesechefsknife.com/blogs/news/grand-master-ichiro-hattori-history-and-pride-kd-cowry-x-Damascus-knives
[16] https://japaneseknifeguide.com/the-hattori-knife-products-and-history/
[17] https://hattoricollector.com/
[18] "Singularity in A Sea of Knives: Koji Hara." *Blade Magazine.* November 2011.
[19] http://www.knifehousehara.com/profile/index.html
[20] https://www.bladegallery.com/shopdisplayproducts.asp?id=452
[21] "Custom Knifemakers Directory." *Knives '96.* 1995.
[22] https://vk.com/wall-88898876_1451
[23] https://japanesechefsknife.com/collections/hattori-kd-series-limited-knife-collections
[24] https://japanesechefsknife.com/pages/product-search-by-blade-steel-japanesechefsknife-com
[25] Okuno, Toshio and Yuji Ito. "Material for tool and parts having high resistance to corrosion and wear." Japanese patent 63-169358. Patented July 13, 1988.
[26] Hitachi, Misasa and Yukinori Matsuda. "Powder cold work tool steel." Japanese patent 2,684,736. Patented July 9, 1990.
[27] Ito, Yuji and Toshio Okuno. "Materials for high corrosion resistant and wear resistant tool parts." Japanese patent 3-277747. Patented December 9, 1991.
[28] Murakawa, Yoshiyuki and Masakazu Ito. "High hardness and corrosion resistant steel for blades." Japanese patent 3894373. Patented March 22, 2007.
[29] Matsuda, Yukinori *et al.* "P/M steel for cutting tool." Japanese patent 11279677. Patented October 12, 1999.
[30] "Uniform organization of stainless steel cutlery." Japanese Patent 2764659, issued June 11, 1998.
[31] "What's New." *Blade Magazine.* March 2004.
[32] https://web.archive.org/web/20061218232535/http://www.surlatable.com/common/products/product_details.cfm?PRRFNBR=15867

36 Wayne Goddard and Knife Testing

[1] Goddard, Wayne. *The Wonder of Knifemaking.* **2000.**
[2] Goddard, Wayne. *Wayne Goddard's $50 Knife Shop.* 2001.
[3] Latham, Sid. "Profile: Wayne Goddard." *American Blade Magazine.* February 1978.
[4] Goddard, Wayne. "Good, Better, Best: Results of Basic Knife Testing." *Blade Magazine.* December 1990.
[5] Bear, Charles and Ronald Koeberer. "Why Forge?" *Knife World.* January 1990.
[6] Wood, William W. "Telling the Truth About Forging." *Knives '95.* 1994.
[7] Harvey, Dave. "Wood's Forging/Stock Removal Story is An Interesting Read." *Blade Magazine.* March 1995.
[8] Goddard, Wayne. "Edge Packing, Part I: History, Theories and Methods." *Blade Magazine.* December 1994.
[9] Goddard, Wayne. "Packing, Part II: Much Ado About Nothing?" *Blade Magazine.* January/February 1995.
[10] Goddard, Wayne. "Packing - Another Name for Normalizing." *Blade Magazine.* March 1995.

37 Ed Fowler and Triple Quenching

[1] Hughes, B.R. "Ed Fowler." *Knife World.* December 1988.
[2] Fowler, Ed. *Knife Talk: The Art & Science of Knifemaking.* **1998.**
[3] Goddard, Wayne. "Follow-Up: Why 52100 Needs Special Heat Treatment." *Blade Magazine.* February 1993.
[4] Petersen, Dan. "Multiple Quench: Fact or Theory?" *Blade Magazine.* February 1994.
[5] "Readers Page." *Blade Magazine.* June 1994.
[6] Verhoeven, John. *Metallurgy of Steel for Bladesmiths & Others Who Heat Treat and Forge Steel.* 2005.
[7] Fowler, Ed. "Ball-Bearing Steel: It Just Keeps On Cuttin'!" *Blade Magazine.* May 2002.
[8] Fowler, Ed. "Blades From Ball Bearings: The 52100 Solution." *Blade Magazine.* April 1992.
[9] Fowler, Ed. "Ball-Bearing Steel: It Just Keeps on Cuttin' (Part II)." *Blade Magazine.* June 2002.
[10] Fowler, Ed. "How To Freeze Quench Your Steel." *Blade Magazine.* June 1995.
[11] Fowler, Ed. "Forging the Fowler Way Part III: Normalizing and Annealing." *Blade Magazine.* July 1997.
[12] Fowler, Ed. "52100 From A Metallurgists's View." *Blade Magazine.* October 1997.
[13] Fowler, Ed. "High Performance 52100: How To Forge It and Why It Works." *Blade Magazine.* July 2002.
[14] Fowler, Ed. "The Benefits of Sharing." *Blade Magazine.* October 2006.
[15] *Blade Magazine.* August 1995: 85.
[16] Breed, Kim. "Barry Gallagher Camp Knife." *Blade Magazine.* April 1996.
[17] Breed, Kim. "Rick Dunkerley Model #8." *Blade Magazine.* December 1996.
[18] "OKCA Winners Get Head Start On BLADEhandmade Awards." *Blade Magazine.* September 1997.
[19] Hughes, B.R. "Sculpted Art From The Forge Of Freedom." *Blade Magazine.* January 1999.
[20] Haskew, Mike. "Pursuit Of The Peerless Period Piece." *Blade Magazine.* June 1999.

[21] Breed, Kim. "Cuttin' The Door Down." *Blade Magazine.* October 2005.
[22] Fowler, Ed. "On The Oregon Cutlery Trail." *Blade Magazine.* September 1999.
[23] Fowler, Ed. *Knife Talk II: The High Performance Blade.* 2003.

38 Differential Hardening and the Hamon
[1] Goddard, Wayne L. "Q&A: Ask Wayne Goddard." *Blade Magazine.* June 1991.
[2] Fowler, Ed. "How To Heat Treat: The Spirit of the Forged Blade." *Blade Magazine.* June 1992.
[3] Goddard, Wayne. "Follow-Up: Why 52100 Needs Special Heat Treatment." *Blade Magazine.* February 1993.
[4] Goddard, Wayne. "Heat Treating: 4 Makers, 4 Different Perspectives." *Blade Magazine.* March 1999.
[5] Kapp, Leon. *Modern Japanese Swords and Swordsmiths: From 1868 to the Present.* 2002.
[6] "The Swordmaker of San Francisco." *American Blade Magazine.* January-February 1975.
[7] Hastings, Don. "Japanese Swordmaking in Dallas." *American Blade Magazine.* August 1980.
[8] Latham, Sid. "Tanto: A Classic Renewed." *Blade Magazine.* February 1984.
[9] https://www.samuraimuseum.jp/shop/product/modern-authentic-japanese-sword-tanto-signed-by-kusan/
[10] Engnath, Bob. "The U.S.-Made Samurai Edge." *Knives '86.* 1985.
[11] Fisk, Jerry. "How to Heat Treat and Get a Temper Line." *Blade Magazine.* March 1996.
[12] Hudson, Robbin. "The Best Method of Differential Tempering." *Blade Magazine.* January 1997.
[13] Haskew, Mike. "Hamon: Mark of a Premier Blade." *Blade Magazine.* November 2006.
[14] Fogg, Don. Phone Interview, 2022.
[15] Fogg, Don. "How to Clay Temper and Obtain a Beautiful 'Hamon.'" *Spirit of the Sword.* 2010.
[16] Haskew, Mike. "Western Knives, Eastern Look." *Blade Magazine.* November 2012.
[17] Hanson, Don. Phone Interview, 2022.

39 Specialty Knife Companies
[1] Shackleford, Steve. "10 Years That Changed The Knife Industry." *Blade Magazine.* May 1998.
[2] Kowalski, David. "Remember The End-Line User." *Blade Magazine.* August 1998.
[3] https://www.recoilweb.com/zeroed-in-sal-glesser-158705.html
[4] "Sal Glesser Enters Shrine of Sharp." *Blade Magazine.* November 1999.
[5] Haskew, Mike. "The Renaissance Knife." *Blade Magazine.* December 2017.
[6] Terzuola, Robert. Phone Interview, 2023.
[7] https://www.bladeforums.com/threads/blast-from-the-past-koncept-knives.336156/#post-3775886
[8] Kertzman, Joe. "Worldwide Web Of Spyderco." *Blade Magazine.* February 1999.
[9] https://www.spydiewiki.com/index.php?title=FB01_Bill_Moran_Featherweight
[10] https://www.spyderco.com/catalog/2008MuleTeam.pdf
[11] "The Blade Show." *Blade Magazine.* October 1983.
[12] Bali-Song 1981-1982 catalog.
[13] Pacific Cutlery 1983-1984 catalog.
[14] Benchmade 1991 catalog.
[15] Benchmade 1993 catalog.
[16] Shackleford, Steve. "Showpieces of the SHOT Show." *Blade Magazine.* June 1987.
[17] Dick, Steven. "Will 1989 Be the Year of the Kukri?" *Blade Magazine.* April 1989.
[18] Benchmade 1990 catalog.
[19] Kowalski, David. "Quality, Value, Performance, and Excitement..." *Blade Magazine.* August 1997.
[20] Voyles, J. Bruce. "The Future is Today at Benchmade." *Blade Magazine.* April 2000.
[21] Haskew, Mike. "Macro-Built to Last." *Blade Magazine.* October 2000.
[22] https://microtechknives.com/microtech-a-look-back/
[23] Kertzman, Joe. "Sharpest Images: State-Of-The Art Designs." *Blade Magazine.* May 2000.
[24] https://www.knifeart.com/emerson-knives-history.html
[25] Kertzman, Joe. "Three Cheers for the Stateside Steel." *Blade Magazine.* July 2000.
[26] Haskew, Mike. "ATS-34: Stainless Steel Of Choice." *Blade Magazine.* April 1999.
[27] Shackleford, Steve. "Factories Finest Strut Their Stuff." *Blade Magazine.* November 1998.
[28] Haskew, Mike. "Knife Collaborations, Hawaiian Style." *Blade Magazine.* August 1999.
[29] Winter, Butch. "King of the Assisted Openers." *Blade Magazine.* August 2004.
[30] Flagg, Doug. Phone Interview, 2023.
[31] Ewing, Dexter. "Collaborations for The Nation." *Blade Magazine.* May 2008.

40 Ralph Turnbull, Phil Wilson, and New Super Steels
[1] Haskew, Mike. "ATS-34: Stainless Steel Of Choice." *Blade Magazine.* April 1999.
[2] "Readers Respond." *Blade Magazine.* June 1999.
[3] Shackleford, Steve. "Ted Dowell: Integrals & Integrity." *Blade Magazine.* September 1999.
[4] https://web.archive.org/web/20030715190347/http://sparksknives.com/app/sparksstory.htm
[5] Winter, Butch. "The Alternative: Vasco Wear." *Knife World.* July 1979.
[6] Lewis, Jack and Roger Combs. "A Feel for Steel." *The Gun Digest Book of Knives, 3rd Edition.* 1988.
[7] Warner, Ken. "Damascus Isn't Everything." *Knives '84.* 1983.
[8] Winter, Butch. "Ralph Turnbull." *Knife World.* October 1985.

[9] Winter, Butch. "Sharpen Up." *Popular Mechanics.* April 1987.

[10] Koper, Terry. "Artisan Expects Knives To Outlast Their Maker." *The Cincinnati Enquirer.* February 26, 1981: 22.

[11] Schroen, Karl. *The Hand Forged Knife.* 1985.

[12] Kindig, Eileen Silva. "The Patience of Papp." *Blade Magazine.* December 1986.

[13] "Knifemaker Showcase." *Blade Magazine.* September 1994.

[14] Wilson, Phil. "CPM-10V." *Seamountknifeworks.com.* 1999.

[15] Wilson, Phil. "Some Thoughts on Knife Blade Steel Impact Toughness." *Seamountknifeworks.com.* 2000.

[16] Wilson, Phil. Email Interview, 2018.

[17] McWilliams, Sean. Phone Interview, 2022.

[18] Hehn, Richard and Norbert Klups. *Messer Profi-Tipps für Benutzer und Sammler.* 2001.

[19] McWilliams, Sean. "Book Review: The Hand Forged Knife by Karl Schroen." *Blade Magazine.* August 1986.

[20] Walker, Michael L. "Composite Cutting Blade and Method of Making the Blade." US Patent 4,896,424. Patented January 30, 1990.

[21] Shackleford, Steve. "Look to the Future With Far-Out Folders." *Blade Magazine.* November/December 1988.

[22] Overton, Mac. "Two Blade Materials; Better Than One?" *Blade Magazine.* June 1990.

[23] Goddard, Wayne. "CPM: The Steel of Tomorrow?" *Blade Magazine.* September 1992.

[24] "What's New." *Blade Magazine.* December 1994.

[25] "A Short Profile of Spyderco." *Blade Magazine.* July 1996.

[26] Winter, Butch. "Best of the CPMs." *Blade Magazine.* May 2000.

[27] Delavigne, Kenneth T. *The Spyderco Story: The New Shape of Sharp.* 2000.

[28] Shackleford, Steve. "Factories Finest Strut Their Stuff." *Blade Magazine.* November 1998.

[29] Wilson, Phil. "CPM 440V Update." *Knives Illustrated.* February 1997.

[30] Haskew, Mike. "Which Corrosion Resistant Steel is Best for You?" *Blade Magazine.* May 1997.

[31] Kerztman, Joe. "The Stainless Steel Stock-Up." *Blade Magazine.* March 2000.

[32] Shackleford, Steve. "A Pirate Looks at 70." *Blade Magazine.* January 2001.

[33] Wilson, Phil. "Preliminary Report on the Newest Particle Metallurgy Blade Material." *Seamountknifeworks.com.* 1996.

[34] Wilson, Phil. "Update on CPM-420V." *Seamountknifeworks.com.* 1998.

[35] "CPMs Get New Names." *Blade Magazine.* December 2000: 49.

[36] Shackleford, Steve. "Cool-Cuttin' Customers." *Blade Magazine.* November 2000.

41 Chris Reeve and CPM-S30V

[1] Harvey, Dave. "Reeve on Reeve." *Blade Magazine.* March 2001.

[2] Bates, Jr., John Davis and James Henry Schippers, Jr. "Lew Booth." *The Custom Knife II.* 1974.

[3] Wilson, Phil. "BG-42! It is Rocket Science." *Knives Illustrated.* October 1998.

[4] Loveless, R.W. "And Testing Retort." *American Blade Magazine.* January-February 1977.

[5] "A Knife Talk With Bob Loveless: Part II." *Blade Magazine.* October 1985.

[6] Shackleford, Steve. "Ted Dowell: Integrals & Integrity." *Blade Magazine.* September 1999.

[7] Goddard, Wayne. "How To Combat The Rising Cost of ATS-34." *Blade Magazine.* September 1995.

[8] Goddard, Wayne. "The Brass-Rod Test: Barometer Of Superior Temper." *Blade Magazine.* February 1997.

[9] Shackleford, Steve. "A Pirate Looks at 70." *Blade Magazine.* January 2001.

[10] https://web.archive.org/web/19990225132121/http://www.admiralsteel.com:80/products/blades.html

[11] Shackleford, Steve. "'97 Guild Show: Viva Las Blades." *Blade Magazine.* December 1997.

[12] Harvey, Dave. "Reeve on Reeve: Part II. " *Blade Magazine.* April 2001.

[13] Shackleford, Steve. "New For '98: Factory Fashions In The Fast Lane." *Blade Magazine.* June 1998.

[14] Kertzman, Joe. "Latest Hunters Are Big Game." *Blade Magazine.* October 1998.

[15] *Blade Magazine.* May 2000: 79.

[16] Shackleford, Steve. "It's All In The Blade." *Blade Magazine.* June 2000.

[17] Shackleford, Steve. "Blades 2000: The Edge Evolution Continues." *Blade Magazine.* March 2000.

[18] Winter, Butch. "Best of the CPMs." *Blade Magazine.* May 2000.

[19] "Readers Respond." *Blade Magazine.* July 2000.

[20] Shackleford, Steve. "Extreme Steels." *Blade Magazine.* April 2001.

[21] "CPMs Get New Names." *Blade Magazine.* December 2000: 49.

[22] Kertzman, Joe. "These Blades Are Boss!" *Blade Magazine.* September 2002.

[23] https://www.bladeforums.com/threads/the-true-history-of-s30v-development.1015831/

[24] https://www.bladeforums.com/threads/chris-reeve-green-beret.384825/page-5#post-3538203

[25] Wilson, Phil. "Breakthrough: A Steel Made Specifically for Knife Blades." *Blade Magazine.* April 2002.

[26] Wilson, Phil. "Best All-Purpose Workhorse Stainless? (Part I)." *Blade Magazine.* July 2002.

[27] Wilson, Phil. "The Best All-Purpose Workhorse Stainless? (Part II)." *Blade Magazine.* August 2002.

[28] Shackleford, Steve. "'02 Factory Knives Are Oh Too Cool." *Blade Magazine.* June 2002.

[29] Ewing, Dexter. "Swing Things." *Blade Magazine.* June 2003.

[30] Haskew, Mike. "The Steel People." *Blade Magazine.* June 2003.

[31] Kertzman, Joe. "Endorsements Roll in for CPM S30V." *Blade Magazine.* September 2003.

42 Competition for Crucible

[1] https://www.microholics.org/post/anyone-here-know-a-lot-about-lcc%E2%80%99s-11351024
[2] "Holly-Day Knives." *Blade Magazine.* December 2003.
[3] https://web.archive.org/web/20050313084240/http://www.timken.com:80/products/specialtysteel/enginee ring/tech_info/knifesteels.asp
[4] https://web.archive.org/web/20051127135836/http://www.timken.com:80/products/specialtysteel/enginee ring/tech_info/knifesteels.asp
[5] McKay, Beren. Email Interview, 2020.
[6] Kertzman, Joe. "Satisfying the Need for New Knives." *Blade Magazine.* October 2005.
[7] "Cover Story." *Blade Magazine.* April 2007.
[8] https://web.archive.org/web/20081120180611fw_/http://www.rickhindererknives.com/aboutme/index.html
[9] https://web.archive.org/web/20090105212031fw_/http://www.rickhindererknives.com/knives/xm-18/index.htm
[10] https://www.rickhindererknives.com/about-rick-hinderer/
[11] Kertzman, Joe. "Is CPM 154 the Ideal Blade Steel?" *Blade Magazine.* April 2006.
[12] Warner, Ken. "Damascus Isn't Everything." *Knives '84.* 1983.
[13] Wright, Timothy. Phone Interview, 2022.
[14] Ewing, Dexter. "CPM M4: The Next Super Steel?" *Blade Magazine.* December 2006.
[15] "Benchmade 2009 Consumer Catalog." 2008.
[16] https://www.spyderco.com/edge-u-cation/mule-team-project/
[17] "Spyderco 2010 Product Guide." 2009.
[18] Haskew, Mike. "The Knife Industry's Steel Team." *Blade Magazine.* July 2007.
[19] Ewing, Dexter. "Perfect Fit For Kershaw?" *Blade Magazine.* April 2009.
[20] Covert, Pat. "Potpurri of the Point Stuff." *Blade Magazine.* July 2010.
[21] Batson, James. Email Interview, 2021.
[22] Farr, Dan. "Mac Daddy of All Carbides Part I." *Blade Magazine.* June 2009.
[23] https://archive.triblive.com/news/crucible-materials-corp-files-for-bankruptcy/
[24] https://archive.triblive.com/news/allegheny-technologies-buys-crucible-units/
[25] https://sbisteel.com/ventures/sb-speciality-metals
[26] https://www.watermill.com/news/crucible-industries-and-latrobe-specialty-steel-distribution-announce-partnership/
[27] https://www.syracuse.com/news/2010/01/crucible_industries_llc_signs.html
[28] https://knifenetwork.com/forum/showthread.php?t=52088
[29] Haskew, Mike. "The Crucible Crunch." *Blade Magazine.* April 2010.
[30] "Trophy Knives." *Blade Magazine.* October 2009.
[31] https://web.archive.org/web/20130711025229/http://www.tactical-life.com/tactical-knives/n690-gain-the-cobalt-edge/
[32] Shackleford, Steve. "Taking the Edge Off." *Blade Magazine.* June 2010.
[33] "Taking the Edge Off." *Blade Magazine.* July 2010.
[34] Kertzman, Joe. "Sweet Deals on Knife Supplies." *Blade Magazine.* January 2011.
[35] https://web.archive.org/web/20100922012944/http://bucorp.com/files/BUC_High-Performance-Steel-for-Knives.pdf
[36] https://web.archive.org/web/20110203113728/uddeholm.com/knife
[37] https://web.archive.org/web/20110918045728/http://www.bucorp.com:80/knives.htm
[38] Wilson, Phil. Email Interview, 2018.
[39] Kawai, Nobuyasu, Katuhiko Honma, Hirofumi Fujimoto, Hiroshi Takigawa, Minoru Hirano, and Masaru Ishii. "Nitrogen containing high speed steel obtained by powder metallurgical process." U.S. Patent 4,121,929, issued October 24, 1978.
[40] Sandberg, Odd, and Leif Westin. "Powder metallurgy manufactured high speed steel." U.S. Patent 6,818,040, issued November 16, 2004.
[41] "Blade's 2010 Knives of the Year." *Blade Magazine.* November 2011.
[42] Ayres, James Morgan. "No Rust for the Wary." *Blade Magazine.* April 2011.
[43] Ejnermark, Sebastian, Thomas Hillskog, Lars Ekman, Rikard Robertsson, Victoria Bergqvist, Jenny Karlsson, Petter Damm *et al.* "Corrosion and wear resistant cold work tool steel." U.S. Patent Application 14/917,521, filed July 28, 2016.
[44] https://www.spyderco.com/catalog/2014_Spyderco_Product_Guide.pdf
[45] Berns, Hans, and Werner Trojahn. "High-nitrogen Cr-Mo steels for corrosion resistant bearings." In *Creative Use of Bearing Steels.* ASTM International, 1993.
[46] https://web.archive.org/web/20021202230600/http://www.benchmade.com/detail.asp?model=100SH2O
[47] https://web.archive.org/web/20041209005049/http://spyderco.com/catalog/details.php?product=40
[48] "Benchmade 2005 Consumer Catalog." 2004.
[49] Perot, Nicolas, Jean-Yves Moraux, Jean-Paul Dichtel, and Bruno Boucher. "X 15 TN:™ A New Martensitic Stainless Steel for Surgical Instruments." *ASTM SPECIAL TECHNICAL PUBLICATION* 1438, 2003: 13-27.
[50] https://mailchi.mp/spyderco/spyderco-byte-may-2022
[51] https://web.archive.org/web/20100918042550/http://cartech.com/news.aspx?id=3716

[52] Novotny, Paul M., Thomas J. McCaffrey, and Raymond M. Hemphill. "Corrosion resistant, martensitic steel alloy." U.S. Patent 5,370,750, issued December 6, 1994.
[53] https://web.archive.org/web/20100918042550/http://cartech.com/news.aspx?id=3716
[54] Haskew, Mike. "The Million Dollar Finish." *Blade Magazine.* April 2012.
[55] https://web.archive.org/web/20150322075852/http://www.cartech.com/news.aspx?id=4058
[56] https://www.businesswire.com/news/home/20110620005821/en/Carpenter-Technology-Acquire-Latrobe
[57] https://knifenews.com/cold-steel-to-transition-from-cts-xhp-to-s35vn-blade-steel/
[58] https://www.crucible.com/History.aspx?c=7
[59] Martin, RJ. Phone Interview, 2022.
[60] Kajinic, Alojz, Andrzej Wojcieszynski, and Maria Sawford. "Corrosion and Wear resistant Alloy." US Patent 7,288,157. Patented March 15, 2007.
[61] http://www.crucible.com/PDFs/DataSheets2010/Datasheet%20CPM%20S110Vv12010.pdf
[62] Wilson, Phil. "A Real Life Knife Test." *Seamountknifeworks.com.* 2009.
[63] Wilson, Phil. "Is CPM S110V Chippy?" *Seamountknifeworks.com.*
[64] https://www.worthpoint.com/worthopedia/kershaw-shallot-with-new-cpm-s110v-super-steel
[65] https://web.archive.org/web/20130306152305/http://www.blademag.com/kitchen-knives/kai-usa-lands-four-knife-of-the-year-awards
[66] "P/M Improves Cold Work Die Steel." *International Journal of Powder Metallurgy* 37, no. 2. 2001: 17.
[67] "New Shock Resistant Cold Work Die Steel." *International Journal of Powder Metallurgy* 37, no. 3. 2001: 13.
[68] "Z-Wear PM." US Trademark serial no. 77,913,387. Registered October 25, 2011.
[69] "Z-Tuff PM. US Trademark serial no. 87,639,519. Registered December 4, 2018.
[70] https://forum.spyderco.com/viewtopic.php?f=2&t=58604&p=835859
[71] https://www.spyderco.com/catalog/details/MT12P/Mule-Team-12-Cruwear/703
[72] https://forum.spyderco.com/viewtopic.php?f=2&t=60927
[73] https://www.bladeforums.com/threads/first-strider-sng-dgg-in-pd-1.1103909/
[74] https://web.archive.org/web/20140402201531/http://www.alphaknifesupply.com/blademateirals.htm
[75] http://www.crucibleservice.com/admin/userfiles/file/CPM4V.pdf
[76] https://knifesteelnerds.com/2019/11/01/crucible-s45vn-steel/
[77] https://catalog.spyderco.com/Reveal4/?page=10
[78] https://knifesteelnerds.com/2021/03/25/cpm-magnacut/

43 Knives and Steel in China and Taiwan

[1] Kowalski, David. "Quality, Value, Performance, and Excitement..." *Blade Magazine.* August 1997.
[2] *Blade Magazine.* April 1998: 113.
[3] Ewing, Dexter. "Beasts of the Far East Part I." *Blade Magazine.* March 2007.
[4] http://ahoneststeel.com/profile.php
[5] https://forum.spyderco.com/viewtopic.php?f=2&t=14579&p=140419
[6] https://knifeinformer.com/kizer-knives/
[7] https://knifeinformer.com/reate-knives/
[8] @rike_knife Instagram.
[9] https://www.weknife.com/pages/we-story
[10] https://artisancutlery.net/pages/about-us
[11] https://www.civivi.com/pages/civivi-story
[12] https://knifenews.com/artisan-cutlery-develops-new-cjrb-brand/
[13] Haskew, Mike. "The Next Great Market for Handmade Knives?" *Blade Magazine.* October 2005.
[14] Haskew, Mike. "Asian Damascus Invasion." *Blade Magazine.* June 2013.
[15] Winter, Butch. "Doctor D-2 & His Cutting Machines." *Blade Magazine.* July 2002.
[16] *Blade Magazine.* March 1993: 29.
[17] Winter, Butch. "Best of the CPMs." *Blade Magazine.* May 2000.
[18] *Blade Magazine.* September 1999: 32.
[19] Krauss, David A. *American Pocketknives: A History of Schatt & Morgan and Queen Cutlery.* 2002.
[20] Kertzman, Joe. "Is D-2 All It's Cut Out To Be?" *Blade Magazine.* March 2004.
[21] https://knifenews.com/artisan-creates-proprietary-powder-metallurgy-steel-for-budget-knives/
[22] https://www.globenewswire.com/news-release/2017/05/12/984489/0/en/ERAMET-group-Change-in-the-partnership-between-ERASTEEL-and-Heye-Special-Steel-Co-Ltd-in-China.html

44 The Family Tree of Knife Steel

[1] https://www.spyderco.com/catalog/details/C229GBN15V/Shaman-reg-Brown-G-10-CPM-reg-15V-reg-Sprint-Run-trade-/2390

INDEX

ABOUT THE AUTHOR

Larrin Thomas is a steel metallurgist specializing in the development of new steel grades and heat treatments. He has a Ph.D in Metallurgical and Materials Engineering from the Colorado School of Mines. Larrin developed his interest in knives and steel through his father, Devin, known for his pattern-welded Damascus steel. Larrin is the author of the blog KnifeSteelNerds.com, the book *Knife Engineering: Steel, Heat Treating, and Geometry*, and he runs the YouTube channel Knife Steel Nerds. Larrin also developed the knife steels MagnaCut and ApexUltra. Larrin lives in Pittsburgh with his wife, two kids, and a dog.

Made in the USA
Las Vegas, NV
17 March 2024

87333857R00280